Annual Conference of the Australasian Corrosion Association 2024

Cairns, Australia
10-14 November 2024

ISBN: 979-8-3313-2966-2

Printed from e-media with permission by:

Curran Associates, Inc.
57 Morehouse Lane
Red Hook, NY 12571

Some format issues inherent in the e-media version may also appear in this print version.

Australasian Corrosion Association Inc.
30a Jessie Street
Preston
VIC 3072
Australia

Phone: 61 3 8608 7900

https://membership.corrosion.com.au/

Additional copies of this publication are available from:

Curran Associates, Inc.
57 Morehouse Lane
Red Hook, NY 12571 USA
Phone: 845-758-0400
Fax: 845-758-2633
Email: curran@proceedings.com
Web: www.proceedings.com

TABLE OF CONTENTS

Bridging the Gap in Coating Performance 1
Matthew Brown

Sheet Pile Wall Protection using a Petrolatum Tape System on Rottnest Island Main Jetty 9
D. Edelman, K. Fortune

Design and Implantation of a New DC Mitigation System for Rail. 17
Richard Brodribb, David O'Brien

Assessment Criteria for the Electrochemical Protection of Steel 29
C. Stone, G. K. Glass

Concrete Asset Inspections on Wastewater Infrastructure 40
J. Rigby

Detecting Corroded Wall Ties Through Non-Destructive Means 51
C. Y. Lam, M. A. Hossain, M. J. Masia, I. A. Chaves, J. Vazey

Effect of Hydrogen on Oxidation of Fe-20Cr-(1Si, 2Mn OR 2Al) Alloys in Water Vapour at 650°C 63
Jiarui Chen, Jianqiang Zhang, David J. Young

Evolution of Corrosion Considerations in the as 3700 Masonry Structures Standard: A Historical Analysis and Implications for Durability Enhancement 79
M. A. Hossain, C. Y. Lam, M. J. Masia, I. A. Chaves

Crevice Corrosion of Mild Steel After Two Years of Exposure to Natural, Nutrient-Dosed, and UV-Treated Seawaters 91
R. B. Petersen, R. E. Melchers

Weathering Steel Durability in Australasia 106
W. L. Mandeno, R. A. Francis, R. El Sarraf

Improving Asset Reliability by Upgrading Base Metal Metallurgy Utilizing True High Velocity Thermal Spray 117
Phil Williams, Peter Johnson

Impact of Corrosion on the Compression Strength of Steel Wall Ties Within a Brick Veneer Wall System 126
L. Terry, S. Gillogly, I. A. Chaves, M. Masia

Protection of Steel at Edges and Scratches by Zinc Coatings 137
R. A. Francis

Durability Plans and Corrosion in Major Infrastructure Projects 146
D. B. McDonald, A. Lu, B. Eliasson

Identification & Control of a Persistent Sulphate Reducing Bacteria (SRB) & Tuberculate Microbial Induced Corrosion (MIC) Consortium from a Mild-Steel Closed-Loop Building Cooling System by a Novel Combination Strategy Relying on Anti-Corrosion & Anti-Microbial Additives. 163
S. Nearhos

Analysis of Dust Levels Created by Centrifugal Abrasive Blasting 176
R. McHendrie, D. Fosdike, D. McDonald, K. Sellars

Superhydrophobic Coatings for Corrosion Control ... 195
David Morton, Liliana Madaleno

Hot Water Pitting and Corrosion Loss Trends for Phosphorous Deoxidised (DHP) and Phosphorous Deoxidised Arsenical (DPA) Copper Alloys ... 204
P. Richardson, R. Melchers, D. Nicholas

Deterioration of Concrete Piles Underwater ... 214
P. Karajayli, S. Odlin, D. Cukierski

Cathodic Protection System Design – Actual Versus Design Outcomes for CP Designs from AEM (Airborne Electromagnetic) Surveys ... 226
A. Sterling

In-Situ Metal Wall Tie Corrosion Inspection of Pre-1990 Brick Buildings ... 237
S. De Prazer, I. Chaves, C. Middleton, B. Smith, L. Van Woensel, M. A. Hossain, M. J. Masia

The Reason for Contrasting Colours ... 247
D. McKeown, B. Lichtendonk

Cairns Main Wharf 1-5 Rehabilitation ... 262
M. Panafieu, N. Berends

Port of Hastings Multifacted Design ... 277
N. Berends, D. Branca, B. McGuiness

Bi-Modal Corrosion Behaviour of Magnesium Alloys ... 290
R. E. Melchers

Observations of Multi-Step Pitting Corrosion of Steels in De-Ionized Water and Seawater ... 299
R. E. Melchers, R. Jeffrey

Recent Progress in Ultra-Low Thermal Conductivity Polymer Materials: A Mini-Review ... 312
Nicholas S. L. Tan, Andrew B. Lowe, Ibukun Olowoye, Thunyaluk Pojtanabuntoeng

Challenges in Condition Assessment and Life Extension Methods for Aged Pipelines with Cathodic Protection & AC Mitigation System ... 321
Bob Phang, Lifu Wang, Anton Bardak

Deep Well Impressed Current Cathodic Protection System for Steel Piles of a New Bridge Over a River ... 332
S. Ali, F. Papworth

Some Microstructural Influences Upon Localised Corrosion ... 340
N. Birbilis

Ports: Anchors of the Energy Transition ... 349
C. Crawshaw, S. Barker

Cathodic Protection and Its Implications for Vessels in the Royal Australian Navy ... 367
K. Dylejko, W. Neil

Evolution of Fouling Release Coatings ... 379
D. Harvey

Sustainably Led Asset Management: A Pacific Case Study Queens Wharf, Lautoka Fiji ... 383
B. McGuiness

Author Index

BRIDGING THE GAP IN COATING PERFORMANCE

Author: Matthew Brown

AkzoNobel International Paint Protective Coatings

ABSTRACT

It can be argued that the protection of structural steel against corrosion using coatings is no better today than 40 or 50 years ago. Despite the advances in protective coating technology and much greater understanding of the requirements and disciplines affording corrosion control, premature corrosion failures are still common. Why are coating failures still occurring when there is so much understanding and information on the subject? The probable answer is in the improper execution of the theoretical knowledge and disciplines resulting from lack of knowledge, lack of training, inadequate inspection, and cost restraints. The result of coating performance can only be realised as a dry film on the surface, where the transformation of the liquid coating from the container to the dry film involves many practical operations. Each of these individual actions can be crucial to the ultimate performance of the coating. Well formulated coatings from reputable manufacturers, correct coating specifications, surface preparation requirements and coating application controls are the prime factors for success. All factors must be carefully addressed, not only individually but also together as an intricate synergistic requirement of the operation. The paper examines and discusses the reasons for many of the variable results evident throughout the industry and how to successfully bridge the gap in coating performance.

Keywords: Protective Coatings, Surface Preparation, Coating Application, Quality Control, Success Factors.

INTRODUCTION

The intention of this paper is to give a personal overview in the use of protective coatings to maximise their performance and to highlight some faults and misconceptions in the various operations, which often lead to coating failures. It is not in the scope to cover the vast technical and practical detail of the individual subjects and operations.

The paper and accompanying presentation therefore basically cover already known information and introduces nothing new. It does however attempt to pull together the available information and give further understanding of the individual decisions and operations that need to be addressed regularly in practice but are not always assessed collectively and executed as planned or specified. It will show the importance of the crucial factors and their influence and respective contribution to the overall job with the objective of mitigating the repeating common problems that do occur.

Coatings are applied to surfaces to serve various functions including protection/preservation, decoration, sanitation etc. In the case of structural steel, the principal reason is for protection against corrosion. The added value a well applied protective coating gives in the asset protection of large and expensive steel structures must never be underestimated and should always be considered and treated appropriately in this respect.

The quality and lifetime of protection afforded by the coating is the consequence of several, important factors involving both chemical and physical operations, which collectively determine the result. Nowadays the terms 'coating' and 'paint' are not technically synonymous and should be viewed separately and specifically rather than indiscriminately. Historically early formulations of paints did not sufficiently differ technically to warrant a distinction between terms, as do their modern-day counterparts.

A coating can be described as a liquid material which when applied correctly to a surface forms a film that progressively transforms, either by chemical and/or physical actions to a solid adhering, cohering, continuous protective layer. Because of the technical nature involved in both the selection and use of coatings for protective purposes, it is important that they are viewed with this distinction and treated and used as engineering materials.

Although many factors need to be fulfilled for the ensured success of protection by coatings, five factors can be considered as having crucial influence, four technical and one commercial, they are:

1. Coating selection.
2. Surface Preparation.
3. Coating Application.
4. Quality Control.
5. Cost.

Diligently addressing and correctly executing the known criteria that affect coating performance can add considerable lifetime value to structures in a cost-effective way. This should be the principal objective for every protective coating project.

HIGHLIGHTING THE IMPORTANT ISSUES:

1. Coating Selection:

There is a belief that sticking to what is known to work is to leave it alone without change. In fact, in many companies and industries, coating specifications have existed and left unchanged for years, following in tradition or simply passed on by predecessors. However, experience does also show that there is often a better way if there is a will to search. This philosophy of continually searching for improvement is the correct approach and is also appropriate for coating selection. It can be worthwhile and advisable to regularly review and change coating specifications if benefits can be realised.

It is not the intention of this paper to go into the vast number of different coating specifications and exposure conditions found in the industry. However, a strong emphasis is placed on the importance to confirm that technically correct coatings as systems are specified for all areas of projects by end users and is one of the key factors to successful coating performance. To help assess and confirm the correct selection particularly in cases of non-expertise, it is recommended that additional professional advice be sought from appropriate technical consultants and reputable coating manufacturers. The following notes give a further insight into the general technology of protective coatings.

1.1 Protective Coatings:

Commercial coatings are produced in the form of chemically complex liquids resulting from the mixing and dispersing, under controlled conditions of specifically selected raw materials from comprehensively tested formulations. (Note: powder coatings are physically different and separate to this discussion). The raw materials available for use in coatings comprise organic and inorganic types, originating from natural or synthetic sources and varying in nature from solid powders and resins to volatile and non-volatile liquids, each incorporated at optimum levels and for specific purposes.

The availability of the numerous raw materials allows the coating chemists to formulate and produce the many differing generic coating types available today. The present range can effectively satisfy all the technical demands of protection and other desired properties including cost effectiveness in the widely diverse fields of the surface coating industry.

1.2 Film Formation Mechanisms:

To further understand the different technical capabilities of the generic types, which ultimately determines their individual usefulness, it is necessary to explain the physical and/or chemical changes that occur with the formation of the film on the surface. As this is such an important feature of the coating properties, protective coatings are classified by the method of film formation. The two classifications are termed 'Convertible' and 'Non –convertible'. The following notes outline the technology involved. The transition from the liquid to the solid film can be either a relatively simple process Involving solvent evaporation (drying) or a more complicated chemical process (curing), or even a combination of both.

The coatings that undergo transformation by solvent evaporation only are termed 'non-convertible' or physically drying types and include Chlorinated Rubbers, Vinyl's, Air-drying Acrylics and Bituminous coatings etc. Conversely those that involve chemical change are termed 'convertible' or chemically curing types that include Epoxies, Polyurethanes, Polysiloxanes, Polyurea's, Silicates, Alkyds etc. To be more specific and to use the correct and accepted terminology 'drying' signifies a physical change (loss of solvent by evaporation) and 'curing' a chemical change (effected by an added curing agent, oxygen, moisture, or heat).

All protective coatings classified as 'non-convertible' and most of 'convertible' types are formulated to include solvents and as such the 'drying' mechanism will occur early in film formation of both types. As the drying process only occurs in non-convertible types, they are all relatively quick drying products. 'Convertible' types, although still going through the drying phase are slower to develop full 'cure' because of the longer time required for chemical change to take place.

There is, however, an anomaly in the special cases of solvent–free coatings, whereby no solvents are incorporated in the coatings at all. As such, a chemical change only occurs during film formation and completion. In this context 'drying' and 'curing' is one and the same mechanism.
Plural pack convertible materials like epoxy coatings have significantly expanded the scope in coating usefulness by allowing the chemist the ability to specifically design or 'tailor-make' coatings with predetermined performance properties by virtue of the known chemical composition created after mixing and application. With this ability in formulation versatility, coatings are made with properties and characteristics for specific, limited, and wide-range use.

Considering the vast and diverse choice of coatings and their individual characteristics. Include the often ambiguous and confusing historical and currently used terminology. Add to this, the glorification of products and performance claims displayed in manufacturers marketing brochures. All of this accumulates to produce an over-abundance of information needing to be analysed and assessed and only adds to the already difficult and daunting tasks faced by specifiers and end users in coating selection. More than ever there is a need to study and have a more detailed understanding of the technology involved in coatings that can add significant better value to the end-user through enhanced performance. Knowledge and skilful selection will preclude many of the failures commonly encountered. Therefore, correct selection of coatings and systems is a key factor to successful coating performance.

2. Surface Preparation:

Most of the evidence available indicates that the quality of surface preparation has the largest single bearing on coating performance. Results showing the effect on the relative performance of lower and higher quality coatings applied to lower and higher quality prepared surfaces can support this, where the former and latter combination respectively performed surprising, better than expected.

The result shows that good surface preparation gives good coating performance by removing intervening contamination that reduces adhesion. A correctly cleaned surface permits closer proximity of the two interfacial surfaces, which in turn allows for the electrically based forces of attraction to take effect. The closer the proximity, the stronger the force and consequent adhesion. The degree and type of contamination left on the surface from poor surface preparation will correspondingly inhibit the formation of the necessary adhesion forces contributing to earlier failure of the applied coating.

2.1: Surface Contaminants:

The principal surface contaminants that are known to affect adhesion of coatings are:

- Oil and grease,
- Corrosion products,
- Soluble salts,
- Dust and dirt,
- Moisture,
- Millscale
- Old loose coatings.

The standard methods currently used for cleaning surfaces can vary in the quality and efficiency of removing the various listed contaminants. It is important that each contaminant is identified, and the most appropriate method (or methods and sequence of use) are chosen for its removal. For instance, abrasive blasting alone will not totally remove oil and grease or soluble salt contamination. Degreasing fluid or solvent cleaning is. the first treatment required for the former and high-pressure fresh water washing for the latter contaminant. The use of abrasive blasting can then follow to achieve the final. quality required. The quality of cleanliness (removal of contaminants) is one major aspect of the surface preparation procedure with the other being creation of surface roughness or profile. A surface profile is necessary for augmenting the adhesion of the coating by increasing the overall surface area (increasing the available points of contact). This is best produced by abrasive blast cleaning whereby a controlled range of roughness is created depending on the blasting media used and the coating type and thickness to be applied.

The quality or standards of surface preparation are fully covered by several national. institutions including the (1) Association for Materials Protection and Performance (A.M.P.P - formally N.A.C.E.) (2) The Society of Protective Coatings. (S.S.P.C.), (3) International Standards Organisation (ISO) etc. It must be remembered that to specify the surface preparation standard is only an intention, it must be achieved and executed correctly in practice. Therefore, creating the conditions for good adhesion by the removal of surface contamination and producing a surface profile are key factors in coating performance.

3, Coating Application:

3.1 Mixing of Coatings:

As explained earlier, coatings are complex heterogeneous mixtures of many raw materials and over time in storage, partial separation can occur. Therefore, to restore uniformity prior to application, all coatings need to be thoroughly physically mixed by power stirring. For plural packaged coatings, in addition to the physical mixing operation, the correct volume mixing ratio of base to curing agent must also be followed to allow the correct chemical reaction to occur between the components. As such, correct physical and volume mixing are key factors in coating performance.

3.2 Dry Film Thickness & Quality of the Film:

Industrial protective coatings are applied to surfaces by various methods including brush, roller, and spray. The spray (airless spray) operation is by far the most widely used method because of the speed of application and the ability to handle high solids, high viscous materials characterised in modern coatings.

The objective of the application procedure is to obtain a consistent and coalesced (continuous) film to produce the required dry film thickness (DFT). The DFT is normally measured in microns (10^{-6} metres) and a typical DFT of a single coating is from 50μm to 500μm. Coatings systems are typically specified between 120μm to 500μm where the former is for low corrosion environments and the latter for long term exposure in aggressive environments. We expect a lot from coating systems: up to 25 years life at only the thickness of a few pieces of paper.

A correctly formulated and applied coating should form a barrier to the substrate that is as physically complex and chemically stable as possible, overall consistent, and free of imperfections like pinholes, bubbles, and other discontinuities. Achieving these criteria will also maximise the cohesive strength of the film, further enhancing performance. The quality of the barrier formed is therefore dependent on the thoroughness of mixing the product and the skill of the operator in application.

Most of the faults found in application revolve around the difficulty in the achievement of both the correct dry film thickness and a consistent coalesced film (without discontinuities). Therefore, correct training to obtain an acceptable level of skill in the manual operation of application is a key factor to coating performance. Environmental conditions also present challenges for the formation of quality films – working at extremes is not only challenging for the applicators, but the coatings as well.

3.3 Over-Coating of Products.

There are some specifications that comprise single coat application only, however most protective coatings specifications involve two to four coats in the system. In multi-coat specifications the main objective is to achieve total system integrity, which essentially means maximising the adhesion between coats. In practice this is accomplished by closely following the minimum and maximum overcoating intervals from the product data information.

The minimum over-coating period is a function of time and temperature involving the physical drying process or removal of solvent from the applied film by evaporation. This criterion applies to both convertible and non-convertible coatings (see earlier notes). The maximum over-coating period is also a function of time and temperature but involves the chemical curing mechanism that is only applicable to convertible coatings. In other words, there is normally no maximum over-coating time given for non-convertible coatings.

Ideally the maximum over-coating time is after the solvent has left the film, but before complete chemical reaction has taken place, and the coating becomes too hard to allow for optimum adhesion to be achieved. However, many products data information shows quite long maximum over-coatings times, even many months. In such cases the coatings have been specifically engineered for long over-coating times that can facilitate the practical scheduling of projects. Therefore, close observation and control of minimum and where applicable, maximum over-coating times to maximise inter-coat adhesion are key factors to coating performance.

3.4 Difficult Surfaces & Stripe-Coating.

As well as the manual controls required for consistent application, other difficulties encountered are from the widely variable physical configurations of surfaces and their accessibility. Coating the complex shapes including corners, edges, weld seams and curvatures pose more difficulties to the operator and are further compounded when application is to be done above ground level where safe access is required from scaffolding, cranes etc.

The most difficult areas to achieve film thickness and consequential protection is on sharp or protruding areas like edges, corners, and rough weld seams as the coating, in the early stages after application naturally flows away from surface peaks. To maximise better DFT control, all edges should have a 2-3mm radius applied and sharp protruding areas removed. To further ensure DFT achievement, additional localised application in the form of stripe coating should be adopted for these areas. The removal of sharp and protruding surfaces and the appropriate application of stripe coating are key factors to coating performance.

3.5: Coating Rheology & Application Technique.

Many protective coatings are purposely designed to give high build (HB) properties to maximise the dry film thickness (DFT) achievable per-coat and to minimise application cost (number of coats for adequate protection). The HB property is achieved by special additives in the coating formulation that change the rheology to allow the attainment and retention of the required DFT on vertical and inclined surfaces. The induced condition technically referred to as thixotropy produces a jelly-like structure into the coating, which remains while in the can and under static conditions.

Under dynamic conditions and shear forces as produced during the application process, the thixotropic nature is broken down allowing the coating to become fluid and able to flow to the required thickness on the surface. Shortly after application and in the absence of shear, the thixotropy is re-established, reducing flow, minimising the formation of runs and sags on vertical and inclined surfaces. Although the thixotropic condition is very beneficial for achieving DFT during vertical surface applications its very nature can cause problems for the subsequent coalescence of the film should poor application techniques be carried out, particularly if the gun to substrate distance is too large or a very fast evaporation spray thinner is used. Severe cases of poor film coalescence, producing considerable discontinuity will cause premature coating failure.

3.6 Human Involvement:

The design and production of coatings as packaged liquids in cans are totally under the control of the manufacturer and trained, qualified staff. However, evidence of coating performance is only realised after application, as a dry film on the surface and after time of exposure. In other words, the manufacturer is only producing and in control of a "partly" finished product with only limited involvement with its application to surfaces. Any subsequent contribution is normally given as advisory information to the user.

Illuminating further on the third-party involvement highlights their importance to the success of coating performance. As mentioned earlier, the other crucial factors identified were surface preparation and application. The practical tasks of blasting and spraying are mainly done by general labour and often with limited qualification or training. Technical understanding of the operations and the importance of carrying out the tasks correctly and in sequence are crucial for ensuring coating performance. As such, the blaster and sprayer's practical skills are as equally important for successful coatings performance to that of the manufacturers formulating/design chemist. Therefore, correct training of the personnel involved in blasting and application operations to give proper understanding of the objectives and to achieve acceptable practical skills are key factors to successful coating performance.

4 Quality Control:

There is no doubt that the adoption of the appropriate quality control procedures are essential parts of any production operation including protective coatings manufacture, surface preparation and coating application.

4.1 Qualifying the Responsibilities.

It is clear, that manufacturing control is the sole responsibility of the manufacturer. However, there are often misconceptions and assumptions made with the responsibility for surface preparation and application control. In many instances it is presumed that the coatings manufacturer takes responsibility because a technical inspector is normally assigned to the job. Often the inspector is designated or categorised, (however wrongly) as a 'supervisor' for the application. This description automatically indicates that he will control and be responsible for achieving the quality of surface preparation and application.

This should not be the case as the manufacturer, or his representative does not have any authority or control over a third-party applicator or contractor. The reason why the manufacturer assigns an inspector is to advise on the practicalities of surface preparation, standards to follow and the application procedure if required by the applicator or the client/owner. The applicator/contractor should therefore be contractually responsible for achieving the required specification and providing in-house quality control in consultation with the advising inspector of the manufacturer. In the operations of surface preparation and application of surface coatings it is vital that quality control procedures are addressed correctly, sequentially and monitored by trained and preferably suitably qualified personnel. Therefore, taking responsibility and adopting the appropriate quality control procedures with mutual consultation and co-operation are key factors in coating performance.

5. Cost:

Whether it's the client/owner, the applicator/contractor or the supplier, cost or price is generally, the major factor to be taken into consideration. What's important and fair to all parties are that the client/owner gets value for money (cost) in terms of protection and performance and the applicator/contractor and the supplier achieve acceptable margins (price) for their products and services. Unfortunately, in practice this does not always work out ideally and often in the negotiation and final securing of contracts, cost and price are forced to a minimum. For the contractor/applicator this can create the need to squeeze costs of the various practical operations involved to maintain overall profitability that may adversely affect quality and subsequent coating performance.

5.1 Cost & Coating Design.

Market and environmental demands are the inevitable driving forces behind present-day coating design and development. The unrelenting competitive climate has effectively channelled the major manufacturers to follow similar routes and policies to produce environmentally friendly, cost-effective coatings to maintain margins and support acceptable business growth. Helping to accomplish the incessant profitability demands, the coating designers' contribution is to focus mainly on producing competitive products with wide-spectrum properties and multi-purpose usage capability that are environmentally acceptable. Add to these, cost engineering and a 'fitness-for-purpose' approach, unnecessarily over-designed coatings are now uncommon entities.

The use of wide-spectrum, multipurpose products does however inevitably lead to compromises with some properties of these coatings. In other words, one coating cannot fulfil 100% of all requirements. For these specific or restricted needs like for instance in internal tank coating, dedicated cargo storage, anti-abrasion properties, high temperature resistance etc, purposely designed products are also available at appropriate prices. Obviously incurring costs for protective-coating projects is unavoidable. However, it's crucial that costs are realistic and appropriate for the performance expectation. Appropriate allocation of costs will be beneficial with added value being provided to projects upon completion and not wasted because of improper analyses, choices, and attention at the onset. Should the latter be the case, premature maintenance can be expected with additional costs incurred. As such, realistic costing from critical analysis of the project is a key factor to coating performance.

Conclusions.

In the paper we have identified numerous key factors that have major bearings on coating performance, to recap they are:

- ✓ Correct selection of coatings and systems.
- ✓ Creating the conditions for good adhesion by the removal of surface contamination and producing a surface profile.
- ✓ Correct training of the personnel involved in blasting and application operations to give proper understanding of the objectives and to achieve acceptable practical skills.
- ✓ Correct physical and volume mixing of the coatings.
- ✓ Close observation and control of minimum, and where applicable maximum over-coating times to maximise inter-coat adhesion.
- ✓ Removal of sharp and protruding surfaces and the appropriate application of stripe coating.
- ✓ Taking responsibility and adopting the appropriate quality control procedures with mutual consultation and co-operation.
- ✓ Realistic costing from critical analysis of the project.

It has been explained that focussing only on one or even several of the factors will not ensure successful coating performance. The key is to understand and make certain that all the key factors are correctly addressed in the overall assessment of the project and costed accordingly.

The project specification encompassing the coating specification should always include most of the key factors for coating performance as part of the requirement. However, it must be remembered that specifications only give written down rules and regulations to follow as an objective or intention to achieve. Only correct and absolute execution of the directions and key factors will accomplish the expected coating performance.

Bibliography

(1) Association for Materials Protection and Performance (A.M.P.P - formally N.A.C.E.) Training Course CIP 1&2

(2) The Society of Protective Coatings. (S.S.P.C.). Surface Preparation Standards for Mechanical, Dry, Wet Abrasive Blasting and Water Jetting.

(3) International Standards Organisation (ISO). ISO 8501 Standards Series; ISO 12944 series.

AkzoNobel internal training courses ProMote and ProStream.

Author Details

Matthew Brown.

Matthew is employed by AkzoNobel Protective & Marine Coatings as a Technical Specialist. Matthew has been with AkzoNobel for 40 years and has been employed in the Protective and Marine Coatings field as a formulation chemist, linings specialist, Technical Manager, Sales Manager, Company Cross-licence Manager and his current role as a Technical Resource and Trainer for the AkzoNobel Protective and Marine company operations in SE Asia, Pacific and Australasian region.

CORROSION & PREVENTION
CAIRNS 2024

THE AUSTRALASIAN CORROSION ASSOCIATION INC.

Sheet Pile Wall Protection Using A Petrolatum Tape System On Rottnest Island Main Jetty

D.Edelman[1,*], K. Fortune[2]

[1]Edelman Projects Pty Ltd, QLD, 4812, Australia

[2] Denso (Australia) Pty Ltd, VIC, 3061, Australia

**Corresponding author email: david@edelman.net.au*

ABSTRACT

A recently developed petrolatum tape marine corrosion protection system for sheet pile walls was used to protect the splash zone on the Rottnest Island main jetty. The system is manufactured by Denso and is known as the SeaShield 2020SP system.

The Rottnest Island Main Jetty was built in multiple sections beginning in the 1960's. The sheet pile wall is approximately 130m long and has experienced severe corrosion in the splash zone resulting in loss of steel thickness and isolated holing of the structure. A corrosion protection system was required to extend the life of the structure.

The system chosen had the advantage of a long expected life, excellent corrosion protection, the ability to be applied above and below the tidal level, and being tolerant to the severe pitting on the steel surface with minimal preparation. The application process included welding of attachment bars to the steel surface, application of petrolatum products to provide corrosion protection and installation of a tensioned HDPE protective covering.

The project demonstrated that the system can be applied successfully above and below water with good productivity rates. The system was tolerant to variations in the sheet pile wall profile and was adapted for several features of the sheet pile wall such as tie-back heads and level changes. The importance of good access and systems of work for applicators was underscored in the installation.

Keywords: steel, seawater, protective coating, petrolatum, sheet pile

Introduction

The petrolatum-tape marine corrosion protection system for sheet pile walls (the system) that is the focus of this paper is known as the Denso SeaShield 2020SP system. It was developed in Australia and has recently been introduced to the market after a period of in-situ testing and development [1]. The system is designed around and based on the highly successful petrolatum tape systems for discrete piles utilising tensioned HDPE jackets [2],[3]. The use of the system on the Rottnest Island Main Jetty is the first commercial application of the system.

The Rottnest Island main jetty provides for all passenger ferry services to Rottnest Island, WA and is therefore critical to Rottnest Island's operations and the local tourism industry. The sheet pile wall section of the jetty was built in multiple sections beginning in the 1960's. The sheet pile wall is approximately 130m long and has experienced severe corrosion and pitting in the splash zone resulting in loss of steel thickness and isolated holing of the structure. A corrosion protection system was required to extend the life of the structure. A section of the sheet pile wall is shown in Figure 1, with passenger ferry at right of image.

This paper will provide a description of the system, the advantages the system provides in relation to the particular case study, and some learnings of the project that potential applicators or specifiers may find useful when planning future projects.

FIGURE 1: ROTTNEST ISLAND MAIN JETTY SHEET PILE WALL.

Description of The System

The system includes the following components:

- Steel bars stitch-welded to the sheet pile wall to attach the system;
- Marine grade petrolatum primer, tapes, and mastic;
- Wall anchor elements (spines) that are constructed of HDPE and clamped to the sheet pile wall via the welded steel bars;
- Rigid foam blocks for profiling the path of the outer covering;
- Outer coverings constructed of high-density polyethylene (HDPE);
- HDPE plates for sealing of the bottom of the system; and
- Components for protection of tie-back heads.

The basic arrangement of the system is shown in Figure 2, with a cross section shown in Figure 3. The petrolatum primer and tape against the steel surface provide the corrosion protection function. The expanded polystyrene foam blocks provide a continuously convex path for the HDPE outer covering. During installation, the outer covering is tightly tensioned with the use of stainless steel fasteners and thus presses and retains the petrolatum tape and primer against the steel surface.

Figure 4 shows the welded steel bars on Rottnest Island main jetty. The bars were stitch welded with 5mm fillet welds of 50mm length every 160mm. Figures 5 shows how the expanded polystyrene foam blocks provide a convex path for the tensioned outer covering. Figure 6 shows the cover for the tie-back heads on the sheet pile wall. The tie-back heads are protected with petrolatum products and then the cover is filled with marine expanding foam. Figure 7 shows a completed section of the wharf.

The system requires an element of custom design. In particular, the foam block profile and width of outer covering was customised to suit the sheet pile profile. Special features of the wall included tie-back heads, level changes and changes of direction. A 3D scan of sections of the wall was conducted using a handheld scanner to characterise the profile and special features of the wall. This custom design process is typical for all installations of the system and special features of other sheet pile walls may include fenders, ladders, drainage points and others. The system can be incorporated with structural repair plates, which was not necessary on this project.

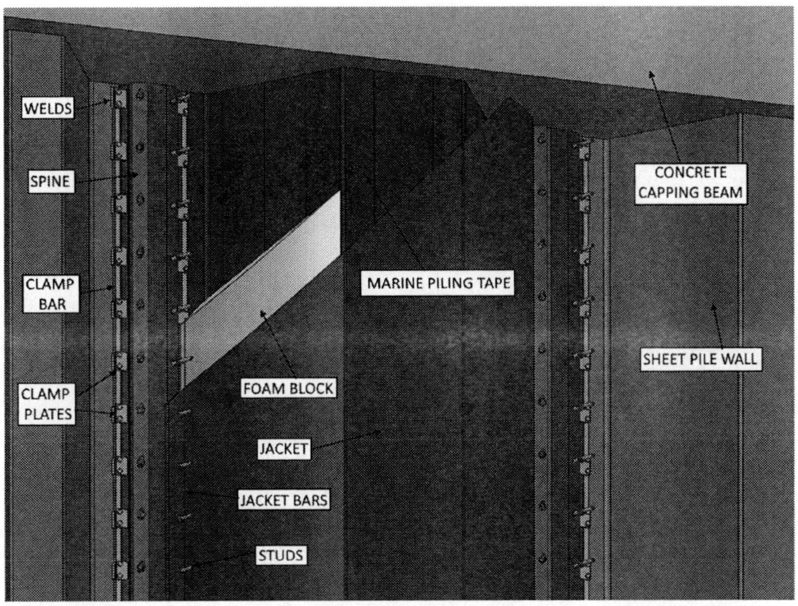

FIGURE 2: THE BASIC ELEMENTS OF THE SYSTEM SHOWN IN A CUTAWAY VIEW.

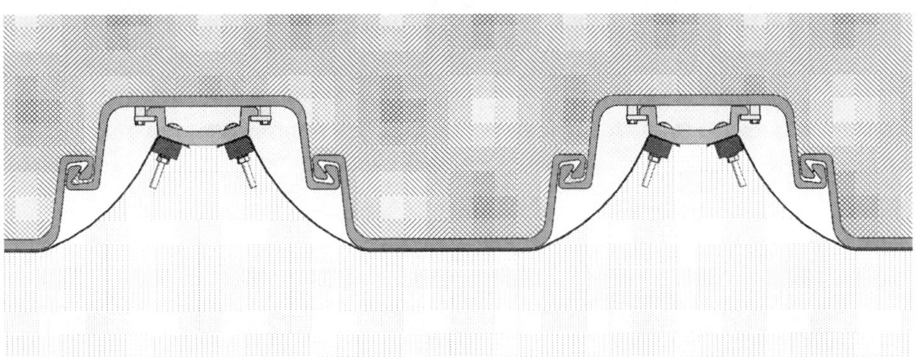

FIGURE 3: A CROSS-SECTIONAL VIEW OF THE ROTTNEST ISLAND MAIN JETTY INSTALLATION

FIGURE 4: SHOWING THE WELDED BAR AND CLAMP PLATES (LEFT), SPINE (CENTRE) AND AN OUTER COVERING (RIGHT) IN THE PROCESS OF BEING TENSIONED.

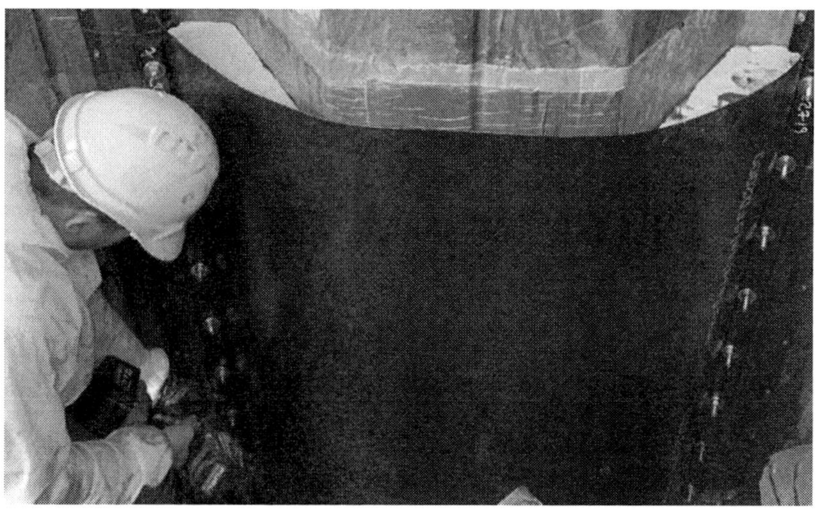

FIGURE 5: SHOWING HOW THE FOAM BLOCKS PROVIDE A CONVEX PATH FOR THE OUTER COVERING (NOTE: NOT AT ROTTNEST ISLAND).

FIGURE 6: COMPLETED TIE-BACK HEAD COVER (FILLED WITH MARINE EXPANDING FOAM)

FIGURE 7: COMPLETED SECTION

ADVANTAGES OF THE SYSTEM AT ROTTNEST ISLAND MAIN JETTY

The system provided the following advantages on the Rottnest Island Main Jetty application:

- Compared to a liquid coating, the system required less surface preparation. The steel surface had extensive deep pitting in the splash zone. A typical liquid coating would require abrasive blasting and then the pits to be thoroughly prepared such as by grinding and/or filling. Even with extensive preparation, it is likely that the pits would cause coating holidays and inclusions that would lead to premature deterioration. In comparison, the system only required preparation in accordance with SSPC-SP2/SP3, achieved by water blasting at 5000psi. The marine grade petrolatum primer that the system uses effectively fills the pits without additional preparation required.
- The ability to apply the system underwater allowed the project to proceed with tight timelines during unfavourable tides. Since the wharf was in use, weekend working hours were not permitted and the project had to be completed through the winter months ahead of the busy summer season. In comparison, it would not have been possible to apply a liquid coating to the height requested since suitable tides were not available to complete an application of a liquid coating during the project period.
- Due to the wharf being active and the works being metres from pedestrian traffic, it would have been extremely restrictive to carry out abrasive blasting due to the noise generated by that activity. In comparison, the system required less intensive tooling and processes.

- Due to the highly effective nature of petrolatum systems it was possible to minimise the extent of welded patching and confine repairs to isolated holes only. In comparison, as a typical liquid coating is likely to break down initially at deeply pitted areas, the risk of further holing and damage to critical areas is very high.

Project Outcome and Learnings

Quality of Application

Despite numerous irregularities to the wall surface, the conformance of the system to the wall profile was observed to be excellent. The conformance of the system to the wall profile is important in ensuring the petrolatum tape is pressed against the wall surface without allowing any voids where water could potentially ingress and degrade performance over time. There was significant variation of the sheet pile wall profile observed from the 3D scan. Further, many sheet piles had remnants of additional thick steel components welded to the face of the sheet pile. These variations result in: 1. differences of path length of the outer covering between fixing points, which could lead to outer coverings being too loose or too tight and, 2. excessive variation in the profile that the foam block is designed to follow, which could result in voids between the foam block and the wall profile.

In the case of outer covering path length, the nominal 2% stretch designed into the HDPE jacket ensured that a sufficient tension was achieved, while over-tensioning was within acceptable limits. For the foam block design, the foam blocks have an inherent amount they will compress, and an additional relieving slot was included in the profile to allow the foam blocks to further conform to the sheet pile profile. Applicators checked for voids during application of each section and were directed to apply additional tape layers if required. However, it was observed that the flexibility in the foam block and ability for the tape to slightly reform in shape under the pressure achieved during outer cover tightening was sufficient to ensure that there were no voids.

Productivity of Application

Following welding of the bars to the wall, it was demonstrated in the first week that a small team of 5-6 applicators could apply around 12 'jacket-widths' per day, this consisted of around 13 spines and 12 outer coverings of height of around 1.5 metres. In this case, this represents a square metreage of around 20 square metres. This indicative square meterage productivity may be increased for a wider sheet pile profile where each jacket would cover a greater area. This is because much of the labour is associated with the attachment and tensioning of the system rather than application of the petrolatum product to larger steel areas, which is relatively rapid.

The welding of the steel bars to the wall had much greater variability of productivity due to the project period requiring a significant portion of the welding to be carried out underwater using divers, rather than scheduling to allow works at low tides. At the time of writing, with the project ongoing, an accurate productivity figure for welding has not been determined. However, welding is likely to be highly dependent on equipment used, condition of the wall and quality of access put in place. Welding of the steel bars is also expected to be relatively easily priced by contractors since it is a standard operation.

Lessons Learnt in Application

A number of key lessons were learnt that led or would lead to greater productivity on site. These are summarised here:

- Improvements to access and the workspace led to increased productivity. In this case improved access was typified by having a sufficiently wide working platform and being able to work in a standing position. On the Rottnest Island main wharf this meant having the working platform situated at a permanently submerged level and having workers work in waders or wetsuits to apply the system. Surprisingly, the need to work at least partially underwater did not result in a significant loss of productivity for the application (excluding the welding).

- Having teams specialise in one task at a time led to productivity gains. On this project, this meant separate teams or separate time periods dedicated to one of applying spines only, applying outer coverings only, or carrying out the specialised components such as tie back head covers. This approach allows the team to have only the materials and tools required for that step on hand and thus minimises the amount of tool changes etc required for the works, while allowing a de-cluttered workspace.
- There was a distinct productivity boost when communication between the 'topside' workers and those applying the product was improved. The productivity was highly dependent on being able to supply tools and materials continuously and smoothly to the applicators at the workface as they were required. Clear communications between applicators and a topside person and between the topside person and those preparing and fetching materials facilitated this process.
- Adopting a 'production mindset' led to productivity gains. This involves investing resources as a deliberate team-based activity to make small improvements for each task. Notable examples on this project included sourcing magnetic nut setters to handle fasteners more easily and setting up dedicated hanging baskets with required tools or materials for each step. In any case, the key finding was to carry this out as a deliberate activity involving members of the team.

CONCLUSIONS

This project demonstrated the benefits and practicality of the system and provided examples and data that can be used to inform and improve future projects.

ACKNOWLEDGMENTS

The Authors would like to thank the Rottnest Island Authority for supporting the system, Wallbridge Gilbert Aztec as the project engineers and contract superintendent; Advanteering Civil Engineers as Principal Contractor, and Shorewater Marine as sub-contract applicators.

REFERENCES

1. Edelman D, Fortune K, Sheet Pile Wall Protection with Denso SeaShield™ 2020SP: A Petrolatum System for Sheet Pile Walls, Proceedings of PIANC APAC 2022, Melbourne, 2022.
2. Smith M, Bowley C, and Williams L. (2002), In-Situ Protection of Splash Zones - 30 Years on, Paper presented at the CORROSION 2002, Denver, Colorado, April 2002.
3. Edelman D. (2017), Project profile: Lucinda Bulk Sugar Terminal, Corrosion and Materials, November 2017 pp 82 - 85.

AUTHOR DETAILS

D. Edelman is the corresponding author of this paper. He is recently an independent consultant with Edelman Infrastructure and Engineering. Prior to this role, he was the Manager Engineering at the Port of Townsville and has delivered capital projects with values up to $30M. He has been a project engineer with Queensland Sugar Limited where he was responsible for corrosion protection on the 5.7km long Lucinda Jetty. He has jointly developed the Denso 2020SP system with Denso Australia.

	K. Fortune is a business development manager at Denso Australia, with 6 years' experience in the corrosion protection industry working across marine, water, pipelines and civil infrastructure projects. He has extensive knowledge of Denso's product range and assists customers to achieve project success. Kyle holds a Bachelor of Science (BSc) and is also a qualified commercial diver (Part 1 & 2).

DESIGN AND IMPLENTATION OF A NEW DC MITIGATION SYSTEM FOR RAIL.

Richard Brodribb[1], David O'Brien[1]

[1]M. Brodribb Pty Ltd, Melbourne

Presenter email address: rbrodribb@brodribb.com.au

1. ABSTRACT

Several railways in Australia use 1500VDC to power locomotives in passenger rail systems. The DC supply circuit comprises a DC generator, an overhead catenary and a return rail. The loco draws current from the overhead and returns this to the generator via the rail. The return current may also find return paths through other metallic structures and this stray current can cause corrosion in those structures. An Electrolysis Drainage Unit (EDU) has been designed to provide varying protective current to foreign structures using a programmable algorithm. The EDU incorporates advanced digital control, data logging and remote interfacing with a 6-pulse thyristor rectifier designed to withstand the arduous DC rail environment.

Keywords: ICCP, DC stray current mitigation, DC rail traction system, DC drainage bond.

INTRODUCTION

Direct Current (DC) current at voltages of 1500V or 750V are used to power trains and tram systems in many countries [1]. In Australia, Victoria, New South Wales and South Australia use train or tram or both train and tram DC powered traction systems.

DC powered traction systems in Australia typically use an overhead feed conductor from the power source to the motor in the locomotive and then via a rail return to the power source.

DC return currents may find return paths back to the power source via other parallel paths to the rail return. These other parallel return currents are often referred to as stray currents.

The propensity of DC stray currents to cause corrosion in other buried metallic structures was observed within a few years of the implementation of DC traction systems [2]

Strategies to mitigate stray current corrosion include [3]

a. Electrical isolation of the DC rail return.
b. Bonding to the rail to provide a single point of discharge
c. Control of discharge current by resistor and diode to prevent high currents or reverse currents
d. Impressed current cathodic protection systems which use electronic control to provide variable dc current to the foreign structure. These devices are commonly called, in Australia, a "TDU" standing for "Transformer Drainage Unit" or a "TRAD" standing for "Transformer Rectifier Assisted Drainage System"

In this paper we will discuss the design, manufacture and implementation of a 480A 75V DC, automatically controlled cathodic protection rectifier, based on the concept of a Transformer Drainage Unit and incorporating a high level of automation and integration with a rail system's data control and collection system. The new rectifier has been called an "EDU" – standing for "ELECTROLYSIS DRAINAGE UNIT".

STRAY CURRENT CORROSION

The flow of current in a typical DC traction system passes from the overhead catenary cable, through the motor of the locomotive and returns via the rail. As the train moves along the rail the potential of a buried metallic structure goes from anodic with respect to the rail to cathodic at the point of current return under the engine and then returns to anodic where the current departs the metallic structure to the substation return, as shown in Figure 1.

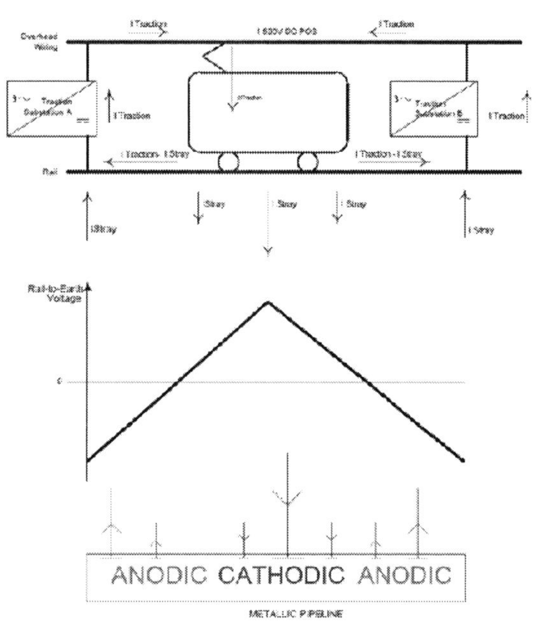

Figure 1 Stray current generation (after [4])

The flow of current from the pipeline (anodic current), back to the substation, may cause corrosion of the pipeline at the area at which this current flow takes place.

MITIGATING STRAY CURRENT CORROSION WITH A TDU.

COMPONENTS AND OPERATION OF A TDU.

A Transformer Drainage Unit (TDU) operates by creating an impressed current system where the cathode is the foreign metallic structure that needs to be protected, and the anode is the steel train rails which the train runs on. The foreign structure becomes more or less cathodic, depending on the level of stray currents that this structure picks up.

The stray current of Figure 1 is related to the traction current drawn by the train motor. If there is no traction current then, intuitively, there can be no stray current effect. When the traction current increases the stray current also increases and this current will flow from the rail to the section of the foreign structure that has become cathodic

The TDU is a DC rectifier with an adjustable output where the rectifier positive is connected to the rail and the rectifier negative is connected to the foreign structure as shown in Figure 2.

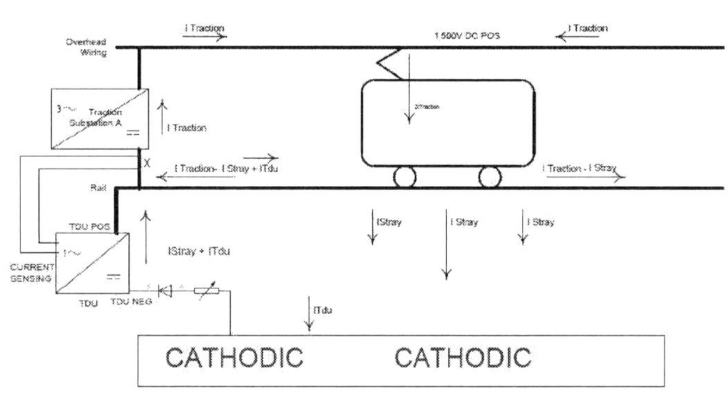

Figure 2 Operation of TDU (after [5])

Stray current (Istray) flows from the rail to the foreign structure, then into the TDU rectifier and from the positive of the TDU back the negative return. Additional current also flows from the structure depending on the condition of its insulating coating – if any, as this structure is also being cathodically protected using the rail as the anode at the same time.

A metallic connection on the foreign structure provides a drain point for the current so that the return current does not leave the foreign structure via defects on its surface.

Control of the output of the TDU can be either from sampling the return current at the substation, as shown in Figure 2 or by measuring the rail to earth voltage and using this to provide increasing output of voltage and current as the rail to earth voltage increases. The TDU can also be connected to multiple structures to provide protection to them.

The output of the TDU can be controller in either voltage mode, where the output voltage remains stable, but the current varies with the resistance of the foreign structure – TDU – rail circuit, or in current mode, where the voltage across the TDU terminals varies but the current drawn remains constant.

The conversion from AC to DC in the TDU is usually done using a single-phase isolation transformer and a thyristor control bridge. The isolation transformer steps down and isolates the AC mains supply and the thyristor control bridge provides the adjustable DC output voltage required to provide protection of the foreign structure. Additional filtering may be provided on the thyristor bridge output to reduce AC ripple in the DC output which may interfere with communications systems and to provide more accurate voltage control.

A typical TDU may have an output voltage of 30V maximum and a DC current rating of 80A [6]

The EDU we will discuss in our paper also operates in voltage mode with sensing available of either the substation negative DC return current or the rail to earth voltage and has the facility to use another control variable leaving giving the opportunity to explore other control strategies for the rectifier output.

The TDU also has an additional blocking diode in the connection from the foreign structure to prevent current flowing from the rail through the TDU and into the foreign structure if the negative terminal of the TDU and structure becomes more negative than the rail. A resistor may also be fitted to limit the maximum current that can flow in this case.

CONTROL RULE OF THE TDU OUTPUT.

The TDU output is determined by either the rail to earth voltage or the substation current, as mentioned in the previous section. The TDU output voltage is adjusted to provide a linear increase of output voltage as the substation DC current increases, often measured from a large current shunt, see Figure 3.

The output voltage is set with a minimum offset value, typically between 1.5 – 7.7V [7], a voltage slope as a proportion of substation current and a maximum voltage to provide safety limits. The current of the TDU is also monitored and set to a maximum value so that the load on the TDU is kept within the rating of the equipment.

The control rules of the TDU output can be characterized as shown in Figure 4 [8].

Figure 3 DC current shunt at substation

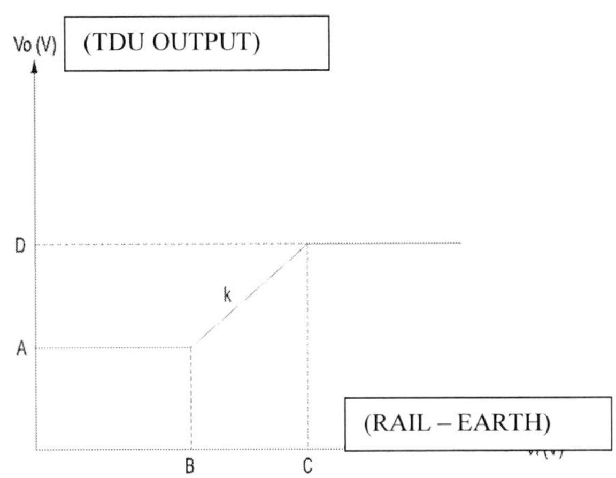

Figure 4 Control Rules for TDU [8]

Rule variables are as follows:

A = offset voltage

B = starting point for TDC Output control

C= Maximum Rail – Earth control value

D = Maximum TDU output voltage.

k = Value of transfer function V0/Vi

The TDU output voltage has 3 separate control rules

Region 1 – Vi<B, Vo = A

Region 2 – B<Vi<D, Vo=k*(Vi-B)+B

Region 3 - Vi>D, Vo=D

A similar control rule was used for the EDU with the choice of rail to earth, substation or other control parameter being able to be substituted for Vi.

DESIGN AND DEVELOPMENT OF THE EDU.

The concept of the operation of the EDU is like that of the TDU discussed above, where the foreign metallic structure is protected by additional voltage developed across the EDU making the structure only cathodic to the rail. However, the new EDU had to be designed to integrate into the client's data and control network, allowing it to be both remotely monitored and controlled using an agreed protocol.

The realization of the system combined our experience with high current DC systems with the need to design the control system to meet new requirements for data monitoring. Input from the client in all parts of the project was essential to help provide an equipment that would be acceptable to a wide range of users.

EDU DESIGN SPECIFICATION.

The specification provided by the client laid out requirements for the following sections of the EDU as shown in Table 1.

Section	Requirement
Power conversion	Dual input 230V Delta or 415V star – using the same rectifier. 0V DC minimum output, 75V DC maximum output 160A DC continuous, 240A DC 1 hour, 480A 1 minute Parallel bypass diode to act as drainage if EDU is off
Controller	Measure rail to earth voltage (+/- 100V DC range) Measure substation return current from external shunt or transducer Adjust DC output with respect to control rule based on rail to earth voltage or substation return current or another control parameter Allow user to input parameter values of controller rule, A, D, B, C, k freely into the rectifier controller via HMI or remotely Remote access to be via the DNP3 communication protocol, which is widely used in power distribution systems.
Display	Show – Rail to Earth voltage, Output voltage, Output current. Show – parameter values of controller rule Record and display – voltage, current visually on Y – T axis graph on HMI Show – active status, faults, alarm history
Store	Measured input control parameter value Output voltage Drainage current Rectifier status.
Mechanical requirements	Indoor mounted Preference for natural convection cooling Specified maximum envelope of 1 180mm W, 780mm D, 2 180mm H. IP 21 touch protection (Test finger not to contact live terminals, drip proof)

	Segregation of test points required by technician
Other requirements	Consideration of electrical safety regarding touch protection. Complete galvanic isolation of rail – earth and substation current circuits from technician accessible components such as control circuit boards. Physical segregation of input, control and output terminations.

Table 1 EDU requirements

7.1 BLOCK DIAGRAM OF EDU

The following illustration shows the key components of the EDU and the interface between the EDU and the client signal sources and SCADA. These elements are described in more detail in Figure 5.

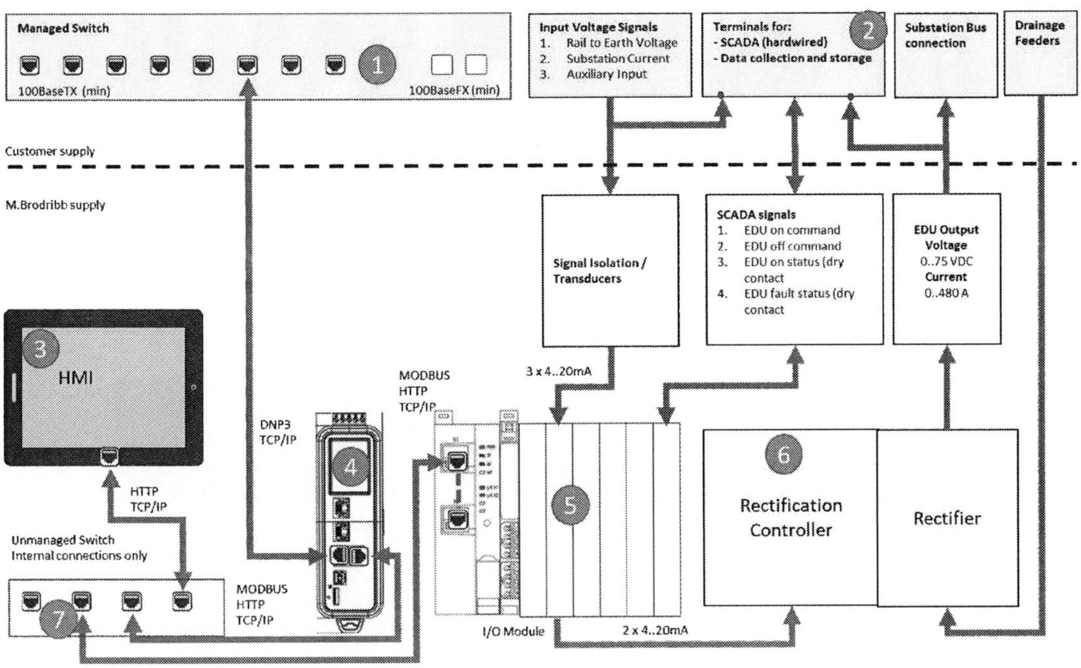

Figure 5 Block diagram of EDU

The components that comprise the EDU are as follows:

1. Managed Network switch (Provided by client, fitted into EDU)
2. Data Collection and Storage module (Proprietary item provided by client, fitted into EDU)
3. HMI Touchscreen (reads web pages served by EDU controller)
4. EDU Controller and DNP3 Gateway. (Calculates output voltage reference from control rule, provides DNP3 Outstation for remote connections and internal webserver for HMI and remote monitoring)
5. IO Modules and Gateway (Interface to internal and external devices including Rail to Earth voltage and Substation current).
6. Rectification controller. (Receives analog signals setting the output from the appropriate IO module and provides trains of gate pulses synchronised to trigger the thyristor bridge rectifier and provide the DC output voltage)
7. Unmanaged Ethernet switch (Provides method for connecting HMI, EDU controller, remote IO)

7.2 EDU CONTROL

The EDU output voltage set point (V0) is calculated by the EDU Controller, item 4 which uses a 3-step linear function of a selectable input parameter (Xi) using the transfer function shown in Figure 4.

All the input parameters for this are entered using the HMI, Component 3.

The EDU controller then provides an 0 – 10V DC analog signal to the Rectifier controller, Component 6. The Rectifier controller regulates the EDU output voltage in proportion to the value of that analog input signal, where 0V input corresponds to 0V output and 10V DC corresponds to 100% output, 75V in this case.

The rectifier controller has independent voltage and current hard limits that are set independent of the controller output. These two limit functions prevent the rectifier exceeding its rated capacity in the event of a controller failure.

The input parameter Xi can be selected by the operator, via the HMI, or via DNP3, from one of three external inputs

1. Rail to earth voltage (-100V to + 100V). The Rail to earth voltage is isolated from the controller input via a 4 – 20mA isolating transducer.
2. Substation current (shunt 0 – 500mV). The substation current is isolated from the controller input via a 4 – 20mA isolating transducer. The shunt value (mV/A) is programmable over a wide range so that the EDU
3. Auxiliary Input. This input could be freely programmed to make the control rule based on an external variable, such as reference potential, realizing the function of a VCDB - Variable Conductance Drainage Bond or to operate as a TRAD - Transformer Rectifier Assisted Drainage system [9] using the same EDU equipment.

These values can all be locally programmed or be programmed through the DNP3 interface.

HUMAN MACHINE INTERFACE (HMI)

The HMI touch screen provides the central control and monitoring interface between the user and the EDU and represented a significant part of the design task.

The HMI has a screen layout intended to provide an overview of operation to any operator and then to logically guide the operator to areas that they are able to access and change, such as voltage settings, depending on their level of authority.

The touch screen view is mirrored on a web page, allowing simple remote control of the HMI.

The EDU default home screen of the HMI is shown in figure 6, indicating rectifier status, trending output and alarm states.

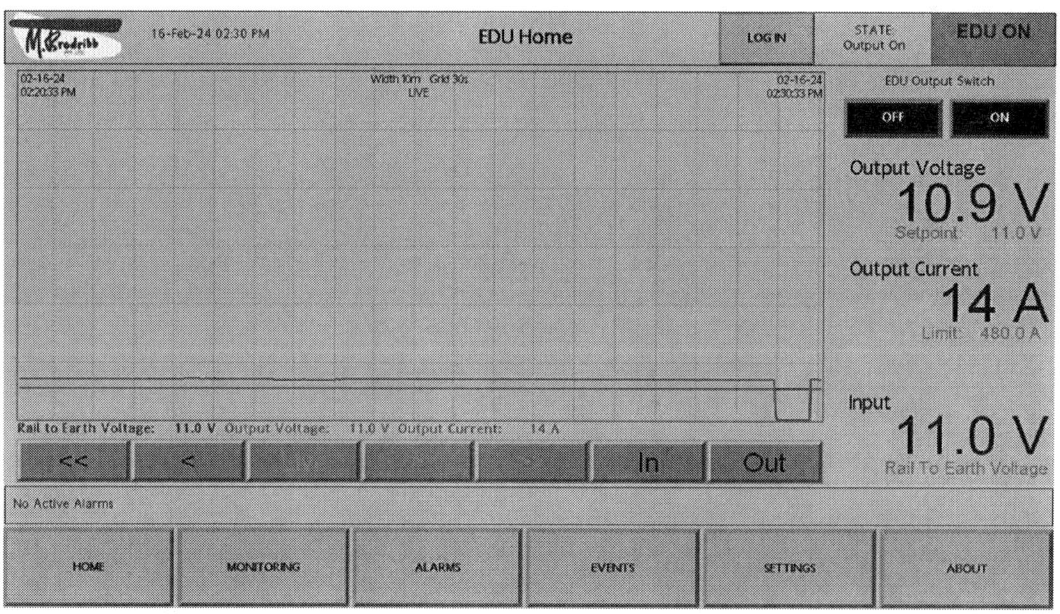

Figure 6 Home Screen of HMI, Output On

The HMI screen display was laid out in a consistent manner between pages, with a status bar at the top, page details of the specific functions in the middle of the page, main navigation buttons at the bottom and subnavigation buttons to the left, where applicable.

The overview of the multiple HMI screens showing content, screen controls and the navigation between the screens is shown in Figure 7.

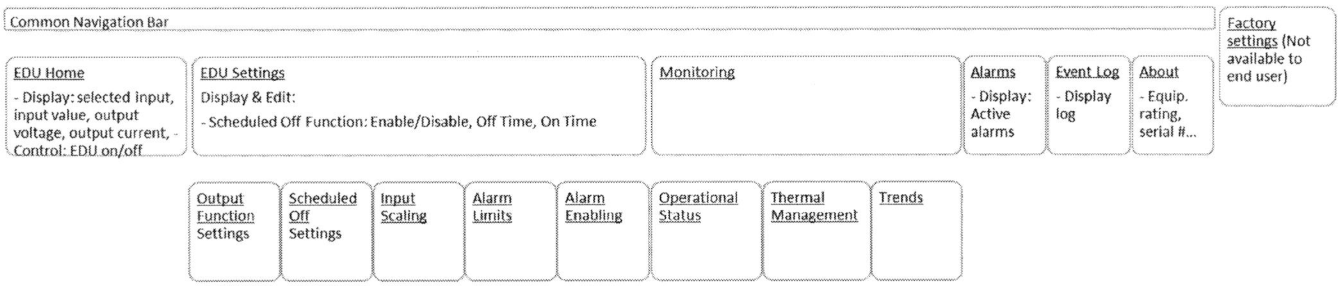

Figure 7 HMI Screen map

HMI screens for settings, alarms and operational status assist the operator in setting the output, checking status and examining the EDU health in detail. A selection of these screens are shown in Figures 8 and 9.

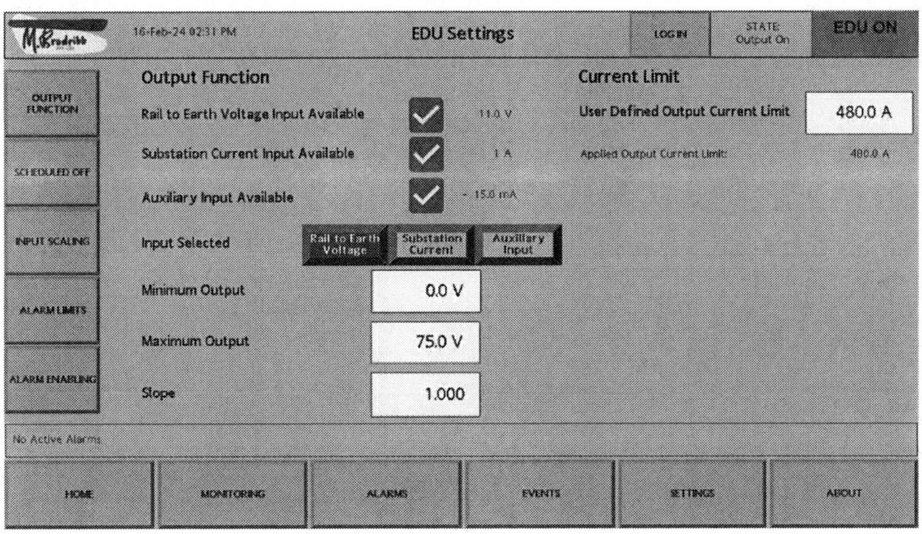

Figure 8 EDU Settings screen

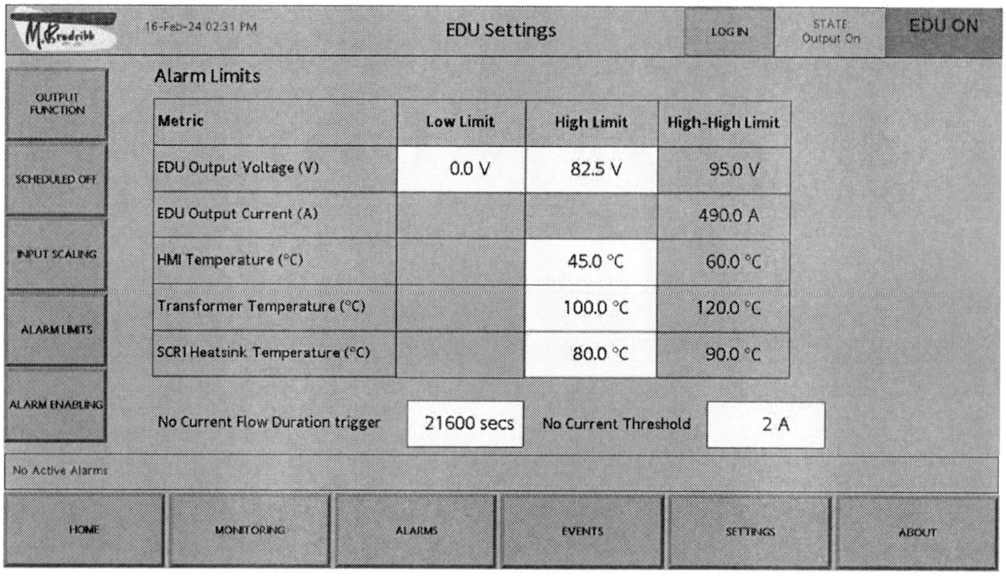

Figure 9 EDU Alarm settings screen

EDU RECTIFICATION CIRCUIT.

The EDU rectification circuit comprised a dual input auto transformer that sets the input to the rectifier transformer to be 415V from either a 230V input or a 415V input, a 415V three phase isolation transformer and a three phase fully controlled rectifier bridge comprising 6 thyristors in a full wave configuration with an additional inductor and capacitor filter to reduce the AC ripple component. The general schematic of the rectifier is shown in Figure 10.

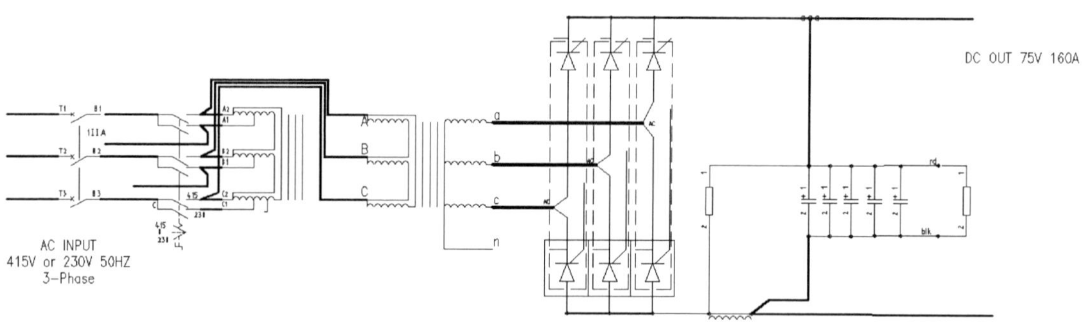

Figure 10 EDU Rectifier schematic

The three phase transformer and rectifier combination provides a more efficient and lower ripple source of DC voltage than the single-phase rectifier system of the TDU and can be more easily designed for higher power outputs, allowing for the protection of more assets. For the EDU of this paper the client specified a continuous rating of 75V 160A, a 1-hour rating of 75V 240A and a 1-minute rating of 75V 480A, this last requirement with a maximum duty cycle of 10%.

The requirement for short term loading required a control rule which could recognize how long the rectifier had been at elevated current and then force the maximum current available down once that time had been exceeded. We chose this method rather than more conventional approaches such as thermal sensing because the heating time of the semiconductor pellets in the thyristor packages at 480A is quick while the heating time of the ceramic thyristor package itself is slow by comparison. This could lead to the situation where the pellet reaches failure temperature inside the thyristor while the outside is still relatively cool. The method of calculation of the time dependent current limit is shown in Appendix 1.

The EDU is also provided with additional inductor capacitor filtering which, combined with the intrinsically lower ripple of three phase rectifier systems compared to single phase systems, performed satisfactorily giving an AC ripple component of approx. 1% rms/dc at 75V 240A load.

The EDU is also designed to provide a drainage function if the foreign structure voltage becomes higher than even if the rectifier is not operating acting as a diode bond drainage without a current limiting resistor.

The EDU has two separate negative outputs, each with a fully rated isolating switch, which enables the asset owner to protect 2 structures, holding them to the same voltage with respect to the rail.

MECHANICAL DESIGN OF EDU.

The mechanical design of the EDU required providing solutions to several problems and both the designers and the client's engineers and field staff worked together to find the result that best suited all parties.

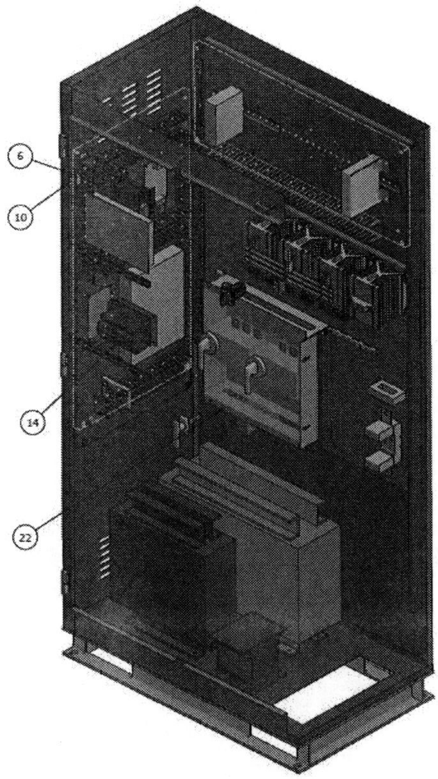

Figure 11 Section of 3D Internal view of EDU

- The EDU cabinet had to fit in the space constraints shown in section 3.1 while incorporating a few bulky components which included the power transformers, rectifier heat sink assemblies and control equipment.
- The cooling of the EDU had to be sufficient that so that fans would not be required except under overload conditions.
- The EDU also had to be safe for technicians to work on. This required shrouding of electrically live components and separation of high voltage or high fault current terminals from low voltage equipment and test terminals.
- The high current cables that connect to the rail and the foreign structure must be accommodated in a way that assists installation.

To assist on solving the problems, 3D modelling techniques were used extensively in the design phase to determine where and whether the components could fit based on the following design order.

- Major components such as transformers and rectifier heatsinks have sizes determined by their power handling capacity and their allowable temperature rise so the design of these components was detailed and 3D blocks of those components were prepared.
- The overall cabinet size was stipulated so the outside cabinet body was designed.
- Terminal strips for high and low voltage were placed in the cabinet at the areas where the external cables could enter or exit
- The remaining smaller control gear could then be positioned in remaining accessible areas.
- The predicted temperature rise in the cabinet was then calculated under load conditions

PERFORMANCE OF EDU

A test program was designed to validate the design and confirm operation of the EDU under normal load and overload conditions, the operation of the control and the performance of the remote interface. The test program included the following elements:

- Electrical safety – the EDU was checked to confirm it was safe to connect to a 415V supply by tests including insulation resistance and high voltage withstand.
- Component checks and correct phasing of the rectifier.
- IO Module tests – conducted via webserver.
- Digital input tests to confirm that the control system recognized connected switches and temperature sensors
- Digital output tests to confirm that the digital outputs corresponding to particular functions such as pilot lights or contactors operated correctly.
- Analog input tests – to confirm that the transfer function used for control recognized rail to earth voltage, current shunt or auxiliary inputs.
- Analog output tests – to confirm that analog signals the EDU provided for remote monitoring were the correct value
- Operation in local and remote control modes
- Heat rise testing at 160A and 240A – using FLIR camera and inbuilt temperature sensors on the transformers and heatsinks
- Short term performance at 480A using a steel bar load – checking that the time based current limit operated.
- Testing of DNP3 interface.

The EDU performance met or exceeded the requirement of the client specification in all cases.

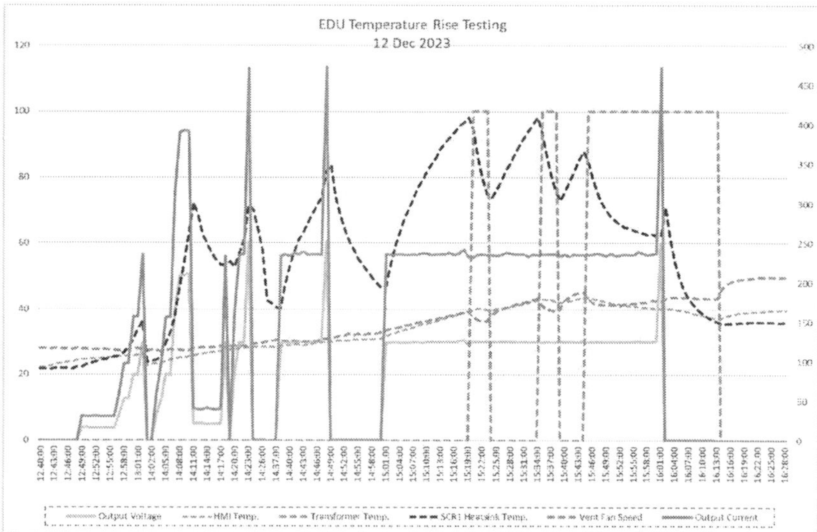

Figure 12 Overload testing at 480A

Figure 12 shows the overload testing being conducted. The left-hand side Y axis is voltage, and the right-hand side Y axis is current. Note the current peaks of 480A are repeated a few times. The operation of the vent fans at overload conditions can also be observed.

This data was presented on the HMI on the EDU and available by the internal webserver in the HMI.

DISCUSSION.

The design and construction of the EDU combined existing technologies that we were familiar with, such as high current DC rectifier design, combined new requirements to provide a high level of automation and integration into the client SCADA Victorian rail network and additional benefits in performance.

Present TDU rectifiers are limited in power to approximately 30V and 80A but the new EDU can source peak currents up to 480A at 75V, allowing more effective mitigation of stray currents as locomotive DC currents increase with more powerful trains and additional DC burdens such as airconditioning.

The TDU also has limited fault reporting and no data gathering capability. The new EDU design addresses this shortfall with both local detailed status monitoring and reporting and remote interfacing to the client SCADA system by an agreed protocol.

The new EDU could also be configured to act as a TRAD for NSW rail networks or a Variable Conduction Drainage Bond where regeneration DC current is an issue.

In the execution of the EDU project we made the following observations.

The design phase needed the following:

- A control platform suitable for rapid prototyping and production use
- Availability of skilled resources familiar with the control platform.
- Robust, reliable and available industrial grade components.

The manufacturing and testing phase raised issues which had to be resolved which included:

- Long supply times post Covid 19 for various electronic components
- Resolution and interpretation of specifications between the stakeholders on the client side and ourselves.
- Suitability of some components to respond to signal changes at the speed required.
- Walk through by client staff who would be using the EDU in the field showed that access to some connection points were not readily accessible, necessitating some rework and design changes
- Designing and executing a test program to cover the operation of the EDU in detail that satisfied the client's requirements.

We also discovered, in discussions with the client's operational staff, that the new EDU would eliminate some of the tasks that they had to carry with separate equipment.

At the time of writing further type tests are being caried out by the client to confirm approval of the EDU on the rail network.

CONCLUSION

A thyristor-based EDU DC power supply was design to replace the function of the existing TDU in reducing stray current corrosion in a rail network. The new EDU operated as required and has further functionality that will increase its utility for the asset owner.

APPENDIX 1 – OVERLOAD TIME CURRENT CALCULATION

The following equations define the time limited overcurrent protection for the EDU to provide protection for the EDU components when repetitive overloads are reapplied at varying time intervals.

OUTPUT CURRENT LIMITS

The following current limits are defined:

- Irated = continuously rated current = 160A
- 1.5 x Irated = 1 hour rated current = 240 A
- Let D1 = Duty cycle @ I > Irated = 10%
- 3 x Irated = 1 min rated current = 480 A
- Let D2 = Duty cycle @ I > 1.5 x Irated = 10%

TRACKING TIME OPERATING ABOVE RATED CURRENT

Two variables are used to monitor and manage the duty cycle when operating above the rated current.

- T1 = operating time (secs) where I > Irated
- T2 = operating time (secs) where I > 1.5 x Irated

T1, T2 are updated every second as follows:

If I <= Irated
T1 = T1 – D1
T2 = T2 – D2
If Irated < I <= 1.5 x Irated
T1 = T1 + 1
T2 = T2 – D2
If 1.5 x Irated < I
T1 = T1 + 1
T2 = T2 + 1
If T1 < 0, T1 = 0
If T2 < 0, T2 = 0

CURRENT LIMIT APPLICATION

If T1 > 1 hour
Output current is limited to Irated
If T2 > 1 minute
Output current is limited to 1.5 x Irated

REFERENCES

Reference number	Source
1	Wikipedia contributors. "List of railway electrification systems.". Wikipedia, The Free Encyclopedia, 13 Aug. 2024. Web. 13 Aug. 2024.
2	Traction Stray Current Mitigation Systems, ACA Conference CP 17, Paper 59, Brodribb, R.
3	ibid, page 3
4	T HR EL 120002 GU Electrolysis from Stray DC Current Version 1.0 01 May 2014. State of NSW, Transport NSW, fig 1
5	VEC Resource Manual, EnergySafe Victoria, July 2014. Melbourne, Victoria, Fig 5
6	ibid, page 22
7	ibid, page 22
8	MTM specification A21111, 2023, Figure 6
9	Traction Stray Current Mitigation Systems, ACA Conference CP 17, Paper 59, Brodribb, R.

AUTHOR DETAILS

Richard Brodribb is the Managing Director of M. Brodribb Pty Ltd.

He has been involved with the design and manufacture of a wide range of DC power supplies for cathodic protection, battery charging and other DC applications for over 40 years.
He has a B.E from Monash (Elec Eng) and a M.B.A from RMIT. He is a lifetime member of IEEE and a member of AMPP.

David O'Brien is an engineer and technology consultant with 30 years of experience in the design, manufacture, installation, and operation of industrial products and equipment. He currently leads the development of advanced controllers for DC power supplies used in cathodic protection, battery charging, and other specialized applications. David holds a B.E. in Mechatronics from Monash University and an M.S. in Computer Science from Georgia Tech.

CORROSION & PREVENTION
CAIRNS 2024

THE AUSTRALASIAN CORROSION ASSOCIATION INC.

ASSESSMENT CRITERIA FOR THE ELECTROCHEMICAL PROTECTION OF STEEL

C. Stone and G.K. Glass

Concrete Preservation Technologies, Long Eaton, UK

garethg@cp-tech.co.uk

ABSTRACT

Cathodic protection of steel is generally assessed by measuring the steel potential or determining the steel potential shift. In this work mixed potential theory is used to evaluate criteria for electrochemical treatments for steel. This study shows that when the kinetics of the cathodic reaction exhibit activation controlled behaviour, mixed potential theory may be used to monitor the development of steel passivity. In general, passive conditions are indicated by low current densities that deliver large potential shifts. Alternatively, when the cathodic reaction is under mass transfer control, the corrosion rate while the current is applied may be estimated. Thus, criteria for monitoring and assessment of electrochemical treatments may be enhanced if potential changes are linked to the current which induces the change. Such assessment might be used to develop new treatments.

Keywords: corrosion, cathodic protection, electrochemical treatment, assessment criteria

INTRODUCTION

The original basis for cathodic protection (CP) is to reduce the force driving metal oxidation (its conversion to positive ions) by shifting the potential of a metal in the negative direction. This is embodied in at least one definition of CP [1]. However, a cathodic current has many other protective effects. These include the removal of aggressive negative ions and the production of inhibitive hydroxyl ions at a steel surface. These effects are already harnessed in other techniques [2]. As a result a cathodic current may promote the formation of a passive film on steel in environments were passivity is not readily induced [3]. Such effects provide an alternative basis for protection.

The original criterion to achieve CP required the achievement of a potential reflecting metal immunity. Another criterion involves the determination of the potential decay or potential shift following the interruption or application of the protection current. The magnitude of the potential decay required is usually 100 mV [4]. The choice of 100 mV has largely been pragmatic; it is generally achievable and experimental data suggests it works in environments like concrete. However some laboratory studies have suggested that the protective effects of such a relatively small negative potential shift may not prevent corrosion in all cases [5,6]. In this work mixed potential theory is used to evaluate protection criteria for steel.

Mixed Potential Theory

Mixed potential theory is a theory used in electrochemistry that considers a corrosion process to result from two or more electrodes supporting distinct electrochemical reactions [7,8]. Each reaction has a distinct potential-current relationship. Potential changes from the natural electrode potential represent a driving force for the electrochemical reaction while the current represents its rate. The main reactions which characterise the corrosion of steel are the oxidation of iron (the anodic reaction) and the reduction of oxygen (the cathodic reaction). Iron has a negative natural potential, while oxygen has a positive natural potential. In the corrosion process, the rate of iron oxidation is balanced by the rate of oxygen reduction.

The corrosion process may be under anodic, cathodic or resistive control depending on the behaviour of the anodic and cathodic reactions as well as the resistance of the electrolyte in the environment. The separate anodic and cathodic reactions are generally considered to be under activation or mass transfer control. The behaviour of the reactions and the effects of resistivity determine the relationship between corrosion potential and corrosion rate as well as the effect of any applied current on the corrosion potential and corrosion rate. The reaction behaviours are well documented and are presented below [9,10].

The applied current density (i_{appl}) is given by the difference between the cathodic and anodic reaction rates (i_c and i_a):

$$i_{appl} = i_c - i_a \tag{1}$$

A cathodic current will stimulate the cathodic reaction and restrict the anodic reaction. The activation controlled contribution to a negative potential shift from the corrosion potential is described by the equations:

$$i_a = i_{corr} exp\left(-\frac{2.3\eta_a}{\beta_a}\right) \tag{2}$$

$$i_c = i_{corr} exp\left(\frac{2.3\eta_a}{\beta_c}\right) \tag{3}$$

where i_{corr}, η_a, β_a and β_c are the corrosion rate, activation induced potential shift (overpotential) and the anodic and cathodic Tafel constants respectively. The corrosion rate in this equation is the corrosion rate in the absence of a current (the open circuit corrosion rate). These equations with be presented slightly differently in the literature dependant on whether an applied cathodic current is considered to be positive or negative and whether the Tafel constant represents a slope on a logarithmic scale with base ten or the natural logarithmic scale. For the purpose of the illustrations in this work, the base 10 logarithmic scale is used and the protection current has a positive value.

Evidence of activation controlled cathodic kinetics on steel in concrete is provided by the reported relationship between the corrosion rate and the corrosion potential. An exponential increase in the corrosion rate is often accompanied by a linear decrease (negative shift) in the corrosion potential. If the variation in corrosion rate results principally from a change in the kinetics of one reaction, the corrosion potential-corrosion rate relationship may provide the Tafel slope of the other reaction. Examples are provided in Figure 1 below [11,12]. The theoretical value of the Tafel slopes is approximately 120 mV. It may vary depending on a symmetry factor also referred to as a transfer coefficient (approximately 0.5) and the number of electrons taking part in the rate determining step of the reaction process (usually one). The whole reaction process may involve several electrons.

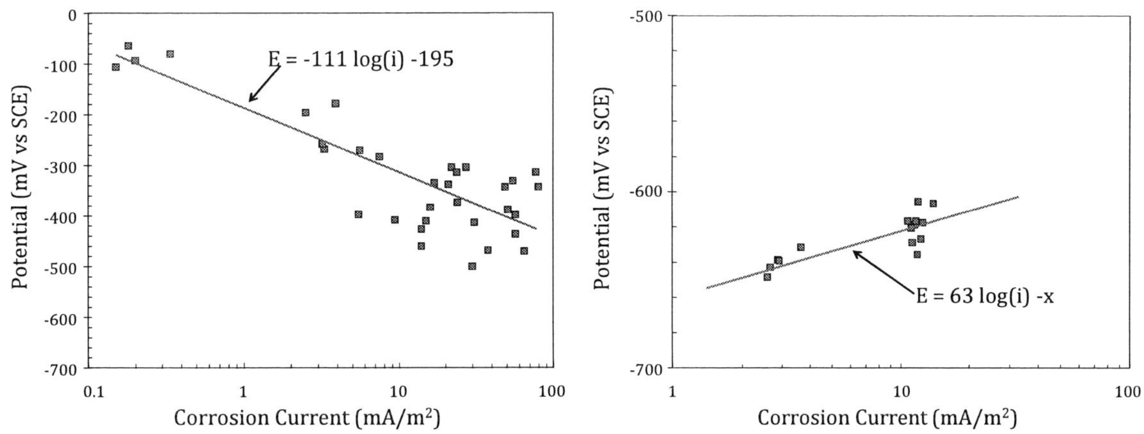

FIGURE 1 EVIDENCE OF ACTIVATION CONTROLLED CATHODIC BEHAVIOUR (LEFT) AND ACTIVATION CONTROLLED ANODIC BEHAVIOUR (RIGHT) IN THE RELATIONSHIP BETWEEN CORROSION POTENTIAL AND CORROSION RATE [12,13].

Mass transfer controlled cathodic kinetics might be expected when the cathodic reaction rate approaches the limiting current for oxygen reduction. As the cathodic reaction rate approaches the limiting current for oxygen reduction (i_L), a mass transfer contribution to the potential shift (η_m) is given by the equation:

$$\eta_m = \frac{\mathrm{RT}}{\mathrm{zF}} \ln\left(1 - \frac{i_c}{i_L}\right) \tag{4}$$

where R, T, z and F are the gas constant, absolute temperature, valence number and Faraday constant respectively,

The potential shift (η) at a cathode is given by the sum of the activation (η_a) and mass transfer (η_m) contributions [13]:

$$\eta = \eta_a + \eta_m \tag{5}$$

The resistive contribution has been neglected as attempts are made to eliminate the effects of the electrolyte resistance (IR drop) when potentials are measured for the purposes of CP assessment.

A potential decay determined relative to the instant off potential (the potential measured on interrupting the current) gives a conservative estimate of the potential shift achieved by a CP system [14]. When oxygen starved conditions exist decays are very slow and potentials are more negative [15]. Thus, a time limit sometimes imposed on the measurement of a potential decay may result in a significant underestimation of the true potential shift.

In oxygen starved conditions (strongly polarised cathodic kinetics) the anodic reaction kinetics are more likely to be under activation control [12]. The corrosion rate may be approximated by the limiting current for oxygen reduction, over a range of potentials in the absence of a protection current. The relationship between the applied current and the potential shift is then defined by the kinetics of the anodic reaction (Equation 2 above).

When activation controlled anodic reaction kinetics are unlikely (e.g. where changes in the corrosion rate result from the formation and breakdown of passive films, a process not represented by Equation 2), the cathodic reaction is not likely to be oxygen starved. Changes in the cathodic reaction rate dominate the applied current-potential shift relationship and give rise to the corrosion potential -corrosion rate relationship of the example in Figure 1 (left).

Basis for Protection Criteria

Immunity

Immunity is achieved when the metal potential is driven to a point where the rate at which the metal oxidises is the same as the rate at which it is deposited back onto the metal surface (also known as the exchange current density). This depends on the concentration and solubility of the metal corrosion products in the environment. It is largely determined by Thermodynamics. For steel, this potential is approximately -850 mV relative to a Copper/Copper Sulphate reference.

Immunity is generally not practically achievable when the cathodic reaction (e.g. oxygen reduction) can occur at very high rates [16]. The cathodic reaction rate determines the required current to achieve the acceptable metal potential. As a consequence, the immunity criterion tends to be limited to submerged and buried conditions. In other cases a more pragmatic potential decay, or potential shift criterion is applied.

Weakly Polarised Cathodic Kinetics

Weakly polarised cathodic kinetics are likely to occur in aerated or atmospherically exposed conditions. Equations 1 to 5 have been solved by iteration to give the relationship between the open circuit corrosion rate, applied current density and potential shift [10]. The equations may also be simplified in some circumstances to give a directly calculated result. Thus if it is assumed that mass transfer effects are negligible, the above equations may be written as:

$$i_{\text{corr}} = i_{\text{appl}} \Big/ \left(\exp\left(\frac{2.3\eta}{\beta_c}\right) - \exp\left(-\frac{2.3\eta}{\beta_a}\right) \right) \tag{6}$$

This is a re-arrangement of the Wagner - Traud equation [7] from which much of the above theory is derived. The assumption that either the anodic or cathodic kinetics are under activation control provides the basis for the non-destructive polarisation resistance method of corrosion rate determination. The widely used Stern – Geary equation for corrosion rate determination is effectively an approximation of the Wagner – Traud equation at small potential shifts [17-19]. It was not practical to use Equation 6 at the time of Stern and Geary because calculators that could solve the exponential function did not exist.

The sensitivity of the assumptions relating to mass transfer and activation control were determined by comparing the iterated solution with the directly calculated result while varying the anodic and cathodic Tafel constants (activation control) and the limiting current (mass transfer control) [20]. The results are provided in Figure 2 below.

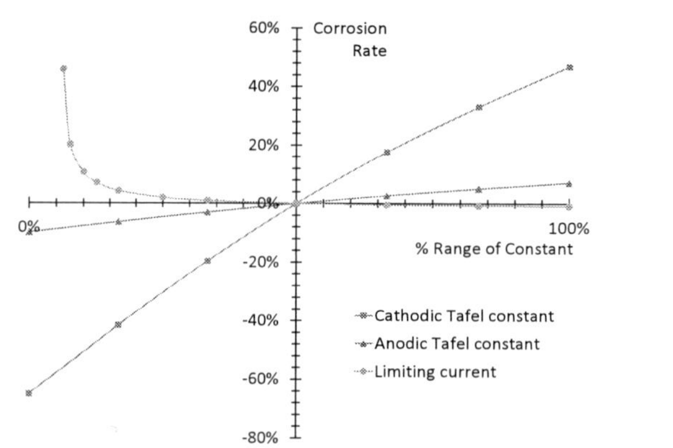

FIGURE 2 SENSITIVITY OF THE CALCULATED OPEN CIRCUIT CORROSION RATE TO CHANGES IN THE LIMITING CURRENT AND ANODIC AND CATHODIC TAFEL CONSTANTS [20].

Figure 2 plots the change in corrosion rate, expressed as a percentage of its value at βa =120 mV, βc = 120 mV and i_L = 60 mA/m^2, as a function of the percentage of the range of each constant (60 to 180 mV for the Tafel constants and 0 to 120 mA/m^2 for the limiting current). In these calculations, 50 mV of polarisation (η = 50 mV) was delivered by an applied current of 5 mA/m^2 (i_{appl} = 5 mA/m^2). The most significant effect arises from the cathodic Tafel constant. The corrosion rate is the calculated corrosion rate in the absence of protection.

To convert corrosion current (rate) to corrosion current density, the area of the corroding steel surface is required. However corrosion current densities are average values and localised corrosion rates may be substantially higher. Various interested parties have taken average corrosion rate data and converted the numbers into a corrosion risk classification [21,22]. The classification adapted from a Concrete Society Report [22] for reinforced concrete is provided in Table 1.

TABLE 1. CORROSION RISK CLASSIFICATION.

Average corrosion current density (mA/m^2)	Average rate of steel section loss (μm/year)	Risk classification
Up to 1-2	Up to 1-2	Very low or passive
2 to 5	2 to 6	Low to moderate
5 to 10	6 to 12	Moderate to high
>10	>12	High

In Table 1, the boundary between passive steel and a low corrosion risk is set at an average steel section loss of 1 mm after 400 to 500 years. The lower boundary of a high corrosion risk is set at a average steel section loss of 1 mm after 80 to 90 years. Both of these average rates are low, but the risk classification relates to the likelihood of the corrosion being localised to a small percentage of the steel surface area and therefore being much higher where it occurs [21,23].

The process of calculating the relationship between corrosion rate, applied current and resulting potential shift is graphically illustrated in Figure 3. In this illustration, the cathodic polarisation behaviour (oxygen reduction) is taken from the experimental data provided in Figure 1 (left). The anodic polarisation behaviour (iron oxidation) is very approximate but has little effect according to the analysis in Figure 2. An applied current of 20mA/m^2 induces a potential shift of 100 mV. This can only occur in the laboratory conditions in Figure 1 (left) when the corrosion rate is about 4 mA/m^2.

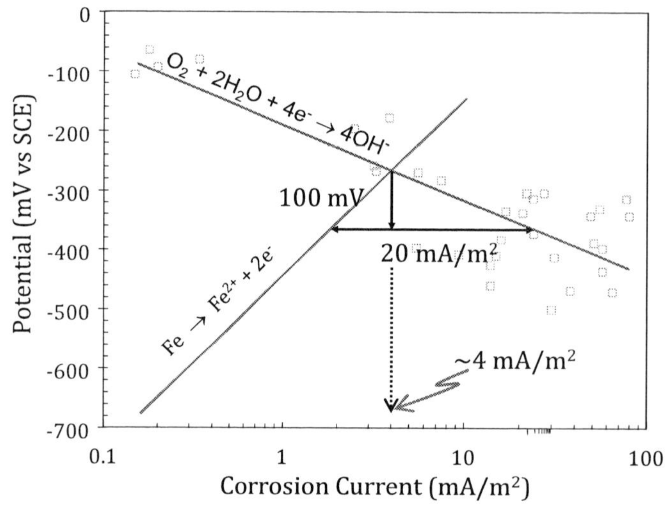

FIGURE 3 GRAPHICAL ILLUSTRATION OF THE PROCESS OF CALCULATING THE RELATIONSHIP BETWEEN CORROSION RATE, APPLIED CURRENT AND POTENTIAL SHIFT

This analysis has been performed to calculate the family of curves in Figure 3 giving the predicted relationship between the open circuit corrosion rate, potential shift and the applied current density [20]. In these calculations the anodic and cathodic Tafel slopes were taken to be 120 mV as this was typically used when developing the corrosion risk classification in Table 1 above.

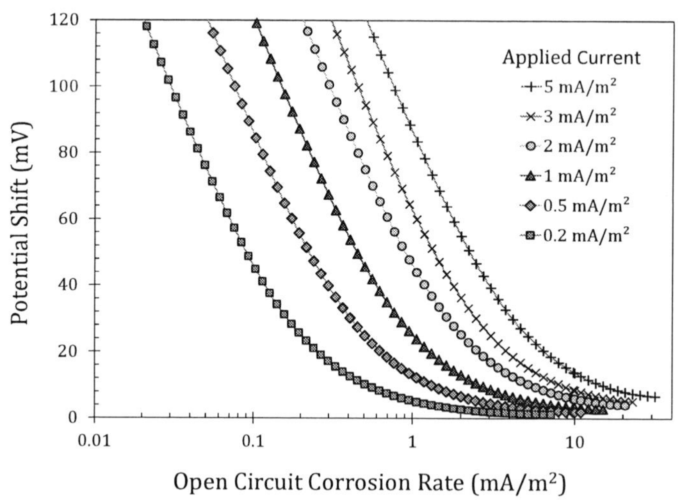

FIGURE 4. THEORETICAL POTENTIAL SHIFT AS A FUNCTION OF OPEN CIRCUIT CORROSION RATE AT VARIOUS APPLIED CURRENT DENSITIES

The interpretation of the data in Figure 4 in the context of the risk classification in Table 1 is provided in Figure 5 [20]. Figure 5 shows the potential shift as a function of the open circuit corrosion rate at an applied current density of 5 mA/m^2. The boundary between passive steel and a low corrosion risk is added as a vertical line that intercepts the x-axis at 2 mA/m^2. At this corrosion rate, about 55 mV of potential shift is required at a protection current of 5 mA/m^2.

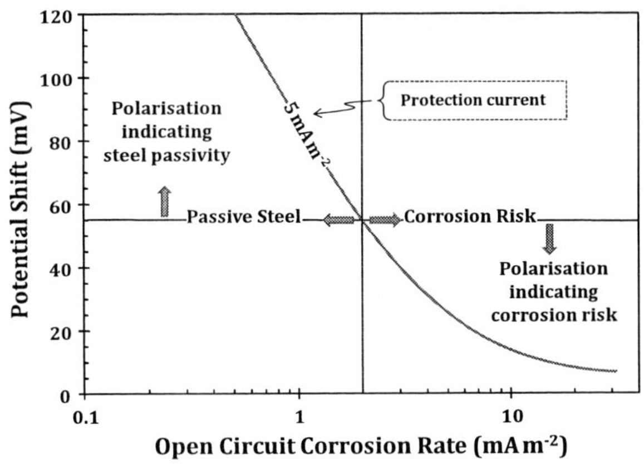

FIGURE 5. INTERPRETATION OF POTENTIAL SHIFT IN TERMS OF CORROSION RISK.

Oxygen Starved Conditions

In terms of CP, oxygen starved conditions represent strongly polarised cathodic kinetics. The dependence of potential shift on corrosion rate changes as the cathodic reaction rate approaches the limiting current for oxygen reduction. When this occurs the anodic reaction kinetics are more likely to reflect activation controlled behaviour (Figure 1 (right) and Equation 2). An anodic Tafel slope of 60 mV will result in a negative potential shift of 100 mV being accompanied by a reduction in the rate of anodic dissolution (the corrosion rate while CP is applied) by a factor of 46. This is embodied in the conventional understanding of CP (to reduce the corrosion rate by making the potential more negative).

A CP criterion based on the achievement of a negative potential shift may however again be enhanced if it is linked to the applied current density. For mass transfer controlled cathodic kinetics, both the open circuit corrosion rate and the cathodic reaction rate may be approximated by the limiting current (i_L). Equation 1 then becomes:

$$i_{app} = i_L \left(1 - exp\left(-\frac{2.3\eta}{\beta_a}\right)\right) \quad (7)$$

This may be used to determine the limiting current for oxygen reduction. Once this is known, the percentage reduction in corrosion rate induced by a negative potential shift provides the protected corrosion rate (anodic reaction rate while the current is applied) [10].

The predicted relationship between the protected corrosion rate (as opposed to the open circuit corrosion rate), the potential shift and the applied current is given in Figure 6 for an anodic Tafel slope of 60 mV [10]. If a potential shift of 100 mV was achieved with a current density of 20 mA/m^2, it can be deduced that the limiting current for oxygen reduction is 20.44 mA/m^2 and, while protection is applied, the protected corrosion rate is only 0.44 mA/m^2.

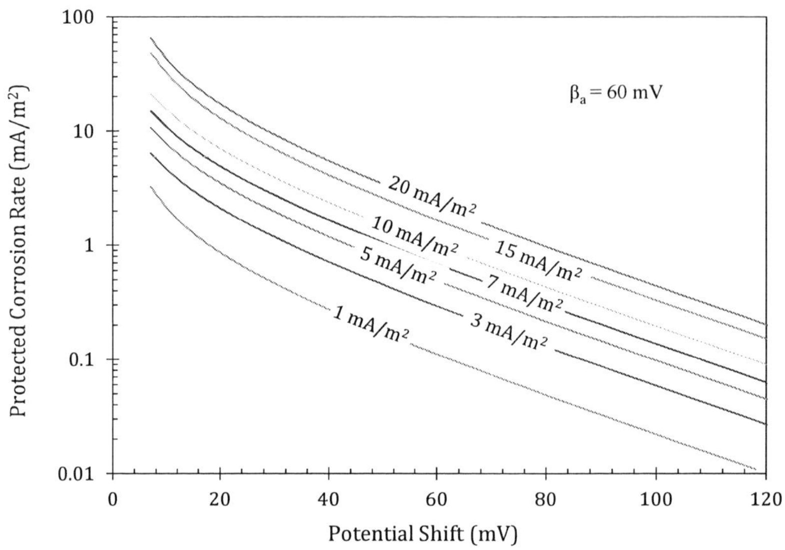

FIGURE 6 - THE PREDICTED RELATIONSHIP BETWEEN THE APPLIED CURRENT, POTENTIAL SHIFT AND PROTECTED CORROSION RATE IN OXYGEN STARVED CONDITIONS [10].

Discussion

The above analysis shows that the achievement of 100 mV of potential shift by the relatively small protection current densities implies that further corrosion induced deterioration will be negligible. This arises either because the steel is passive or because it has a very low protected corrosion rate.

For example, when 100 mV is achieved by 20 mA/m^2 in aerated conditions (weakly polarised cathodic kinetics), the open circuit corrosion rate is calculated to be under 4 mA/m^2. 20 mA/m^2 is seldom exceeded as a design current density for above ground reinforced concrete and many are designed at lower current densities. The borderline corrosion rate for a low to moderate corrosion risk in concrete is considered to be a corrosion rate of about 5 mA/m^2 (Table 1).

Evidence of CP induced passivation has previously been provided by a positive shift in the monitored steel potentials as the duration of CP increases in above ground concrete. Indeed a reduced current demand accompanied by a gradual shift in protection potentials (instant off potentials measured during CP) to values that were more positive than the corrosion potential determined prior to the application of CP, has been observed in both laboratory and field studies [24]. This occurs despite the fact that a cathodic current displaces the potential in the negative direction.

However, continued corrosion has been observed in some laboratory experiments when 100mV of potential shift was achieved. The reason for this becomes evident when the high applied current required to achieve this level of potential shift is noted. Thus in one case corrosion was observed when currents of the order of 100 mA/m^2 were used to achieve only 100 mV of potential shift [5]. This is consistent with the above hypothesis which suggests that the unprotected corrosion rate in this case was very high and therefore more protection current would have been required. Thus it can be helpful to consider the local applied current density when assessing performance.

A theoretical basis for the 100 mV decay criterion in above ground concrete is that, when it is achieved by relatively low applied current densities, passive or near passive conditions have been established on the reinforcing steel [10]. By the same analysis, another protective treatment might only achieve 55 mV with an applied current of 5 mA/m^2. This is also indicative of passive steel even though it may not comply with cathodic protection standard practice [20].

The direct protective effects of a negative potential shift will be limited when the cathodic reaction occurs with ease. This is because the cathodic reaction kinetics prevent substantial potential shifts from being achieved. On the other hand when the cathodic reaction rate approaches the limiting current for oxygen reduction (mass transfer control), a substantial reduction in an initially high corrosion rate may result. When the cathodic reaction kinetics are under mass transfer control and the anodic reaction kinetics are under activation control, the magnitude of the potential shift and the current density may be used to estimate the corrosion rate during protection. A negative potential shift of 100mV will only be induced by relatively low protection current densities if the resulting rate of anodic dissolution (the corrosion rate during protection) is low.

This provides another basis for the 100 mV decay criterion that is applicable when the cathodic reaction kinetics are under mass transfer control, although potential decays are more difficult to measure in these conditions.

Conclusions

When the cathodic reaction kinetics are predominantly under activation control (aerated or atmospherically exposed), the negative potential shift produced by a known current density may be used to estimate the open circuit corrosion rate of the protected steel (the corrosion rate in the absence of protection). A substantial potential decay will only be achieved at relatively low protection current densities if the open circuit corrosion rate is low. In this case the restoration of passivity provides a basis for electrochemical protection.

When the cathodic reaction kinetics are under mass transfer control (buried or submerged conditions) and the anodic reaction kinetics are under activation control, the magnitude of the potential shift and the current density may be used to estimate the corrosion rate during protection. A substantial negative potential shift will only be induced by relatively low protection current densities if the resulting rate of anodic dissolution (the corrosion rate during protection) is low.

Potential shifts and decays may be linked to the local applied current density to improve existing protection criteria and provide criteria for electrochemical treatments that fall outside of existing standard practice. This is particularly true when the cathodic reaction occurs with ease (weakly polarised cathodic kinetics).

References

1. Shreir LL, Electrochemical Principles of Corrosion - A Guide for Engineers, National Corrosion Service, Teddington, U.K.: Department of Industry (Publishers), 1982).
2. Mietz J (Ed.), Electrochemical Rehabilitation Methods for Reinforced Concrete Structures – A State of the Art Report, European Federation of Corrosion Publication Number 24, IOM Communications (Publishers), London, 1998.
3. Glass GK, The effect of a change in surface conditions produced by anodic and cathodic reactions on the passivation of mild steel, Corrosion Science 26(6) (1986) 441-454.
4. BS 7361: Part 1 : 1991, Cathodic Protection - Code of Practice for Land and Marine Applications, British Standards Institute (Publishers), London, 1991.
5. Takewaka K, Cathodic protection for reinforced-concrete and prestressed-concrete structures, Corrosion Science 35(5-8) (1993) 1617-1623.
6. Funahashi M, Bushman JB, Technical review of 100 mV polarization shift criterion for reinforcing steel in concrete, Corrosion 47(5) (1991) 376-386.
7. Flitt H, Will G, and Green W, Electrokinetic review and appraisal of steel reinforcement corrosion in concrete, ACA Conference: Corrosion & Prevention 2015, Paper 016.
8. McCafferty E. Introduction to Corrosion Science, Springer (Publishers), New York, 2010, 140-146.
9. Bard AJ, Faulkner LR, Electrochemical Methods, Fundamentals and Applications, John Wiley & Sons (Publishers) New York, 1980, 100-110.
10. Glass GK, The 100-mV potential decay cathodic protection criterion, Corrosion 55(3) (1999) 286-290.
11. Glass GK, Hassanein AM and Buenfeld NR, Monitoring the passivation of steel in concrete induced by cathodic protection, Corrosion Science 39(8) (1997) 1451-1458.

12. Zayed AM, Sagues AA, Corrosion at surface damage on an epoxy-coated reinforcing steel, Corrosion Science 30(10) (1990) 1025-1044.
13. Evans UR, The corrosion and oxidation of metals: Scientific principles and practical applications, Edward Arnold (Publishers) Ltd., London, 1960.
14. Bennett J, Broomfield JP, Analysis of studies on cathodic protection criteria for steel in concrete, Materials Performance 36(12) (1997) 16-21.
15. Glass GK, Analysis of data on a reinforced concrete cathodic protection system, Materials Performance 35(2) (1996) 36-41.
16. Ashworth V, Booker CJL, Cathodic Protection, Theory and Practice, Institute of Corrosion and Ellis Horwood (Publishers), Chichester, 1986.
17. Stern M and Geary AL, Electrochemical Polarization: I . A Theoretical Analysis of the Shape of Polarization Curves, Journal of the Electrochemical Society 104 (1) (1957) 56-63.
18. Stern M, A method for determining corrosion rates from linear polarization data, Corrosion 14(9) (1958) 60-64.
19. Frankel GS, Wagner-Traud to Stern-Geary; Development of Corrosion Kinetics, Corrosion science: a retrospective and current status symposium, Electrochemical Society (Publishers), Philadelphia, May, 2002.
20. Glass GK, Davison N and Roberts AC, Assessment of cathodic protection applied to above ground reinforced concrete, Materials and Corrosion 70(3) (2019) 503-510.
21. Andrade C, Alonso C, Gulikers J, Polder R, Cigna R, Vennesland O, Salta M, Raharinaivo A, Elsener B, Recommendations of RILEM TC-154-EMC: Electrochemical techniques for measuring metallic corrosion, Materials and Structures 37(273) (2004) 623-643.
22. Concrete Society Technical Report No 60, "Electrochemical tests for reinforcement corrosion", The Concrete Society (Publishers), Camberley, U.K. 2004.
23. Dodds W, Christodoulou C, Goodier C, Atkins C and Preston J, Discussion: Hybrid anode concrete corrosion protection – independent study, Proceedings of the Institution of Civil Engineers - Construction Materials, 170(CM5) (2017) 265-270.
24. Glass GK and Chadwick JR, An investigation into the mechanisms of protection afforded by a cathodic current and the implications for advances in the field of cathodic protection, Corrosion Science 36(12) (1994) 2193-2209.

Author Details

	Christian has both a Masters in Physics and a Masters in Material Science and Engineering. He is a technical expert in RAAC and cathodic protection of concrete, and spends his time between CPT and Loughborough University. Christian works with many prestigious clients including NHS trusts and Pension Funds, sits on the Technical Committee of the CPA/CRA, and contributes to international bodies helping to guide engineers in corrosion management, as well as supporting our partners outside the UK.
	Gareth Glass is the corresponding author of this paper. He is a Director at Concrete Preservation Technologies, a position he has held since the inception of the company in 2005. He has extensive experience in materials technology, durability and rehabilitation of structures with over 100 publications to his name in the area of corrosion protection. He obtained his PhD from the Corrosion and Protection Centre, University of Manchester, in 1984.

Concrete Asset Inspections On Wastewater Infrastructure

J. Rigby[1]

[1]*Remedy Asset Protection, Brisbane, 4000, Australia*

Justin.Rigby@RemedyAP.com.au

ABSTRACT

Environmental exposure, poor concrete quality and aspects of structural design will affect the initiation and propagation of steel reinforcement corrosion. A relatively short time is required between the occurrence of the first visible crack along the reinforcement and actual structural damage. This paper describes concrete degradation prior to corrosion occurring on the reinforcing steel and explanation of how a diagnostic approach to asset inspection can be employed to meet and extend asset lifecycles. Discussion is provided on the embodied carbon of concrete materials and a case study of an asset inspection to a 90-year-old wastewater well.

Keywords: concrete diagnostic, concrete inspection, remedial concrete, concrete asset condition, concrete investigation.

Introduction

The processes of concrete degradation, resulting in corrosion of steel reinforcement, are dependent on exposure condition and concrete quality.

Exposure conditions cannot be controlled without relocation or change of processes. However, concrete quality can be controlled because it is influenced by the mix design, quality of the constituents, placement, compaction, curing conditions, design and detailing. The water-to-binder ratio together with the total binder content of the mix design, and the extent of curing afforded to the concrete, are key factors in controlling permeability and, hence, concrete durability.

When concrete is correctly placed, and exposure conditions remain static, the interaction between exposure and concrete quality will show a near linear deterioration over the lifecycle until the time that passivation of the steel reinforcement is lost and corrosion can initiate.

Using both as-built and in-service inspection data, modelling can be used throughout the structure's service life to identify the predominate degradation mechanism and predict asset service life. This information provides understanding of when interventions are most appropriate to prevent premature concrete failure or to ensure an extended service life.

An inspection of a wastewater infrastructure is discussed in this case study.

Relevant Concepts

Concrete Degradation

Concrete quality and aspects of structural design affect the initiation and propagation of steel reinforcement corrosion.

The deterioration reactions occur mainly because of the transport of aggressive substances either from the environment to the concrete or internally within the concrete matrix. Moisture is required for most of the chemical reactions which cause concrete to deteriorate. By itself, moisture or water does not normally degrade concrete, rather it transports deleterious solids or gases into concrete and serves as an electrolyte to enhance chemical reactions.

The structural and crack control requirements in steel-reinforced concrete design propose that the steel be placed as close to the concrete surface as possible, where tension forces are high, to control crack formation. However, an adequate cover of good quality concrete to the reinforcement is needed to serve as a physical barrier against the ingress of aggressive substances.

The key environmental factors that initiate corrosion of steel are carbonation and chloride. Other factors which may influence either the initiation or rate of reinforcement corrosion include cracks in concrete, temperature, moisture, oxygen, chemical attack and inadequate concrete quality or cover.

Concrete Quality

The solution contained in the pore structure of concrete is highly alkaline as a result of cement hydration, with a pH of about 13 for Portland cements. The alkaline pore solution is particularly beneficial to steel reinforcement as it provides chemical protection in the form of a passive iron oxide film. This film is adherent and protects against corrosion.

High quality concrete can significantly retard corrosion initiation. This is because good quality concrete, made with low water-to-cement content ratio and suitable materials, well compacted and cured, has low permeability and hence a greater resistance to the ingress of liquids and gases.

Adequate concrete quality and cover to steel reinforcement is necessary because the transport of chlorides, atmospheric carbon dioxide and other aggressive substances occurs at rates that typically show a parabolic curve. This implies that if the concrete cover is halved, the time to corrosion initiation is reduced to about a quarter of the time [1].

The effect of cracks on the risk and severity of corrosion of reinforcing steel is determined by the origin, width, intensity and orientation of the cracks. CO_2 and chlorides tend to penetrate more readily towards the reinforcement along cracks compared to sound concrete.

Where the crack is perpendicular to the reinforcement, corrosion is usually confined to a small area. Cracks aligned parallel to reinforcement bars present a larger exposure area and thus a higher risk of causing damage from spalling of the concrete.

Exposure Environment

Environmental exposure will affect the initiation and propagation of steel reinforcement corrosion. Some of these are summarised below:

- Carbonation deteriorates reinforced concrete structures by reducing the alkalinity of concrete
- The diffusion of chloride ions into concrete from external sources is dependent on concrete quality, cement type, cover and exposure conditions. Alternating wet and dry exposure conditions severely enhances corrosion rates

- Chemical attack from sulphate, soft water or acidic solutions.

Furthermore, within AS 3735, four basic exposure classifications for exposure to chemical or penetrating agents are defined in terms of their resultant effect on a concrete member as follows [2]:

- A – Where the concrete is in a non-aggressive environment or is protected from aggressive agents. This classification may be appropriate for surfaces protected or isolated from the attacking environment or where a lower level of durability is applicable
- B – Where the concrete is in an aggressive environment but only subjected to agents to which normal concrete of adequate quality is resistant. This is the lowest category applicable to concrete members in contact with water or condensation
- C – Where aggressive agents will attack the concrete, but provision of a superior quality will enable the member to remain serviceable for the required design life
- D – Where the concrete is subject to an environment that will attack the concrete to such an extent that the required design life cannot be met, i.e. cannot retain or exclude the liquid in an acceptable manner.

During design, most water and wastewater structures fall into exposure classification B, concrete bunds for storage of sodium hypochlorite and other chemicals are group C. However, in-service exposure conditions are often dynamic and variable, and areas of early degradation occur where exposure is greatest.

Furthermore, biogenic corrosion and soft water attack need to be considered.

Biogenic Corrosion

Domestic sewage, both fresh and stale, is non-aggressive to concrete. The main risk is the anaerobic generation of sulphides within the sewage, which occurs particularly in slowly moving or stagnant systems. This results in the formation and release of gaseous hydrogen sulphide (H_2S) into the space above the liquid. H_2S combines with oxygen to form sulphuric acid and attacks the cementitious material of the concrete which leads to eventual structural failure.

Soft Water Attack

Soft water is water which has relatively low concentration of calcium carbonate and other ions and dissolves the calcium hydroxide $Ca(OH)_2$ from the hardened cement paste, increases the porosity and reduces the alkalinity and strength of the concrete and therefore may initiate the corrosion of reinforcement.

Where $Ca(OH)_2$ is leached from the concrete matrix, the concrete becomes weak and friable.

To determine potential for waters to leach $Ca(OH)_2$, an approximate value of Langelier Saturation Index (L_I) may be obtained from the equation (Clause 4.2, AS 3735):

$$L_I = pH\ of\ water - pH\ when\ in\ equilibrium\ with\ calcium\ carbonate$$

$$= pH - 12.0 + log10\,[2.5 \times Ca^{2+}\,(mg/L) \times total\ alkalinity\ (as\ CaCO_3\ mg/L)]$$

A negative value for L_I means the water has a demand for calcium carbonate ($CaCO_3$).

Lifecycle

Within major water retaining structures a design life is usually specified in accordance with AS 3600 which nominates up to 50 years ± 20%, and owners may seek greater design life for reservoirs and other structures [3].

Often the design life is not realised, primarily because of aggressive environments, combined with deficiencies in design, detailing, supply or construction, leading to corrosion of steel reinforcement which substantially reduces the service life of many structures.

In terms of the life of a concrete structure, concrete durability issues can be divided into four phases [4]:

- Phase A: Design, construction and concrete curing
- Phase B: Corrosion initiation processes are underway, but propagation of damage has not yet begun
- Phase C: Propagating deterioration has just begun
- Phase D: Propagation of corrosion is advanced, with extensive damage manifesting.

In these phases, $1 extra spent in Phase A is equivalent to saving $5 of remedial expenditure in Phase B, or $25 remedial expenditure in Phase C, or $125 remedial expenditure in Phase D [4].

A relatively short time is required between the occurrence of the first visible crack along the reinforcement and actual structural damage in the form of large prominent cracks running parallel to the steel reinforcement. When corrosion reaches an advanced stage, the concrete cover to the reinforcement simply delaminates and spalls off.

It is therefore advisable to manage degradation prior to corrosion occurring on the reinforcing steel. This requires a diagnostic approach to identify chemical degradation prior to physical signs of distress.

Sustainability

On the path to net-zero carbon emissions asset owners should consider the production of embodied carbon (CO_2e) from asset maintenance versus run to failure.

Embodied Carbon in Reinforced Concrete

There have been several studies looking at the correlation between the concrete grade and its embodied carbon factors.

One example is the Institution of Civil Engineers (ICE) freely available carbon calculator (inventory of carbon and energy) equation. This was used by the author to measure CO_2e using a concrete mix common to local industry [5] for 32 MPa, 40 MPa and 50 MPa concrete mixes and each assuming 200 kg/m^3 of reinforcing steel where:

$$\sum EC = (ms)(ECFs) + (mc)(ECFc)$$

where:

- $\sum$EC = total embodied carbon (kg)
- ms = reinforcement steel mass (kg)
- mc = concrete mass (kg)
- ECFs = embodied carbon factor for steel (kg·CO_2e/kg)
- ECFc = embodied carbon factor for concrete (kg·CO_2e/kg).

An embodied carbon value is calculated for reinforced concrete as 531, 579 and 581 kg·CO_2e/m^3 respectively.

This includes carbon emissions during the extraction, processing, transportation and manufacture of materials up to the point where they leave the factory gate [6].

Material suppliers are starting to publish environmental product declarations (EPD) that measure the CO_2e of their materials and as an industry we will see society demanding CO_2e be considered within future feasibility and design stages of infrastructure works.

Likely maintaining structures for longer will produce less CO_2e in relation to use of new materials such as concrete and reinforcing steel that would otherwise be needed for replacement of old structures hence, there is a renewed focus on improved design and design for durability.

Case Study - Reinforced Concrete Wastewater Pump Station

Considerable expertise is required for proper diagnosis and remediation of concrete since the chemical processes of deterioration are complex, involving several contributory factors which give rise to very localized or extended forms of damage.

An experienced assessor was used to review all the information provided including:

- analysis of exposure conditions provided by the owner
- in-situ testing by the inspector
- laboratory testing of samples retrieved by the inspector
- visual defects recorded by the inspector.

The following case study is discussed to illustrate the scientific approach necessary to accurately assess and diagnose a concrete structure.

This is a real-life condition assessment of a reinforced concrete wastewater pump station built in circa 1930. The pump station is situated on a traffic island at the intersection of two busy streets in Melbourne.

The pump station comprises of two wells – dry and wet. The wet well is the focus of this case study [7]. The wet well is divided into three levels by intermediate concrete slabs as shown in Figure 1.

Given the difficult roadside conditions, a visual check using pole inspection cameras was undertaken by the owner. The visual check revealed spalling concrete and widespread coating breakdown.

The owner approached Remedy Asset Protection (RemedyAP) as they assumed needing to replace the structure and required a condition assessment to support development of a feasibility study.

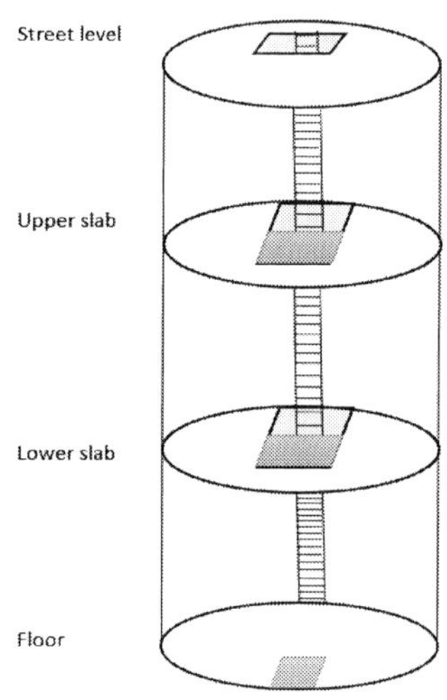

FIGURE 1: SKETCH INDICATIVE OF THE WET WELL LAYOUT.

FIGURE 2 - ACCESS INTO THE WELL

FIGURE 3 - GENERAL CONDITION OF THE WELL

Condition Assessment Planning

A Level 3 Condition Assessment includes investigation of physical and chemical properties of the concrete structure. These are recommended in the following scenarios [8]:

- The asset is approaching one third of its design life
- There are no previous reports on asset condition
- The asset is showing signs of distress
- The asset has a change of exposure or function.

For this structure, all four prerequisites were applicable and so, RemedyAP developed a level 3 inspection plan for the pump station.

The main objectives of the condition assessment were agreed as follows:

- To establish and record the current physical and functional condition of the structure
- To identify likely future material and/or structural issues and the approximate timing of those issues
- To determine and measure the type and extent of the maintenance needs
- To establish a history of material performance and concrete chemistry; or
- To provide feedback to design, construction and maintenance engineers in relation to residual service life.

The inspection plan considered the relevant history of the structure, including engineering design and drawings, construction practice, previous inspections, maintenance and repair records. The highest corrosion rates tend to occur in concrete subjected to periodic wet and dry cycles, at the waterline within liquid retaining structures. The test plan therefore included testing locations in relation to the waterline.

There are several forms of deterioration or distress signals which were considered, including:

- specific attention to isolate the types and forms of cracks, particularly crack widths, location, orientation, depth, length and classification as either affecting the structural properties of the structure or non-structural
- wherever possible, structural cracks needed to be further identified as flexure, shear or direct tension
- signs and details of corrosion of steel reinforcement, including rust stains and extent and amount of lost cross-section
- evidence of loose, corroded or otherwise defective access systems, electrical items and mechanical items including lifting points used to hoist the pumps

- surface defects such as spalling, scaling, honeycombing and efflorescence
- condition of penetrations and seals
- signs of differential settlement, heave, water leakage, aggressive chemical attack or deterioration, poor drainage and related distress.

Procedures were adopted for sampling/coring and an overall program of the inspection and testing were integral to the objectives of the investigation. Other inspection requirements, such as ease of accessibility, equipment, power and water supplies and personnel requirements were essential however are not discussed in this case study.

The conditions of exposure, in relation to aggressiveness of the environment, drainage, and climatic conditions were considered during the assessment stages of the investigation to identify potential sources of observed deterioration.

Both sophisticated and simple equipment were used in this investigation and the results generated, analysed by our specialists who have the expertise and experience to complete the condition assessment that includes: defects notification, results of in-situ testing, sampling and laboratory testing, condition assessment, estimation of residual service life (RSL), preservation and intervention options, sustainability considerations and repair recommendations.

In situ Testing

Testing in place (In situ) is required to test and measure the physical and chemical properties of the reinforced concrete. The data returned from in situ testing is used to assess the concrete and its condition. Tests used in this investigation included:

- Delamination survey
- Rebound hammer
- Crack mapping
- Cover meter
- Ground penetrating radar
- Depth of erosion and carbonation/neutralisation.

The inspector could provide the following observations:

- The well was lined, and widespread coating failure was observed throughout
- Several areas of significant spalling of concrete with exposed reinforcement was observed to the intermediate slabs. The spalling was concentrated around the slab openings (interfaces between concrete and metal grates)
- Areas of exposed and corroding reinforcement were observed in these slabs
- Steel pipes inside the well were showing signs of coating failure and general corrosion
- A neutralisation depth of approximately 30 mm was observed on the well walls, with a concrete cover over steel at 55 mm
- Reinforcing steel was visually observed in the walls by means of concrete breakout, with no signs of corrosion.

Sampling

Core samples are the preferred method for concrete sampling and provide a means of determining the strength of the in situ concrete as a basis for evaluating structural integrity.

Procedures were adopted for sampling/coring and an overall program of the inspection and testing was integral to the objectives of the investigation.

Other inspection requirements, such as ease of accessibility, equipment, power and water supplies and personnel requirements were essential however are not discussed in this case study.

Laboratory testing

Laboratory tests for concrete chemistry included cement content and concentration of chlorides within the concrete.

The results of the laboratory testing were as follows:

- The estimated the cement content in the concrete mix (295 to 384 kg/m3) was considered relatively high for the age of the well and the type of construction.
- The concentrations of chlorides measured inside the well were low and significantly below the corrosion threshold level commonly assumed as 0.4% by weight of cement (0.049% by weight of concrete assuming cement content of 295 kg/m^3). Hence the risk of chloride-induced corrosion of reinforcement was considered low to very low.

Condition Assessment Findings

Concrete elements

In general terms, the condition of the concrete elements in the well ranged from 'poor' (intermediate slabs), 'fair' (invert slab) to 'good' (walls).

The significant deterioration of the intermediate slabs in the well was associated with corrosion-induced delamination from spalling around the openings. The corrosion, in turn, is related to low concrete cover. Although the history of the protective coating applied in the dry well was not known, it likely was applied quite some time after construction of the well. This allowed moisture ingress (potentially from ingress from the street level) into the slabs which then led to neutralisation of concrete and onset of corrosion of reinforcement and steel angles embedded in the slabs around the openings. The slabs in the well were in a poorer condition and were built using a lower quality concrete than the wall of the well.

Structural analysis concluded that the loss to reinforcement was not significant enough for needing steel replacement.

Load testing of the lifting points confirmed they were serviceable and were retagged.

The recorded concentrations of chlorides were significantly below the threshold level. Therefore, the risk of chloride-induced corrosion of reinforced was considered low to very low.

The current condition of the protective coating applied to all concrete elements was poor with up 80% of the lining failed (blistered or missing). Although the cause of failure of the coating had not been investigated in detail, it was likely to be attributable to surface preparation and/or application of the product.

Presence of the protective coating is likely to have reduced the rate of neutralisation of the concrete. The measured depths of neutralisation are considered low to moderate for an approximately 90 years-old structure.

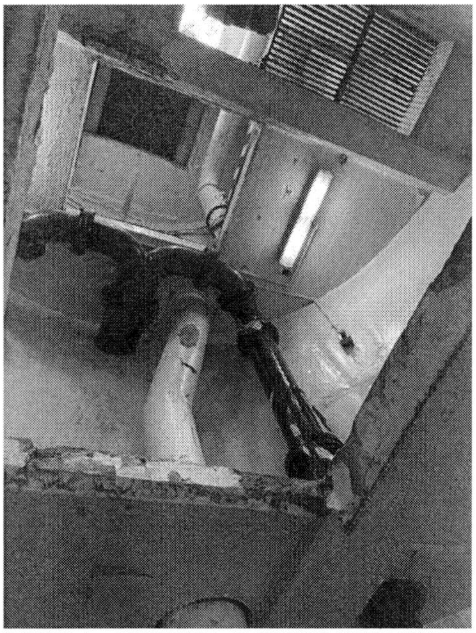

FIGURE 4 - CONDITION AS VIEWED FROM WITHIN THE WELL BASE

FIGURE 5 - INSPECTION OF REBAR AT AREAS OF SPALLING

Steel elements

The steel pipes at the lower section of the well displayed extensive failure of the coating. These pipes appear to have been coated in-situ using a similar coating system as the one used on the concrete components. Although the cause of failure of the coating has not been investigated in detail, it is likely to be attributable to product selection, surface preparation and/or application of the product.

The upper parts of the steel pipes remained intact and were protected using the original (factory applied) fusion bonded epoxy coating system.

The steel angles supporting the steel grates within the intermediate slabs were significantly corroded and required replacement.

Residual Service Life

Once all the information was reviewed, the assessor considered primary and secondary degradation mechanisms that may affect RSL. An exact or definitive RSL can never be calculated, however a reasonable prediction was made based on the visible evidence, relevant laboratory results, modelling results and existing literature regarding deterioration of reinforced concrete.

The assessment concluded RSL would be extended up to 15 years if the repair recommendations were adopted. Further extension of RSL could be achieved by maintaining the protective coatings and operating conditions within the well.

Defects Notification

Where extreme distress was observed such as the access platforms, these were reported immediately rather than waiting for the pending condition assessment report.

Repair Recommendations

The correct specification of a repair system is very much dependent on the proper examination of the structure and an accurate diagnosis of the causes of deterioration.

For this structure the recommended remedial works were to include repair to all areas of spalled concrete and refurbishment of protective coating to both steel and concrete elements summarized as follows:

- Remove all defective coating from concrete elements
- Break out all delaminated and unsound concrete
- Clean steel reinforcement and replace if necessary
- Remove and replace angles and steel grates
- Reinstate removed concrete using high quality, proprietary repair material and introduce adequate slopes to allow moisture to drain off the platforms and into the well base
- Apply a new protective lining system to all concrete elements
- Remove all defective coating from steel elements
- Apply a new protective coating system to all steel elements.

The remedial actions were further assessed to inform the owner:

- Expected service life of the proposed remedial works (time to next major maintenance) was 10 to 15 years.
- At an estimated cost of AUD $84,000 with a ±30% accuracy [9].

Conclusions

In this case study the assessment showed the structure could achieve an extension of service life by performing maintenance and, by performing maintenance, would have nothing left to do for 10 to 15 years. This was a significant finding for the owner who first believed the asset to have minimal residual life and shows the value of a detailed concrete condition assessment.

Key learnings are:

- Replacement of the structure may cost over $3.5M however the condition assessment that cost the client under $50k identified how $84k of maintenance would provide an extension of service life.
- Testing of material condition after construction, establishes a reference from which to measure and monitor deterioration (establishing a benchmark)
- Building a register of asset condition provides an owner with baseline data for future inspections
- Having a proper assessment can provide the necessary information to determine specific preservation and maintenance interventions.

On the path to net-zero carbon emissions asset owners will start considering the production of embodied carbon (CO_2e) from asset maintenance versus run to failure.

References

[1] Standards Australia, "HB 84 Guide to concrete repair and protection," 2018.

[2] Standards Australia, *AS 3600 Concrete Structures for Retaining Liquids,* 2001.

[3] Standards Australia, *AS 3735 Concrete Structures,* 2018.

[4] W. de Sitter, "The Law of Fives," in *CEB-RILEM Workshop on Durability of Concrete Structures*, Copenhagen, 1984.

[5] Sydney Water, "Sydney Water Technical Specification - Civil v.10".

[6] Institution of Civil Engineers, "Inventory of carbon and energy," 2019.

[7] Remedy Asset Protection, "1168 Wet Well - Concrete Assessment Report Rev0," 2023.

[8] Institute of Public Works Engineering Australasia, *Condition Assessment & Asset Performance Guidelines - Practice Note 7: Water Supply & Sewerage,* 2016.

[9] Association for the Advancement of Cost Engineering (AACE International), "Recomended Practice No. 56R-08," 2020.

AUTHOR DETAILS

	Justin Rigby is the Principal at Remedy Asset Protection (RemedyAP) and has extensive experience in the corrosion industry developed over 31 years and is an ACRA Board Member, an AMPP lecturer, Certified Corrosion Technologist and Protective Coatings Specialist, Level 2 Bridge Inspector and serves as Chairperson of the Australasian Corrosion Associations (ACA), Coatings Technical Group along with being active in various Standards committees both in Australia and internationally. Justin has authored numerous papers, peer reviewed for publication in prestigious journals and has presented at international conferences and in Australia with regard to protection of concrete and steel.
Links	www.facebook.com/remedyap www.linkedin.com/company/remedy-asset-protection www.instagram.com/pro_coatings_guy

CORROSION & PREVENTION
CAIRNS 2024

THE AUSTRALASIAN CORROSION ASSOCIATION INC.

DETECTING CORRODED WALL TIES THROUGH NON-DESTRUCTIVE MEANS

C.Y. Lam[1,*], M.A. Hossain[1], M.J. Masia[1], I.A. Chaves[1], J. Vazey[2]

[1]*Centre for Infrastructure Performance and Reliability – The University of Newcastle, NSW, 2308, Australia.*

[2]*Structural Health Monitoring, EngAnalysis Ptd Ltd, NSW, 2300, Australia.*

**Corresponding author email: CheeYin.Lam@uon.edu.au*

ABSTRACT

Corroded wall ties within masonry veneer and cavity wall systems are sometimes difficult to detect without using an intrusive and destructive method. This experimental study presents a non-destructive wall tie assessment method based on vibration measurements obtained from the impact hammer test. The modal parameters of a masonry veneer wall with undamaged and deteriorated wall ties were compared and analysed. When the wall ties were unscrewed, simulating a severe corrosion that completely disconnected the outer brick wall from the timber frame, a reduction of natural frequencies was observed. It was observed that the decreased natural frequencies were more apparent when the deteriorated wall ties occurred close to the top of the masonry veneer wall, causing a larger stiffness loss of the veneer wall and highlighted a potential need for more rigorous inspection work to be done near the area. In addition, mode shape-based analysis using the Parameter-based Method showed that it is effective in approximating the locations of deteriorated wall ties within the masonry veneer wall except when the deteriorated wall ties caused much less stiffness reduction near the bottom support of the wall. The findings contribute to the development and application of non-destructive methods to detect corroded wall ties within masonry veneer wall systems and enhance the understanding of the impact of corrosion on the structural integrity of existing masonry buildings.

Keywords: Wall ties, masonry structures, impact hammer test

INTRODUCTION

Masonry construction systems are historically significant in Australia, especially the masonry veneer and cavity wall systems. Both walling systems rely on the wall ties to connect the outer brick wall and a timber frame for the veneer wall systems, and another leaf of brick wall for the cavity wall systems respectively. Wall ties are critical to ensure the horizontal loads that act on the outer brick wall can be safely transferred back to the load-bearing elements in a strong wind or earthquake loading conditions. Typical construction details of the two masonry construction systems are shown in Figure 1.

In the early stage of constructing veneer and cavity walls, the corrosion of metal wall ties within veneer and cavity walls were not studied extensively and steel wall ties were commonly used [1]. The corrosion of steel wall ties produces rusts that cause the expansion of the wall tie and subsequently induce cracks in the mortar joint, hence weakening the strength of mortar joints. The corrosion process is especially severe when ties are embedded in permeable mortars due to rapid carbonation of mortar. Exposure to extreme marine and wet environments further accelerates the corrosion process significantly [2].

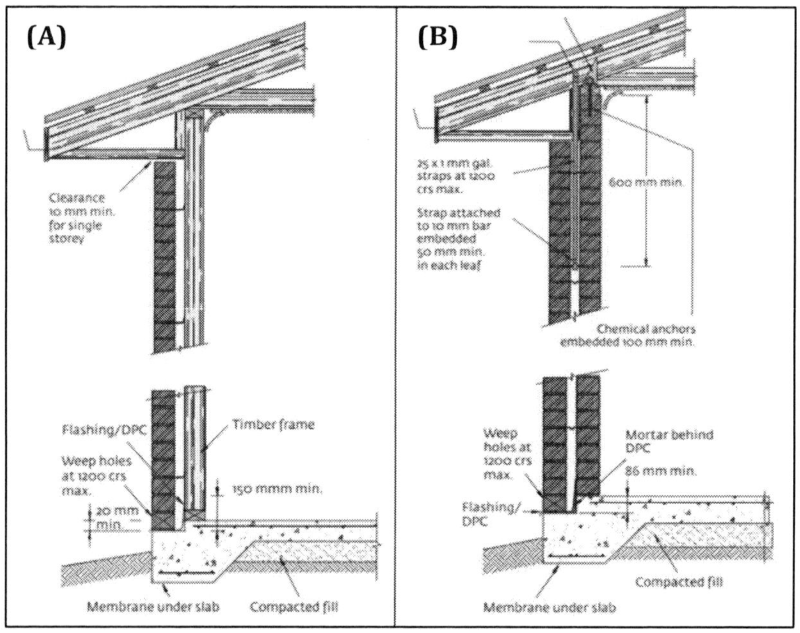

FIGURE 1: THE TYPICAL DESIGN DETAILS OF MASONRY (A) VENEER AND (B) CAVITY WALLS ON RAFT SLABS [3].

WALL TIE INSPECTION TECHNIQUES

For the case of corroded steel wall ties in masonry veneer and cavity walls, visual inspection is frequently used to pick up signs of corrosion such as the presence of rust stains, horizontal cracking and bulging brickwork due to the expansion of the rust products [4]. However, corrosion is not the only threat to the wall ties in masonry veneer and cavity wall systems. The loss of wall tie connection due to poor detailing and workmanship are also critical, and visual signs are no longer effective in identifying the absence of wall tie connections, which may have serious consequences such as the collapse of the outer brick wall. Therefore, other inspection techniques have been developed to obtain more generalised information of wall tie conditions when visual signs are not apparent. Examples of the commonly used wall tie inspection techniques include radar, thermography and borescope methods.

Radar method allows the high-frequency electromagnetic radiation to pass through the targeted wall, and analysis of the reflected signal correlates to the conditions and properties of the wall ties. However, this method has proven effective only if the properties of the target location appear to be strongly contrasted to the surrounding material, and deep defects were not easily detected compared to surface defects [5]. Thermography method on the other hand utilises infra-red radiation to establish a temperature gradient between the outer and inner walls [6]. Disconnected wall ties can be detected based on the thermal conductivity, but the method is limited by the environmental influences, such as the weather, temperature and humidity, which may reduce the accuracy of thermography. If the results of previous inspection methods are inconclusive, a more intrusive borescope method may be used. The borescope penetrates the outer layer of the brick wall through a drilled hole, and the wall tie inspection is carried out by analysing the video or images obtained by the borescope [6]. However, this method becomes less effective if insulation material within the cavity blocks the view and increase the difficulty of the penetration of a borescope. Although the borescope method can identify the loss of wall tie connection reliably, the cases of corroded wall ties in a cavity are more challenging as many researchers have observed that the corrosion of wall tie is more severe at and within the bed joints of the outer brick wall [7-9]. This phenomenon is difficult to detect using the abovementioned inspection techniques which prompted the use of non-destructive and vibration-based detection methods. An example of the mentioned corroded wall ties is shown in Figure 2, where severe corrosion was observed at and within the mortar that caused the outer brick wall to collapse during the Newcastle earthquake in 1989 [8].

FIGURE 2: THE WALL TIE CORROSION IN MASONRY CAVITY WALL SYSTEM BEING REVEALED IN THE NEWCASTLE EARTHQUAKE WHICH CAUSED THE COLLAPSE OF THE OUTER BRICK WALL [8].

Non-destructive and Vibration-based Testing on Masonry Structures

Vibration-based testing methods have been gaining interest for structural health monitoring purposes. Instead of focusing on the corrosion or loss of wall tie connections, the global structural behaviours are analysed, and the wall tie conditions are inferred directly. The main idea is to vibrate the structure and measure its response accordingly to obtain the modal parameters such as the natural frequencies, mode shapes and damping ratios. If the non-destructive nature of the operation is prioritised, less intense vibration sources from the environment may be utilised such as the vibration generated by the wind, traffic and pedestrian activity. To minimise disruption or lower the risk of damaging the masonry structures, further non-contact measurement systems such as a scanning laser vibrometer can be utilised to record the vibrating motion from a distance without coming into contact to the test subject instead of using the traditional accelerometers or strain gauges approach. Many researchers have utilised vibration-based methods for damage identification and structural health monitoring of masonry structures, especially on heritage masonry buildings, masonry arches and multi-leaf masonry walls [10-12]. However, the applications of this methodology for investigating the damage caused by the wall tie deterioration in masonry veneer and cavity walls are limited, which is the motivation of this paper.

The absence of wall tie connection is thought to reduce the out-of-plane stiffness of the overall structure which generally results in a reduction of natural frequencies. For this study, the natural frequencies of the veneer wall under various configurations of deteriorated wall ties were measured and compared to testify the effectiveness of natural frequency as a damage indicator. Although the change in natural frequencies can serve as a damage indicator to flag the presence of deteriorated wall ties, information regarding the locations of the deterioration may not be obtained, as natural frequency is a global property of a structure. For identifying the damage locations, the usage of mode shape-based analysis is often needed. Mode shape represents the relative displacement of all parts of the structure at a given natural frequency [13], and it contains local information that can be used in damage localisation.

For example, comparing the mode shapes between the undamaged and damaged states of a structure provides an indication of damage locations where large changes are often observed around the damage region. Several cases of crack and delamination of brick walls have been successfully identified using various mode shape-based methods [14], and this approach was extended in this study to detect deteriorated wall ties in masonry veneer wall. In particular, Parameter-based Method (PM) was employed where the absolute change of mode shapes under the influence of natural frequency difference was calculated [15, 16]. The equation is shown in Equation 1, where m is the number of vibration modes included in the analysis, j is the mode number, φ is the mode shape vector, ω is the natural frequency, the superscript u represents the undamaged state while d represents the damaged state of the structure respectively.

$$\Delta\varphi = \sum_{j=1}^{m} \left| \varphi_j^d \left(\frac{\omega_j^u}{\omega_j^d}\right) - \varphi_j^u \right| \quad (1)$$

EXPERIMENTAL PROGRAM

In this experimental study, a single-storey masonry veneer wall was built, which consisted of a 2.4m by 2.4m outer brick wall that was connected to four timber studs that were at a horizontal spacing of 600mm using five rows of wall ties, and each row consisted of four wall ties. One end of the wall tie was embedded into the mortar and the other end was side fixed to the timber studs using screws, and a cavity of 50mm was set. The installation of wall ties started at the third course from the bottom of the wall, and each row of the ties was spaced vertically at approximately 600mm with the last row of ties being one course above the previous row. This wall tie configuration was designed in accordance with AS3700 [17] since the outer brick wall is much stiffer than the timber studs, and the wall ties at the top of the wall bear more load generally. The overall dimension of the veneer wall was chosen as it is representative of the common masonry veneer wall constructions encountered in practise [18].

In terms of construction materials, the outer brick wall was made up of cored clay bricks with the Standard Australian sizes of 230mm long by 110mm wide by 76mm high alongside the conventional M3 mortar in accordance with AS3700 [17]. On the other hand, the light duty stainless steel wall ties with a R4 durability classification in accordance with AS2699.1 [19] were used. The timber studs were machine-graded pine (MGP10) and had a cross-section of 90mm by 35mm. The construction of the veneer wall is shown in Figure 3.

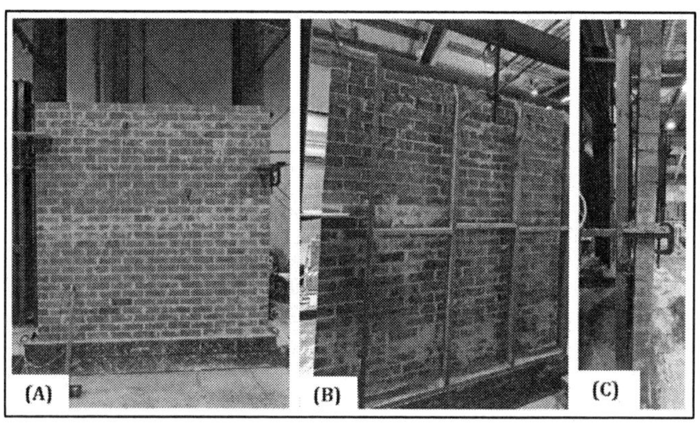

FIGURE 3: PHOTOS OF THE (A) FRONT VIEW, (B) BACK VIEW AND (C) SIDE VIEW OF THE VENEER WALL.

The vibration test commenced after the veneer wall was cured for more than 14 days. A PCB Piezotronics 086D50 impact hammer was explicitly selected to vibrate the wall due to its wide availability and the ability to achieve a higher signal-to-noise ratio compared to ambient vibration testing, which in turn provides a more accurate extraction of modal parameters [11]. The impact and measurement locations were carefully chosen to ensure all the vibration modes of interest are excited accordingly, and the vibration measurements can be recorded without aliasing effect [20]. A diagram of the impact and wall tie locations is detailed in Figure 4.

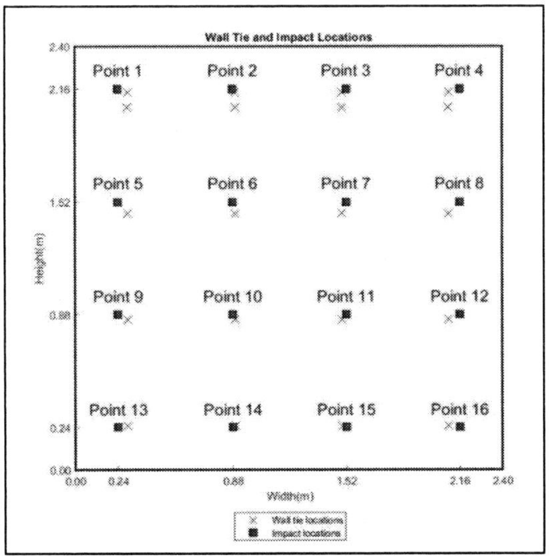

FIGURE 4: DIAGRAM OF THE IMPACT AND WALL TIE LOCATIONS.

The vibration of the veneer wall was recorded using two accelerometers which were fixed at Points 2 and 7 respectively. Each impact location was hit five times in the out-of-plane direction to obtain an averaged vibration response of the veneer wall. The natural frequencies and mode shapes were extracted using MATLAB Least Square Rational Function (LSRF) algorithm [21] with the vibration measurements as the input. One disadvantage of using the impact hammer is the difficulty of maintaining a constant swing throughout the whole test. Although the force was kept at a relatively constant level during the impact hammer test, the damping ratios are known for its sensitivity to the varying force levels, hence damping ratios are not further analysed and discussed in this study.

Once the vibration measurements were recorded for the veneer wall under the condition of all fully functional wall ties, four wall tie deterioration cases were tested by unscrewing the wall ties at different locations to simulate the disconnection of wall tie due to severe corrosion. The orientation of the building is speculated to affect the severity of wall tie corrosion, depending on the time of exposure to the sun. The wall ties that are exposed to more sun and therefore less wet environment show a less severe corrosion compared to the wall ties at other locations within the same structure [7]. Hence, this study mainly focuses on investigating the differences in vibration responses of the veneer wall due to deteriorated wall ties at four different heights. The details of each wall tie deterioration case are summarised in Table 1.

TABLE 1: SUMMARY OF THE WALL TIE DETERIORATION CASES.

Test case	Description of wall tie deterioration
VW1a (Reference)	All ties are fully functional.
VW1b	Top two rows of wall ties were unscrewed.
VW1c	Middle row of wall ties was unscrewed.
VW1d	Second last row of the wall ties was unscrewed.
VW1e	Bottom row of wall ties was unscrewed.

RESULTS AND DISCUSSION

NATURAL FREQUENCY AS DAMAGE INDICATOR

The first four natural frequencies of the veneer wall for all five test cases are summarised in Table 2. The relative natural frequency compared to the reference state of the veneer wall VW1a are visualised in Figure 5, where the natural frequencies of the deterioration cases are all expressed in relative values using the reference case as the benchmark result.

TABLE 2: THE FIRST FOUR NATURAL FREQEUNCIES FOR ALL THE TEST CASES.

Test case	Natural frequency (Hz)			
	Mode 1	Mode 2	Mode 3	Mode 4
VW1a (Reference)	9.12	20.52	30.54	56.66
VW1b	5.64	17.94	28.17	55.58
VW1c	8.64	19.79	29.32	55.72
VW1d	8.69	19.93	29.52	56.12
VW1e	8.68	19.95	29.54	56.09

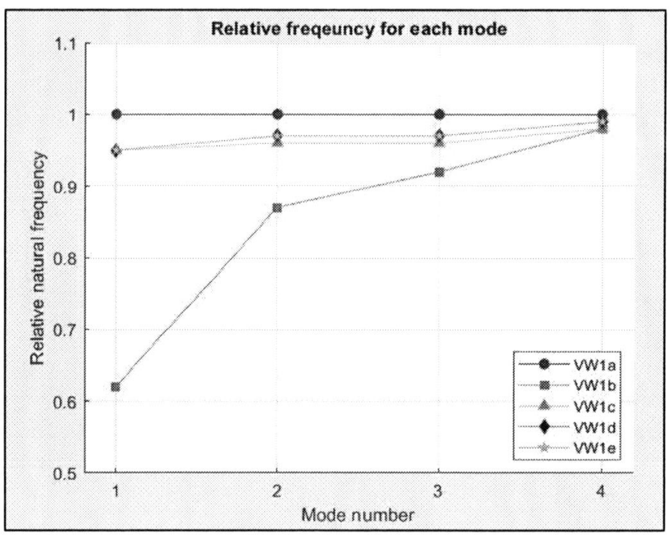

FIGURE 5: RELATIVE FREQEUNCY FOR EACH DETERIORATION CASE WITH RESPECT TO THE REFERENCE TEST CASE VW1A.

Several observations and discussions are presented below based on the natural frequency results.

- Wall tie deterioration caused the natural frequencies of the veneer wall to reduce accordingly for all four wall tie deterioration cases, irrespective of the damage locations. The clear trend of natural frequency reduction shows the potential to flag the presence of wall tie deterioration using a vibration-based method.
- Out of the four wall tie deterioration cases, the case with the most severe reduction of natural frequencies is VW1b, where approximately 38% of natural frequency loss occurred for the first vibration mode. The same trend was observed for the remaining three modes. The observation indicates the loss of connection of wall ties near the top of the veneer wall had a more significant implication in terms of the out-of-plane

stiffness of the veneer wall compared to other deterioration cases. This information may further inform the inspection work of wall ties to be carried out more rigorously near that area.

- The magnitude of natural frequency reduction for test cases VW1c, VW1d and VW1e was similar where an average of only 3% of natural frequency loss across all four vibration modes were observed, which further highlights the importance of the wall ties near the top of a masonry veneer wall.
- Larger differences were observed for the first natural frequency for all four deterioration cases compared to the remaining vibration modes. The difference indicates that perhaps analysing just the first natural frequency provides enough information in terms of damage identification of deteriorated wall ties, which may save computational and time effort in the data processing process.
- No local information is available from the natural frequency results, localisation of the deteriorated wall ties was not achieved based on natural frequency results alone.

Mode Shape-based Analysis

Given that the natural frequencies alone were not fully capable of locating the damage locations, mode shapes were included in the subsequent analysis. First of all, the corresponding mode shapes from the four vibration modes for VW1a are shown in Figure 6. Since the vibration measurements were only taken on the outer brick wall, only the mode shapes of the outer brick wall were estimated, while fulfilling the assumption that no access to the timber studs was provided during the testing stage. This assumption was made to ensure the effectiveness of the non-destructive vibration test was evaluated fairly, since back access is not provided in the first place.

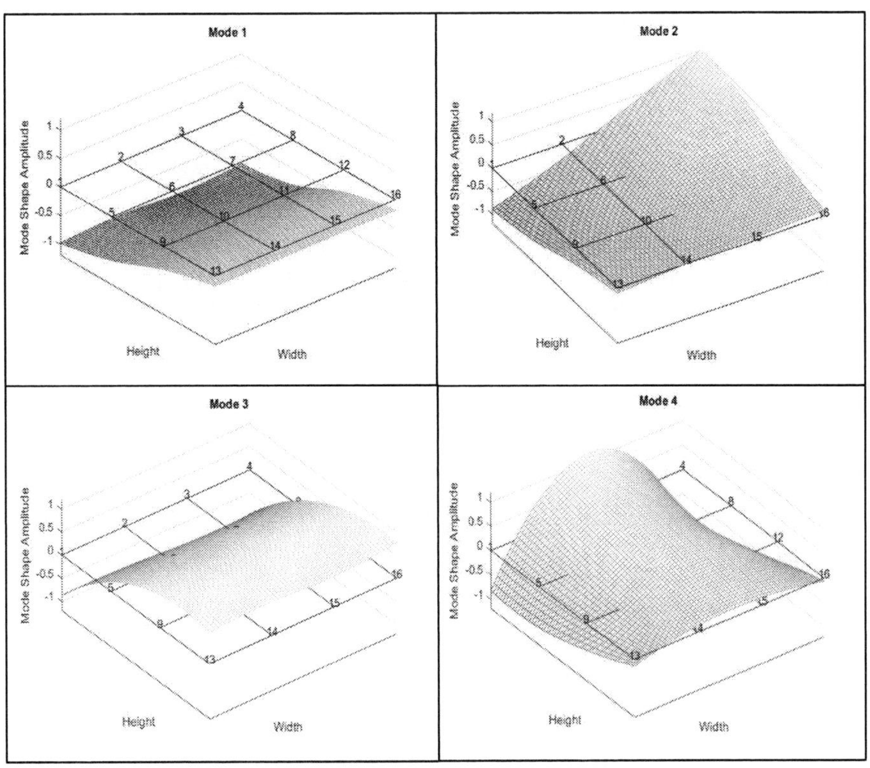

FIGURE 6: THE FIRST FOUR MODE SHAPES OF VW1A.

Based on Equation 1, the absolute differences of mode shapes that are scaled by the change of the natural frequencies between the reference VW1a and the rest of the deterioration cases were calculated. Since the first vibration mode was significantly affected by the deterioration of wall ties, only the first vibration mode was considered in the calculation. The absolute differences were calculated at each impact location, and the results are plotted as vertical bar charts at each impact location and shown in Figure 7.

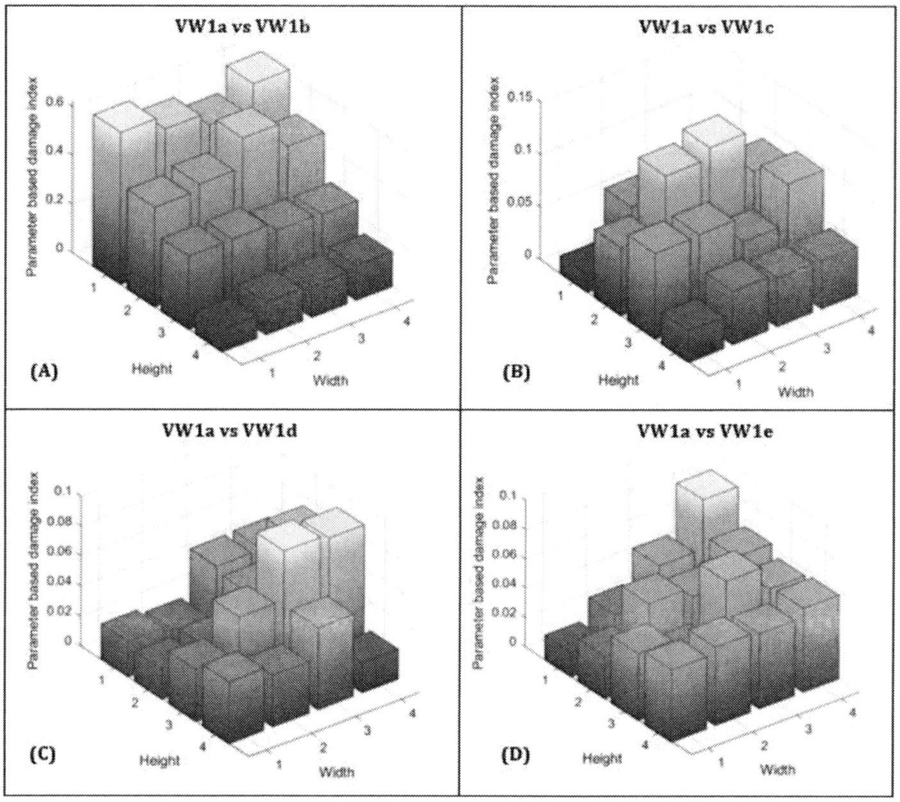

FIGURE 7: RESULTS OF THE PARAMETER-BASED METHOD USING THE FIRST VIBRATION MODE ONLY, (A) VW1A VS VW1B, (B) VW1A VS VW1C, (C) VW1A VS VW1D AND (D) VW1A VS VW1E.

Several observations were drawn from the results of the Parameter-based Method.

- Large absolute differences in mode shapes were observed near the damage regions for VW1b, VW1c and VW1d as shown in Figure 7(A), (B) and (C) respectively. The peaks of the damage indices correspond to the locations of the deteriorated wall ties, indicating the potential for locating the deteriorated wall ties within a veneer wall by combining the information of natural frequency and mode shapes.
- False positive result was obtained for VW1e case when the deteriorated wall ties were simulated at the bottom of the veneer wall. Given that the nature of the first vibration mode that exhibits a smaller mode shape amplitude at the bottom of the veneer wall, the magnitude of the calculated absolute difference in mode shape was small as well. The nature of the first vibration mode combines with the measurement errors and the uncertainties in the mode shape estimation resulted in the false positive result. Further studies involving more vibration modes in the analysis are planned.
- For the first three deterioration cases, although damage locations were identified as peaks at the correct locations, not all the damage locations were identified simultaneously.

CONCLUSIONS

This paper summarises the results of the impact hammer test for investigating the influence of wall tie deterioration within a masonry veneer wall. Vibration measurements provided useful information regarding the structural dynamics of the veneer wall. A total of four deterioration cases were simulated by unscrewing the wall ties at different locations, covering the potential cases of severe wall tie corrosion due to height variations. The presence of wall tie deterioration was captured accurately by the trend of decreased natural frequencies. Furthermore, mode shape-based analysis using the Parameter-based Method is able to approximated the locations of deteriorated wall ties. The method was especially effective except when the deteriorated wall ties were close to the fixed end of the masonry veneer wall. The findings showed the importance of the wall ties near the top of the veneer wall for contributing to the out-of-plane stiffness of a masonry veneer wall system.

Further investigation of different wall tie deterioration test cases will be conducted by unscrewing wall ties at other locations to ensure various deterioration cases are fully covered. In addition, excitation and measurement systems that are more hands off should be investigated, such as utilising the ambient vibration, a shaker excitation system and a scanning laser vibrometer measurement system. In conclusion, vibration-based techniques have potential to investigate wall tie deterioration issues within masonry veneer and cavity walls, especially in the cases of wall tie corrosion when other inspection techniques are less effective.

ACKNOWLEDGMENTS

The financial support provided by The University of Newcastle through the UNIPRS and UNRS Central Scholarships and the Australian Research Council under the Linkage Project - LP220100028 is gratefully acknowledged. The authors would also like to acknowledge the generosity from EngAnalysis for providing the impact hammer and the assistance provided by the civil laboratory staff in The University of Newcastle to setup various testing equipment.

References

1. Chaves, I., Masia, M., Terry, L., Lam, C.Y., Hossain, M.A., Smith, B., de Prazer, S., Vazey, J., Melchers, R. and Chen, W., 2024. Estimating stability and resilience of ageing masonry walls for enhanced infrastructure management and public safety. Australian Journal of Structural Engineering, pp.1-10.
2. de Vekey R C. Corrosion of steel wall ties: history of occurrence, background and treatment. *BRE Information Paper* IP12/90. Garston, BRE, 1990.
3. Think Brick Australia, "Detailing of clay masonry," Think Brick Australia, NSW, Australia, Tech. Rep. TBA 09, January 2020.
4. de Vekey R C. Corrosion of steel wall ties: recognition and inspection. *BRE Information Paper* IP13/90. Garston, BRE, 1990.
5. M. Clark, D. McCann, and M. Forde, "GPR as a tool for the characterisation of ballast," in Proceedings of the international conference and exhibition railway engineering, 2003.
6. Building Research Establishment. Installing wall ties in existing construction. *BRE Digest* 329. Garston, BRE, 2000.
7. Chaves, I.A., de Prazer, S., Jardim do Nascimento, B., Flowers, G., 2021. Empirical coastal atmospheric corrosion of masonry metal wall ties. Corrosion and Materials Degradation 2, 657–665.
8. Page, A.W., 2019. The newcastle earthquake and the masonry structures code as3700, in: Proceedings of 2019 AEES conference. Australian Earthquake Engineering Society, Newcastle.
9. Maurenbrecher, A., Brousseau, R., 1993. Review of corrosion resistance of metal components in masonry cladding on buildings. National Research Council Canada, Institute for Research in Construction.
10. Soleymani, A., Jahangir, H. and Nehdi, M.L., 2023. Damage detection and monitoring in heritage masonry structures: Systematic review. Construction and Building Materials, 397, p.132402.
11. Ramos, L.F., 2007. Damage identification on masonry structures based on vibration signatures. PhD thesis. University of Minho, Portugal.

12. Boscato, G., Reccia, E. and Cecchi, A., 2018. Non-destructive experimentation: Dynamic identification of multi-leaf masonry walls damaged and consolidated. Composites Part B: Engineering, 133, pp.145-165.
13. Avitabile, P., 2018. Modal testing: a practitioner's guide. Wiley, Hoboken, New Jersey.
14. Fan, W., Qiao, P., 2011. Vibration-based damage identification methods: A review and comparative study. Structural Health Monitoring 10, 83–111. doi:10.1177/1475921710365419.
15. Ho, Y., Ewins, D., 2000. On the structural damage identification with mode shapes, in: Proceedings of the European COST F3 conference on system identification and structural health monitoring, Universidad Politecnica de Madrid Madrid, Spain.
16. Dong, C., Zhang, P., Feng, W., Huang, T., 1994. The sensitivity study of the modal parameters of a cracked beam. Proceedings of the 12th International Modal Analysis 2251, 98.
17. Standards Australia, 2018. Masonry structures. Technical Report AS3700:2018. Standards Australia. Sydney, Australia.
18. Li, J., Stewart, M., Masia, M., 2019. Probabilistic modeling of unreinforced masonry walls subjected to lateral out-of-plane loading. Elsevier. pp. 91– 140.
19. Standards Australia, 2020. Built-in components for masonry construction, Part 1: Wall ties. Technical Report AS2699.1:2020. Standards Australia. Sydney, Australia.
20. Lam, C. Y., M. J. Masia, and I. A. Chaves. 2023. "Application of Non-Destructive Testing Techniques for Condition Assessment of Wall Ties in Masonry Construction." In Proceedings of the 26th Australasian Conference on Mechanics of Structures and Materials (ACMSM26), Auckland, New Zealand.
21. Ozdemir, A.A., Gumussoy, S., 2017. Transfer function estimation in system identification toolbox via vector fitting. IFAC-PapersOnLine 50, 6232– 6237.

Author Details

	Chee Yin Lam is the corresponding author of this paper. He is a PhD student at the University of Newcastle since 2022 and a Civil engineer who has been awarded a prestigious Australian Government Research training program scholarship for his academic excellence. His PhD research interest lies in non-destructive testing of wall ties in masonry constructions.
	Md Akhtar Hossain is a research associate at The University of Newcastle, Australia. He has over 10 years of applied structural engineering research and development experience. His research interests include the experimental and numerical modelling of masonry structures.
	Mark Masia is a Professor in Structural Engineering, and Director of the Centre for Infrastructure Performance and Reliability, School of Engineering at The University of Newcastle, Australia. He has a special interest in the structural performance, assessment and the design of masonry structures.

	Igor Chaves is an Associate Professor of Civil Structural Material Science at The University of Newcastle, Australia. He is a civil engineer, holds a Master degree in structural engineering and a PhD for his studies in corrosion analysis and prediction. His research interests also include structural reliability and risk assessment of infrastructure.
	John Vazey is the managing director of EngAnalysis Ptd Ltd who is serving as the lead partner investigator in this project's non-destructive endeavours. John is a highly skilled mechanical engineer and MEngSci with over 18 years of experience consulting to a wide range of industries. Underpinning John's broad skill base is a depth of knowledge and experience in physical testing, data analysis, vibration analysis, and computational mechanics.

CORROSION & PREVENTION
CAIRNS 2024

THE AUSTRALASIAN CORROSION ASSOCIATION INC.

EFFECT OF HYDROGEN ON OXIDATION OF Fe-20Cr-(1Si, 2Mn OR 2Al) ALLOYS IN WATER VAPOUR AT 650°C

Jiarui Chen, Jianqiang Zhang*, David J. Young

School of Materials Science and Engineering, University of New South Wales, Sydney 2052, Australia

**Corresponding author email: j.q.zhang@unsw.edu.au*

ABSTRACT

Hydrogen production and utilisation technologies will involve water vapour at high temperatures. However, the effect of hydrogen on chromia-forming heat resistant alloys in water vapour is still unclear. In this research, model ferritic alloys, Fe-20Cr with and without 1 wt% Si, 2 wt% Mn or 2 wt% Al, were exposed in Ar-xH_2-10H_2O gases (vol%, x = 2, 10 and 20) at 650°C for up to 310 hours to investigate the hydrogen effect. The undoped alloy experienced breakaway oxidation in all H_2-containing gases. Adding Mn in the alloy did not significantly change the oxidation behaviour of the alloy. Although Al doped alloy was also in breakaway in gases with 2% and 10% H_2, it was protected in Ar-20H_2-10H_2O gas. Adding silicon remarkably increased alloy corrosion resistance. Increasing H_2 concentration lowered oxidation weight gains for all alloys except Si-containing alloy where weight gains were very low.

Keywords: Fe-Cr alloys, hydrogen, high temperature corrosion

INTRODUCTION

Hydrogen is a very promising green energy carrier in the world because water is the only product of its combustion. Solid oxide electrolyser cells (SOEC) provide a high-efficient and cheap technology for hydrogen production at high temperatures [1]. Hydrogen could be used in electricity generation and other industry fields with water as a byproduct. However, water vapor present in a high-temperature environment could lead to severe corrosion for Fe-based alloys due to their rapid breakaway oxidation [2-4], eventually causing material failure. The role of hydrogen has been investigated for decades, and several mechanisms have been proposed; for example, hydrogen could result in the finer grain size and faster growth of Cr_2O_3 [5]. Increasing H_2 only slightly decreases the oxidation rate of steel 304L in steam at 600°C [6]. Besides, additional alloying elements (Si, Mn and Al) significantly increase the corrosion resistance of alloys by forming protective oxide layers in dry gases [3, 7, 8]. However, their effects in H_2-containing gases are rarely reported. Although hydrogen embrittlement caused by hydrogen dissolution and enrichment at grain boundaries is widely reported [9], understanding of its effects on oxide scale formation are still limited.

The aim of this work is to systematically investigate the corrosion behaviour of Fe-20Cr alloy and the effect of alloying elements (Si, Mn, and Al) in Ar-xH_2-10H_2O gas mixtures with different hydrogen concentrations at 650°C.

METHODS AND MATERIALS

Four Fe-based ferritic model alloys, Fe-20Cr, Fe-20Cr-1Si, Fe-20Cr-2Mn and Fe-20Cr-2Al (all in wt%), were prepared by arc melting pure metals (Fe 99.98%, Cr 99.995%, Si 98.5%, Mn 99.9%, Al 99.95%) under a protective Ar-5H_2 (vol%) gas. The cast alloy ingots were annealed at 1100°C under flowing Ar-5H_2 for 50 h for homogenization. Then they were cut into rectangular alloy coupons with the dimensions of (8~14) mm × (5~8) mm × (0.6~1.5) mm and then drilled with a 1.5 mm diameter suspension hole. All sample surfaces were ground to a 1200-grit finish and polished to a 3 μm finish, and then electro-polished in a 15% hydrochloric solution to remove the cold work zone to avoid the formation of fine recrystallized structures during heating. Electro-polished samples revealed a columnar grain structure with the width of (0.2~0.8) mm and the length of (2.8~6.4) mm. All samples were ultrasonically cleaned in ethanol after each step and dried for the reaction.

Isothermal corrosion experiments were carried out in a horizontal furnace with an alumina tubular reactor under flowing reaction gases with the flow rates of 2 cm/s at 650°C. The reaction gases were Ar-2H_2-10H_2O, Ar-10H_2-10H_2O and Ar-20H_2-10H_2O (all in vol%), and the oxygen partial pressure (pO_2, atm) in different gases and the dissociation/equilibrium oxygen partial pressure ($pO_{2,eq}$, atm) of iron oxides calculated by FactSage are listed in Table 1. Water vapor was generated by peristaltic pumping of the required volume of demineralized water into a heated vaporizer where Ar gas was introduced to carry water vapor into the reactor.

TABLE 1: OXYGEN PARTICAL PRESSURE (pO_2, atm) OF REACTION GASES AND EQUILIBRIUM OXYGEN PARTIAL PRESSURE ($pO_{2,eq}$) FOR IRON OXIDE FORMATION AT 650°C.

Gases	Ar-2H_2-10H_2O	Ar-10H_2-10H_2O	Ar-20H_2-10H_2O
pO_2	1.31×10^{-21}	5.26×10^{-23}	1.31×10^{-23}
$pO_{2,eq}$	Fe_2O_3	Fe_3O_4	FeO
	2.8×10^{-14}	2.4×10^{-23}	6.83×10^{-24}

Before reaction, all samples were suspended in the cold zone of the alumina reactor, which was then flushed with argon gas for 1h before heated to 650°C. After temperature was stabilized (approx. 20 min), the sample holder was inserted into the hot zone for varied reaction durations (20 h, 100 h and 310 h). For each set time, new alloy samples were used for the reaction. After reaction, the sample holder was carefully pulled out from the hot zone to the cold zone for cooling.

An analytical balance with 0.01 mg accuracy (Mettler Toledo XP205) was used for the weight measurement of samples before and after reaction. Reacted samples were analysed by X-ray diffraction (XRD; Empyrean II, Co source) and then mounted into cold-setting resin to obtain cross-sections. Cross-section morphologies were observed and captured by optical microscopy (OM; Nikon 600) and scanning electron microscopy (SEM; FEI Nova NanoSEM 450) equipped with energy dispersive X-ray spectroscopy (EDS; Oxford). Oxide scales were further identified by Raman spectroscopy (Renishaw inVia 2 Raman Microscope; wavelength 532 nm with argon laser).

RESULTS

WEIGHT GAIN

The weight gains of Fe-Cr alloys in three H_2-containing gases at 650 °C are shown in Figure 1. Assuming the weight gain kinetics follows a parabolic rate law, the rate constants (k_w) of oxidation in all gases were calculated and are listed in Table 2. Comparison of the oxidation rates of each test alloy in all three H_2-containing gases indicates a decrease in weight gain with increased H_2 concentration. Increasing H_2 concentration from 2% to 10% led to a slight decrease in weight gain. However, the oxidation rate dropped dramatically when alloys reacted in 20%H_2 gas. In addition, in all three gases, the oxidation rate of Fe-20Cr alloys after 310h reaction was the highest among all the test alloys, followed by Mn-bearing and Al-bearing alloys, and alloying with Si yielded the lowest oxidation rate in all conditions.

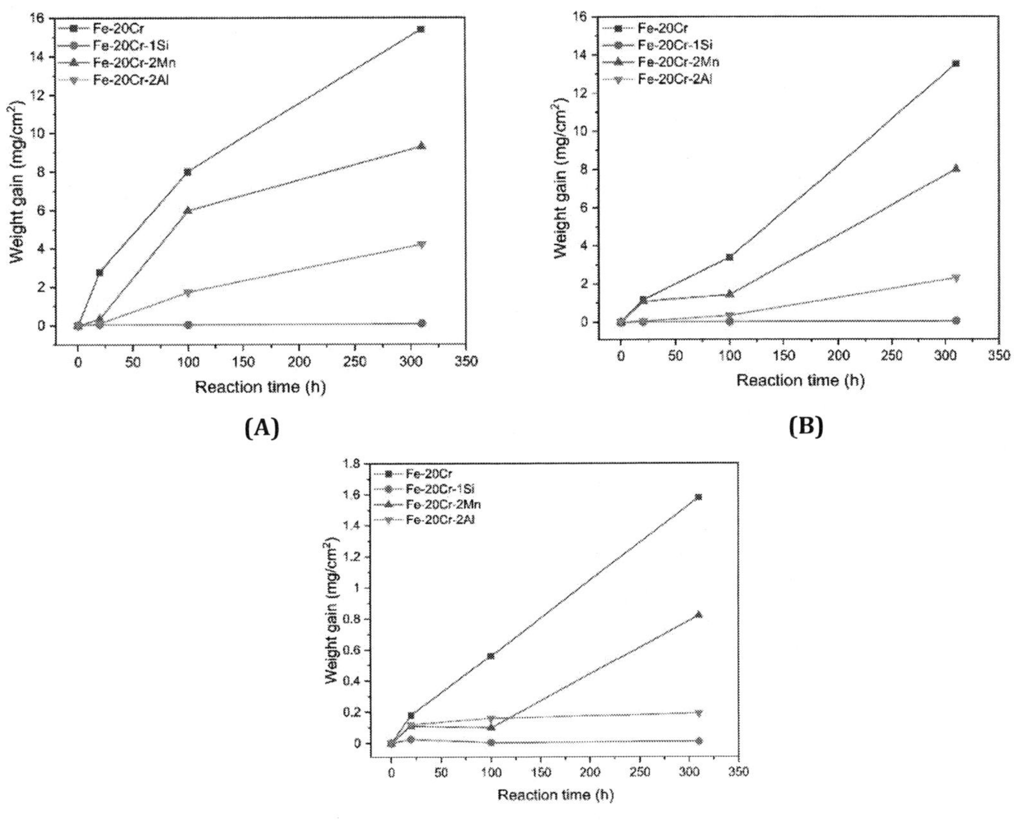

FIGURE 1: WEIGHT GAIN KINETICS OF Fe-BASED FERRITIC ALLOYS AT 650°C IN (A) Ar-2H_2-10H_2O, (B) Ar-10H_2-10H_2O AND (C) Ar-20H_2-10H_2O.

TABLE 2: PARABOLIC RATE CONSTANT, k_w, FOR Fe-20Cr-(1Si, 2Mn, 2Al) REACTED IN Ar-2H_2-10H_2O, Ar-10H_2-10H_2O AND Ar-20H_2-10H_2O GASES AT 650°C.

Alloys	k_w, (10^{-10}, $g^2 \cdot cm^{-4} \cdot s^{-1}$)		
	Ar-2H_2-10H_2O	Ar-10H_2-10H_2O	Ar-20H_2-10H_2O
Fe-20Cr	1.1	0.88	1.2×10^{-2}
Fe-20Cr-1Si	-	6.9×10^{-6}	-
Fe-20Cr-2Mn	0.4	0.31	3.2×10^{-3}
Fe-20Cr-2Al	0.07	2.3×10^{-2}	0.1×10^{-3}

Reation product cross-sections

Fe-20Cr

Figure 2 shows the cross-sectional morphology of Fe-20Cr after 310 h reaction in wet gases containing 2 and 10 vol% H_2. A thick oxide scale was formed in these two gases (Figs. 1A-C), consisting of similar multi-layered Fe-rich oxides (Fe_3O_4 and FeO outer layers and the inner spinel and internal oxidation zone (IOZ) layers) identified by Raman spectra (Fig. 2D).

Further analysis by TEM of the inner layer formed in Ar-10H_2-10H_2O (not shown) after 310 h showed that the inner layer contains two phases of FeO and spinel. In the internal oxidation zone, there were Cr-rich internal spinel precipitates in the metal matrix.

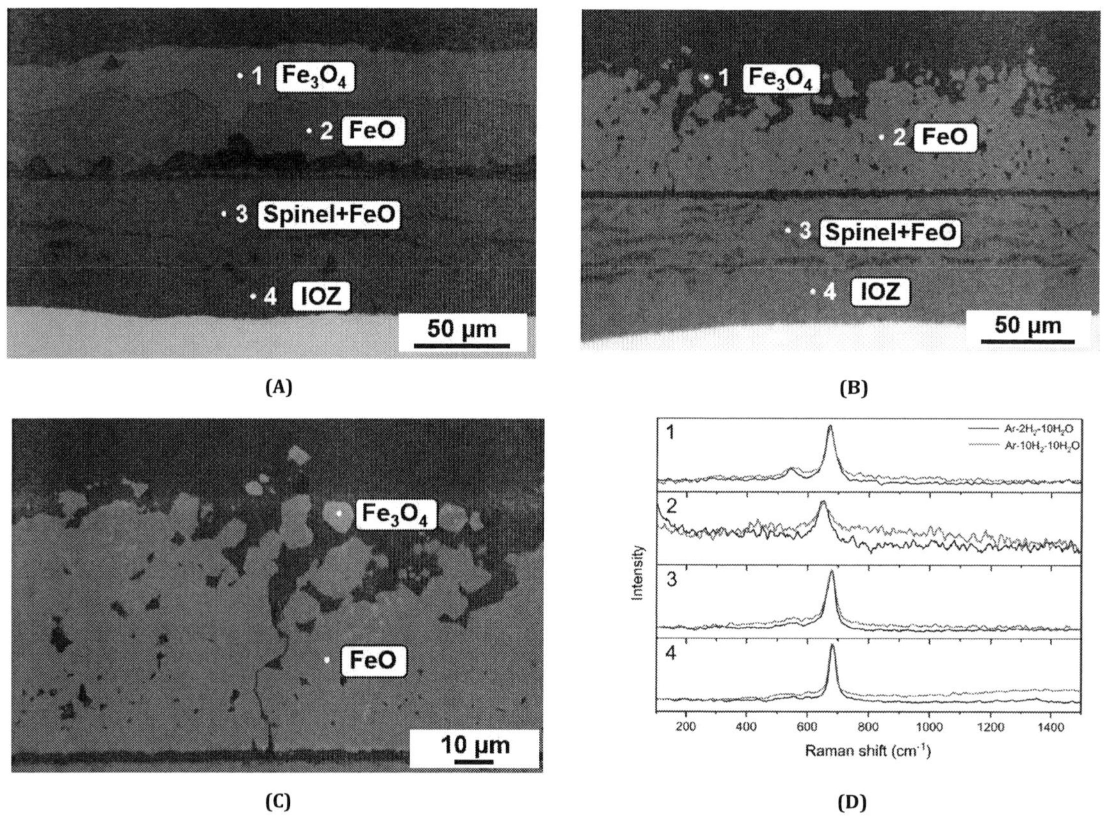

FIGURE 2: CROSS-SECTION MORPHOLOGY OF Fe-20Cr AFTER 310 h REACTION AT 650°C IN (A) Ar-2H_2-10H_2O, (B) Ar-10H_2-10H_2O, (C) ENLARGED VIEW OF THE OUTER LAYER IN (B) AND (D) RAMAN SPECTRA.

Figure 3A shows a cross-section of Fe-20Cr in Ar-20H_2-10H_2O after 310 h reaction. The scale composition was similar to that in 2H_2 and 10H_2 gases (Fe_3O_4, FeO, inner FeO + spinel and IOZ). Although Fe_3O_4 was visually not observed, the XRD result of the outer layer shown in Fig. 3B indicates that the outer layer contains both FeO and Fe_3O_4. Figure 3C shows SEM cross section of the outer layer, and there were several iron-rich metal particles observed and identified by EDS point analysis. Clearly, Fe_3O_4 and FeO formed in 2H_2-10H_2O gas (Fig. 1A) were continuous and dense layers, but Fe_3O_4 was discontinuous and dispersed on the top of the FeO layer when increasing H_2 to 10% and 20% (Figs. 2B and C and Fig. 3A).

The thickness of the oxides scale changed with varying H_2 concentrations (Table 3). The total thickness of the scale decreased with increasing H_2 concentration. Similarly, the thicknesses of the outer and inner layers were also decreased with the increase in H_2 concentration. However, the depth of IOZ had an opposite trend, increasing H_2 resulting in an increased IOZ depth.

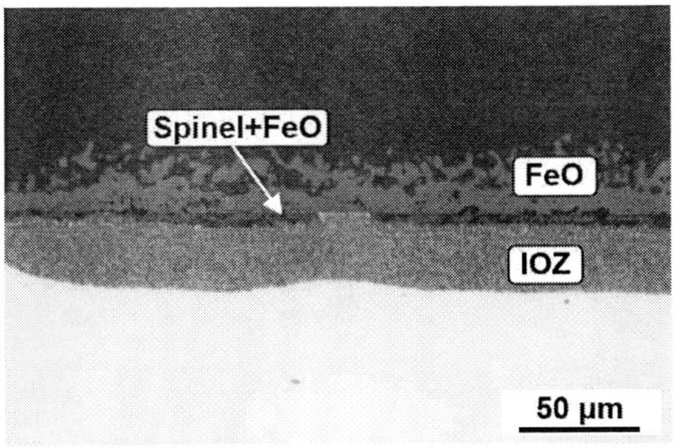

(A)

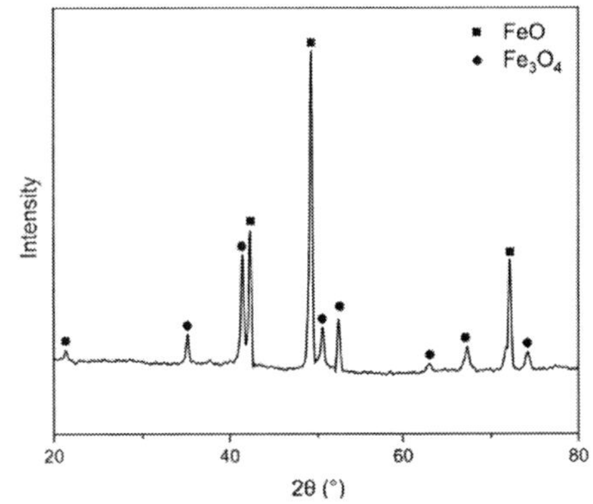

(B)

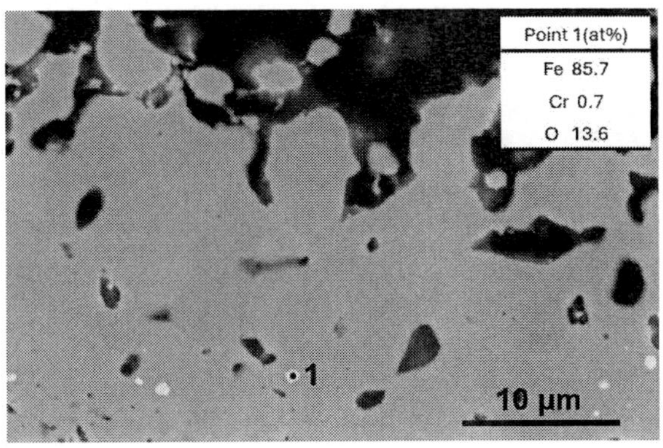

(C)

FIGURE 3: (A) CROSS-SECTION MORPHOLOGY OF Fe-20Cr AFTER 310 h REACTION AT 650°C IN Ar-20H_2-10H_2O, (B) XRD RESULTS OF THE OUTER LAYER, AND (C) SEM CROSS-SECTION OF THE OUTER LAYER WITH EDS POINT ANALYSIS.

TABLE 3: THICKNESS OF THE OXIDE SCALE LAYERS (μm) FORMED IN Fe-20Cr AFTER 310 h IN H_2-CONTAINING GASES.

Oxide layer	Ar-2H_2-10H_2O	Ar-10H_2-10H_2O	Ar-20H_2-10H_2O
Fe_3O_4	26±4	Small islands (not measured)	Small islands (not measured)
FeO	41±11	50±5	35±6
Inner layer	46±9	44±2	13±5
IOZ	23±4	27±1	50±6
Total	136	121	98

Si-BEARING ALLOY

The addition of 1wt% Si to the binary alloy greatly improves its resistance to oxidation, resulting in a notable decrease in weight gain compared to Fe-20Cr (refer to Fig. 1). In three H_2-containing gases, the scale morphology and constituent were similar (Figs. 4A-C), featuring a thin protective Cr_2O_3 layer with a thinner dark contrast SiO_2 layer beneath it. The protective scale thickness after 310 hours of reaction did not show significant differences in three H_2-containing gases, at about 0.27 - 0.28 μm.

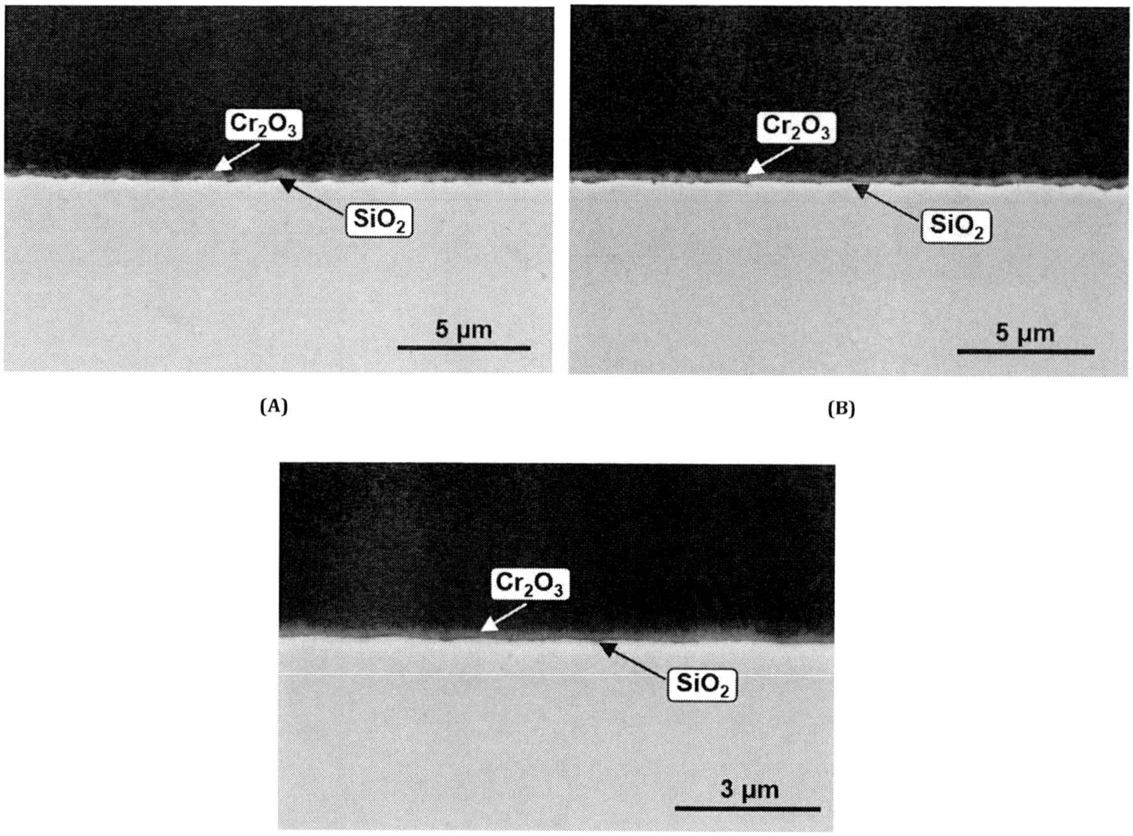

(A) (B)

(C)

FIGURE 4: HIGH MAGNIFICATION BSE-SEM CROSS-SECTIONS OF Fe-20Cr-Si AFTER 310 h REACTION IN (A) Ar-2H_2-10H_2O, (B) Ar-10H_2-10H_2O AND (C) Ar-20H_2-10H_2O.

Mn-BEARING ALLOY

The reaction morphology in 2 and 10% H_2 gas mixtures was not significantly altered by alloying with Mn, as shown in Figure 5. Fast-growing type iron-rich oxide layers/nodules were observed in Ar-2H_2-10H_2O and Ar-10H_2-10H_2O gases, comprising the outer Fe_3O_4 and FeO, and the inner FeO + spinel layer (confirmed by Raman, not shown), which are similar to that of Fe-20Cr (Fig. 2). BSE-SEM of the scale formed in Ar-10H_2-10H_2O and its EDS mapping results are shown in Fig. 5D, indicating that Mn was randomly distributed in the inner layer.

In Ar-20H_2-10H_2O, however, only a part of the scale formed breakaway-type oxide layers (Fig. 6A) with the same constituent as that in 2% and 10% H_2 gases. There were numerous iron-rich metal particles in the outer layer (identified by EDS point analysis in Fig. 6B). Similar phenomena were observed in both binary and Mn-containing alloys, where the morphology of the top of outer FeO layer was changed from dense layer in 2H_2-10H_2O (Fig. 5A) to broken and dispersed particles in 10 and 20 vol% H_2 gases (Fig. 5B and Fig. 6A). Apart from the localized breakaway-type Fe-rich oxides, about 50% of the alloy surface developed a protective, but somewhat porous chromia layer in Ar-20H_2-10H_2O gas (Fig. 6C). EDS-line scan result indicated that Mn was enriched and distributed in the chromia layer (Fig. 6D).

Overall scale and layer thicknesses are listed in Table 4. When increasing H_2 concentration from 2 vol% to 10 vol%, the thickness of each layer was comparable. However, the thicknesses of FeO and the inner layer in the unprotected region decreased remarkably in Ar-20H_2-10H_2O.

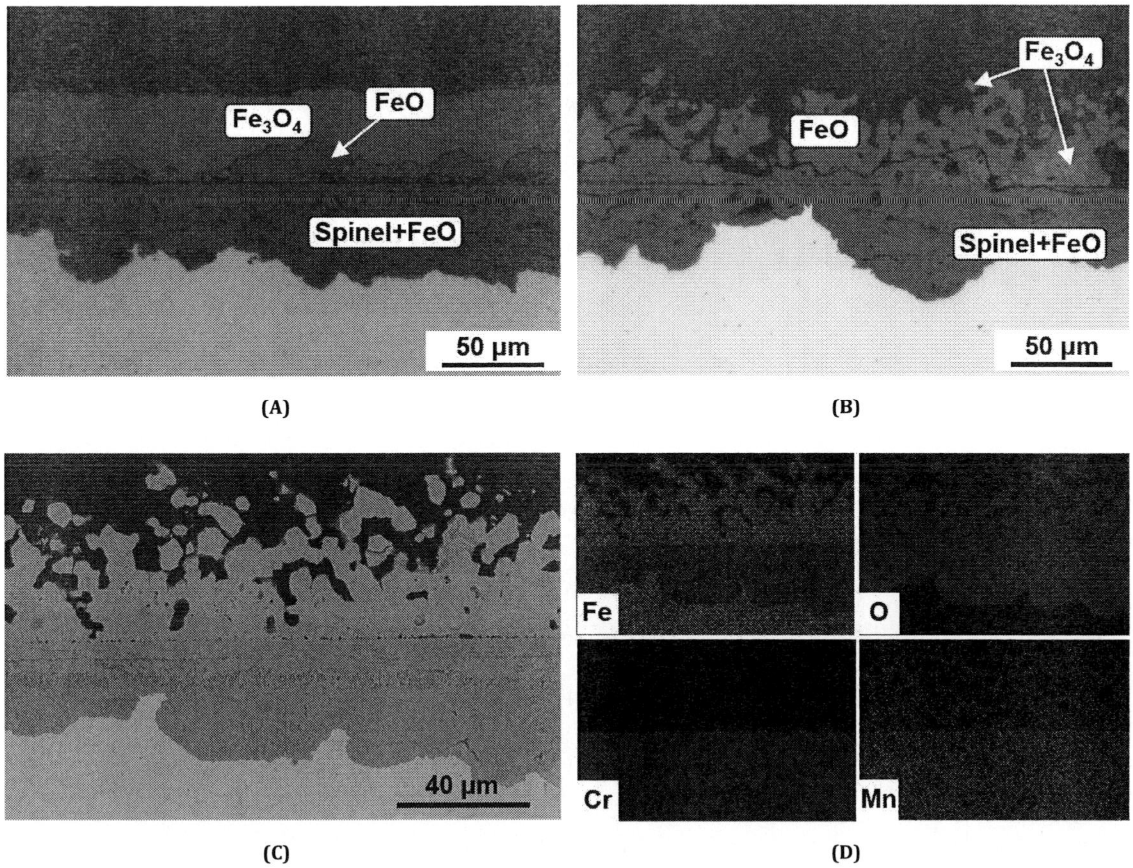

FIGURE 5: OPTICAL CROSS-SECTION OF Fe-20Cr-2Mn AFTER 310 h REACTION IN (A) Ar-2H_2-10H_2O, (B) Ar-10H_2-10$_2$HO AND (C) BSE-SEM IMAGE OF CROSS-SECTION IN Ar-10H_2-10H_2O WITH ITS EDS-MAPPING RESULTS IN (D).

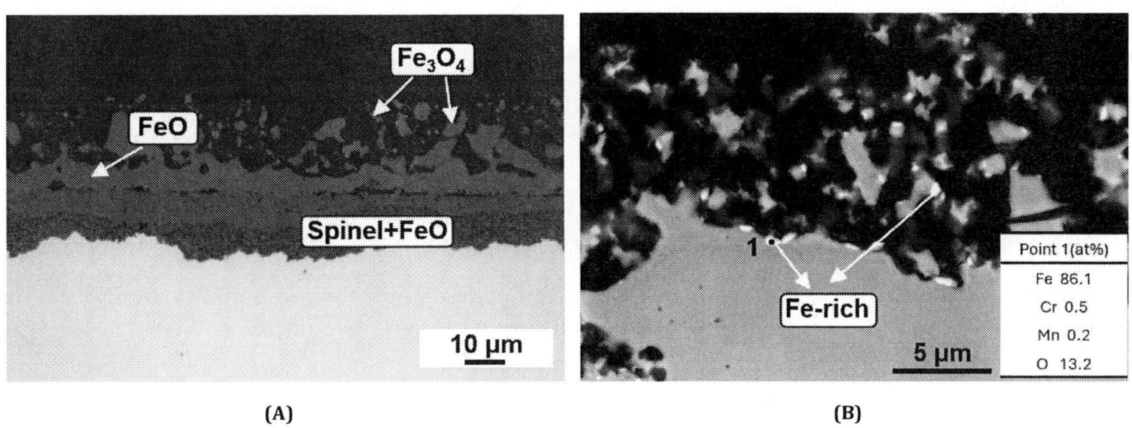

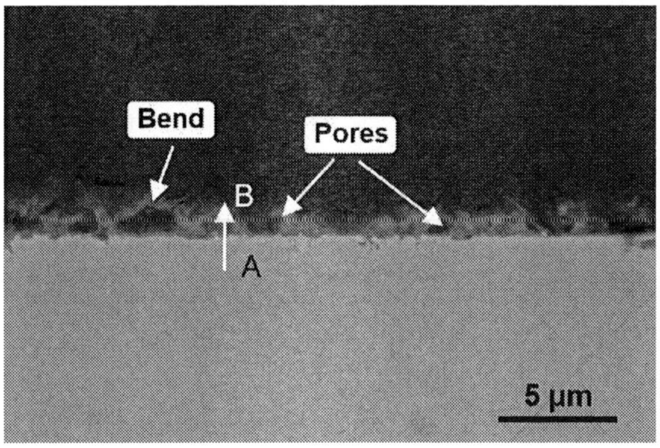

(C)

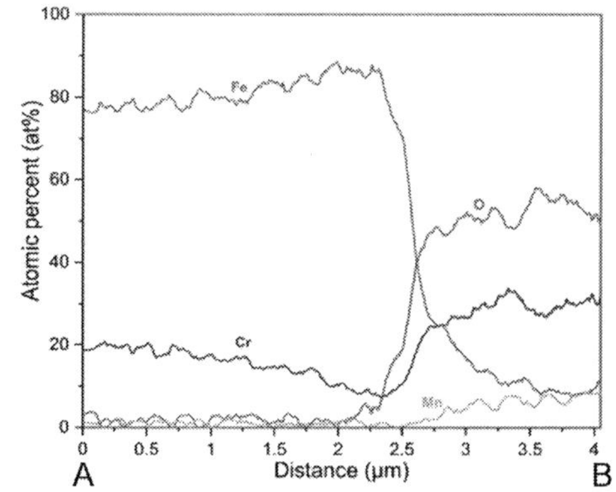

(D)

FIGURE 6: OPTICAL CROSS-SECTION OF Fe-20Cr-2Mn AFTER 310 h REACTION IN Ar-20H_2-10H_2O IN (A) UNPROTECTED REGION AND HIGH MAGNIFICATION BSE-SEM OF THE OUTER LAYER (B) WITH POINT ANALYSIS AND THE PROTECTIVE REGION (C) AND ITS EDS LINE SCAN IN (D).

TABLE 4: THICKNESS OF THE OXIDE SCALE LAYERS (µm) FORMED IN Fe-20Cr-2Mn AFTER 310 h IN H_2-CONTAINING GASES.

	Ar-2H_2-10H_2O	Ar-10H_2-10H_2O	Ar-20H_2-10H_2O	
			Non-protective	Protective
Fe_2O_3	-	-	-	-
Fe_3O_4	32±7	Very limited, Not measured	Very limited, Not measured	-
FeO	Islands, not measured	29±1	13±3 (dispersed oxides not included)	-
Inner layer	36±13	38±14	16±4	-
IOZ	-	-	-	-
Chromia	-	-	-	0.8±0.5
Total	68	67	29	0.8

Al-BEARING ALLOY

In 2H_2-10H_2O and 10H_2-10H_2O gases, the oxide scale (Fig. 7) was unprotective, consisting of the outer Fe_3O_4 and FeO layer, the inner FeO + spinel mixed layer and IOZ. Increasing H_2 concentration (2 vol% to 10 vol%) also led to dispersed oxides at the surface of the outer layer which is similar to that of Fe-20Cr. The scale had two regions: the thick region accounting for 30-40% of the total surface, and a relatively thin region (shown in 7C) over the majority of the surface. Aluminium was found and distributed in the inner layer by EDS line scan (Figs. 7D and E), but the degree of enrichment was small. In contrast, a higher H_2 concentration (20 vol%) resulted in an outstanding corrosion resistance, forming a fully protective scale (Fig. 8). High magnification BSE-SEM cross-sections with its EDS-line scan are shown in Figs. 8 B and C. The constituents of the protective scale were an external chromia layer with an Fe-rich metal layer beneath and Al-rich IOZ.

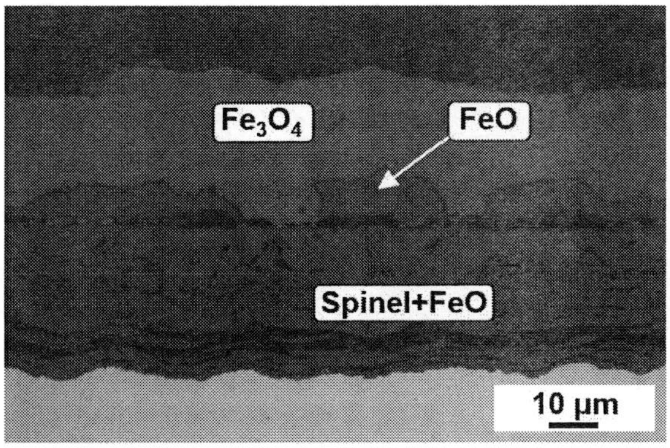

(A)

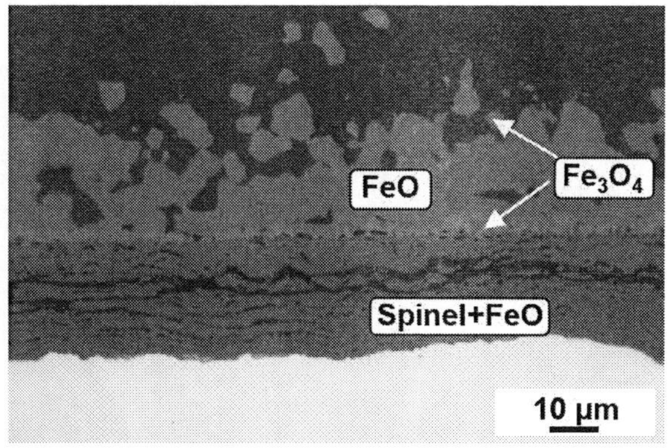

(B)

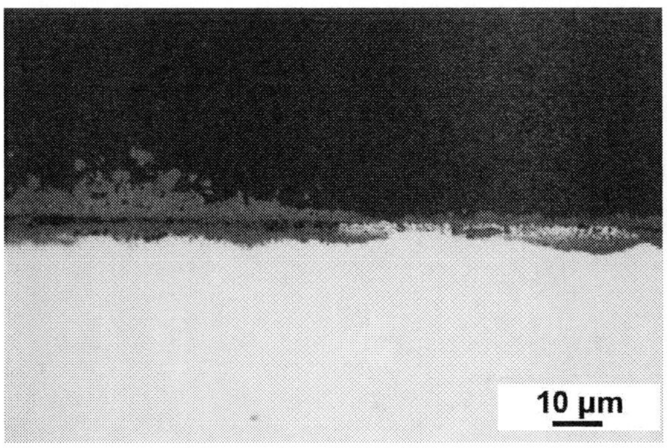

(C)

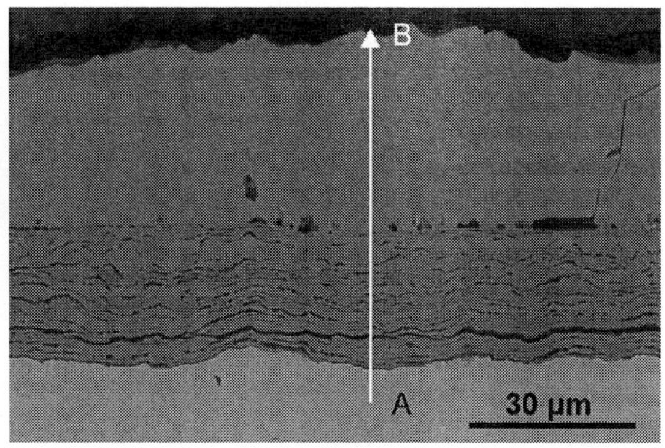

(D)

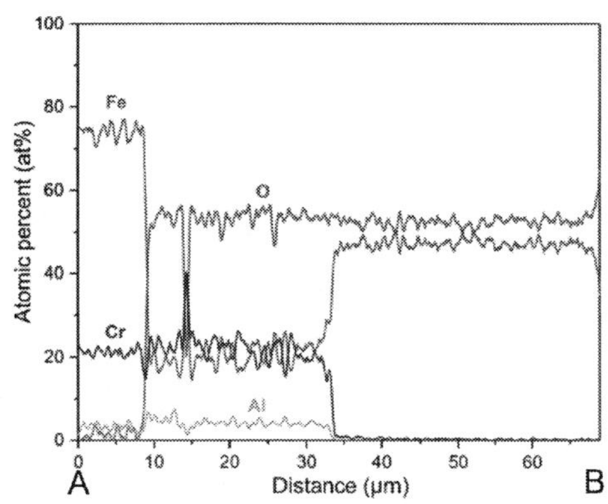

(E)

FIGURE 7: CROSS-SECTIONS OF Fe-20Cr-2Al AFTER 310 h REACTION IN (A) Ar-2H_2-10H_2O, (B) THICK REGION INTHICK REGION IN Ar-10H_2-10H_2O AND (C) THIN REGION IN Ar-10H_2-10H_2O, (D) BSE-SEM IMAGE OF THE SCALE IN Ar-2H_2-10H_2O AND EDS-LINE SCAN IN (E).

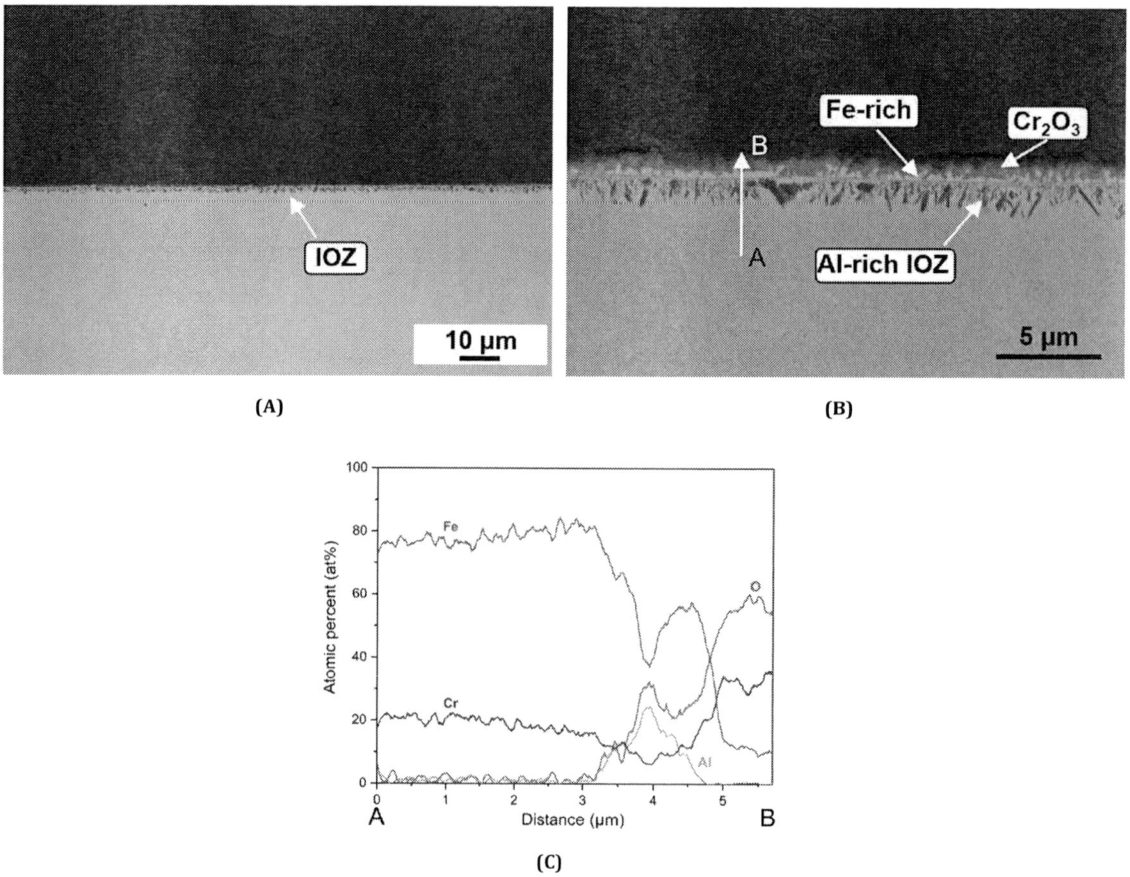

FIGURE 8: (A) OPTICAL CROSS-SECTION OF Fe-20Cr-2Al IN Ar-20H_2-10H_2O AFTER 310 h REACTION AND (B) HIGH MAGNIFICATION BSE-SEM IMAGE OF THE SCALE WITH ITS EDS-LINE SCAN IN (C).

Oxide layer thicknesses formed in Fe-20Cr-2Al after 310 h reaction are listed in Table 5. In the case of Ar-10H_2-10H_2O, there was a slight decrease in the thickness of the outer and inner layers in the thick region, but a remarkable decrease in the thin region compared to that in Ar-2H_2-10H_2O. Importantly, increasing H_2 to 20 vol%, led to protective behaviour and a much thinner scale.

TABLE 5: THICKNESS OF THE OXIDE SCALE LAYERS (µm) FORMED IN Fe-20Cr-2Al AFTER 310 h IN H_2-CONTAINING GASES.

	Ar-2H_2-10H_2O	Ar-10H_2-10H_2O		Ar-20H_2-10H_2O
		Thick layers (30%)	Thin layers (70%)	
Fe_2O_3	-	-	-	-
Fe_3O_4	28±2	-	-	-
FeO	Not measured	20±6	4±2	-
Inner layer	31±1	24±3	8±2	-
Chromia	-	-	-	0.6±0.2
IOZ	-	-	-	1.5±0.3
Total	59	44	12	2.1

Discussion

EFFECTS OF pO_2 AND HYDROGEN ON THE FORMATION OF THE OXIDE SCALE

Ferritic (Fe-20Cr) alloys were found to form a multi-layered structure when reacting in H_2-contianing wet gases. In H_2-containing gases (e.g., Figs. 2 and 3), the outer layer was composed of Fe_3O_4 and FeO, with no Fe_2O_3. The latter being formed in water vapor only condition [10]. This is due to the difference in oxygen partial pressure between H_2-containing and H_2O-only atmospheres. The gas pO_2 of Ar-2H_2-10H_2O and Ar-10H_2-10H_2O is higher than the dissociation oxygen partial pressure of Fe_3O_4 and FeO but lower than that of Fe_2O_3 (Table 1). In Ar-20H_2-10H_2O, the gas pO_2 is slightly lower than the $pO_{2,eq}$ for Fe_3O_4 formation, creating a marginal situation. This is consistent with the observation of limited Fe_3O_4 in Ar-20H_2-10H_2O, being detected only by XRD but not metallographically.

In addition, the outermost layer morphology in different H_2-containing gases is significantly different for all alloys except Fe-20Cr-1Si. Iron oxide layers had a relatively flat surface when it formed in Ar-2H_2-10H_2O (Figs. 2A). However, they became dispersed and discontinuous when H_2 concentration was increased to 10 and 20% (Figs. 2B and 3A). This phenomenon could be related to hydrogen embrittlement of oxides. According to the previous research [9], a compact tension test was conducted to compare the fracture surface of alloys with and without H_2 pre-charging, indicating the intergranular fractures in the H_2 pre-charged sample. Besides, a recent research done by Chen et al. [11] revealed that deuterium was detected and identified by atom probe tomography (APT), showing that hydrogen is accumulated at grain boundaries and carbon-rich dislocations in ferritic and martensitic steels. Therefore, this kind of hydrogen enrichment might also be applicable to the oxide scale, in which hydrogen could enter the oxide scale via grain boundaries leading to the dispersed and discontinuous appearance of the external oxide layer.

Moreover, the presence of hydrogen may increase the alloy oxygen permeability. The direct evidence is that the depth of IOZ was increased gradually when adding hydrogen into water vapor (Table 3) from 23 μm in Ar-5H_2-10H_2O to 50 μm in Ar-20H_2-10H_2O. Commonly, the kinetics of internal oxidation are described as

$$X_i(X_i - X_c) = 2k_p^i t \quad (1)$$

Where X_i and X_c represent the depth between the internal oxidation front and the original surface and that between the scale-substrate interface and the original surface, respectively. k_p^i is the parabolic rate constant of internal oxidation when assuming $(\frac{k_p^i}{2D_O})^{1/2} \ll 1$ and $X_c < X_i$, and t is the oxidation duration. The values of X_i and X_c were measured and are plotted in Fig. 9. In addition, the relationship between the parabolic rate constant and oxygen permeability can be expressed as follows [12]

$$k_p^i = \frac{N_O^{(S)} D_O}{\nu N_{Cr}^0} \quad (2)$$

where $N_O^{(S)}$ and D_O are the solubility of oxygen in local equilibrium at the scale-matrix interface and the diffusivity of oxygen, respectively. The calculated values of k_p^i and oxygen permeability ($N_O^{(S)} D_O$) values derived from them are listed in Table 6.

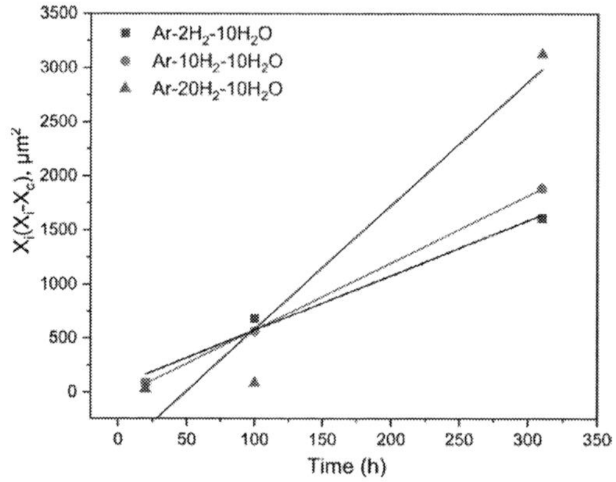

FIGURE 9: INTERNAL OXIDATION KINETICS IN Ar-xH$_2$-10H$_2$O.

TABLE 6: INTERNAL OXIDATION RATE CONSTANT (k_p^i, μm^2·h^{-1}) AND OXYGEN PERMEABILITY ($N_O^{(S)}D_O$, 10^{-14} cm^2/s^{-1}) FOR Fe-20Cr IN DIFFERENT GASES.

Gases	Ar-2H$_2$-10H$_2$O	Ar-10H$_2$-10H$_2$O	Ar-20H$_2$-10H$_2$O
k_p^i	2.54	3.11	5.72
$N_O^{(S)}D_O$	0.68	0.84	1.54

For Equation (2) to be valid, the oxygen permeability should satisfy $N_O^{(S)}D_O \gg N_{Cr}\widetilde{D}_{Cr}$ where $\widetilde{D}_{Cr}$ is the Fe-Cr alloy diffusion coefficient, taken from the literature [13]. The value of $N_{Cr}\widetilde{D}_{Cr}$ is 7.5×10^{-15} cm^2·s^{-1}, which is significantly less than oxygen permeability. Surprisingly, adding hydrogen could apparently increase the oxygen permeability. Hydrogen can penetrate the oxide scale, reaching the underlying alloy. This can occur as gas species (H_2 or H_2O) passing through cracks and pores [14] or possibly grain boundaries, or as protons bonded to oxide ions [2, 15, 16]. Just how dissolution of O in the metal matrix of the IOZ accelerates oxygen diffusion is as yet unknown.

EFFECT OF CHROMIUM CONCENTRATION ON THE FORMATION OF CHROMIA LAYER

The failure to form fully protective chromia scales on most of the tested alloys in H_2-containing gases might be due to the insufficient chromium supply from the alloys. Wagner's theory [17] was used for predicting the formation of chromia and its maintenance. N_{Cr}^{crit} is denoted to be the critical chromium concentration needed to form an external chromia layer rather than an IOZ. Wagner's analysis yield

$$N_{Cr}^{crit} = \left(\frac{\pi g}{2v} N_O^{(S)} \frac{D_O V_M}{\widetilde{D}_{Cr} V_{CrO_v}}\right)^{1/2} \tag{3}$$

here g is the critical volume fraction of oxide in the transition from internal to external oxidation, about 0.3 [18], v the stoichiometric coefficient of oxide CrO_v, V_M and V_{CrO_x} the mole volumes of alloy and oxide, D_O the oxygen diffusion coefficient in alloy, and $\widetilde{D}_{Cr}$ the Fe-Cr alloy diffusion coefficient. Both D_O and $\widetilde{D}_{Cr}$ were extrapolated from literature [19] and [13], respectively.

The value of $N_O^{(S)}$ is calculated based on Sievert's equation

$$N_O^{(S)} = K \cdot pO_2^{\frac{1}{2}} \tag{4}$$

In this equation, K is a temperature dependent value which was calculated for α-Fe from literature [20]. pO_2 is the interfacial equilibrium partial pressure, determined by

$$\mathrm{Fe} + \frac{1}{2}\mathrm{O}_2 = \mathrm{FeO} \tag{5}$$

The critical chromium concentration and all related values are listed in Table 7.

TABLE 7: CRITICAL CHROMIUM CONTEMTRATION WITH RELATED PARAMETERS.

V_M	V_{CrO_v}	$N_O^{(S)}$	D_O (cm^{2}s^{-1})	$\widetilde{D}_{Cr}$(cm^{2}s^{-1})	N_{Cr}^{crit}	N_{Cr}
7.3	14.6	1×10^{-7}	2.5×10^{-8}	3.6×10^{-14}	0.1	0.21

The calculation for Fe-20Cr (N_{Cr}=0.21) suggests that it should be able to form an external chromia layer, but this prediction failed for Fe-20Cr in all gases (Figs. 2 and 3). This phenomenon could be due to the enhanced oxygen permeability because of the presence of water vapor and hydrogen, increasing N_{Cr}^{Crit}.

In the case of Fe-20Cr-2Mn, a discontinuous chromia layer did form but was interrupted by iron-rich oxide nodules in Ar-20H_2-10H_2O (Fig. 6C). An external chromia layer was also observed for Fe-20Cr-2Al in Ar-20H_2-10H_2O (Fig. 8B), due to the formation of Al-rich IOZ. In addition, the scale formed on Fe-20Cr-1Si is a fully protective scale of chromia and SiO_2 (Fig. 4) in H_2-containing gases. The effects of these additional elements are now considered.

EFFECT OF ALLOYING ELEMENTS ON CORROSION OF Fe-20Cr

Si EFFECT

According to the weight gain kinetics, adding silicon significantly lowered the oxidation rate in all H_2-containg gases. Owing to its lower dissociation pO_2, silica is formed preferentially at the scale-alloy interface, where it slows outward chromium diffusion and inward oxygen diffusion. Therefore, the critical silicon concentration (N_{Si}^{crit}) for silica formation should be considered.

$$N_{Si}^{crit} = \left(\frac{\pi g}{2v} N_O^{(S)} \frac{D_O V_M}{\widetilde{D}_{Si} V_{SiO_2}}\right)^{1/2} \tag{6}$$

where $\widetilde{D}_{Si}$ is the diffusion coefficient of silicon in α-Fe, taken from [21]. Because silica was formed underneath the protective chromia, the oxygen solubility is determined by the reaction

$$\frac{2}{3}\mathrm{Cr(s)} + \frac{1}{2}\mathrm{O_2(g)} = \frac{1}{3}\mathrm{Cr_2O_3(s)} \tag{7}$$

and relevant parameters are listed in Table 8.

TABLE 8: CRITICAL SILICON CONTENTRATION RELATED PARAMETERS.

V_M	V_{SiO_2}	$N_O^{(S)}$	D_O (cm^{2}s^{-1})	$\widetilde{D}_{Si}$(cm^{2}s^{-1})	N_{Si}^{crit}	N_{Si}
7.3	22.6	2.5×10^{-12}	2.5×10^{-8}	3.1×10^{-13}	3.5×10^{-4}	0.02

In the current work, the protective silica was formed beneath the chromia layer in all H_2-containing gases (Fig. 4) which is consistent with the prediction. As a result, chromia grows slowly and internal oxidation is prevented.

Mn EFFECT

Manganese has little effect on corrosion resistance in all gases. The kinetic plots (Fig. 1) for Fe-20Cr-2Mn exhibited relatively lower weight uptakes than undoped alloys in the same conditions, corresponding to the thinner outer and inner layers. A similar phenomenon was observed in wet CO_2 where a significant weight gain decrease for in Fe-20Cr-2Mn was due to a protective Cr, Mn-rich scale in dry CO_2 but only a slight decrease resulted in wet CO_2, as the breakaway scale was little changed [7]. The effects of Mn alloying on corrosion resistance are rather small, they are not discussed further here.

Al EFFECT

Alloying with aluminium also decreased the oxidation rate in three gases (Fig. 1). Although the scale constituents in the gas containing 2% H_2 (Fig. 7) was similar to that of Fe-20Cr in the same conditions (Fig. 2), the thicknesses of the oxide layers were decreased (Tables 3 and 5). Evidently, the presence of Al in the inner layer (Fig. 7E) slows the diffusion of iron and oxygen [22, 23]. More importantly, the alloy exposed to the 10% H_2 gas forms a scale which is locally protective (Fig. 7C).

In Ar-20H_2-10H_2O, Al-rich oxides were formed internally rather than as a protective layer. The critical aluminium for forming an external alumina scale is considered.

$$N_{Al}^{crit} = \left(\frac{\pi g}{2v} N_O^{(S)} \frac{D_O V_M}{\widetilde{D}_{Al} V_{AlO_v}}\right)^{1/2} \tag{8}$$

Here $\widetilde{D}_{Al}$ is the diffusion coefficient of aluminium in Fe-Cr alloy, taken from [24]. The oxygen solubility was determined by Equation (5). The values of all relevant parameters are listed in Table 9.

TABLE 9: CRITICAL ALUMINUM CONCENTRATION WITH RELATED PARAMETERS.

V_M	V_{AlO_v}	$N_O^{(S)}$	D_O (cm^2s^{-1})	$\widetilde{D}_{Al}$(cm^2s^{-1})	N_{Al}^{cirt}	N_{Al}
7.3	12.8	1×10^{-7}	2.5×10^{-8}	1.8×10^{-13}	0.05	0.04

No external alumina was observed in all wet gases which is consistent with the prediction. However, in Ar-20H_2-10H_2O, a protective chromia scale (Fig. 8B) was observed rather than the iron-rich oxide layers formed in Al-free alloy (Fig. 3A). The formation of internal alumina in preference of chromia (Al_2O_3 more stable than Cr_2O_3) leads to more Cr diffusion to the surface and formation of the protective layer that was observed (Fig. 8B).

In addition, internal Al-rich IOZ caused volume expansion, leading to the ejection of Fe-rich nodules/layer [25, 26], which is unable to be oxidized due to the very low oxygen partial pressure under the thin Cr_2O_3.

CONCLUSIONS

Ferritic model alloy, Fe-20Cr, was reacted in three gases with different hydrogen concentrations (Ar-2H_2-10H_2O, Ar-10H_2-10H_2O and Ar-20H_2-10H_2O) at 650°C. The effects of hydrogen and oxygen partial pressures on alloy corrosion were investigated by weight gain kinetics and the oxide scale analysis. Additionally, alloying effects (Si, Mn and Al) on the alloy corrosion were also investigated in these H_2-containing wet gases.

For Fe-20Cr alloys, the oxide scale was a typical multi-layered structure consisting of outer Fe_3O_4 + FeO layers, an inner spinel + FeO layer, and IOZ in all H_2-containing gases. Lower pO_2 values led to a decrease in the amount of Fe_3O_4 formation in H_2-containing gases.

Increasing hydrogen content in the gas decreased the pO_2, and changed the oxide formation and morphology. The observation of discontinuous and dispersed outer iron oxide layer in the gas containing 10% and 20% hydrogen could be related to hydrogen accumulation at oxide grain boundaries, causing hydrogen embrittlement of the oxide. Furthermore, the presence of hydrogen enhanced the effective oxygen permeability, resulting in a thicker IOZ.

Silicon has significant effects on improving corrosion resistance of the alloy by forming a SiO_2 sublayer as a barrier and facilitating chromia formation. The presence of manganese did not have significant effects on oxidation rate or scale morphology. In addition, aluminium had limited effects on oxidation rate and the scale formation in Ar-2H_2-10H_2O and Ar-10H_2-10H_2O gases, but significantly reduced the oxidation rate in

Ar-20H_2-10H_2O due to the formation of an alumina IOZ which prevented the internal oxidation of Cr, allowing its Cr diffusion outwards to form a protective chromia on the surface.

ACKNOWLEDGMENTS

Financial support from the Australian Research Council's Discovery Program is gratefully acknowledged.

REFERENCES

1. Ni M, Leung MKH, and Leung DYC, *Electrochemical modeling of hydrogen production by proton-conducting solid oxide steam electrolyzer.* International Journal of Hydrogen Energy, 2008. **33**(15): p. 4040-4047.
2. Essuman E, Meier GH, Żurek J, Hänsel M, and Quadakkers WJ, *The Effect of Water Vapor on Selective Oxidation of Fe–Cr Alloys.* Oxidation of Metals, 2008. **69**(3): p. 143-162.
3. Jianian S, Longjiang Z, and Tiefan L, *High-temperature oxidation of Fe-Cr alloys in wet oxygen.* Oxidation of Metals, 1997. **48**(3): p. 347-356.
4. Chen J, Shen Z, and Zhang J, *Effect of Water Vapor and Oxygen Partial Pressure on Oxidation of Fe and Fe–Cr Alloys.* High Temperature Corrosion of Materials, 2023. **99**(1): p. 15-46.
5. Nguyen TD, Zhang J, and Young DJ, *Effects of H2 on microstructures of Cr_2O_3 scales grown in water vapour and consequences for breakaway.* Corrosion Science, 2024. **236**: p. 112265.
6. Hooshyar H, Jonsson T, Hall J, Svensson JE, Johansson LG, and Liske J, *The Effect of H_2 and H_2O on the Oxidation of 304L-Stainless Steel at 600 °C: General Behaviour (Part I).* Oxidation of Metals, 2016. **85**(3): p. 321-342.
7. Nguyen TD, Zhang J, and Young DJ, *Effects of cerium and manganese on corrosion of Fe–Cr and Fe–Cr–Ni alloys in Ar–20CO_2 and Ar–20CO_2–20H_2O gases at 650°C.* Corrosion Science, 2015. **100**: p. 448-465.
8. Nguyen TD, Zhang J, and Young DJ, *Effects of Silicon on High Temperature Corrosion of Fe–Cr and Fe–Cr–Ni Alloys in Carbon Dioxide.* Oxidation of Metals, 2014. **81**(5): p. 549-574.
9. Mills WJ, Lebo MR, and Kearns JJ, *Hydrogen embrittlement, grain boundary segregation, and stress corrosion cracking of alloy X-750 in low-and high-temperature water.* Metallurgical and Materials Transactions A, 1999. **30**(6): p. 1579-1596.
10. Yuan J, Wang W, Zhu S, and Wang F, *Comparison between the oxidation of iron in oxygen and in steam at 650–750°C.* Corrosion Science, 2013. **75**: p. 309-317.
11. Chen Y-S, Lu H, Liang J, Rosenthal A, Liu H, Sneddon G, McCarroll I, Zhao Z, Li W, Guo A, and Cairney JM, *Observation of hydrogen trapping at dislocations, grain boundaries, and precipitates.* 2020. **367**(6474): p. 171-175.
12. Wagner C, *Reaktionstypen bei der Oxydation von Legierungen.* Zeitschrift für Elektrochemie, Berichte der Bunsengesellschaft für physikalische Chemie, 1959. **63**(7): p. 772-782.
13. Braun R and Feller-Kniepmeier M, *Diffusion of chromium in α-iron.* physica status solidi (a), 1985. **90**(2): p. 553-561.
14. Rahmel A and Tobolski J, *Einfluss von wasserdampf und kohlendioxyd auf die oxydation von eisen in sauerstoff bei hohen temperaturen.* Corrosion Science, 1965. **5**(5): p. 333-346.
15. Tuck CW, Odgers M, and Sachs K, *The oxidation of iron at 950°C in oxygen/water vapour mixtures.* Corrosion Science, 1969. **9**(4): p. 271-285.
16. Quadakkers W and Żurek J, *Oxidation in steam and steam/hydrogen environments.* 2010.
17. Wagner C, *Beitrag zur Theorie des Anlaufvorgangs.* 1933. **21B**(1): p. 25-41.
18. Rapp RA, *The transition from internal to external oxidation and the formation of interruption bands in silver-indium alloys.* Acta Metallurgica, 1961. **9**(8): p. 730-741.
19. Takada J and Adachi M, *Determination of diffusion coefficient of oxygen in α-iron from internal oxidation measurements in Fe-Si alloys.* Journal of Materials Science, 1986. **21**(6): p. 2133-2137.
20. Young DJ, *Chapter 2 - Enabling Theory*, in *High Temperature Oxidation and Corrosion of Metals (Second Edition)*, D.J. Young, Editor. 2016, Elsevier. p. 31-84.

21. Borg RJ and Lai DYF, *Diffusion in α-Fe–Si Alloys.* Journal of Applied Physics, 1970. **41**(13): p. 5193-5200.
22. Wolff IM, Iorio LE, Rumpf T, Scheers PVT, and Potgieter JH, *Oxidation and corrosion behaviour of Fe–Cr and Fe–Cr–Al alloys with minor alloying additions.* Materials Science and Engineering: A, 1998. **241**(1): p. 264-276.
23. Kvernes I, Oliveira M, and Kofstad P, *High temperature oxidation of Fe-13Cr-xAl alloys in airH_2O vapour mixtures.* Corrosion science, 1977. **17**(3): p. 237-252.
24. Yoneda S, Hayashi S, Saeki I, and Ukai S, *The Effect of Cr on the Lifetime of Al-Rich Amorphous Oxide Layer Formed on Fe–Cr–Al Alloys at 650 °C.* Oxidation of Metals, 2017. **88**(5): p. 669-686.
25. Jiang C, Zhang J, and Young DJ, *The Effect of Water Vapor on NiO Formation by Ni–Cr Alloys at 650 °C (HTCPM Focus Issue, FNS-111).* Oxidation of Metals, 2021. **96**(1): p. 57-68.
26. Xie Y, Zhang J, and Young DJ, *Water vapour effects on corrosion of Ni-Cr alloys in CO_2 gas at 650 °C.* Corrosion Science, 2018. **136**: p. 311-325.

	Chen is a PhD student in the School of Materials Science and Engineering in UNSW Sydney since 2023.
	Zhang is the corresponding author of this paper. He is a professor in the School of Materials Science and Engineering in UNSW Sydney.
	Young is an emeritus professor in the School of Materials Science and Engineering in UNSW Sydney.

CORROSION & PREVENTION
CAIRNS 2024

THE AUSTRALASIAN CORROSION ASSOCIATION INC.

EVOLUTION OF CORROSION CONSIDERATIONS IN THE AS 3700 MASONRY STRUCTURES STANDARD: A HISTORICAL ANALYSIS AND IMPLICATIONS FOR DURABILITY ENHANCEMENT

M. A. Hossain[1,*], C. Y. Lam[1], M. J. Masia[1], I. A. Chaves[1]

[1]The University of Newcastle, New South Wales, Callaghan 2308, Australia

**Corresponding author email: Akhtar.a.Hossain@newcastle.edu.au*

ABSTRACT

The AS 3700 Masonry Structures Standard has undergone significant evolution. It incorporates various corrosion-related considerations to enhance the durability and longevity of masonry (veneer and cavity) structures. This paper explores the historical development of corrosion considerations within the AS 3700 Standard, tracing the evolution of relevant provisions and methodologies. Beginning with early standard iterations, which may have lacked specific corrosion-focused provisions, the study examines how subsequent revisions addressed corrosion concerns. Key aspects (mainly focusing on wall ties) such as material selection, protective coatings, design detailing, and maintenance practices are analyzed in the context of corrosion prevention and mitigation strategies. Additionally, the paper discusses advancements in corrosion-resistant materials and technologies that have influenced the evolution of the AS 3700 Standard. By documenting the historical progression of corrosion considerations within the AS 3700 Masonry Structures Standard, this study aims to provide valuable insights for practitioners, researchers, and policymakers for metal wall ties involved in masonry construction and corrosion management.

Keywords: AS 3700, Masonry code, corrosion, protective coating

INTRODUCTION

The Australian Standard AS 3700, commonly known as the "Masonry Structures Code," has a rich history highlighting its fundamental role in defining masonry construction practices in Australia; the latest version was published in 2018 [1]. It was initially developed to offer detailed guidelines and specifications for the design and construction of masonry structures, AS 3700 has consistently adapted to address the evolving needs of the construction industry. This evolution reflects advancements in building materials, construction techniques, and technological innovations. By continually updating its provisions, AS 3700 ensures that masonry construction remains robust, efficient, and aligned with modern standards, thereby supporting the integrity and durability of masonry structures across various applications.

A notable advancement within AS 3700 has been the incorporation of corrosion-related provisions. These provisions have significantly influenced masonry construction practices, making builders and engineers increasingly aware of the need to select appropriate materials and employ meticulous design detailing to mitigate corrosion risks. This proactive approach has markedly improved the durability and longevity of masonry structures, especially in coastal and other corrosive environments. Additionally, AS 3700 requires compliance with AS 2699.1, in which durability considerations are a large focus. To maintain performance criteria, regular inspections and timely repairs to address early signs of corrosion are required, which will contribute to preserving masonry structures and reducing long-term repair and replacement costs.

The historical evolution of the AS 3700 Standard, particularly concerning wall ties, reflects an enhanced understanding of masonry construction practices and the critical role of wall ties in maintaining structural integrity. Wall ties are crucial for the stability of masonry panels, and proper selection, spacing, and installation are essential to prevent damp penetration, distortion, cracking, or collapse. Factors influencing wall tie selection include the type of brick/block, cavity width, building type and height, location, and design life. It is important to consult relevant documents and be aware of recent regulatory changes. The regulations governing wall ties are addressed in two distinct documents. AS 3700 focuses on the structural aspects of masonry, detailing provisions for cavity wall construction, including the spacing of ties within walls, near openings, and the embedment depth of ties in mortar joints. This standard ensures that ties are effectively integrated into the wall design to uphold structural integrity. Conversely, AS 2699.1 pertains explicitly to the wall ties themselves [2]. It details the types of ties, the materials from which they are made, and the required protective coatings to prevent corrosion. Additionally, it includes provisions for marking ties to ensure correct usage according to the specified durability standards. Together, these standards offer a comprehensive framework for selecting, installing, and maintaining wall ties in masonry structures, balancing the need for structural integrity with material longevity.

Understanding the historical construction codes and standards is crucial for assessing and restoring historic masonry structures. These codes offer valuable insights into the design principles, material specifications, and construction techniques used for wall ties. Appreciating historical context and relevant regulations allows restoration specialists to make informed decisions that preserve the structure's integrity and authenticity while ensuring its stability and longevity in contemporary settings.

Overall, the AS 3700 Masonry Structures Standard remains a cornerstone of the Australian construction industry, governing the design and construction of masonry structures. Its evolution, particularly in addressing corrosion concerns, has significantly enhanced the durability and longevity of masonry structures, especially veneer and cavity walls. This paper explores the historical development of corrosion considerations within the AS 3700 Standard (incorporating amendments), highlighting how these provisions have advanced over time to improve masonry construction practices. Some key aspects of metal wall ties based on Australian standards are also discussed to illustrate these advancements.

Historical Development of Corrosion Provisions

In the initial versions of AS 3700, corrosion was not a primary concern for wall ties. When the AS 3700 Standard was first introduced, the focus was primarily on general masonry construction practices. Wall ties were recognized as essential components for the stability of cavity walls, connecting the inner and outer leaves of masonry. However, early versions of the standard provided basic guidance on wall tie use, primarily concerning their placement and spacing within the wall assembly. The emphasis was on ensuring that ties were positioned correctly to resist lateral loads and prevent differential movement between the leaves. However, as instances of wall tie failures due to corrosion became more prevalent, there was a growing recognition of the need to include specific provisions for corrosion prevention.

Early Iterations of the Standards

The initial versions of AS 3700 were influenced by earlier masonry codes and practices from both Australia and other countries, particularly the UK. The need for a standardized approach to masonry construction became evident as Australia's construction industry grew, particularly in the post-World War II era, when the demand for durable, cost-effective building materials surged [3]. Early versions of the code focused primarily on traditional materials like clay brick and concrete block, with guidelines centred around load-bearing capacities, metal ties, mortar types, and construction methods typical of the mid-20th century. One of the earliest responses to this issue was the introduction of guidelines for selecting materials in environments prone to corrosion. The standard began to recommend the use of galvanized or stainless steel wall ties in areas with high exposure to moisture or saline environments, where the risk of corrosion was significantly higher.

In May 1953, the Standards Association of Australia released the first specification for wall ties, designated as SAA Int. 324, under the title "Interim Specification for Metal Wall Ties for Brickwork" [4]. This specification was designed to establish a minimum quality standard for wall ties used in cavity brick walls and brick veneer construction, ensuring that the outer brick leaf was securely anchored to the inner leaf. It also provided specific guidelines on manufacturing metal ties, as shown in Figure 1. It is also noted that there is a 'drip" in all the metal ties proposed in Figure 1 to prevent moisture movement along the ties and across the cavity, and length of all ties is 7.5" (~190mm) ± 0.5"(~12mm).

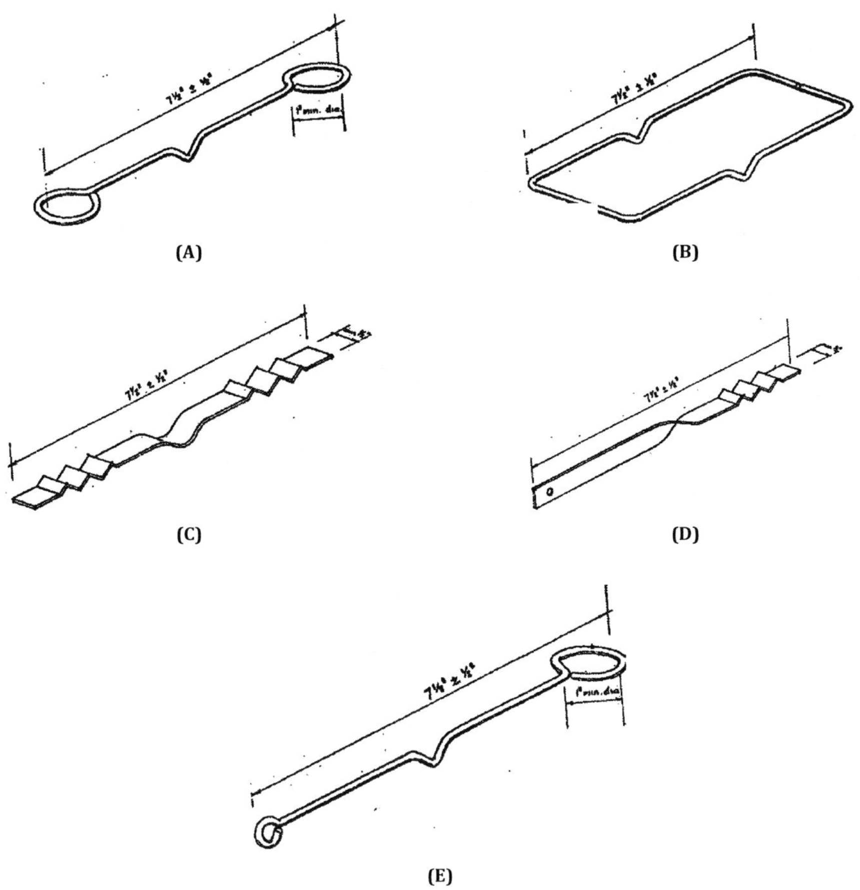

FIGURE 1: FIRST INTRODUCED METAL WALL TIES (A), (B) AND (C) USED FOR DOUBLE WALL OR CAVITY MASONRY CONSTRUCTION, AND (D) AND (E) FOR BRICK VENEER CONSTRUCTION, ADOPTED FROM [4]

According to the specification, wall ties must be made from one of the following materials: mild steel coated with zinc, 99.5% pure copper, or copper alloys with at least 55% copper content. It also required that the coating for mild steel wall ties contain a minimum of 98.5% pure zinc, as tested per A.S. No. K. 53-1942. This document was withdrawn in October 1984 after the release of AS 2699-1984. In 1963, the Australian Standard Rules for the Use of Concrete Blocks in Masonry Construction (AS CA32-1963) provided the first comprehensive guidelines for masonry structures in Australia.

Subsequent Revisions

As the industry evolved, so did the standard. In the late 20th century, AS 3700 underwent significant revisions to incorporate advances in material science, including the use of reinforced and post-tensioned masonry [7]. The introduction of new materials and construction techniques necessitated updates to the code to ensure that structures built using these methods would be safe, durable, and fit for purpose. The revisions also reflected a growing understanding of the behaviour of masonry structures under different environmental conditions, such as seismic activity and wind loads.

A significant milestone in the evolution of AS 3700 was the adoption of performance-based design principles. For instance, in 1998, a design method for wall ties was introduced [8], providing greater flexibility in the design process. This shift empowered engineers to employ innovative solutions while maintaining compliance with the safety and performance requirements established in the standard. This transition represented a move away from the prescriptive approach of earlier versions, which focused more on specific methods and materials. Additionally, the introduction of AS 2699, first edition in 1984, a companion standard, further refined these requirements by specifying the types of materials suitable for wall ties, including galvanizing and stainless steel, which offered enhanced corrosion resistance. Subsequent revisions to AS 3700 provided more detailed guidance on the use of wall ties in masonry construction. The standard expanded to include specific recommendations on tie embedment depth, minimum spacing near openings, and tie spacing across different masonry units. These updates were driven by experimental research and field observations, underscoring the critical role of proper tie placement in preventing structural failures and ensuring effective load distribution. The key modifications related to AS 3700 and AS 2699.1 are presented in Table 1 and Table 2, respectively.

TABLE 1: CHRONOLOGICAL ORDER OF DEVELOPING AS 3700

Publication Year	Key modifications related to metal wall ties, cavity details and corrosion
1963	Code: AS CA32-1963 [6] ✓ The first pivotal standard outlined the rules for using concrete blocks in Masonry Construction in Australia. ✓ Announced a new Australian standard on metal wall ties is in the course of preparation. ✓ Regarded as tentative code and opened for constructive comments. ✓ Spaced metal wall ties at intervals not exceeding 16"(~400mm), horizontally and vertically (cavity construction). ✓ Provided additional ties near the opening not more than 12" (~300mm) apart around the perimeter. ✓ Maximum cavity width 2.5" (~63mm).
1967	Code: AS CA32-1967 [9] ✓ Revised edition of AS CA32-1963. ✓ The previously announced standard on metal wall ties was still under development. ✓ Designed to simplify and clarify previous rules. ✓ Introduced the clauses related to cavity width, metal tie spacing (horizontal and vertical), and attachment requirements to the timber frame for masonry veneer construction. ✓ Cavity width (veneer)- minimum 1" (~25mm) and maximum 2"(~50mm).

	✓ Spaced metal wall ties at intervals not exceeding 16"(~400mm) in vertically and 48" (~1200mm) in horizontally (veneer construction). ✓ Secured metal ties to the timber frame using a minimum of two 1-inch (~25mm) by 8 s.w.g. (4.064mm) galvanized clouts.
1969	Code: AS CA47-1969 [10] ✓ Provided the first comprehensive guidelines for brickwork in masonry construction. ✓ Adopted SAA Int 324 [4] (brickworks) and British Standards BS 1234 [11] (cavity wall construction) as metal wall ties standard. ✓ Enforced the minimum thickness of leaf for cavity walls is 3" (~75mm). ✓ Introduced additional ties at discontinuities in the brickwork (e.g., control joints). ✓ Proposed the tie spacings and tie types corresponding to various leaf thicknesses and cavity widths in a tabular form. ✓ Proposed uniform spacing of ties for both cavity and veneer construction includes a maximum tie spacing of 18" (~450 mm) horizontally and vertically for cavities up to 2" (~50 mm). For cavity widths up to 3" (~75 mm), the horizontal spacing may be increased to 36"(~900 mm). ✓ Also introduced minimum tie embedment in the mortar 2" (~50mm) in each leaf. ✓ Modified the ties fixing requirement (veneer construction). It was proposed **one** 1-inch (~25mm) by 8 S.W.G. (4.064mm) galvanized clouts or other approved equivalent method of fixing instead of **two** 1-inch (~25mm) by 8 S.W.G. (4.064mm) galvanized clouts.
1974	Code: AS 1640-1974 [12] ✓ Developed a metric version of AS CA47-1969. ✓ Intended for application to both non-loadbearing and loadbearing structures constructed in Australia using clay, calcium silicate, or concrete bricks. ✓ Modified the minimum cavity width requirement (40mm for cavity walls and 25 mm for veneer construction). ✓ Modified the wall tie spacings. For cavity widths of 60 mm or less, the maximum horizontal spacing was set at 400 mm, with vertical spacing at 600 mm, applicable to both cavity walls and brick veneer construction. ✓ Updated the fixing requirement for brick veneer (with a 25mm x 2.5 mm galvanized clout or other approved equivalent method of fixing).
1977	Code: AS 1475.1-1977 [13] ✓ Developed a separate code for blockwork. ✓ Superseded the previous version AS CA47-1969. ✓ AS 1475 is comprised of two parts, with the first part addressing the design of buildings featuring unreinforced concrete block walls. ✓ Modified the wall tie spacings. For cavity widths of 80 mm or less, the maximum horizontal spacing was set at 600 mm, with vertical spacing at 600 mm, applicable to cavity construction. The ties shall be spaced not more than 600 mm apart vertically and horizontally for veneer construction. ✓ Cavity width for cavity walls shall be within the range of 40 to 160 mm. ✓ Maximum and minimum cavity widths for brick veneer were set at 25mm and 50 mm, respectively. ✓ Requirement for metal wall ties to be corrosion-resistant was introduced for the first time, though no specific details were provided. Previously, earlier standards had limited corrosion considerations to metal lintels, materials for flashings and weathering, etc. ✓ Specified that no part of the ties should be positioned closer than 15 mm to a wall face exposed to the weather.
1988	Code: AS 3700-1988 [14] ✓ Presented brickwork and blockwork requirements as one standard. ✓ Introduced performance-based specifications for masonry construction.

	✓ For the first time, the consideration of lateral loads, including wind and seismic forces, was introduced in masonry design. ✓ Wall ties shall comply with AS 2699. ✓ Introduced exposure classifications for steel in masonry based on surface and exposure environment. ✓ Modified the wall tie spacings: For general applications, ties shall be spaced at 600 mm centres in each direction for cavity wall and brick veneer construction. ✓ Introduced corrosion resistance rating for steel products in masonry based on the type and/or coating of steel. ✓ Adopted galvanized steel (coated with hot dip zinc, coating class Z275 and Z600, or aluminium/zinc), cadmium-coated steel and stainless steel (grade 316) for wall ties.
1998	Code: AS 3700-1998 [8] ✓ Increased adoption of performance-based specifications for requirements. ✓ Inclusion of the virtual work method for designing unreinforced masonry subjected to wind or earthquake loads. ✓ Introduction of additional requirements for structures exposed to earthquake actions. ✓ Implementation of a new design method for wall ties. ✓ Classified wall ties categories, including veneer, cavity, collar joint, and head and column ties. Additionally, developed a classification system (Type A and Type B) based on seismic considerations in tie design. Ties are also classified as light, medium and heavy duty based on different characteristics strength in tension and compression. ✓ Introduced capacity reduction factor for wall ties in tension or compression. ✓ Enforced the minimum width of cavity is 40 mm for both cavity and veneer walls. ✓ Removed corrosion resistance rating details from AS 3700-1988, and did not include coating specifications for steel products within this code.
2001	Code: AS 3700:2001 [15] ✓ Introduced a new section for simplified design of masonry for small buildings. ✓ Minimum clear width of a cavity shall be 25 mm for veneer construction and between 35 mm and 75 mm for cavity construction. ✓ Included atmospheric corrosivity classification based on ISO 9223 [16].
2011	Code: AS 3700:2011 [17] ✓ Added a separate section for the design of cavity walls subjected to lateral loading. ✓ Implemented mandatory screw fixing for face-fixed masonry veneer ties in certain locations. ✓ Enhanced strength design provisions for wall ties. ✓ Removed Section 12 for simplified design and developed a new standard specifically for simplified design. ✓ Revised the wall tie categories, reducing the classification from four types to two (veneer ties and cavity ties). ✓ Adopted screw and nail fixing methods for brick veneer. ✓ Eliminated the atmospheric corrosivity classification.
2018	Code: AS 3700:2018 [1] ✓ Reinforced the minimum cavity width requirement to 40 mm for both cavity and veneer walls. ✓ Included a new appendix providing informative details on the relationship between ISO 9223 corrosivity categories and durability classes.

TABLE 2: CHRONOLOGICAL ORDER OF DEVELOPING AS 2699.1

Publication Year	Key modifications related to metal wall ties, and protective coatings
1953	Code: SAA Int. 324 [4] ✓ First publication in Australia related to metal wall ties. ✓ Proposed three types of wall ties for cavity construction and two types for veneer construction, as shown in Figure 1. ✓ Suggested three materials for wall ties: mild steel, copper, and copper alloys. ✓ Specified that strip metal ties should have a minimum width of 0.5 inches and a thickness of at least 20 S.W.G. (0.914mm) for mild steel or 22 S.W.G. (0.711mm) for copper or copper alloys. ✓ Stipulated that the total cross-sectional area of wire wall ties must be no less than that provided by a 10 S.W.G. wire (8.3mm2) for steel and a 12 S.W.G. wire (5.4mm2) for copper or copper alloy. ✓ Coated mild steel with more than 98.5% of pure zinc.
1978	Code: BS 1234 [11] ✓ Six materials are specified for metal wall ties: low carbon steel strip, low carbon steel wire, copper, copper alloys, stainless steel wire, and stainless steel strip. ✓ Proposed three wall tie types for wall cavity construction: Butterfly type, Double-triangle type, and Vertical-twist type. ✓ Recommended minimum diameters for ties: Butterfly type (3.15±0.08 mm) and Double-triangle type (4.5±0.08 mm) for low carbon steel, with a diameter range of 3 mm to 3.83 mm for Vertical-twist type. ✓ Highlighted the need for additional precautions, such as bitumen coating, when wall ties are used with black ash mortar or marine aggregates. ✓ Proposed tie marking based on the weight of the wall ties. ✓ Steel elements should be galvanized or primed with an anti-corrosive primer to prevent corrosion.
1984	Code: AS 2699-1984 [5] ✓ SAA Int. 324 was withdrawn on October 1, 1984. ✓ Superseded BS 1234 in Australia. ✓ Officially cancelled the adoption of both SAA Int 324 and BS 1234 as standards for metal wall ties. ✓ First code dedicated for wall tie specifications only in Australia. ✓ Introduced a duty classification for ties (light, medium, heavy, and extra-heavy). ✓ Implemented color coding for each wall tie or package. ✓ Proposed three types of masonry veneer ties: face-fixed, flange-fixed, and side-fixed.
2000	Code: AS/NZS 2699.1 [18] ✓ Provided manufacturers with guidelines for classifying wall ties into three duty ratings. ✓ Proposed two types of wall ties: Type A for cavity and veneer walls, classified by characteristic strength, and Type B for seismic-resistant veneer ties, evaluated based on stiffness, serviceability, ultimate strength, and ductility. ✓ Included provisions for remedial (retrofitted) ties to ensure performance and durability. ✓ Tabulated coating requirements for metal ties based on durability classification. ✓ Included stainless steel grades AS 1449/316 or AS 1449/316L as acceptable materials for wall ties.
2020	Code: AS 2699.1:2020 [2] ✓ Established a minimum design life of 50 years for wall ties. ✓ Introduced durability criteria based on proximity to surf and non-surf coasts. ✓ Specified coating thickness for galvanized steel ties, replacing the previous notation (weight per unit surface area).

	✓ Proposed that tie markings include both durability class and duty classification. ✓ Presented the relationship between the durability class of wall ties and atmospheric corrosivity categories.

KEY ASPECTS OF CORROSION MANAGEMENT IN THE AS 3700 STANDARD

The issue of wall tie corrosion was first officially recognized in the 1960s in South Wales, UK. Initially, it was believed that corrosion was due to inadequate protection of the ties and prolonged exposure to harsh environmental conditions [19]. However, it became evident that many early wall ties and the mortars used in construction were inherently prone to corrosion. Early mild steel ties were either left unprotected or coated with bitumen. By the early 1930s, zinc coatings or galvanizing became the standard protection [20]. The most common types of mild steel ties under BS 1234 included strip ties (vertical twist or fishtail) and wire ties (butterfly or double triangle). Despite their widespread use, these traditional ties were vulnerable to corrosion, risking the structural integrity of masonry buildings, as shown in Figure 2. Modern wall ties, in contrast, are meticulously engineered for strength and durability, using advanced materials and protective coatings to significantly reduce corrosion risk. This evolution reflects a deeper understanding of corrosion factors and a commitment to the long-term stability and safety of masonry structures.

As the Australian standard evolved, particularly from the 1980s onward, the impact of environmental conditions on wall ties' performance and longevity became increasingly recognized. Corrosion, especially in coastal areas or high-humidity regions, became a significant concern due to moisture infiltration. This prompted AS 3700 to include more detailed provisions on material composition and protective coatings for wall ties.

FIGURE 2: CORROSION OF METAL WALL TIES, TAKEN FROM [21]

MATERIAL SELECTION

The AS 3700 Standard has placed significant emphasis on material selection, particularly for wall ties, as a key strategy in preventing corrosion. The standard introduced distinctions between various grades of galvanized steel and stainless steel, offering specific guidelines on their use in different environmental conditions. For instance, in highly corrosive environments, stainless steel wall ties are required due to their superior resistance to corrosion compared to galvanized steel. In coastal areas, where the risk of corrosion is particularly high, stainless steel is now exclusively used for manufacturing wall ties, restraint fixings and masonry support systems in masonry construction. As the industry moves towards higher specifications and longer service life, stainless steel presents a cost-effective, long-term solution for various architectural applications. However, simply specifying "stainless steel" is not enough to prevent corrosion; understanding the correct specification and application of stainless-steel products is essential to ensure their durability and performance over time.

Protective Coatings

The evolution of the AS 3700 Standard highlighted the importance of protective coatings as a critical measure for enhancing the durability of metal components in masonry construction, particularly in corrosive environments. The standard provided detailed specifications for the type and thickness of coatings to be applied to metal components, such as wall ties, to create a protective barrier against corrosive agents like moisture, salt, and pollutants. By doing so, these coatings helped to extend the lifespan of the metal, preventing direct exposure to elements that could accelerate degradation. The emphasis on protective coatings reflects a broader understanding within the construction industry that material longevity depends not solely on the base material but also on the quality and application of the protective layers that shield it from environmental stressors.

In addition to specifying coatings, research into the real-world performance of zinc-coated wall ties became increasingly important, especially following incidents where the failure of these ties led to structural collapses. A notable case was the collapse of a high-rise wall at Plymouth Polytechnic in 1983 [22], which underscored the vulnerabilities of zinc-coated ties under extreme weather conditions, such as high winds. This event, among others, prompted a closer examination of the durability of zinc coatings under service conditions, leading to further refinements in the AS 3700 Standard. These refinements aimed to ensure that wall ties and other metal components could withstand the demands of their operational environments, thereby safeguarding the structural integrity of masonry constructions over the long term.

Maintenance Practices

Maintenance of metal wall ties is crucial due to the inherent susceptibility of mild-steel ties to corrosion over time. The Building Research Establishment (BRE) in the UK has highlighted this issue since the 1970s, particularly in its publications Digests 329 and 401, emphasizing that the lifespan of masonry often exceeds that of the metal ties used within it [22-24]. As a result, there is a need for regular inspection and potential replacement of these ties to ensure the structural integrity of the building. Corrosion damage is typically identified by regular horizontal cracks in bed joints, occasionally accompanied by step cracking. It is important to distinguish these signs from other forms of masonry deterioration, such as vertical cracks caused by moisture expansion, drying shrinkage, or foundation movement, and sulfate attack, which affects the Portland cement mortar but not black ash or lime mortars [23]. Recognizing these symptoms early is vital for timely maintenance and prevention of further structural damage.

To address the issue of corroding wall ties, remedial wall ties, also known as retrofit or replacement ties, have been developed and standardized in AS/NZS 2699.1, introduced in 2000. These ties, designed for both seismic (Type B) and non-seismic (Type A) conditions, are used to replace defective ties or those that have been omitted during initial construction. Remedial ties, whether cavity or masonry veneer types, are essential for restoring and maintaining the structural stability of buildings, particularly as the original ties corrode or lose effectiveness over time. These modern solutions ensure that the integrity of the building is upheld, providing a durable and secure means of addressing the long-term challenges posed by wall tie corrosion. Regular maintenance and timely replacement of wall ties are, therefore, critical practices for preserving the safety and stability of masonry structures.

Advancements in Corrosion-Resistant Materials and Technologies

The AS 3700 Standard has been significantly influenced by continuous advancements in corrosion-resistant materials and technologies. The introduction of new alloys, improved galvanization techniques, and more effective protective coatings have greatly enhanced the standard's provisions for corrosion prevention. Additionally, the development of more accurate and reliable methods for assessing the corrosion potential of materials has led to better-informed decisions in material selection and design detailing. These advancements have been incorporated into the standard, ensuring it remains relevant and effective in addressing the challenges of corrosion in masonry construction.

In recent updates, AS 3700 mandates compliance with AS 2699.1, incorporating advancements in corrosion-resistant technologies and materials. The standard now includes provisions that reflect the development of new coatings and the increased use of stainless steel ties in high-risk environments, ensuring that masonry structures are better equipped to withstand corrosion over their intended lifespan. It also emphasizes the durability considerations including potential degradation of masonry components including wall ties. Regular maintenance and inspection of wall ties, particularly in older buildings, are critical to identify potential degradation issues before they cause significant structural damage. This holistic approach to wall tie performance, from material selection to maintenance, underscores the evolving understanding of these components' critical role in masonry construction. As research in material science and corrosion technology progressed, the AS 3700 Standard underwent several revisions to incorporate new knowledge and best practices, including the critical update of requiring protective coatings on metal components to enhance their lifespan. The standard also stressed the importance of proper design detailing, such as ensuring adequate drainage and ventilation in cavity walls, to reduce moisture retention and mitigate the risk of corrosion.

CONCLUSIONS

The evolution of corrosion considerations within the AS 3700 Masonry Structures Standard reflects the construction industry's growing understanding of the challenges posed by corrosion in masonry structures. From its early iterations, which primarily overlooked corrosion issues, to its current form, which incorporates advanced materials and technologies, the standard has significantly advanced in addressing the durability and longevity of masonry structures. This study provides valuable insights for practitioners, researchers, and policymakers involved in masonry construction and corrosion management by documenting this progression. The history of AS 3700's provisions regarding wall ties illustrates the growing sophistication of masonry construction standards in Australia. From essential guidance on tie placement to comprehensive requirements addressing material selection, corrosion protection, and installation practices, the evolution of AS 3700 demonstrates the standard's commitment to enhancing the durability and safety of masonry structures. The incorporation of modern materials and technologies ensures that the standard remains relevant in the face of changing environmental conditions and construction practices.

ACKNOWLEDGMENTS

The authors gratefully acknowledge the financial support provided by the Australian Research Council and the Western Australian Government Department of Energy, Mines, Building Regulation and Safety (DEMIRS) through Linkage Project LP220100028. The authors would like to acknowledge Emeritus Professor Adrian Page for his valuable assistance in providing access to historical standards.

References

1. AS 3700:2018, Masonry Structure, Fifth Edition, Standards Australia, 2018, Sydney, Australia.
2. AS 2699.1:2020, Built-in components for masonry construction, Part 1: Wall ties, Standards Australia, 2020, Sydney, Australia.
3. Piroozfar P. and Farr ER, Evolution of nontraditional methods of construction: 21st century pragmatic viewpoint. Journal of Architectural Engineering 19(2) (2013) 119-133.
4. SAA Int. 324, Interim Specification for Metal Wall Ties for Brickwork, Standards Association of Australia, 1953, Sydney, Australia.
5. AS 2699-1984, Wall Ties for Masonry Construction, Standards Australia, 1984, Sydney, Australia.
6. AS CA32-1963, Australian Standards Rules for the Use of Concrete Blocks in Masonry Construction known as the SAA Code for Concrete Block Masonry, First Edition, Standards Association of Australia, 1963, Sydney, Australia.
7. Ewing BD, Performance of Post-Tensioned Clay Brick Masonry with Openings, 2008.
8. AS 3700-1998, Masonry Structure, Second Edition, Standards Australia, 1998, Sydney, Australia.
9. AS CA32-1967, The Use of Concrete Blocks in Masonry Construction known as the SAA Code for Concrete Block Masonry, Revised Edition, Standards Association of Australia, 1967, Sydney, Australia.

10. AS CA47-1969, SAA Brickwork Code (Clay Brick and Concrete Bricks), Standards Association of Australia, 1969, Sydney, Australia.
11. BS 1243, Metal Ties for Cavity Wall Construction, Fourth Edition, British Standards Institution, 1978, London, England.
12. AS 1640-1974, Brickwork in Building known as the SAA Brickwork Code (Metric Units), Standards Association of Australia, 1974, Sydney, Australia.
13. AS 1475 Part 1-1977, Rules for Concrete Blockwork in Buildings known as the SAA Blockwork Code Part 1 – Unreinforced Blockwork, Standards Association of Australia, 1977, Sydney, Australia.
14. AS 3700-1988, Masonry in Building known as the SAA Masonry code, Standards Australia, 1998, Sydney, Australia.
15. AS 3700:2001, Masonry Structure, Third Edition, Standards Australia, 2001, Sydney, Australia.
16. ISO 9223, Corrosion of metals and alloys—Corrosivity of atmospheres—Classification, determination and estimation, Second Edition, International Organization for Standardization, 2012, Geneva, Switzerland.
17. AS 3700:2011, Masonry Structure, Fourth Edition, Standards Australia, 2011, Sydney, Australia.
18. AS/NZS 2699.1-2000, Built-in components for masonry construction Part 1: Wall ties, Standards Australia and Standards New Zealand, 2000, Sydney-Australia and Wellington-New Zealand.
19. Addis B, Concrete and steel in twentieth. Structure and Style: Conserving Twentieth Century Buildings, E & FN Spon, 1997, 103-109, London, UK.
20. Way AGJ, Popo-Ola SO, Biddle AR and Lawson RM, Durability of Light Steel Framing in Residential Building, Sci Publication P262 (2009).
21. Corroded mild steel cavity wall ties see: https://actionwallties.co.uk/defect/cavity-wall-ties-corroding-failing/page/2/ Last accessed 11 August 2024.
22. de Vekey RC, Ties for cavity walls and masonry cladding, Structural Survey 8(4)(1990), 384-392.
23. de Vekey RC, Installing wall ties in existing construction, BRE Digest 329, 2000, Watford, UK.
24. Digest 401, Replacing wall ties, BRE Digest Concise reviews of building technology, 1995, Watford, UK.

AUTHOR DETAILS

M. A. Hossain is the corresponding author of this paper. He is a Research Associate at Centre for Infrastructure Performance and Reliability, The University of Newcastle, Australia. He assists with all aspects of this project's experimental and field research. Dr Hossain has over 12 years of applied structural engineering research and development experience. He specializes in experimental and numerical modelling analysis of structural response.

	C. Y. Lam is a PhD student at the University of Newcastle since 2022 and a Civil engineer who has been awarded a prestigious Australian Government Research training program scholarship for his academic excellence. His PhD research interest lies in non-destructive testing of wall ties in masonry constructions.
	Professor M. Masia is the Director of the Centre for Infrastructure Performance and Reliability at The University of Newcastle, Australia. He isspecializes in the structural performance, assessment, and design of masonry structures. He has a longstanding reputation with Standards Australia, the Australian Research Council, and national and international professional masonry societies.
	Associate Professor I. A. Chaves is the Deputy Director of the Centre for Infrastructure Performance and Reliability at the University of Newcastle, Australia. He is a chartered materials professional academic specializing in Civil/Structural/Material Science. He holds a MEngSci degree in structural engineering and a PhD for his studies in corrosion analysis and prediction. His research interests also include structural reliability and risk assessment of energy infrastructure.

CORROSION & PREVENTION
CAIRNS 2024

THE AUSTRALASIAN CORROSION ASSOCIATION INC.

CREVICE CORROSION OF MILD STEEL AFTER TWO YEARS OF EXPOSURE TO NATURAL, NUTRIENT-DOSED, AND UV-TREATED SEAWATERS

R.B. Petersen[1,*], R.E. Melchers[1]

[1]The University of Newcastle, NSW, 2308, Australia

**Corresponding author email: robert.petersen@newcastle.edu.au*

ABSTRACT

Localised corrosion occurring in crevices at joints, fittings, and other discontinuities in low-alloy steel infrastructure has important practical implications. It is considered herein for bolted mild steel assemblages with different crevice gaps, crevice lengths and outstand lengths, exposed for up to 2 years in natural Pacific Ocean seawater, and seawater dosed with calcium nitrate to inhibit early abiotic corrosion, and seawater treated with UV-radiation to inhibit bacterial corrosion. Contrary to expectations, there was little evidence of severe localised corrosion within the crevices on any test samples. Instead, corrosion was more severe on the coupon surfaces outside of the crevice. These observations are related to findings from the literature and the practical implications of the work are discussed. It is also shown that pitting both inside and outside the crevice follows a bimodal trend with time.

Keywords: crevice corrosion, mild steel, seawater, pitting, long-term

INTRODUCTION

Crevice corrosion of low alloy steels is a significant concern in the offshore oil and gas industry. This form of localised corrosion occurs in regions of restricted flow where a metallic surface contacts a small volume of trapped, stagnant solution, while the remaining material surface is exposed to the bulk environment [1]. Crevices may form due to the design of a component, such as at joints, fittings, or other discontinuities in steel structures. They may also result from surface deposits, including waterborne sediments, microbiological material, or rusts [1, 2]. Crevice corrosion leads to highly localised corrosion inside and around the mouth of the crevice [1, 3], and has been observed in practice (Figure 1).

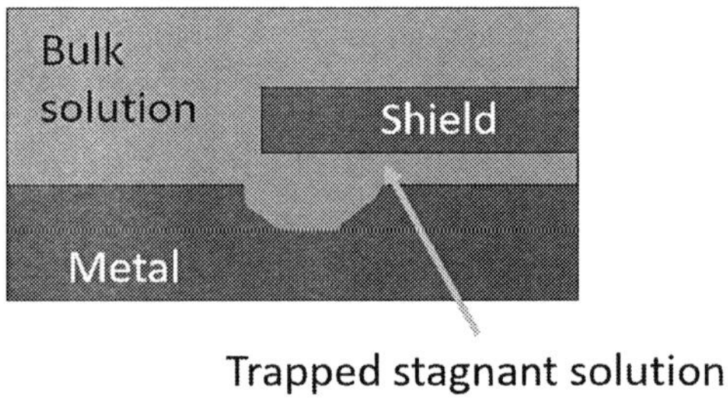

(A)

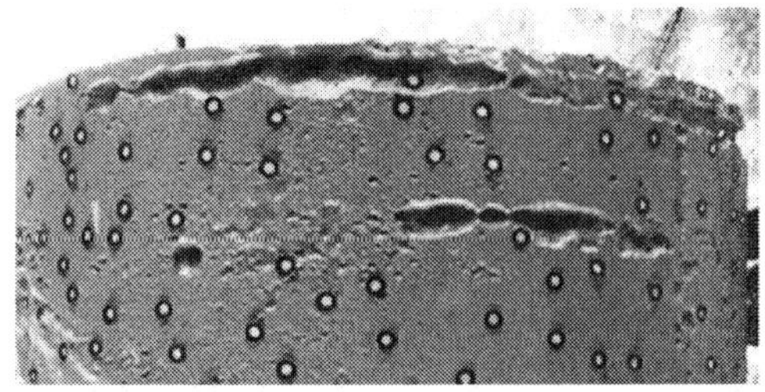

(B)

FIGURE 1 (A) SCHEMATIC OF CREVICE CORROSION; (B) CREVICES ON 300 MM DIAM. PIPE NEAR JUNCTION

Corrosion within and adjacent to crevices can eventually lead to perforation or structural failure of metallic components. For asset owners and managers, predicting the time to failure (or remaining service life) is crucial for asset management, including condition assessment, maintenance, and replacement schedules, to avoid potentially costly failures. Increasingly, it is recognised that quantitative estimation of expected remaining service life requires models for the spatial extent and rate of corrosion growth. Models are especially important for crevice corrosion, as these cases are often difficult to locate and inspect. Crevice corrosion may also take decades to become partially evident. This study aims to investigate the crevice corrosion of mild steel assemblages using experiments, with the goal of developing predictive models for corrosion losses and pit depths as a function of geometrical and environmental parameters.

Most research on crevice corrosion has focused on passive metals in aerated chloride environments, such as those found in the food, chemical, and oil industries [1]. In these applications, the initiation of crevice corrosion can be a practical and aesthetic problem. Accordingly, most research has focused on the events leading up to the initiation of corrosion, the factors influencing the time to initiation, and the susceptibility of different alloys to crevice corrosion [1, 2]. Typically, in these applications, initiation is considered the failure state. In contrast, little attention has been given to the rate of growth or the extent of crevice corrosion [1]. However, for mild and low alloy steels, initiation is less of an issue, and the development of crevice corrosion is likely more practically important.

The development of crevice corrosion of mild and low alloy steels over time is unknown, but it likely shares similarities with the development of general corrosion loss and pitting corrosion on metal surfaces without pre-defined crevices. In these cases, it has been shown that the development of corrosion losses and pit depths over time follows a nonlinear bimodal trend [4] (Figure 2). This behaviour results from a change in rate-controlling processes as corrosion products build up on the corroding surface [5]. Figure 2 also shows the transition point (ta) indicating the changeover point between the rate controlling modes, and the influence of the addition of dissolved inorganic nitrogen (DIN) on corrosion loss in mode 2.

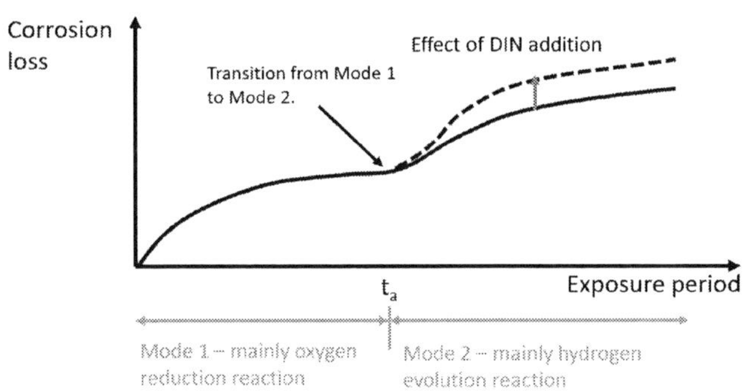

FIGURE 2 SCHEMATIC OF BIMODAL TREND FOR MILD AND LOW-ALLOY STEELS SHOWING MODES, TRANSITION POINT TA, AND THE EFFECT OF ADDING DISSOLVED INORGANIC NITROGEN (DIN) ON MODE 2 CORROSION

Several parameters influence crevice corrosion, including material type, the conductivity of the bulk solution, the chemistry of the bulk solution, the geometry of the crevice, the area of the free metal surface outside the crevice (the cathodic area), and the possibility of microbiological corrosion [1, 2, 6-8]. For crevice corrosion of mild steel in seawater, parameters such as material type, conductivity, and chemistry of the bulk solution may be ignored and are not considered herein.

The next section describes a long-term, field-based experimental program studying the crevice corrosion of mild steel assemblages. Preliminary results are provided, including data obtained, initial analysis, and interim findings. The discussion considers the relation of the current findings with those described in the literature. Practical implications are noted, and areas of potential further work are described. A previous paper outlined results after one year of testing [9]. This paper presents results after two years of testing.

EXPERIMENTAL PROGRAM

The experimental study is being undertaken at The University of Newcastle's marine exposure testing facility at The Port Stephens Fisheries Institute at Taylors Beach, NSW, Australia. The (on-going) testing program consists of mild steel samples with artificially created crevices being continuously immersed in seawater for periods of time up to four years. The present paper reports observations for coupons exposed for up to two years.

The samples used for the tests consisted of three mild steel pieces, bolted together with nylon bolts, to create four (4) gaps where crevice corrosion may occur (Figure 3). The crevice spacer, the width of the sample (i.e. the length of crevice) W, and the length of the outstand (L) varied between the samples. This allowed the effect of these variables on the corrosion behaviour of the crevices to be observed.

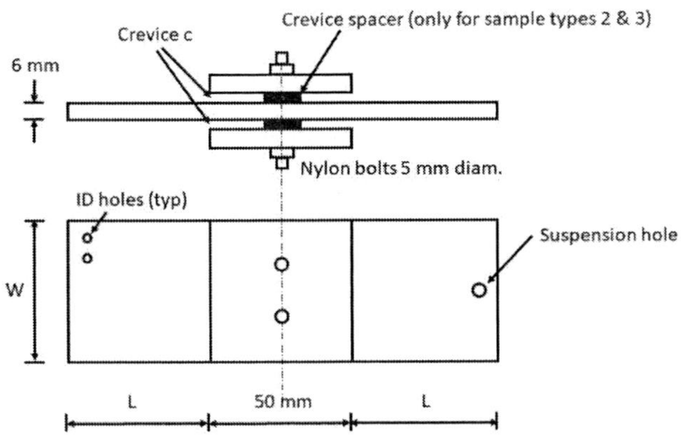

FIGURE 3 GENERAL GEOMETRY OF TEST SAMPLE FOR CREVICE CORROSION IMMERSION TESTS

Based on practical considerations and experience as reported in the literature, six (6) different sample geometries were selected for testing (Table 1). Sample type 1, adopted as the standard case, was tested in three different seawater tanks (natural, nutrient added, and UV treated) as described further below. Sample types 2-6 were tested only in natural seawater. Sample types 1,3-6 were bolted together directly. This was to attempt to replicate as closely as possible the conditions, for example as would occur when flange plates are bolted together. Sample types 2 and 3 were used to investigate the influence of the size of the crevice gap. This was done by using layers of adhesive tape, applied in the position shown in Figure 1. The size of the crevice gap c was measured using stainless steel feeler gauges in multiple locations. The average sizes are reported in Table 1.

The outstand length L for most samples was selected at 50 mm, considered sufficient to avoid anode/cathode area limitation effects. To check on this possibility sample type 4 had a larger outstand of 100 mm. Sample type 5 was designed to represent crevices at flanges (with a zero outstand length), and Sample type 6 was designed to investigate the effect of length of crevice W.

TABLE 1. TEST PROGRAM

Sample Characteristics		Sample Code					
		1	2	3	4	5	6
Crevice length W	(mm)	50	50	50	50	50	100
Outstand L	(mm)	50	50	50	100	0	50
Crevice gap c	(μm)	30	100	200	30	30	30
No. of Samples:	SW (natural)	10	8	8	8	8	8
	Nutrient added	10	-	-	-	-	-
	UV treated	8	-	-	-	-	-

All samples were constructed from 6mm thick mild steel plates, guillotined from one metal sheet to minimize material variability. For identification purposes, the central part of each assembly was hole-drilled to provide a unique code for comparison against a numbered guide. Prior to assembly each part of the sample was cleaned to remove mill scale. This was done by submerging them for 1 minute in an inhibited hydrochloric acid solution, in accordance with ASTM G1-03 [10]. The mild steel parts were then hand washed in tap water with a scouring nylon sponge and stored in an oven at 60°C for an hour. The mild steel parts were then each weighed to determine their pre-exposure mass. The thickness of the parts was also measured using digital calipers, with several readings at random locations over the surface. For each test sample the parts were then assembled using threaded nylon bolts and washers to hold the parts together. The completed test samples were stored in plastic containers with desiccant beads. Within 24 hours of assembly all samples were taken to the Taylors Beach test site and immersed in the seawater test tanks.

A flow diagram of the Taylors Beach corrosion facility is shown in Figure 4. It is located within the NSW Fisheries facility that requires high quality seawater for fish-breeding purposes and that closely monitors the seawater. The seawater taken from their pumped inflow system is split three ways. One branch leads to sediment settling tanks used to ensure removal of sediment from the seawater. From these tanks the water is then pumped through a series of filters to remove very fine particles, and then through UV radiation filters used to remove as much as possible all living materials such as bacteria. Previous studies have shown that this arrangement removes some 99.9 % of the living biotic material in the seawater [11]. The filtered and UV treated seawater is then released slowly into a UV protected sample exposure tank. The second branch goes straight to a similar, but not UV protected, exposure tank, while the third branch goes to an exposure tank fitted with nutrient dosing equipment.

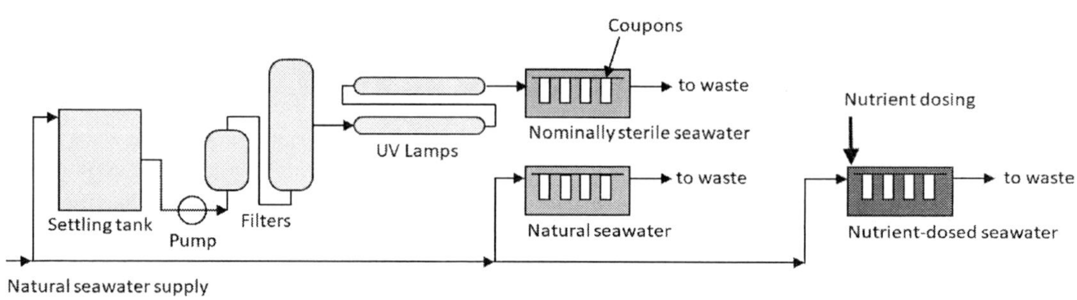

FIGURE 4. SCHEMATIC OF THE TAYLORS BEACH CORROSION TESTING FACILITY

The system is programmed in such a way that the seawater in all tanks is replaced every 24 hours in an attempt to maintain seawater oxygen levels at near saturation. Water velocity when flushing the tanks was adjusted to a very low flow rate to avoid disturbing the samples. For the nutrient-dosed seawater tank, a pre-determined amount of calcium nitrate tetrahydrate ($Ca(NO_3)_2$) was added to the seawater every time it was replenished so as to achieve a constant 0.15 mg nitrogen/L concentration of dissolved nitrogen. This value was chosen to represent a realistic pollution level, and based on previous observations that it increases corrosion of mild steel immersed in seawater by a factor of 2-3 compared to fresh (non-nutrient dosed) seawater [12] for exposure periods of 2-5 years. Recent work has shown that dosing with calcium nitrate reduces corrosion losses for exposure periods less than 2 years [13]. In that work, the added nitrate inhibited corrosion loss in the short term. In the current experiments the calcium nitrate was stored in a large 100 L plastic container and electronically discharged by an automated wall-mounted pump system. Over the duration of testing, sediment settled within the natural seawater tank and the nutrient dosed tank. This sediment build-up in both tanks was periodically removed manually.

Each test sample was suspended from PVC half-pipes using cable ties in such a way that each sample was fully immersed in seawater (Figure 5). This system of exposure ensured minimal sediment deposition on the crevices and adjacent sample areas.

(A) (B)

FIGURE 5. PHOTOGRAPHS SHOWING INITIAL DEPLOYMENT: (A) SAMPLE TYPES 1 AND 4 SUSPENDED FROM PVC HALF-PIPE BEFORE IMMERSION; (B) SAMPLES IMMERSED IN NATURAL SEAWATER TANK.

The number of samples placed in each tank was sufficient for sample recoveries at (nominally) 3, 6, 12, 18, 24, 36, and 48 months with some additional samples as spares. In each recovery one sample of each type was removed from the tanks. Herein only observations and results from the first five recoveries (up the 24 months of exposure) are presented.

Some complications in the exposure experiments were experienced not long after initial exposure. A short time into testing, the filtration and UV-sterilisation system failed. This failure occurred during the NSW state-wide Covid lockdown period towards the end of 2001. It was difficult to repair the issue quickly, so it was decided to continue the experiments in the natural and nutrient dosed tanks but to leave the experiments in the UV-treated tank in an offline mode. This meant that water was not replenished in the UV-treated tanks during this period. The failure occurred 4 months after the experiments started. It was repaired and the UV-treated tanks were operational again some 5 months after the original failure occurred (9 months into testing).

After the desired exposure time, samples were removed from the tanks, placed in a plastic bag, and delivered immediately to the University of Newcastle laboratories. The samples were then dismantled, and each mild steel piece of the sample was cleaned by firstly flushing it lightly with water, and then drying it with a paper towel. The sample pieces were then stored in an oven at 60 degrees Celsius until they were further cleaned and examined as outlined below.

Prior to examination, the mild steel pieces were subjected to a further detailed cleaning using electrolysis and an ultrasonic technique, both in accordance with ASTM G1-03 [10]. This cleaning procedure allowed detailed cleaning of the micro-sized corrosion pits. It was performed, in most cases, on the day of recovery. The mild steel pieces were cleaned using electrolysis in rounds of 2 hours in duration, and then they were scrubbed with a lightly abrasive nylon cloth removing any loosened rust products. The pieces were then dried thoroughly using paper towels and then placed back in the oven (temperature of 60°C) for at least an hour to dry further. After drying in the oven, the coupons were removed, photographed and a number of measurements and observations were made on each coupon (see below). When not being assessed the coupons were stored in a plastic container containing desiccant beads.

For corrosion loss and pit depth analysis the surface area of each coupon was divided into a rectangular grid pattern. As an example, the grid pattern used for Samples 1-3 are shown in Figure 6. A similar grid pattern was used for Sample 4, except an additional 2 rectangular divisions measuring 25 x 20 mm were included on each outstand. Sample 5 included no outstand. Two additional rows were included for Sample type 6 to account for the additional width, W, of the coupon. Three (3) measurements of remaining thickness and pit depth were taken in each grid area, with locations selected randomly for thickness and for the deepest pit for pit depth. This allowed (after some analysis) the creation of corrosion loss and pit depth profiles along the length of the coupon. The remaining thickness was measured using digital calipers (accuracy of ± 0.02 mm, and average deviation of ± 0.01 mm). These values were subtracted from the initial thickness to obtain thickness loss, which was then divided by two to obtain an estimate for general corrosion loss (on each side of the coupon).

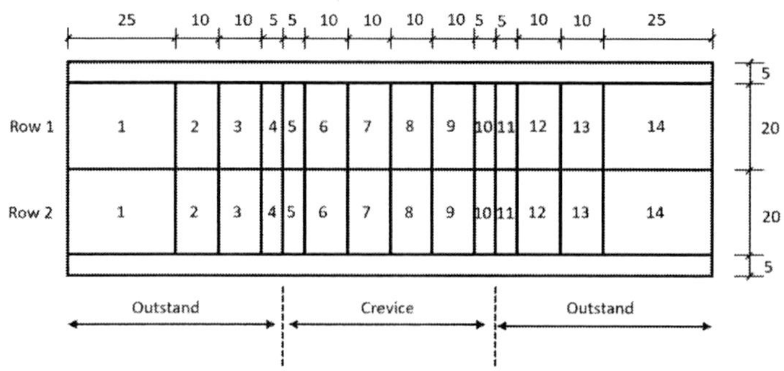

FIGURE 6. SCHEMATIC OF SAMPLE 1 SHOWING DIVISIONS USED FOR THICKNESS AND PIT DEPTH MEASUREMENTS

Pit depths were measured using the standard differential focusing technique with an optical microscope (Zeiss Axio Imager A1M, Oberkochen, Germany). This procedure involves first focusing the microscope on the metal surface at the outside perimeter of the pit and then raising the microscope stage to focus on the lowest point of the pit. The distance that the stage moved through (from pit mouth to lowest point) is equal to the pit depth relative to the corroded surface. Based on the previous experience of the authors, this technique has a measured absolute error of 0.003 mm, and average deviation of ± 0.003 mm. The measured pit depths were added to the estimates of general corrosion for one side of the coupon to obtain estimates for absolute pit depth.

EXPERIMENTAL OBSERVATIONS

Typical corrosion patterns on Sample 1 exposed to natural seawater for periods of 4-, 8-, 12-, 19- and 25 months are shown in Figure 7. Minimal corrosion damage was observed across the entire surface after 4 months of exposure. After 8 months of exposure, regions of deep pitting were observed on the outstand surfaces of the coupon (i.e., on the surfaces outside of the crevice zone). Inside the crevice zone the corrosion damage was minimal, relative to the damage observed on the outstand surfaces. Similar observations were made for the coupons exposed for 12-, 19- and 25 months; however, corrosion on the outstand surfaces was more severe. Also, some pitting was observed inside the crevice, but this pitting was shallow and covered less area than the pitting observed on the outstand surfaces.

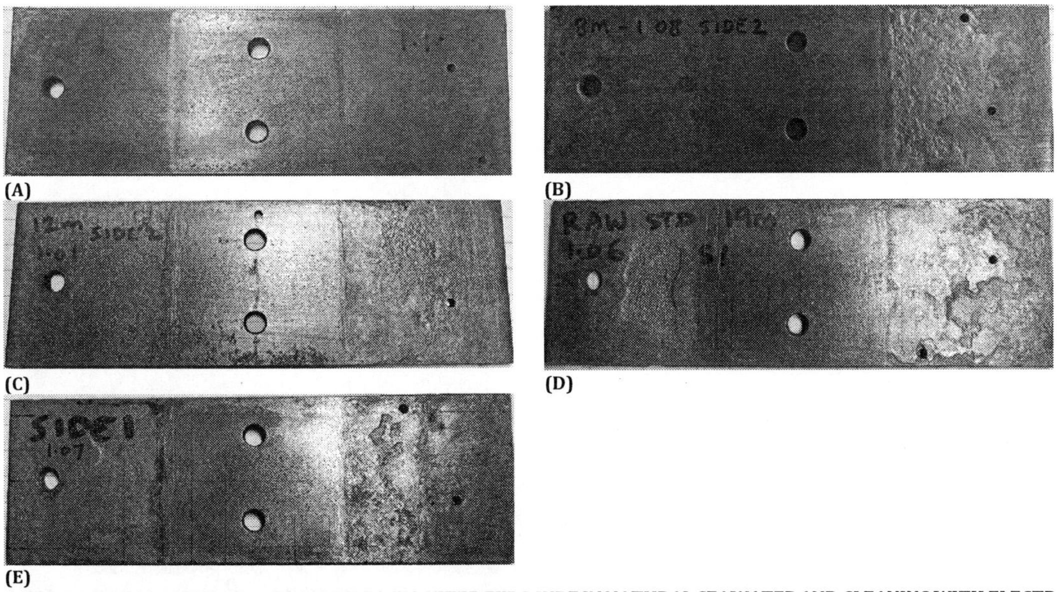

FIGURE 7. PHOTOGRAPHS OF COUPON SURFACES AFTER EXPOSURE IN NATURAL SEAWATER AND CLEANING WITH ELECTROLYSIS AND ULTRASONICS, FOR SAMPLE TYPE 1 IN NATURAL SEAWATER, AND EXPOSED FOR: (A) 4 MONTHS; (B) 8 MONTHS; (C) 12 MONTHS; (D) 19 MONTHS; (E) 25 MONTHS.

Corrosion patterns on the other types of coupons are shown in Figure 8. To save space only photographs of coupons exposed for 25 months are shown. The corrosion damage after 25 months of exposure was more severe on the outstand surfaces, with very little corrosion damage observed on the surfaces within the crevice zones. The depth and extent of corrosion in the outstand zones appeared similar across all sample types exposed to the natural seawater. Corrosion losses on the outstand surfaces in nitrate dosed seawater appeared similar to corrosion losses on the outstand surfaces in natural seawater. The corrosion losses on the exposed surfaces of coupons immersed in filtered, UV-sterilized seawater were lower than those on the other coupons.

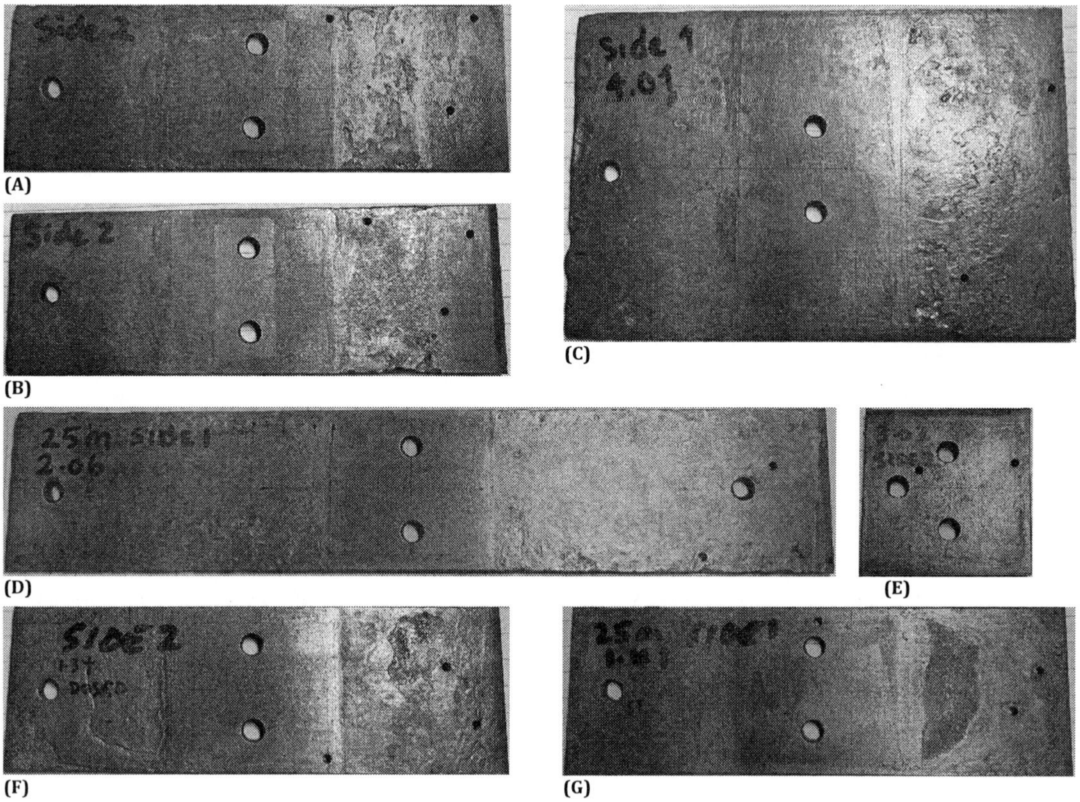

FIGURE 8. PHOTOGRAPHS OF COUPON SURFACES AFTER 25 MONTHS OF EXPOSURE AND CLEANING WITH ELECTROLYSIS AND ULTRASONICS, FOR THE FOLLOWING TEST COUPONS AND EXPOSURE CONDITIONS: (A) SAMPLE TYPE 2 (C=100); (B) SAMPLE TYPE 3 (C=200); (C) SAMPLE TYPE 6; (D) SAMPLE TYPE 4; (E) SAMPLE TYPE 5; (F) SAMPLE TYPE 1 IN DIN-DOSED SW; (G) SAMPLE TYPE 1 IN FILTERED AND UV-TREATED SW.

Interpretation of experimental observations

Corrosion loss and pit depth profiles

Corrosion profiles measured on the coupon exposed to natural seawater (Sample type 1) are shown in Figure 9 for increasing exposure periods (4, 8, 12, 19, 25 months) and for general corrosion loss, average pit depth (average of the 3 maximum values) and maximum pit depth. The circles indicate the measured data points, and the smooth lines are curves fitted using the Stineman function for best-fit smooth trends in the Kaleidograph graphing software program[14]. The crevice and the outstand zones are as marked. In all cases more corrosion had occurred outside the crevice, that is in the outstand zone, compared with inside the crevice. Evidently, this is more pronounced for the longer exposures, i.e. after 12, 19 and 25 months. Corrosion profiles for the remaining coupons showed similar overall trends.

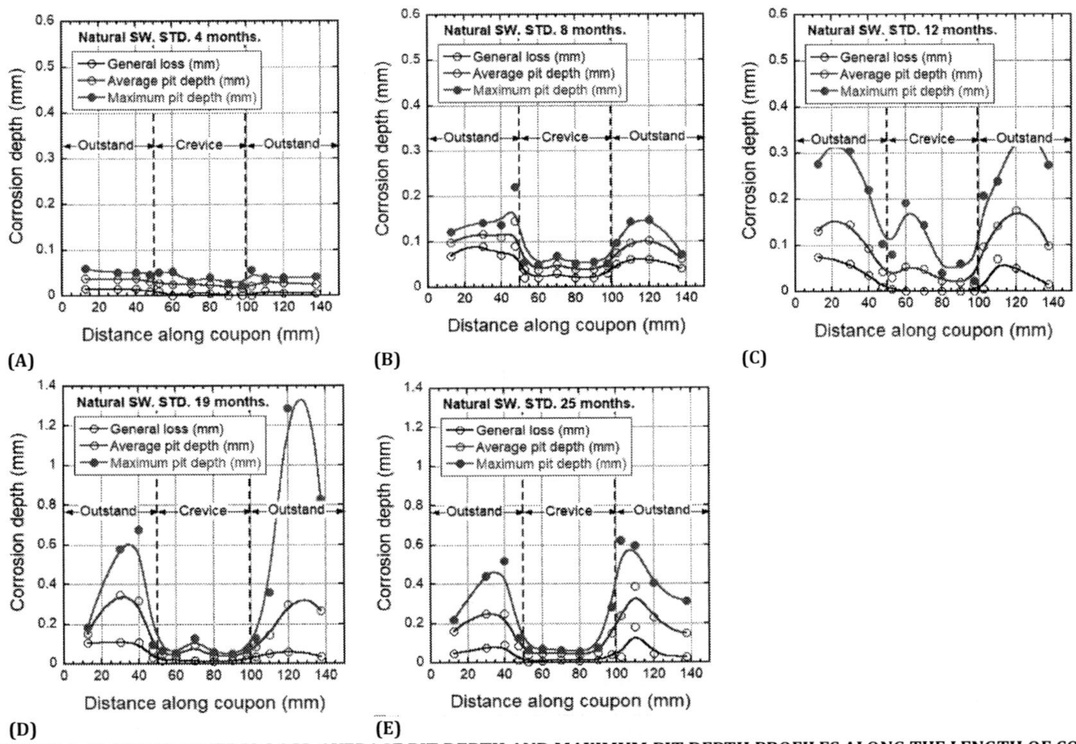

FIGURE 9. PLOTS OF GENERAL LOSS, AVERAGE PIT DEPTH AND MAXIMUM PIT DEPTH PROFILES ALONG THE LENGTH OF COUPONS IN NATURAL SEAWATER, FOR SAMPLE TYPE 1 IN NATURAL SW AND EXPOSED FOR: (A) 4 MONTHS; (B) 8 MONTHS; (C) 12 MONTHS; (D) 19 MONTHS; (E) 25 MONTHS. THE DASHED LINES INDICATE THE LOCATION OF THE INTERFACE BETWEEN CREVICE ZONE AND OUTSTAND. CIRCLES DENOTE MEASURED DATA POINTS. THEY APPLY FOR DATA OBTAINED FROM BOTH SIDES OF THE COUPON. THE SMOOTH LINES ARE BEST-FIT SMOOTH TREND LINE FITS TO THE DATA USING A SOFTWARE PROGRAM. NOTE THE DIFFERENT Y-AXIS SCALES BETWEEN A-C AND D,E.

Progression of pit depth with time

The main focus of the test program was to try to obtain information about the progression of corrosion with time, for locations inside the crevice, outside the crevice and around the crevice mouth, and also the dependence of various geometrical and environmental factors. In this paper the progression of absolute pit depths with time are considered. The progression of general losses with time, although not shown here, follow similar trends to pit depth with time. Evidence of the link between pit depth and general loss can be seen in the corrosion damage profiles in Figure 9.

Figure 10 shows trends for the absolute pit depth as a function of time for the following regions on each coupon: outside of the crevice, the crevice mouth (5 mm outside of the crevice gap), the crevice mouth (5 mm inside the crevice gap), and inside of the crevice gap. Each coupon contained 4 crevice gaps with their associated crevice mouth regions and outstand areas (Figure 3). Each data point in the plots is the average of the deepest pit measured for each of the 4 crevice gaps (and associated outstand areas). For example, each data point in the "outside crevice" data subset was the average of the deepest pit from each of the 4 outstand surfaces on each coupon. Similarly, each data point in the "inside crevice" data subset was the average of the deepest pit from each of the 4 crevice gaps. Also shown are trend lines through the data points at any exposure time. All trend lines were obtained using the smooth curve-fitting algorithm in Kaleidagraph [14]. In all cases and for all exposure periods pitting was worse on surfaces outside of the crevice. This is consistent with the corrosion profiles and photographs of observed corrosion damage (Figure 7, Figure 8) and with the measured corrosion profiles (Figure 9). The mean value trend lines for pit depth progression with time is nonlinear, and discounting cases where pit depths decrease with time, the trends are consistent with the bi-modal model for corrosion in marine environments. The influence of parameters on pit depths are considered in the following sections.

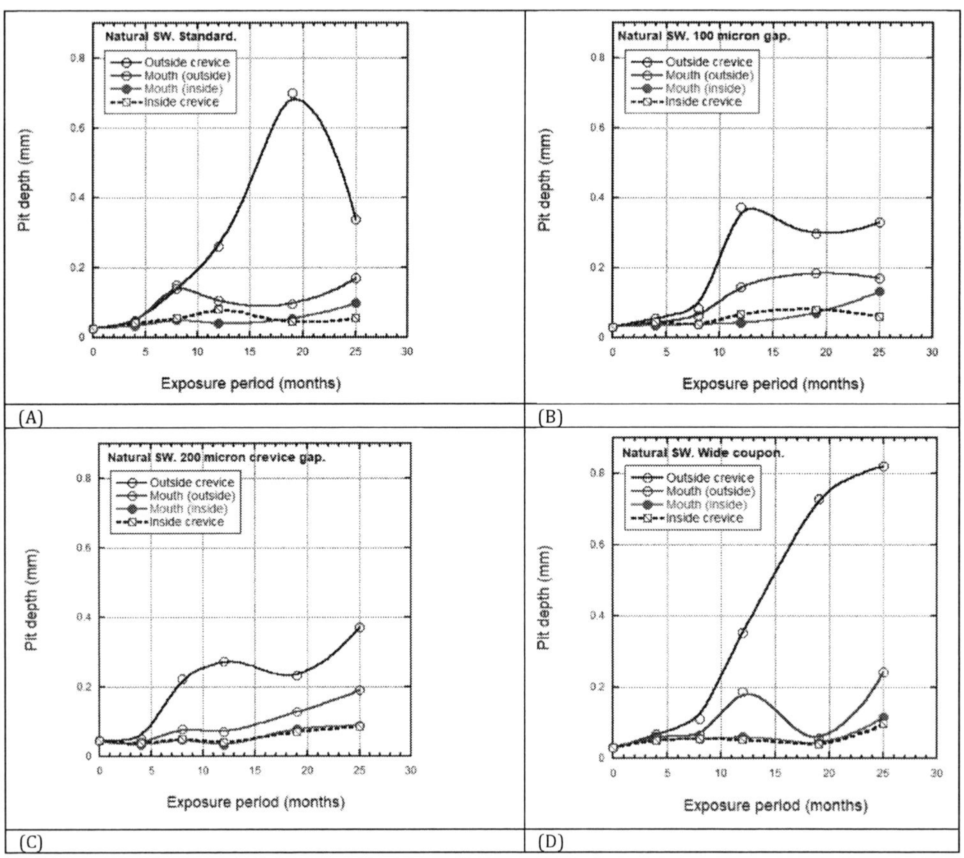

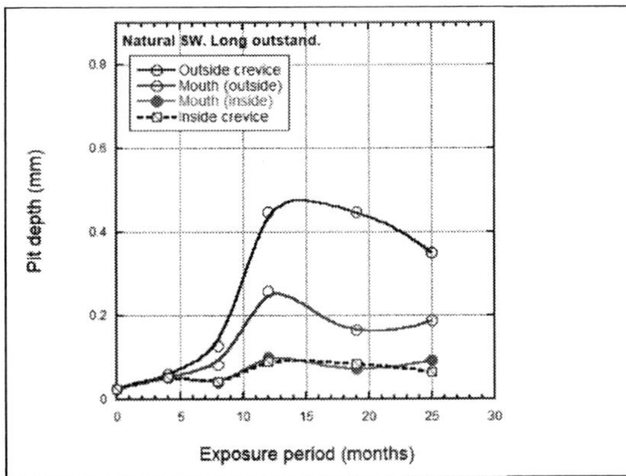

(E)

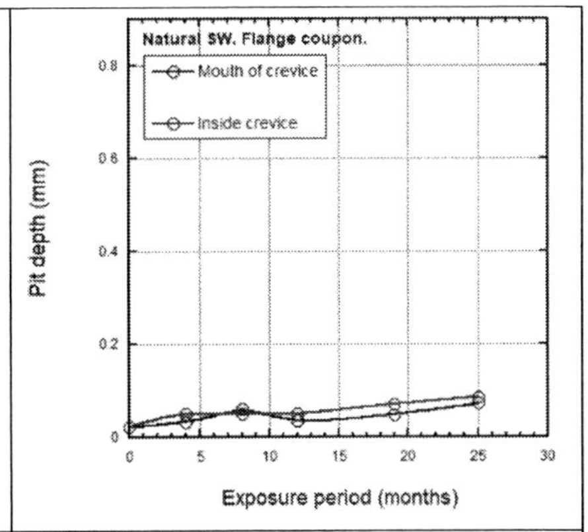

(F)

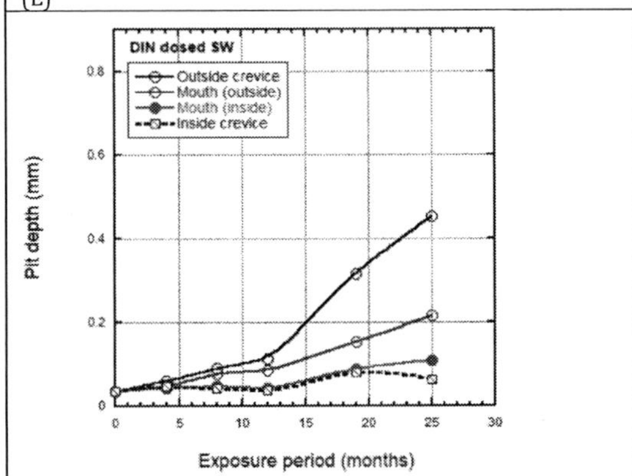

(G)

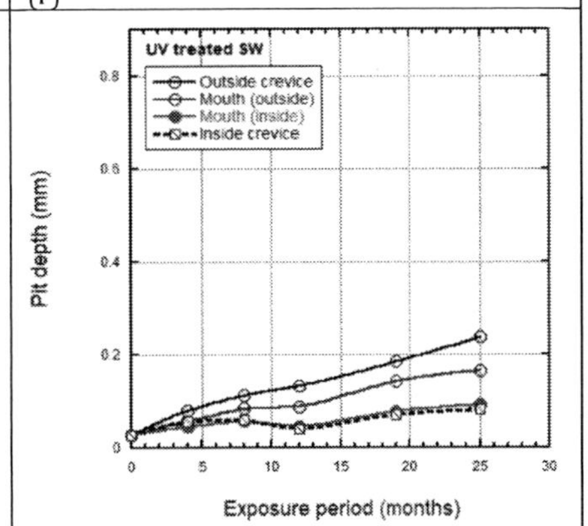

(H)

FIGURE 10. MEAN VALUE DATA FOR THE 4 DEEPEST PITS, TAKEN FROM EACH OF THE 4 CREVICE GAPS, AS FUNCTIONS OF TIME, SHOWING ALSO TREND LINES, FOR THE FOLLOWING TEST COUPONS AND EXPOSURE CONDITIONS: (A) SAMPLE TYPE 1; (B) SAMPLE TYPE 2; (C) SAMPLE TYPE 3; (D) SAMPLE TYPE 6; (E) SAMPLE TYPE 4; (F) SAMPLE TYPE 5; (G) SAMPLE TYPE 1 IN DIN-DOSED SEAWATER (SW); (H) SAMPLE TYPE 1 IN UV-TREATED SW.

INFLUENCE OF GEOMETRIC FACTORS

A comparison of Figure 10 a, b & c shows that the size of the crevice gap (within the tested range 30-200 micron) does not have a large influence on pit depths within the crevice, or at the mouth. Depth of pitting inside the crevice and at the mouth was approximately 0.1 mm. The size of the crevice gap also does not appear to have an effect on the pit depths observed on the outstand surfaces, after discounting the one large reading at 19 months for the coupon with a crevice gap of 30 microns (Figure 10 a). Pit depths on the outstand after 25 months of exposure measured approximately 0.35 mm in all cases.

The width of the coupon also does not appear to influence the depth of pitting within the crevice and at the crevice mouth, which measured approximately 0.1 mm for both the standard width case (W = 50 mm) (Figure 10a) and the longer width case (W=100 mm) (Figure 10d). The pit depths on the outstand surface of the wide coupon were large, measuring 0.8 mm after 24 months, in comparison to pit depth of 0.35 mm on the standard size coupon.

The observations made to date suggest that a longer outstand does not significantly change the pit depths inside the crevice. This can be seen by comparing plots of corrosion for Sample type 1 (Figure 10a) with Sample type 4 (with the long outstand) (Figure 10e). After 25 months the maximum pit depths in the crevice and the outstand surfaces were 0.1 mm and 0.4 mm, respectively.

Pit depths on the surface of coupons with no outstand (Sample type 5) (Figure 10f) were similar to pit depths measured on the standard coupon (including an outstand) after 25 months. The measured pit depths were again 0.1 mm.

Pit depths are observed to increase over time in the flange type coupons (Sample type 5) (Figure 10f). After 25 months of exposure the depth of pitting is 0.1 mm and is consistent with the pit depths inside the crevices of other coupons with outstands that were also exposed to natural seawater. This indicates that, for the coupons tested herein, the presence of an outstand does not influence pitting within the crevice zone.

INFLUENCE OF NUTRIENT DOSING

Dosing with calcium nitrate appears to have had little effect on pit depths inside the crevice region (as evident in comparing Figure 10a with Figure 10g). Dosing with calcium nitrate does, however, appear to have reduced early-stage pitting (up to 12 months) on the outstand surfaces. At 25 months of exposure, pit depths appear to be similar, both around 0.4 mm, for the dosed and natural seawater exposures.

INFLUENCE OF FILTRATION AND UV-STERILISATION

Longer-term pit depths inside the crevices of coupons immersed in the filtrated/ UV-treated seawater (Figure 10h) were similar to long-term pits depths in the crevices of coupons immersed in natural seawater (Figure 10a) and DIN-dosed seawater (Figure 10g). Pit depths in crevice zones at 25 months were approximately 0.1 mm for all seawater types. Pit depths on the outstands of the surfaces in the filtrated/UV treated seawater were lower than those observed on the outstand surfaces of coupons exposed to natural and DIN-dosed seawater. This particular finding is consistent with previous results for the effect of UV treatment on the corrosion of steel coupons without crevices [11].

As noted above, the UV-treated seawater tank was not in operation for a significant time period (between months 4-9). However, this is only likely to have influenced the short-term results, with the influence of this interruption becoming less noticeable for longer testing durations.

DISCUSSION

The findings from this study indicate that crevice corrosion, characterised by intensified corrosion inside and around the crevice's mouth, is not observed in the tested coupons. Although corrosion is present on the surfaces within the crevices, it is significantly more severe on the free surfaces outside the crevices. These observations align with the results presented after one year of exposure in [9].

A potential explanation for the absence of crevice corrosion in these specimens was provided in [13], drawing from Evans' observations [15]. Evans proposed that low alloy steels typically possess numerous 'sensitive spots' on their surface susceptible to anodic attack. These 'spots' are likely the Manganese Sulphide (MnS) inclusions, identified by researchers such as Wranglen [16], as necessary in initiating localised corrosion and pitting.

In the context of a crevice, MnS inclusions outside the crevice might develop more rapidly due to the proximity of necessary cathodic zones and the ready availability of oxygen. This suggests that corrosion initiation points outside the crevice could 'out-compete' those within crevices. This scenario could also apply to corrosion-resistant alloys (CRAs). However, CRAs heavily depend on oxygen availability to rapidly re-form the protective diffusion barrier once impaired, and such oxygen is likely less readily available within tight crevices. On the other hand, for mild steels the lower availability of oxygen within tight crevices tends to reduce the rate of any corrosion that might initiate.

The pit depth versus time plots (Figure 10) exhibit a clear bimodal trend over time, both inside and outside the crevice. The data implies that the transition time (ta), where a change in the rate-controlling process and an increase in corrosion rate occurs (Figure 2), is similar for pitting development both inside and outside the crevice for each coupon.

The plots in Figure 10 could be used to predict pit depths after extended exposure periods. However, due to the variability in pit depths (for example Figure 10a), more accurate predictions require additional data, particularly for longer exposure durations. As noted, further coupon recoveries are planned at 36 and 48 months of exposure. Incorporating this data into the plots will enable more precise predictions of long-term pit depth rates.

In addition to the exposure tests discussed here, electrochemical investigations have been conducted at the University of Newcastle laboratories. These investigations used wire beam electrode technology [17] to study non-uniform corrosion behaviour under crevices in similar naturally sourced seawater and crevice geometries. This work is currently continuing and will involve comparison to the experimental results presented herein.

A further study, involving Deakin University has employed a wire beam electrode system integrated into a probe to examine the evolution of crevice corrosion under marine immersion field conditions. This work will be reported in due course. Preliminary findings suggest that for crevice corrosion of mild steel, corrosion appears to commence within the crevice. However, within a few weeks to months, corrosion becomes more severe outside the crevice, broadly aligning with the findings reported in the present study.

CONCLUSIONS

For bolted mild steel connections with formed crevices exposed to natural, DIN dosed, and UV-treated seawater for up to 2 years, with crevice gaps ranging from less than 30 micron up to 200 micron and for different outstand geometries, the following conclusions were drawn:

1. Corrosion is significantly more severe outside of the crevice zone, compared to inside the crevice zone.
2. The tested parameters influenced the depth of corrosion inside the crevices by an insignificant amount.
3. Pit depth progression on surfaces inside and outside of the crevice follow a bimodal trend with time.

ACKNOWLEDGMENTS

This project was funded by ARC project DP200100746. The authors acknowledge the support of the laboratory staff at the University of Newcastle and also post-doctoral researcher Peter Richardson for assistance in preparing coupons for testing and setting up and maintaining the experimental testing facility at Taylors Beach. The authors also acknowledge the continued support of The Port Stephens Fisheries Institute at Taylors Beach.

References

1. Combrade, P., *Crevice corrosion of metallic materials*, in *Corrosion Mechanisms*, P. Marcus, Editor. 2012, CRC: Roca Baton. p. 449-498.
2. Fontana, M.G., *Corrosion engineering*. 1988, New York: McGraw-Hill. 556 p. : ill.
3. Harro, H.H. and R.D. Port, *The Nalco guide to cooling water systems failure analysis*. 1993, New York: McGraw.
4. Melchers, R.E. and R. Jeffrey, *The Transition from Short- to Long-Term Marine Corrosion of Carbon Steels: 1. Experimental Observations.* Corrosion, 2022. **78**(5): p. 415-426.
5. Melchers, R.E. and R. Jeffrey, *The Transition from Short- to Long-Term Marine Corrosion of Carbon Steels: 2. Parameterization and Modeling.* Corrosion, 2022. **78**(5): p. 427-436.
6. Ijsseling, F.P., *Electrochemical Methods in Crevice Corrosion Testing - Report Prepared for the European-Federation-of-Corrosion-Working-Party Physicochemical Testing Methods of Corrosion - Fundamentals and Applications.* British Corrosion Journal, 1980. **15**(2): p. 51-69.
7. Oldfield, J.W. and W.H. Sutton, *Crevice Corrosion of Stainless-Steels .1. Mathematical-Model.* British Corrosion Journal, 1978. **13**(1): p. 13-22.
8. Oldfield, J.W. and W.H. Sutton, *Crevice Corrosion of Stainless-Steels .2. Experimental Studies.* British Corrosion Journal, 1978. **13**(3): p. 104-111.
9. Petersen, R.B. and R.E. Melchers, *Crevice corrosion of mild steel under extended exposures in natural, nutrient-dosed, and UV-treated seawaters*, in *Corrosion and Prevention 2023*. 2023, The Australasian Corrosion Association: Perth, Australia.
10. ASTM, *ASTM G1-03 Standard Practice for Preparing, Cleaning, and Evaluating Corrosion Test Specimens.* 2017, ASTM International: West Conshohocken, PA, United States. p. 9.
11. Melchers, R.E. and R.J. Jeffrey, *Long-Term Corrosion of Mild Steel in Natural and UV-Treated Coastal Seawater.* Corrosion, 2014. **70**(8): p. 804-818.
12. Melchers, R.E., *Long-term immersion corrosion of steels in seawaters with elevated nutrient concentration.* Corrosion Science, 2014. **81**: p. 110-116.
13. Petersen, R.B. and R.E. Melchers, *Development of Pitting Corrosion for Mild Steel Exposed to Particulate Media and Natural and Nutrient-Dosed Seawaters.* Corrosion, 2023. **79**(1): p. 121-131.
14. Stineman, R.W., *A consistently well-behaved method of interpolation.* Creative Comput., 1980.
15. Evans, U.R., *The corrosion and oxidation of metals; scientific principles and practical applications.* 1960, London,: E. Arnold. 1094 p.
16. Wranglen, G., *Pitting and sulphide inclusions in steel.* Corrosion Science, 1974. **14**: p. 331-349.
17. Tan, Y.J., *Wire beam electrode: A new tool for studying localised corrosion and other heterogeneous electrochemical processes.* Corrosion Science, 1999. **41**(2): p. 229-247.

AUTHOR DETAILS

	Robert Petersen is the corresponding author of this paper. He is a Lecturer at The University of Newcastle, Australia. He holds a BE (Civil Engineering, 2005) and a PhD (2009), both from the University of Newcastle. His current research interests are the long-term corrosion of mild steel surfaces and assemblages in marine environments. Recently he has been awarded the prestigious Guy Bengough award in recognition for the practical relevance of his research outcomes.
	Robert Melchers is an Emeritus Professor of Civil Engineering at The University of Newcastle, Australia. His qualifications are: BE and MEngSc (Monash University), PhD (University of Cambridge, UK), Fellow, Australian Academy of Technology and Engineering, Honorary Fellow, The Institution of Engineers Australia. His awards include the 2009 ACA Corrosion Medal, 2012 Jin S Chung Award (ISOPE), 2013 John Connell Gold Medal (The Institution of Engineers, Australia), 2014 Eminent Speaker, Structural College, The Institution of Engineers, Australia, 2006 and 2018 Guy Bengough Award (IOM3).

CORROSION & PREVENTION
CAIRNS 2024

THE AUSTRALASIAN CORROSION ASSOCIATION INC.

WEATHERING STEEL DURABILITY IN AUSTRALASIA

WL. Mandeno [1*], RA. Francis [2], R. El Sarraf [3].

[1] *Mandeno Consulting Ltd, Wellington 6035, New Zealand*

[2] *Retired Consultant, Victoria, 3147, Australia*

[3] *WSP NZ, Christchurch, 8011, New Zealand*

**Corresponding author email: wlmandeno@gmail.com*

ABSTRACT

The atmospheric corrosion rates for Australian-made weathering steels, as predicted by ISO 9224, are compared with historic 15-year weight loss results from Australian exposure sites, and current international design guides. Examples of weathering steel bridges in Australia and New Zealand are also presented.

Keywords: weathering steel, durability, atmospheric testing.

INTRODUCTION

The purpose of the paper is to share unpublished historic data on the durability of weathering steel panels that were exposed in 1968 at different sites around Australia and monitored for fifteen years. The paper also compares these results with the durability predicted by ISO 9224 for these and a newly available Australian-made weathering steel in different atmospheric corrosivity environments in Australasia. Finally, it reviews the sacrificial thickness allowance recommended by various international bridge design guides for these environments.

It is hoped that this paper will give greater confidence to Australasian bridge designers to specify the use of weathering steel instead of conventional coated carbon steel in non-coastal environments.

WEATHERING STEEL

Weathering steel (WS), or to use its technical title of "structural steel with improved atmospheric corrosion resistance", is a high-strength low-alloy steel that, in suitable environments, may be left unpainted because it forms an adherent protective rust "patina" to prevent significant further corrosion. The added alloying elements comprise only 2% of the steel, with copper and chromium being the two most important, along with silicon and in some cases phosphorus. These do not diminish the structural capability of the steel which has strength, ductility, toughness and weldability suitable for bridge construction [1, 2].

All structural steel rusts at a rate which is governed by the access of the metallic iron to moisture and oxygen. As this process continues, the oxide (rust) layer becomes a barrier restricting further ingress of moisture and oxygen to the metal, and the rate of corrosion slows down.

The rust layer that forms on most conventional carbon-manganese structural steels is relatively porous and it regularly detaches from the metal surface after which the corrosion cycle commences again. Hence, rusting progresses as a series of incremental curves approximating to a straight line, the slope of which depends on the aggressiveness of the environment. This is illustrated schematically in Figure 1.

The weathering steel rust patina is initiated in the same way. However, due to the alloying elements in the steel, it produces a more stable rust layer that adheres to the base metal and is much less porous. This layer develops under conditions of alternate wetting and drying to produce a protective barrier, which impedes further access of oxygen and moisture. Eventually, if the rust layer remains sufficiently impervious and tightly adhering, the corrosion rate may diminish virtually to zero. The resulting reduction in corrosion rate is illustrated in Figure 1. In a suitable environment, this stable condition may be reached within 8 years or so, and the metal is then protected from significant future corrosion by the rust patina [2].

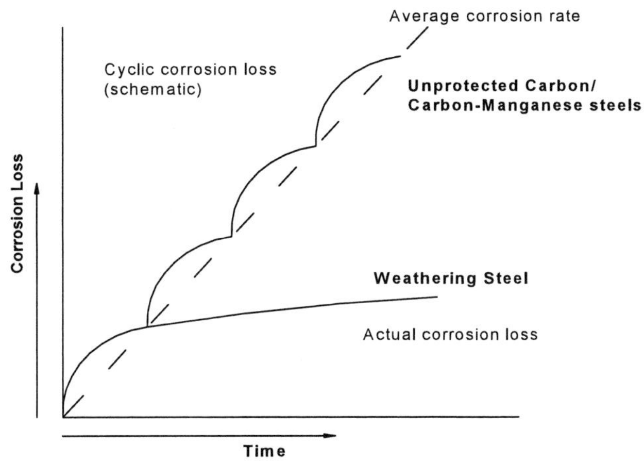

FIGURE 1. SCHEMATIC COMPARISON BETWEEN THE CORROSION LOSS OF WEATHERING STEEL AND CARBON STEEL [2]

ATMOSPHERIC TESTING

In 1968, the Corrosion Research Department of Australian Iron & Steel Pty Ltd (AI&S), at the time part of Australian steel making firm BHP Pty Ltd, commenced a comprehensive test program to compare the atmospheric corrosion performance of *AusTen Type A*, their proprietary low alloy weathering steel (WS), with a plain carbon steel (CS). Twenty-three exposure sites were located throughout Australia in a broad range of environments from heavy industrial/marine at Port Kembla to the very dry arid environment of Mt Newman in WA. Test panels were exposed, removed and analysed at regular intervals over 15 years and reported, but not published, by Badger and Wallace [3].

This current paper uses the AI&S data to estimate the expected WS metal loss after 100 years exposure within different ISO 9223 atmospheric corrosivity categories. These losses are then compared to the current allowances given in Table 3.2 of the Australian design guide [2] and Table 3.2 in the latest European design guide [4].

ESTIMATE OF 100-YEAR METAL LOSS

The 100-year metal loss can be determined by analysing the long-term results in a number of steps.

Step 1: WS Corrosion Rates for ISO Corrosivity Categories

The ISO 9223 Atmospheric Corrosivity Categories [5] are based on first year CS corrosion rates. WS in a given category will show a reduced corrosion rate, but there is no simple formula relating first year CS steel rates to first year WS rates. Figure 2 plots the first-year corrosion rates for the two steel types from the AI&S report.

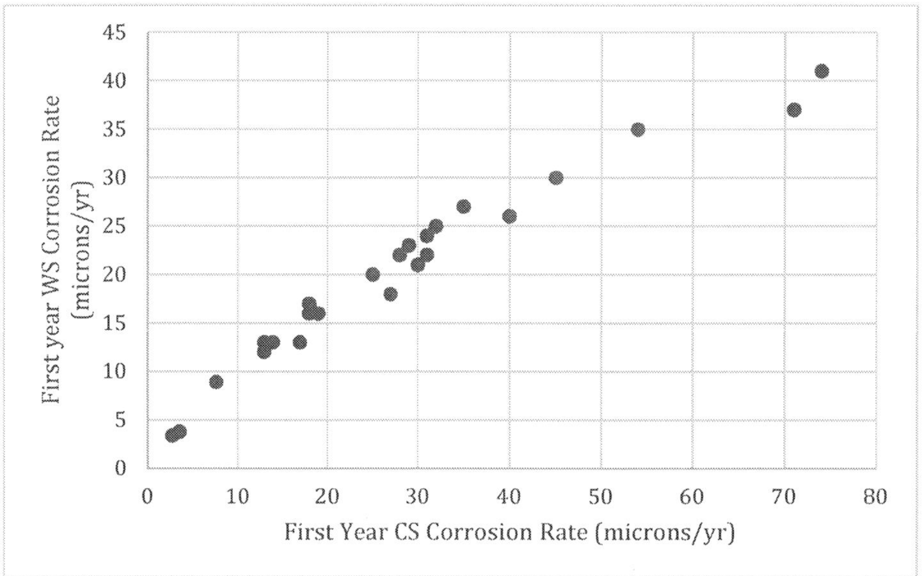

FIGURE 2: FIRST YEAR WEATHERING STEEL CORROSION RATES COMPARED TO CARBON STEEL RATES FROM [3].

Figure 2 shows significant scatter, but there is clearly a relationship between the rate in CS and WS. Table 1 presents suggested first year WS corrosion rates for the standard ISO Corrosivity categories based on the relationship shown in Figure 2.

TABLE 1: FIRST YEAR CORROSION RATES FOR WEATHERING AND CARBON STEELS FOR ISO CATEGORIES

ISO 9223 Category	1st yr CS Corrosion Rate (microns/yr)	1st yr WS Corrosion Rate (microns/yr)
C2	1-25	1-20
C3	25-50	20-35
C4	50-80	35-45

Step 2: Long Term Rates from Trends

Figure 3 shows the changes in corrosion rate with exposure time from six selected sites with 15 years exposure, which cover a wide range of results and show a relatively smooth transition. (Other results, and those from similar overseas investigations, show more varied behaviour, but do not seriously contradict these examples).

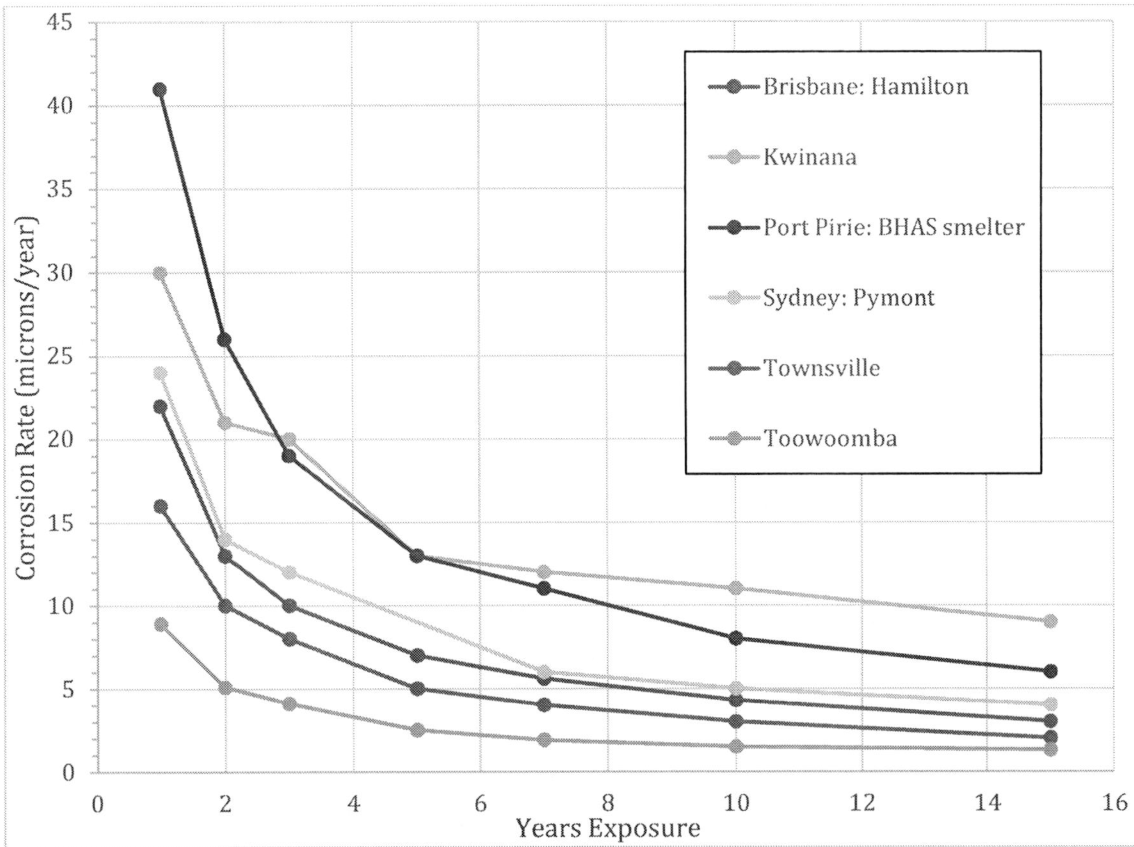

FIGURE 3: DECREASE IN WEATHERING STEEL CORROSION RATE WITH TIME FROM SIX SITES.

From these trends, a first year WS rate of 20 microns per year (C2/C3 border) shows a 15-year rate of between 2–3 microns per year. A first year WS rate of 35 microns per year (C3/C4 border) shows a 15-year rate of between 6–9 microns per year. The more conservative (greater) figure can be considered a long-term rate and has been used to calculate the 100-year corrosion allowance. These are given in columns 3 and 4 of Table 2.

Step 3: Check Mathematical Relationship

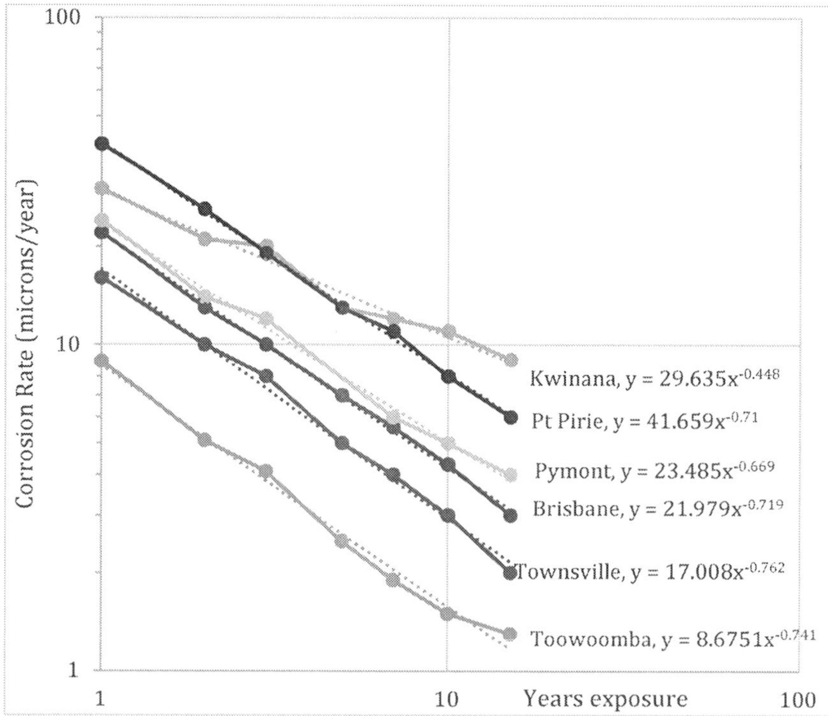

FIGURE 4: FIGURE 3 SITES PLOTTED USING LOGARITHMIC CO-ORDINATES, WITH TRENDLINES AND MATHEMATICAL RELATIONSHIPS.

Figure 4 shows that weathering steel, at least in these environments, does appear to follow a definite mathematical relationship, as per Section 4 of ISO 9224 [6], of the form:

$$[\text{Rate in year } t] = [\text{First year rate}] \text{ x } [t] \text{ x } [e^{(-b)}]$$

where the constant *b* depends on the environment for a given metal. As the value of *b* becomes more positive, the difference between the short and long-term rate decreases. For the examples given, the value of *b* is approximately -0.75 for C2 environment, -0.7 for C3 and -0.5 for C4 (following the example of Kwinana). As the environment becomes more corrosive, it would appear that the value of *b* approaches zero and the long-term rate approaches the first-year rate, similar to the behaviour observed with carbon steel [7].

Using these exponents and the suggested boundaries noted in Step 1, we can calculate a 20-year corrosion rate using ISO 9224. Assuming the 20-year rate is a good approximation of the long-term constant rate, the loss over 100 years can also be calculated. These are given in columns 6 and 7 of Table 2 and give similar results to the extrapolated values.

TABLE 2: LONG-TERM WS RATES AT ISO BORDERS FROM 15 YEAR RATE, AND EXTRAPOLATION.

1	2	3	4	5	6	7
ISO Category	**Calculated from AI&S data**			**Calculated using ISO 9224**		
	1st yr. WS rate (µm/yr)	**15-year WS rate (µm/yr)**	**100yr WS loss (mm)**	**Exponent *b***	**Long-term Rate (µm/yr)**	**100yr WS Loss (mm)**
C2/C3	20	3	0.3	-0.7	2.5	0.25
C3/C4	35	9	0.9	-0.5	8	0.8

Step 4: Compare with Rates Predicted by ISO 9224

According to ISO 9224, total corrosion depth (D) for less than 20 years may be predicted using Equation 1:

$$D = r_{corr}\, t^b$$

For >20 years, Equation 3 of ISO 9224 is used:

$$D(t > 20) = r_{corr}[20^b + b\,(20^{b-1})\,(t - 20)]$$

Annex C of ISO 9224 is used to calculate (b) for an alloy. The specific value of (b) in a non-marine environment = b_a

$$b_a = 0.569 + \sum b_i w_i$$

where:

b_a is the alloy-specific value of *b* in non-marine exposures;

b_i is the multiplier for the *i*th alloying element;

w_i is the mass fraction of the *i*th alloying element.

Table 3 presents the $b_i w_i$ calculated for *AusTen Type A* WS from the element multipliers given in ISO 9224 Table C.1

TABLE 3. ELEMENT MULTIPLIERS

Element	*AusTen Type A* (%)	ISO 9224 Multiplier (b_i)	*Calculated* $b_i w_i$
C	0.11	-0.084	-0.00924
Mn	0.91	-	-
S	0.019	1.44	0.0274
Si	0.42	-0.163	-0.0684
Ni	0.25	-0.066	-0.0165
Cr	0.25	-0.124	-0.031
Cu	0.26	-0.069	-0.0179
P	0.09	-0.49	-0.0441

An estimate of the 100-year loss was made using Equation 3 of ISO 9224 and the steel chemistries for *AusTen Type A* and the recently Australian developed conventional weathering steel, *REDCOR®*, manufactured by BlueScope. The results are presented in Table 4, along with the estimated 100-year loss for WS presented in this paper.

Table 4 uses the data provided in ISO 9224 to predict the 15 year and 100-year losses for *AusTen Type A* WS, using a $\sum b_i w_i$ of -0.15988 and b_a of 0.40912. Also shown are the predicted 100-year losses for REDCOR W24W1 (High Phosphorus Type WR350A) with a b_a of 0.34258 and REDCOR W25W7 (Low Phosphorus Type WR350B) with a b_a of 0.41514.

TABLE 4. ISO 9224 PREDICTED 15 AND 100-YEAR LOSSES FOR WS

Atmospheric Corrosivity Category	C2/C3	C3/C4
r_{corr} (for Carbon Steel)	25 μm/yr	50 μm/yr
D(15) for REDCOR®	63 to 77 μm	126 to 154 μm
15yr rate for AusTen Type A using AI&S data	4 μm/yr	7 μm/yr
D(100) for AusTen Type A	**0.23mm**	**0.45mm**
D(100) for REDCOR®	**0.17 to 0.23mm**	**0.33 to 0.46mm**
D(100) for AusTen Type A using AI&S data	**0.25mm**	**0.8mm**

Step 5: Comparison with the Recommended Corrosion Allowances

Table 5 compares the calculated 100-year loss for WS in this work to the corrosion allowances in the two published guides. The allowance given in the guides would be expected to include the 100-year loss plus a margin of error, so the figures from the ECCS Guide for C2 and C3 could probably be adopted. However, there are insufficient exposure test figures that would justify adoption of an allowance for the C4 environment.

TABLE 5: ALLOWANCES COMPARED TO WS GUIDES [2] [4]

ISO Category	100 yr Loss (mm)	Allowance (mm)	
	This work	Aust. Guide	ECCS Guide
C2	0.25 – 0.3	1	0.8
C3	0.8 – 0.9	1.5	1
C4	NA	NA	1.5

COASTAL ENVIRONMENTS

Exposure to high concentrations of chloride ions, deposited from airborne marine salts in aerosol originating from breaking waves at sea or on the shoreline, or from salt fogs, will negatively affect the patina formation on WS. The hygroscopic nature of salt can prevent surfaces from fully drying when relative humidity is elevated and thus stop the formation of the rust patina, thereby resulting in the weathering steel continuing to corrode at a rate similar to carbon steel. Figure 5 shows weathering steel exposed to a severe marine environment, showing formation of rust delamination at the edges.

FIGURE 5: WEATHERING STEEL IN A SEVERE MARINE ENVIRONMENT

Evidence of a higher corrosion rate (similar to carbon steel) and a delayed, or even absent, formation of the protective patina has been identified by Morcillo [8] for unwashed and sheltered surfaces (i.e. microclimate effects), as well as in crevices, on weathering steel structures in coastal locations. McKenzie [9] reported that severe pitting developed on weathering steel bridge members in sheltered marine environments.

Therefore, when determining the suitability of using weathering steel in a given location, the atmospheric corrosivity assessment needs to take into account not just the macroclimate, in accordance with AS 4312 [10] or SNZ TS 3404 [11] and/or the CSIRO Corrosion Mapping App [12]; but also the microclimate effects, as discussed by Francis [13] and described in the Australian Steelwork Corrosion and Coatings Guide [14].

Details regarding humidity levels, wind strengths and wind directions, which assist in determining the macroclimate and microclimate for a specific location, can be obtained from the Australian Bureau of Meteorology website [15].

Based on the findings of the Morcillo review of weathering steel performance data [8], it is recommended that conventional weathering steel should only be used where its steady state corrosion rate is less than 6 microns/year, or in areas where the maximum first year corrosion rate (taking into account both the macroclimate and microclimate effects on sheltered surfaces) of mild steel is less than 50 μm/yr. This is equivalent to a rain-washed surface in atmospheric corrosivity category C3 (Medium) to ISO 9223 or AS 4312. Therefore, if the determined surface-specific atmospheric corrosivity is C4 (High) or C5 (Very High), weathering steel should not be used. Higher nickel grades of weathering steel have been developed in Japan which are claimed to resist these more corrosive environments [16] but long-term experience is lacking.

AS 4312 suggests the C4 corrosivity boundary commences about 1km inland at exposed surf beaches, and 0.1km inland on calm shores, but a conservative approach to these distances is advisable to allow for microclimate effects on sheltered surfaces. Generally, weathering steel can be used in locations that are more than 2km from the ocean seacoast, where the maximum first year corrosion rate (taking into account both the macroclimate and microclimate effects) of mild steel is less than 50 μm/yr. On sheltered bays without breaking surf, WS that is freely exposed to rain washing may be acceptable up to 500m from the shoreline.

In the case that conventional weathering steel is being considered for sites where the microclimate corrosivity is borderline high C3 (Medium)/low C4 (High), or less than 2km from the open seacoast; then determination of the actual site-specific corrosivity environment (including those of unwashed and sheltered surfaces), with a minimum of 1-year record, is required. This assessment should be undertaken by an experienced and qualified corrosion engineer or materials scientist. It is recommended that the 'coupon weight loss' technique, to ASTM G1 [17], be used to determine the site-specific corrosivity, rather than testing for salt and sulphur dioxide levels in the atmosphere. Guidance for conducting the coupon weight loss test is given in the Australian WS bridge guide [2]. Alternatively, use of the 3 monthly qualitative 'wire-on-bolt' test method, to ASTM G116 [17], could be considered.

In summary, when assessing the site atmospheric corrosivity category in Australia (as per Figures A1 to A6 and Table 4.1 of AS 4312) to determine the suitability for the use of weathering steel, the following assessment guidelines should be used [2]:

- Any location in a C4, C5 or CX category or within 500 metres of the shoreline, whether the seas are temperate semi-sheltered, quiet (calm) or tropic quiet (calm), weathering steel should not be used.
- Any location within 5km of temperate surf (C4, C5 or CX), weathering steel should not be used.
- Weathering steel should be acceptable for all other locations (C2, C3) in Australia if at least 500m inland.
- However, if actual measurements or local features show high corrosivity, then this evidence should override these guidelines that apply to both boldly exposed macroclimates and sheltered micro-climates.

EXAMPLES OF WS BRIDGES IN AUSTRALASIA

BERWICK ROAD BRIDGE, INLAND NEAR DUNEDIN, NZ (1985)

MERCER RAIL OVERBRIDGE, SOUTH OF AUCKLAND, NZ (2006)

NZ RAIL REPLACEMENT BRIDGE (2012)

NZ RAIL REPLACEMENT BRIDGE (2012)

REPLACEMENT KUROW BRIDGE, NZ (2014)

KARAPIRO VIADUCT, NZ (2015)

PICTON RIVER BRIDGE, TASMANIA (1981)

JAMES RUSE DRIVE LIGHT RAIL BRIDGE, PARRAMATTA, NSW (2021)

FIGURE 6: EXAMPLES OF WEATHERING STEEL BRIDGES

Conclusions

The historic exposure trials have confirmed that weathering steel can be specified with confidence for steel structures exposed to the atmosphere in C2 and C3 corrosivity zones, provided they are detailed and located in accordance with the relevant guide for Australia [2] or New Zealand [19]. The recommended corrosion allowance given in these guides at 1mm for C2 and 1.5mm for C3 has been shown to be conservative and allows for more severe environments due to microclimatic effects that may exist. These include underdeck bridge girders being sheltered from the removal of wind-deposited contaminants by rain washing, and being sheltered from sunlight which would otherwise aid evaporation of condensed moisture if steel temperatures drop to below dewpoint overnight.

Acknowledgments

The authors acknowledge the support of the NZ Heavy Engineering Association and BlueScope in preparing their guides for the use of weathering steel in Australian and New Zealand bridges. We also thank BlueScope for their permission to use the historic BHP test data [3].

References

1. El Sarraf R, Mandeno W, Hicks S and Killmore C, The use of Weathering Steel in Australian Bridges, 10th Austroads Bridge Conference, Melbourne, Australia, 2017, Paper 163.
2. El Sarraf R, Mandeno W, and Hicks S, Weathering Steel Design Guide for Bridges in Australia, BlueScope Steel Ltd (Publishers), Melbourne, Australia, 2017.
3. Badger GT and Wallace NE, The Long Term Exposure Testing of Plain Carbon and AusTen Steels, Including Fifteen Year Site Results and Related Statistical Extrapolation, Australian Iron & Steel Pty Ltd Corrosion Research Department (unpublished) Report No. PK/TD/88/001, 1988.
4. Ungermann D and Hatke P. European Design Guide for the Use of Weathering Steel in Bridge Construction. In ECCS-European convention for constructional steelwork, 2021.
5. ISO 9223:2012, Corrosion of metals and alloys – Corrosivity of atmospheres – Classification, determination and estimation.
6. ISO 9224:2012, Corrosion of metals and alloys – Corrosivity of atmospheres – Guiding values for the corrosivity categories.
7. Francis RA, Long-term corrosion rates: Guide values in codes compared to actual results, C&P23, Perth, Paper ACA2023-19921, 2023.
8. Morcillo, M., Chico B, Diaz I, Cano H and de la Fuente D, Atmospheric Corrosion Data of Weathering Steels. A Review. Corrosion Science Vol.77 pp 6-24, December 2013.
9. McKenzie M, The corrosion performance of weathering steel in highway bridges, Atmospheric Corrosion pp 717-736, The Electrochemical Society, New York, 1982.
10. AS 4312: 2019, Atmospheric corrosivity zones in Australia
11. SNZ TS 3404:2018, Durability requirements for steel structures and components
12. Cole IS, Patel MJ, Chu C, Ganther WD, Harrison D, Golding P and Jose J, Corrosion Map of Australia Re-Imagined, C&P23, Perth, Paper ACA2023-20181, 2023.
13. Francis RA, Corrosive effects of micro-environments and design features, On-line C&P20, Paper 15, 2020.
14. Clifton GC, El Sarraf R, Mandeno W, Golding P, Sheehan A, Australian Steelwork Corrosion and Coatings Guide. Australian Steel Institute, Sydney, Australia, 2013.
15. http://www.bom.gov.au/climate/data/
16. Kihara H. and Kimura M, Advancements of Weathering Steel Technologies in Japan, *Corrosion* Vol. 67, No. 9, NACE International, Houston, USA, September 2011.
17. ASTM G1-03(2017) Standard Practice for Preparing, Cleaning, and Evaluating Corrosion Test Specimens.
18. ASTM G116-99(2020) Standard Practice for Conducting Wire-on-Bolt Test for Atmospheric Galvanic Corrosion.
19. El Sarraf R, Mandeno W, and Karpenko M, New Zealand Weathering Steel Guide for Bridges, HERA Report R4–97:2020, Manukau, New Zealand, 2020.

AUTHOR DETAILS

Willie Mandeno is the corresponding author of this paper. He is a civil engineer who retired in 2017 from WSP Opus as their Technical Principal Materials & Corrosion Engineering and is a now a Coatings Consultant based in Wellington, New Zealand.

Willie is a Chartered Professional Engineer, a Fellow of Engineers NZ, an ACA Life Member and Corrosion Medallist, and an AMPP Corrosion Specialist and CIP3 Peer.

Rob Francis is a recently retired corrosion and coating consultant with over 40 years of academic, industrial and consulting experience in corrosion and protective coatings. Dr Francis has a B.Sc. in metallurgy and a Ph.D. in corrosion science. He has authored or co-authored over 40 technical papers or presentations on corrosion and coatings. He was awarded the Victor Nightingall award in 2014 and made an ACA Life Member in 2016.

Raed El Sarraf is the Technical Principal Asset Integrity Management with WSP NZ based in Christchurch, New Zealand, and is an ACA Corrosion Technologist and a Fellow of Engineers NZ.

Raed is the principal author of the BlueScope publication Weathering Steel Design Guide for Bridges in Australia and HERA Report New Zealand Steelwork Corrosion and Coatings Guide. He has also contributed to the production of several specifications including SNZ TS 3404, NZTA S9 and NZTA S10.

CORROSION & PREVENTION
CAIRNS 2024

THE AUSTRALASIAN CORROSION ASSOCIATION INC.

IMPROVING ASSET RELIABILITY BY UPGRADING BASE METAL METALLURGY UTILIZING TRUE HIGH VELOCITY THERMAL SPRAY

Phil Williams & Peter Johnson

Integrated Global Services

Richmond, Virginia, 23237 USA

ABSTRACT

First patented in 1914, Thermal Spray technology is not new, but it is only in the last few decades that it has been developed into a robust, impermeable protective barrier suitable for immersion service. There are many factors for consideration when upgrading base metal metallurgy of critical assets. These include, but are not limited to; material development, the delivery / conveyance system being used, the application process and the cost and time to complete. Weld Metal Overlay (WOL) is a common, much used technology in providing metallurgical upgrades to less noble materials, but application factors mean it is best suited to new build equipment where it can be completed in the shop, rather than in-situ. In Brownfield applications, especially during turnarounds where time and speed of application is more critical, additional consideration must also be made for the requirements of other alternative protective systems where environmental conditions, cure time, stress relieving, etc., all play a part in the application process. This paper addresses how True High Velocity Thermal Spray (HVTS) overcomes the additional requirements associated with these other systems, allowing the rapid application of an erosion-corrosion resistant barrier which is now widely regarded as one of the most reliable metal protection systems available today.

INTRODUCTION

When erosion and corrosion lead to metal wastage, threatening the integrity of the equipment, asset owners and operators are faced with an immediate challenge of addressing these issues to remain operational. A metal upgrade with a Corrosion Resistant Alloy (CRA) barrier is often chosen as the best solution and there are several viable options available. The method and extent of the CRA system used will largely be determined by the shutdown time available to complete the repairs.

Weld Metal Overlay (WOL) remains a tried and tested option for this purpose but can be slow and difficult to apply in-situ. In contrast, continued development, extensive laboratory testing and thorough performance validation carried out over several decades, coupled with numerous field applications and subsequent site inspections has confirmed that True High Velocity Thermal Spray (HVTS) alloy cladding systems provide enhanced performance, with additional benefits to the end user.

Reducing Maintenance and Increasing Uptime

Mission critical assets (process vessels, towers, and columns) are an integral part of every Oil & Gas and Petrochemical facility worldwide, with their operation and uptime playing a vital role in maintaining continued levels of production. The media within these vessels is often corrosive in nature, with the corrosive effects further exacerbated by the process (high temperature / high pressure), particularly where sour service conditions exist.

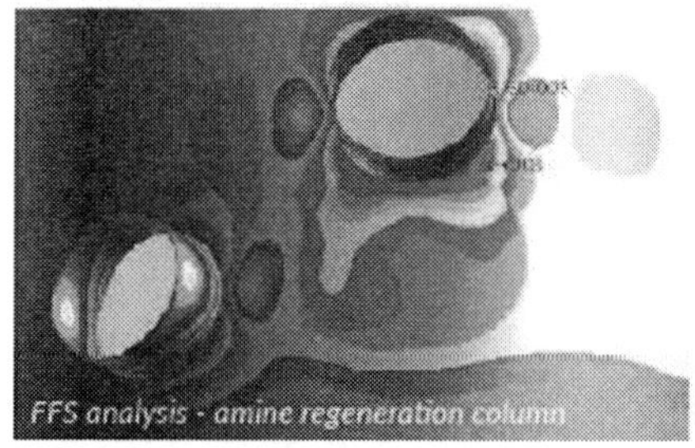

FIGURE 1 - FITNESS FOR SERVICE ANALYSIS

For example, process equipment in amine service are exposed to several potentially aggressive corrosion mechanisms. In the lower section of a regenerator, condensation of water vapor could cause the formation of carbonic acid which can attack the carbon steel surfaces not wetted by the amine solution, leading to aggressive corrosion rates.

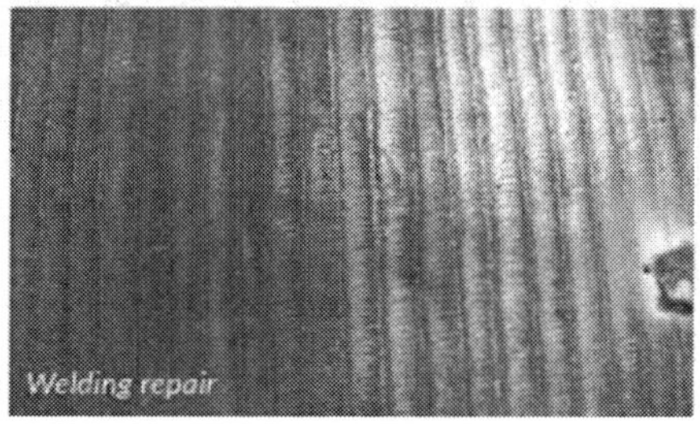

FIGURE 2 - WELDING REPAIR

In such situations, acid-gas ratio, choice of amine, contaminants, two-phase flow, flashing, high velocities, vessel design and insulation are all factors which require consideration.

Corrosion ultimately leads to metal wastage and shell thickness loss. Plant and process shutdowns caused by issues with these critical assets can quickly run up losses of tens of millions of dollars.

FIGURE 3 - WELDING REPAIR APPLICATION

According to the US Department of Energy's Energy Assurance Daily, there were over 1700 refinery shutdowns from 2009 to 2012 with 23% of them caused by maintenance.

Depending on the unit affected and the work scope, the duration of a shutdown can range from a couple of hours to several weeks, with over 90% of them being unplanned. Any shutdown, however long, is a disruption in the refining process which comes at high cost to the business.

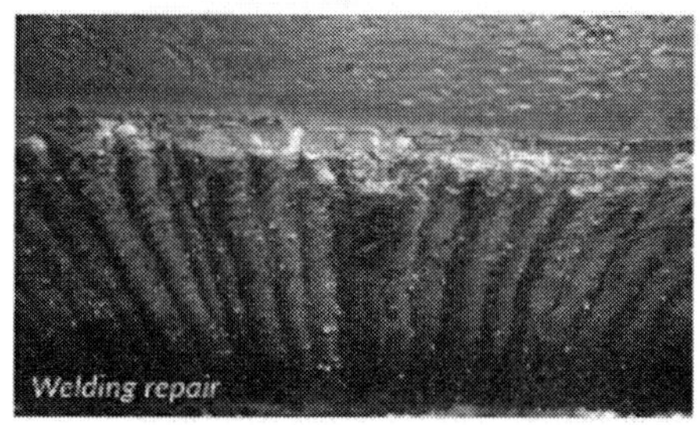

FIGURE 4 - WELDING REPAIR

An unplanned shutdown has even larger implications and can cost companies millions of dollars in lost production.

Once it becomes evident that a piece of equipment has suffered wall thickness loss due to corrosion and / or erosion, the countdown begins. If allowed to continue unchecked, wall thickness loss may continue beyond the corrosion allowance. When this occurs, the choice of available repair options becomes significantly narrowed, and oftentimes, the only choice is between either major mechanical repairs, or replacement of the equipment. However, if intervention takes place before this time, or when Fitness for Service (FFS) evaluations deliver a favourable result (Figure 1) and the vessel still maintains its operational integrity, there are several other options that may be considered.

Welding has long been an industry solution (Figure 2) for both rebuilding areas of wall thickness with the parent metal alloy, and in providing a reliable corrosion resistant alloy (CRA) barrier. In the main, it is chosen because it can also add strength back to the substrate (Figure 3), restoring pressure retention capabilities (Figure 4), meaning it can be applied even when the vessel is below its corrosion allowance.

When applied correctly, the metallurgical bond between the base substrate and weld metal is strong. Welding is also resistant to mechanical damage and shock, and where necessary, localized damage can also be repaired, providing that appropriately skilled welders are readily available.

For these reasons, multiple asset owners and operators depend on welding to protect their assets from further corrosion. But, as with most technologies, welding application is not without its downsides.

One major issue, although rare in occurrence, is the catastrophic distortion of the shell due to overheating during the welding process. A high degree of stress gets added, especially on thinner vessels. In vertical columns, welding can lead to shrinkage of approximately 1%. In a 5m diameter column this could mean shrinkage of up to 50mm – a significant level of distortion.

There is also the question of time and cost. Welding is a relatively slow process with an application rate of approx. 0.5 – 1.5m^2 per welding head/shift, so can also cause additional delays in bringing the asset back into service. Working conditions inside the vessel from high temperature preheat move from uncomfortable to near impossible. The additional time, cost and resources required for Post Weld Heat Treatment (PWHT) to stress relieve Heat Affected Zones (HAZ), where mandated, may generally render this option non-viable and if not done properly, can also create localized zones of high hardness that may lead to cracking.

Temper bead welding has been offered in place of PWHT but is difficult to execute, especially on larger scopes and lends itself more toward reliable automation than manual application. Another time-consuming exercise is the removal and reinstallation of support rings and downcomers. This alone can double the duration of the project. It is also worth mentioning that welding is only as good as your chosen process, procedure, technician, and quality program.

Evaluating Alternatives – Thermal Spray

Thermal spray technology has been utilized for the application of a CRA barrier since the 1980s, spraying metals widely used in the welding process. However, it was quickly noted that the thermal spray process itself can negatively affect the condition of the material being sprayed. The resulting cladding, when applied using traditional 'off the shelf' metal alloys and commercially available thermal spray equipment, was highly stressed and inherently permeable.

This permeability coupled with the internal stresses and a lower bond strength between the cladding and substrate creates a path for corrosion and premature failure. These early failures resulted in an understandable and rather universal distrust of early iterations of commercially available, low velocity thermal spray technology. At the same time, this evoked a great question.

'Is it possible to reduce the permeability, porosity, and internal stress of the thermal spray applied coating, and improve bond strength?'

A handful of engineers and material scientists have successfully developed a solution to this problem by redesigning both the equipment used to apply the metal cladding, the process technology and the alloy of the feedstock material. (Figure 5)

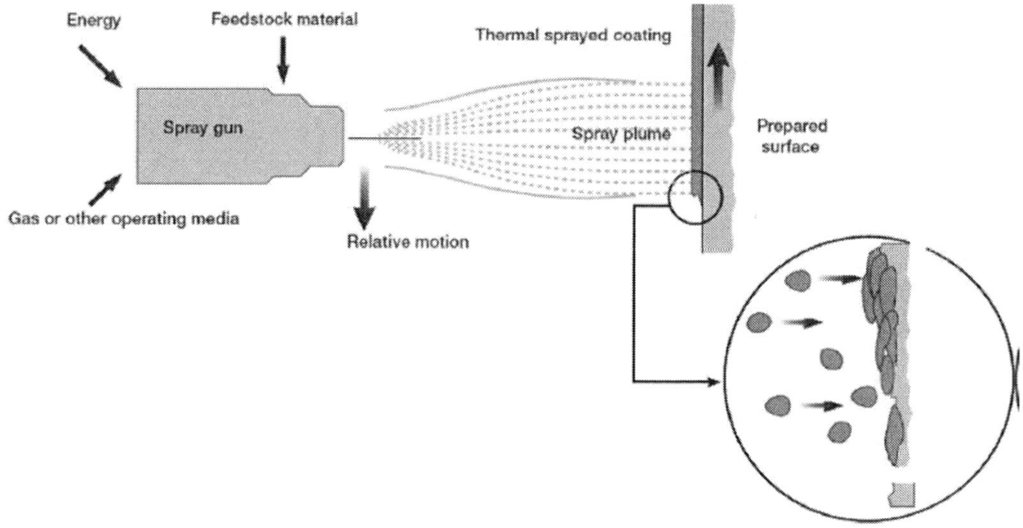

FIGURE 5 - THERMAL SPRAY EVOLUTION - HIGH VELOCITY THERMAL SPRAY (HVTS) ALLOY CLADDING

Evolution of True High Velocity Thermal Spray Cladding

The problem of bond strength, both between the applied metal particles and with the substrate was solved by increasing velocity and improving the quality of the substrate surface preparation.

Higher velocities provide improved atomization of the feedstock, resulting in a reduced particle size. The increase in speed also creates higher impact velocities leading to better embedment of the particles in the prepared substrate (Figure 6)

Improved surface preparation methods provide a cleaner, rougher surface, and a better anchor profile for the cladding to bond to.

When the molten metal particles hit the rough, profiled, white metal substrate at speeds in excess of Mach-1, they splat like pancakes and become embedded in the rough surface to form tight mechanical bonds. In addition, the particles themselves have a rough outer skin; an important feature that further promotes excellent inter splat adhesion, helping to lock the particles together. Multiple, additional overlapping passes of the thermal spray torch create a thick cladding layer (typically 500-micron thick) with excellent adhesion throughout. Dolly pull-off adhesion testing measures bond strengths between the High Velocity cladding and substrate of between 30 - 60MPa. (Figures 7 & 8)

Figure 6 – HVTS Alloy Bond Strength

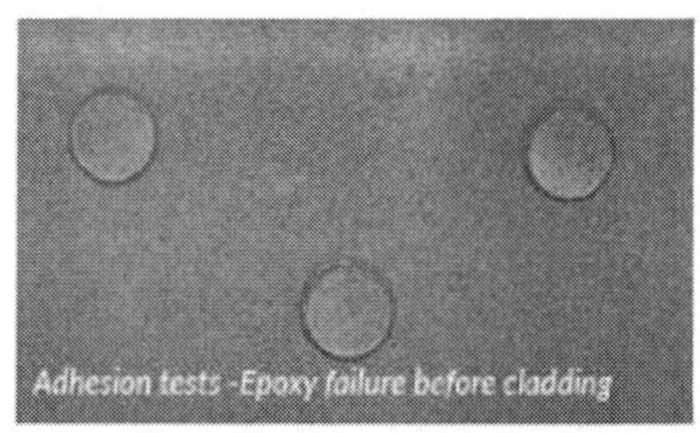

Figure 7 - Adhesion Tests
100% Epoxy Adhesive Failure

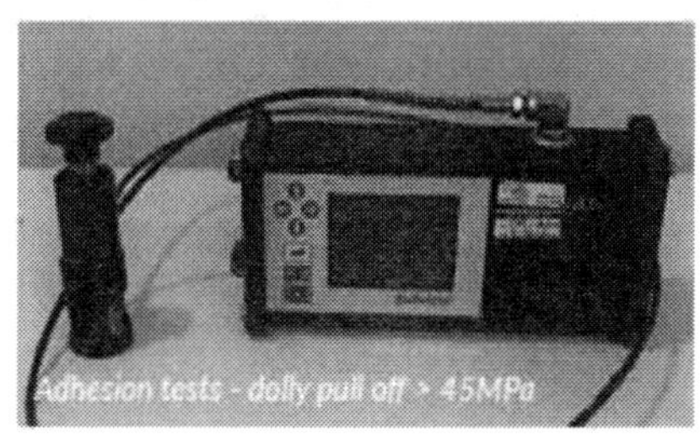

Figure 8 - Adhesion Tests
Dolly Pull off >45MPa

The second issue to address was the inherent permeability of the conventional thermal spray cladding. When the sprayed metal droplets leave the gun, during their flight to the surface, they are exposed to the atmosphere and the corresponding 21% oxygen level.

As a result, and due in part to their composition, these high temperature droplets experience a phenomenon referred to as 'in-flight oxidation'.

If not properly controlled, when the metal droplets lands on the substrate, the metal is deposited with an accompanying layer of metal oxide.

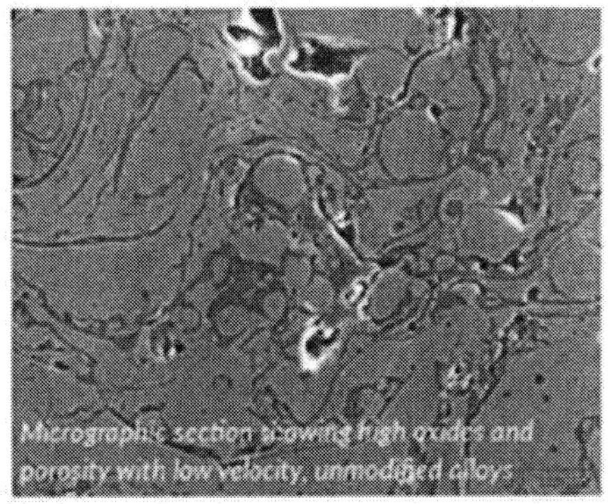

FIGURE 5 - MICROGRAPHIC SHOWING HIGH OXIDE & POROSITY WITH UNMODIFIED ALLOYS

Creating an Impermeable Protective Barrier

Under a microscope, the low velocity thermal spray cladding can appear dense, but in reality, the larger particle sizes combined with the oxide structures formed in flight create a pathway for permeation to the substrate through the metal cladding (Figure 9). Even though these high-alloy systems cannot themselves corrode or thin over time, corrosive media can permeate through the oxide layer and cause delamination by reacting with or corroding the underlying substrate (Figure 10).

FIGURE 10 – UNMODIFIED ALLOY FEEDSTOCK

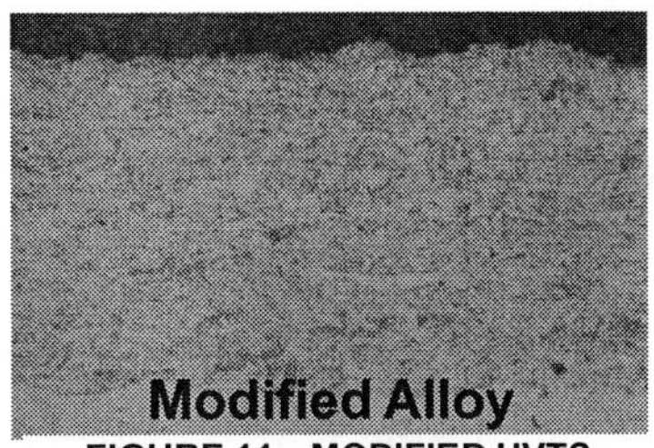

FIGURE 11 – MODIFIED HVTS ALLOY FEEDSTOCK

Currently only one company has solved this issue of macro-porosity with their High Velocity Thermal Spray Alloy Cladding. Off-the-shelf materials have been modified with unique alloying components to minimize in-flight oxidation. By mitigating oxide formation, the corrosion performance and permeability of the protective barrier has been greatly improved (Figure 11).

Quality Assurance and Control

Several workshop thermal spray vendors ventured into on-site mission critical asset applications and their inconsistent results have arguably damaged the reputation of "thermal spray applied cladding" as a whole.

FIGURE 13 – HVTS QUALITY CONTROL PLANNING

Welding, however, has been successfully applied for decades as an engineered solution, with relevant Quality Control requirements for both the technician/operator (welder), the welding process and the welding alloy. As a result, there are multiple procedures, standard practices, and controls that a client can follow to ensure the application is successful.

In Situ, High Velocity Thermal Spray Alloy Cladding on the other hand, is relatively new. Consequently, to ensure their applications would not follow the "thermal spray" experience of the past, it was necessary to develop in house engineering standards with multiple quality control systems governing the equipment, materials, process, procedures, performance qualification and manpower training, adopted from the template of the welding industry (Figure 13),

Proven Performance – Amine Regeneration Column

FIGURE 14 - AMINE REGENERATION COLUMN

An amine regeneration column in the Middle East (Figure 14) had clear indications that metal thinning was taking place. Thickness mapping caried out during the turnaround showed the problem to be more severe (Figure 15). With only two weeks of turnaround time left, a very significant part of the shell was found to be below its corrosion allowance. Following a comprehensive Fitness for Service Assessment (FFS), it was determined that although the vessel was below its minimum thickness in some areas, there was enough residual strength for it to be able to continue to operate, if it could be preserved in its current condition.

FIGURE 15 - CONDITION BEFORE CLADDING

The vessel was marked for replacement, but needed to operate for another four years, until at least the next turnaround for this to happen. The only technology believed to be available to get this vessel back into serviceable shape before the plant could go back online was weld overlay.

However, weld overlay would have required circumferential bands to be welded. Due to the sour service conditions of the vessel, post weld heat treatment would be required with the likelihood that the vessel would also need to be stabilized with cranes during the operation. The weld overlay and subsequent PWHT processes would have extended the turnaround by two to three weeks, resulting in significant costs from related production losses.

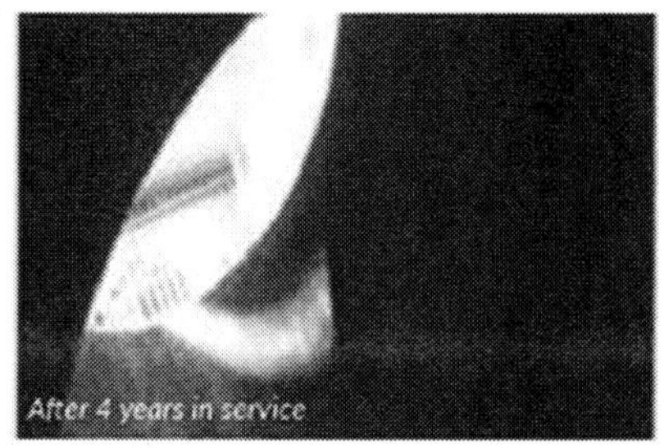

FIGURE 16 - INSPECTION AFTER 4 YEARS IN SERVICE

The facility owners became aware of high velocity thermal spray cladding technology. Within the space of four to five days the corroded areas had been cladded and protected. The vessel was promptly put back into service without any extension to the turnaround. The cost savings associated with the speed of application, production savings and the vessel life cycle were substantial.

After two years, there was an opportunity for inspection during a short turnaround. The corrosion condition had essentially been frozen (Figure 16). Inspection after four years in service further confirmed that the cladding was performing as expected, stopping any attack to the shell (Figure 17).

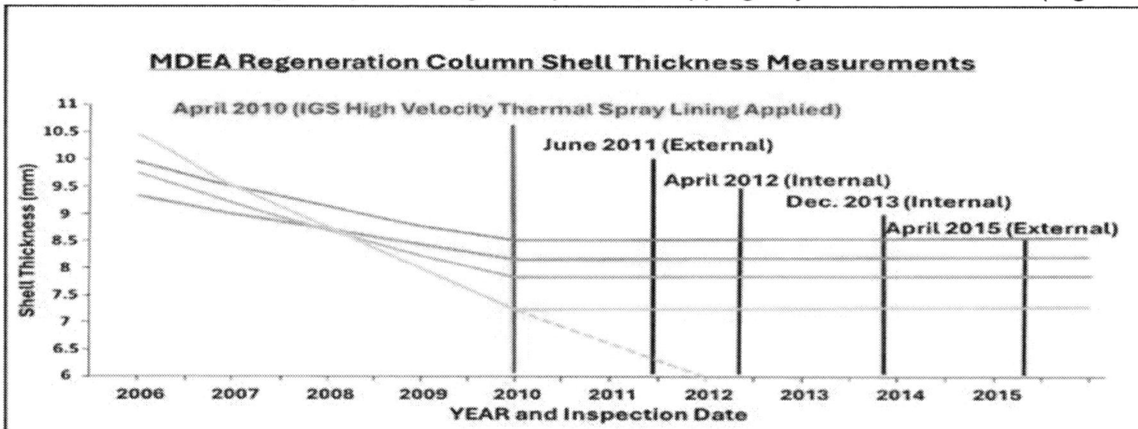

FIGURE 17 - WALL THICKNESS MEASUREMENTS BEFORE AND AFTER HVTS

Case Study – De-ethaniser Column

The de-ethaniser column made from a non-Hydrogen Induced Cracking (HIC) resistant grade of carbon steel was placed in service in 1975 (Figure 18). Turnaround & Inspection (T&I) for the bottom section of the column was completed in 2013. In total, 722 cracks were found in ring supports of 51 trays.

The majority of these cracks were found at tray ring support welds. These cracks were confirmed to be caused by Sulphide Stress Cracking (SSC) at previously weld-repaired locations that had high hardness levels, thought to have been caused by poor Post Weld Heat Treatment (PWHT).

Advanced UT inspection revealed the second significant point of the investigation in identifying hydrogen-induced cracking blisters in the top and bottom heads of the column as well as the 24" piping of the bottom product cycle.

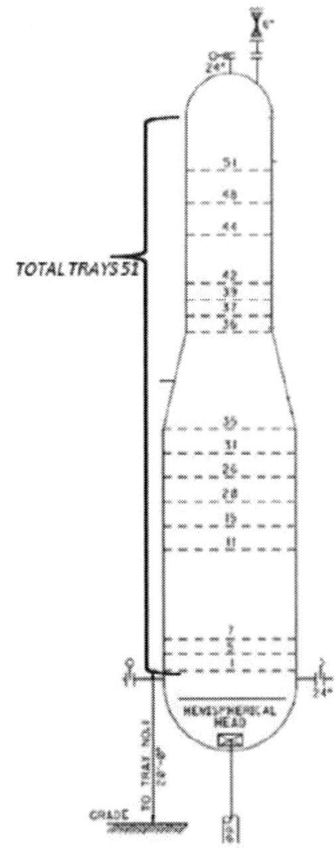

FIGURE 18 – SKETCH OF DE-ETHANISER COLUMN

Vessel Repair Alternatives

A risk matrix was developed to consider different alternatives for vessel repair. Due to the large surface areas involved (the vessel was ~5 metres across and 70 metres tall), internal weld geometries, and low heat input, the thermal spray method was deemed the most feasible of the CRA options.

A corrosion test cell, with a computerized potentiostat scan and Tafel analysis of the results to determine effective corrosion rates, demonstrated that the conventional twin wire arc spray applied using a standard alloy material exhibited deleterious performance due to an inability to effectively shield the carbon steel substrate.

However, a specifically modified NiCrMoW-XX alloy, applied using the high velocity thermal spray process resulted in a corrosion performance similar to that of wrought UNS N10276 plate (Figure 19).

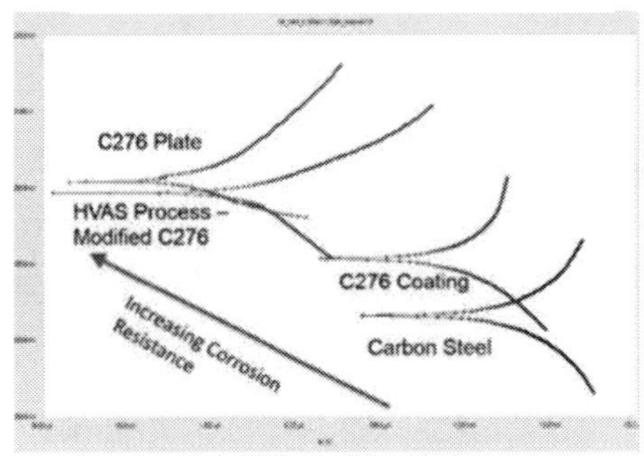

FIGURE 19 – CORROSION RESISTANCE QUALIFICATION TESTING

Application

As a result, a modified NiCrMoW cladding was applied to the entire internal shell (approx. 635m^2) using a proprietary modified Twin Wire Arc Spray process.

In July 2014, the vessel was taken out of service and cleaned with an approved hydro blasting procedure for a subsequent T&I.

An inspection was conducted by a combined vendor and client technical team to determine the integrity of the applied coating. No indications of SSC or corrosive coating delamination were observed. The vessel was returned to service with continued internal and external monitoring.

A further inspection was carried out in 2018 and again in 2023. No damage or hydrogen induced cracking was noted on the shell, welds or tray supports of the column, and the modified metal spray cladding showed no signs of wear or deterioration in any form.

CONCLUSIONS

When faced with metal wastage, asset owners and operators have several viable options available to them in addressing the corrosion mechanisms using a Corrosion Resistant Alloy (CRA). The chosen technique used to apply this CRA will in the main be determined by the shutdown time available to carry out the application.

While weld metal overlay continues to be a reliable option, laboratory testing and performance validation, coupled with field application and subsequent site inspections of High Velocity Thermal Spray Alloy Cladding projects, have confirmed this solution can provide performance "in the same league" as weld overlay, but with several added benefits.

When selecting "thermal spray" to apply the CRA barrier however, one needs to be aware of the vendor-to-vendor differences in the design and application of such protection systems. Suitable references, operator experience, the correct process and material technologies and performance test reports need to be thoroughly reviewed to ensure the applied CRA barrier will perform as expected.

ACKNOWLEDGEMENTS

Colin Bateman
Marina Silva
Iain Hall
Integrated Global Services

REFERENCES

1. Patrick J Christensen, William H Graf and Thomas W Yeung, "Refinery power failures: causes, costs and solutions," https://www.digitalrefining.com/article/1000773/refinery-power-failures-causes-costs-and-solutions#.YShegYgzY2w (27/08/2021).
2. T. Shrestha, I. Hall, "Corrosion Mitigation Using Thermal Spray Coating", NACE Corrosion2014, San Antonio, TX, USA
3. I. Hall, R. Muir, T. Shreshtha, T. Sabti, S. Dossary, S. Raghunathan, S. Shehri, "Application of HVAS Cladding to mitigate Sulfide Stress Cracking and Hydrogen Induced Cracking in a De-ethanizer Column" NACE Corrosion 2015, Dallas, TX, USA
4. NACE Standard MR0175/ ISO 15156-1:2009, "Petroleum and natural gas industries -Materials for use in H2S-containing environments in oil and gas production.", 2009Edition, NACE International, Houston, TX, USA
5. NACE Technical Report 1F192 (1993 Revision), Use of Corrosion Resistant Alloys in Oilfield Environments, NACE, Houston, TX, USA.

Author Details

	Phil Williams is the corresponding author of this paper. Phil Williams is the Director of Business Development – Oceania a position he has held since November 2020. Working for Integrated Global Services, Phil is a AMPP Senior Certified Coating Inspector and has over 35 years of technical sales experience, with more than 20 years in corrosion and surface technology management. He is responsible for business development and client facing services across Oceania and the Philippines, providing asset integrity and thermal efficiency solutions for mission critical equipment in the oil and gas, power, petrochemical and pulp and paper industries.
	Peter Johnson is Director – APAC Business Development, a position he has held since July 2020. Working for Integrated Global Services, Peter is a AMPP Senior Certified Coating Inspector with more than 15 years' experience in the corrosion and surface technology sector. He is responsible for client facing services in Southeast Asia, providing asset integrity and thermal efficiency solutions for mission critical equipment in the oil and gas, power and petrochemical industries

CORROSION & PREVENTION
CAIRNS 2024

THE AUSTRALASIAN CORROSION ASSOCIATION INC.

Impact of Corrosion on The Compression Strength of Steel Wall Ties Within a Brick Veneer Wall System

L. Terry[1,*], S. Gillogly[1], I. A. Chaves[1], M. Masia[1]

[1]*Centre for Infrastructure Performance and Reliability – The University of Newcastle, Australia, 2308*

**lyndsey.terry@uon.edu.au*

ABSTRACT

Steel wall ties are an important structural component of masonry veneer and cavity brick walls, connecting the external leaf of masonry to the internal load-bearing frame and transferring lateral forces from wind and earthquakes. Corrosion of steel wall ties diminishes their effectiveness by reducing their cross-sectional area, compromising their strength when subject to lateral forces. The experimental work conducted for this study entailed the compression testing of artificially corroded wall tie samples with known section loss. The artificially corroded wall ties were constructed into a brick couplet and timber subassembly system to replicate a veneer wall. The findings reveal a decline in strength and change in the failure mechanism of the wall tie compared to existing studies, with increasing corrosion-induced section loss, highlighting the clear impact that corrosion of steel wall ties has on the performance of brick veneer wall systems.

Keywords: steel wall ties, compression strength, brick veneer, corrosion

Introduction

Masonry veneer wall systems have been used around the world, primarily due to their aesthetic, thermal and durability functional features [1]. A veneer wall system consists of a non-structural exterior masonry wall designed to support its self-weight, connected to a structural backing system with steel wall ties across an air cavity. The connection between an external leaf of masonry and the internal framework, such as timber, steel or masonry, is critical to prevent failure of the structure when exposed to lateral forces. For example, strong winds or an earthquake will subject the brick veneer wall ties to both tension and compression loading, along with cyclic loading. The work presented in this paper reports on the behaviour and strength of brick-tie-timber subassemblies subjected to compression testing. Figure 1 (A) illustrates the brick veneer wall as a subassembly, including a single connection comprising the timber frame, wall tie, brick couplet and mortar joint. Figure 1 (B) displays the configuration used in this experimental study, featuring metal wall ties that were artificially corroded.

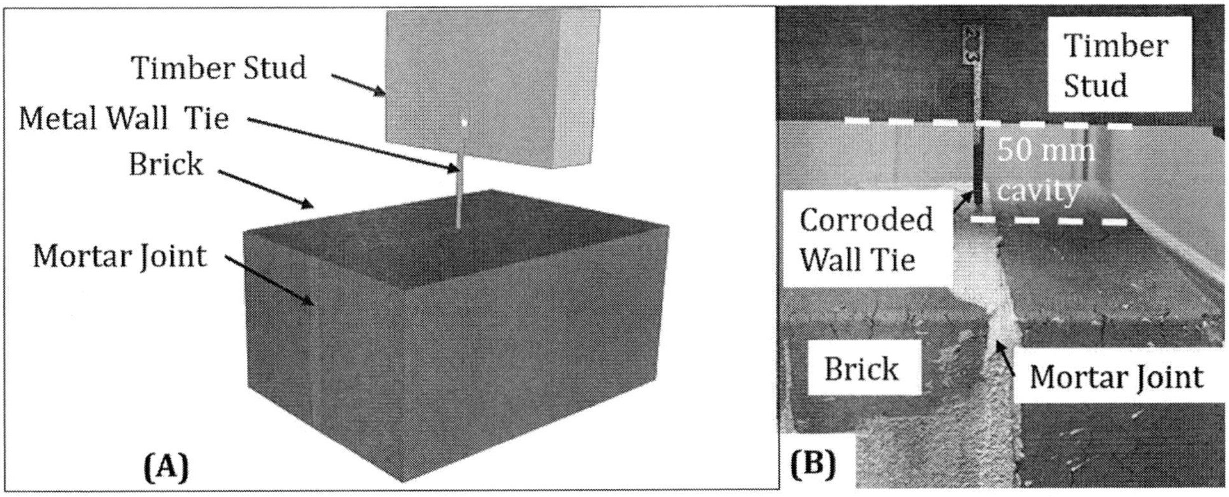

FIGURE 1: BRICK VENEER SUBASSEMBLIES (A) COMPRISING BRICK COUPLET, WALL TIE AND TIMBER STUD [2] AND (B) INCLUDING BRICK COUPLET, ARTIFICIALLY CORRODED WALL TIE AND TIMBER STUD.

Evaluating the condition of brick veneer wall ties can be difficult and is typically only possible through destructive techniques. Without proper condition maintenance of wall ties, corrosion remains undetected and will lead to the deterioration of the wall ties. If wall ties in a brick veneer wall system are compromised due to corrosion or even poor construction, the connection between the external leaf of masonry and the internal timber framework may become ineffective [3]. Such failure in brick wall systems can have serious consequences. After the 1989 Newcastle earthquake, Melchers and Page documented numerous instances of failed brick wall systems where the external leaf of masonry was disconnected from the internal framework, causing collapse. The wall ties in the failed brick walls investigated after the earthquake were found to have either experienced severe corrosion losses at the mortar interface zone or had been installed incorrectly [4, 5]. Further, in 2015 a brick building in Bar Beach, Newcastle failed after being subject to severe wind loading which caused the wall to collapse. In the debris of this collapsed wall, it was found that corrosion staining was present on the mortar, signifying that corrosion had led to the failure of this system [6]. The findings from the 1989 Newcastle Earthquake and recent works indicate that severe wall tie corrosion is evident in the mortar interface zone [7]. In a brick veneer wall, this zone refers to a 30 mm section of the wall tie that extends from the embedment zone in the external leaf of masonry to the air cavity. By concentrating on the mortar interface zone of the wall tie, the current study emphasises the effects of localised steel wall tie corrosion on the compressive strength of brick veneer wall subassemblies.

Previous studies by Muhit (2021) and Choi et. al. (2004) have examined wall ties under compression in subassembly systems similar to that shown in Figure 1. Their findings indicate that the primary failure mechanism in compression are the buckling of the wall tie and nail pull-out. Muhit found that in static compression the dominant failure pattern in 23 out of 25 specimens was wall tie buckling near the timber connection, while some wall ties failed by a combination of wall tie buckling and nail pull-out from the timber. The wall ties used were R4 stainless steel light duty side fixing wall ties and the results showed a mean peak compression load of 1.04 kN and corresponding buckling displacement of 3.08 mm [2]. In contrast to Muhit's later work, Choi et. al. analysed brick-tie-timber subassemblies using wall ties with different embedment lengths and thicknesses. Their research revealed that the compressive strength of wall ties depended on their thickness, with thicker ties yielding higher average buckling loads in compression [8].

Understanding the impact that corrosion has on the compressive capacity of wall ties in a brick-tie-timber subassembly is important to understanding the capacity of brick veneer walls when subject to lateral loading. This paper presents data relating to the corrosion of wall ties in a subassembly system similar to that of Muhit's (2021) study (see Figure 1). This work is an extension of previous findings regarding the impact of corrosion on the tensile strength of wall ties. Terry et. al. (2023) conducted a study involving brick-tie-timber subassemblies equipped with galvanised R2 light-duty side-fixing wall ties. The galvanised wall ties used were subjected to artificial corrosion at the mortar interface zone and subsequently tested under tension as part of the subassembly. The results indicated a significant reduction in tensile strength as the localised section loss increased at the mortar interface region of the wall tie [9].

The primary failure mechanism for wall ties under compression loading is axial buckling near the timber connection or nail pull-out [2]. However, it was hypothesised that introducing corrosion at the mortar interface zone, which leads to localised section loss of the wall tie, may shift this failure mechanism. Specifically, it was proposed that the mortar interface zone that has been artificially corroded in this study would become the weakest point in the system. The goal of this study is to determine the percentage of localised section loss necessary for wall ties to fail in compression at the mortar interface, rather than buckling at the timber connection or pulling out from the timber stud. Additionally, this research aims to quantify the strength reduction of the wall tie resulting from the accelerated corrosion induced section loss.

METHODS AND MATERIALS

ELECTROLYSIS

Electrolysis is a form of accelerated corrosion that utilises an electrolytic current to corrode metal in a solution of sea water. The wall ties used for this study were galvanised R2 corrugated light-duty side-fixing wall ties. These wall ties match the shape and size of the R4 stainless steel wall ties previously tested by Muhit, however, are different in their material. The galvanised wall ties were selected for this study as the electrolysis was unlikely to corrode the stainless-steel material as significantly as the galvanised material. The methodology employed to artificially corrode the wall tie samples using electrolysis was replicable to that completed in the previous work by Terry et. al. (2023). Electrolysis was used as a form of accelerated corrosion using a Tenma Dual DC 0-30 volts regulated power supply, aligning with ASTM G71-81 standards [10]. The method of electrolysis is known to have some inconsistencies, particularly regarding the effectiveness of replicating natural corrosion mechanisms, a phenomenon confirmed by Nascimento [6]. ASTM G31-21 also indicates that electrolysis can sometimes be misleading because metals and alloys do not all respond similarly to the factors that affect corrosion, resulting in the tests providing "indicative results" only [11]. In this way, the electrolysis method was not employed to replicate natural corrosion mechanisms and was instead used as a method of stripping the wall tie of its mass or section at a localised region along the wall tie. Additionally, other means such as mechanically or chemically stripping the wall tie of cross section were not feasible and not as traceable as electrolysis was. The test set-up specified in ASTM G71-81, and illustrated in Figure 2 (A), was utilised for all tests setting the power supply at 3 amps for all samples. The R2 lightly galvanised wall tie samples were used to represent the anode, while aluminium was chosen as the cathode, producing the lowest standard potential when in the presence of galvanised coating of the wall tie (zinc) in the galvanic cell [12].

The mortar interface region of a wall tie has been found to show severe levels of corrosion in existing or failed structures [4-6]. A study investigating the impact of inter-cavity corrosion on metallic wall ties in masonry structures through utilising accelerated corrosion testing with an electrolysis corrosion cell, also reports on greater corrosion at the ~30 mm mortar interface zone of the wall tie [7]. For this experimental study, a trial wall tie was first exposed to electrolysis for 30 minutes. It was found that when the entire wall tie is exposed to corrosion through electrolysis it is still localised what is hypothesised to be the weakest point along the wall tie (within the mortar-interface zone as circled in Figure 2(B)). From the knowledge of existing literature and the trial electrolysis test, the wall tie samples in this study were then corroded using electrolysis at the mortar interface zone only. Figure 2 (A) shows an example of the electrolysis cell with the wall tie being artificially corroded. Figure 2 (B) provides an example of the results for three of the wall tie samples exposed for 15 minutes in electrolysis showing the 30 mm mortar interface zone that was corroded. As seen in Figure 2 (A), the mortar embedded zone was wrapped in electrical tape to avoid section loss from electrolysis in this region.

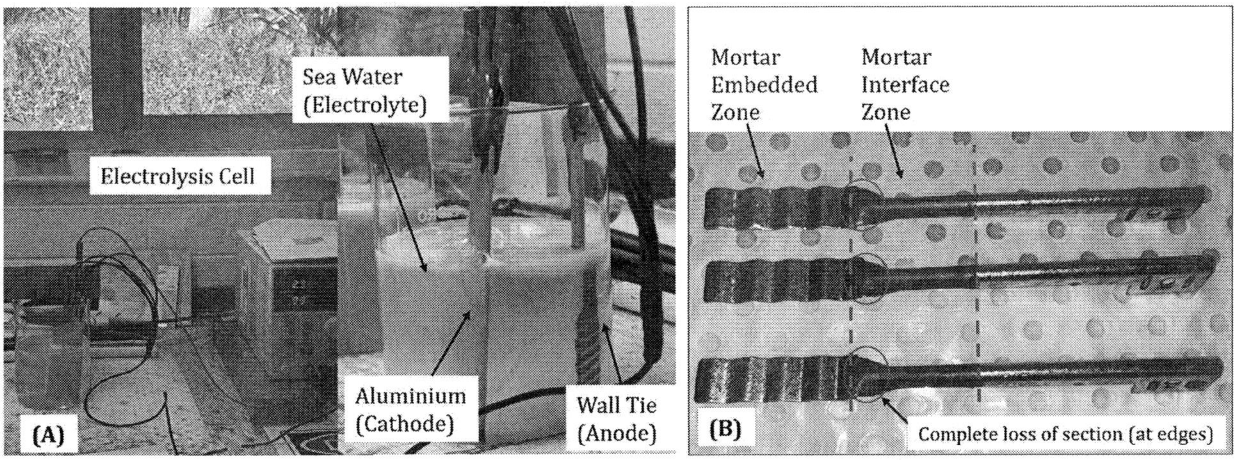

FIGURE 2: (A) ELECTROLYSIS CELL SET-UP AND (B) EXAMPLE OF RESULTS OF ELECTROLYSIS FOR THREE WALL TIE SAMPLES (ID 10, 11, AND 12) SHOWING LOSS OF ZINC COATING AND SECTION LOSS IN THE MORTAR INTERFACE ZONE OF THE WALL TIE WITH COMPLETE LOSS OF SECTION (CIRCLED) AT THE EDGES.

Based on the results of the previous study that also used the technique of electrolysis to artificially corrode wall ties in a brick-tie-timber subassembly [9], it was hypothesised that 15 minutes of electrolysis at the interface zone would likely cause a reduction in compressive strength and a changed characteristic failure mode of the wall ties. The wall ties used in this study were grouped based on a range of planned exposure times in the electrolysis cell, with increments of 2.5 minutes to allow for a systematic comparison of compressive strength as the section loss increased. Following electrolysis, the wall ties were constructed into brick-tie-timber subassemblies, referred to as "samples" in this paper. Each exposure time was linked to a specific 'level' of corrosion, with results detailed in Table 1, which shows the average localised section loss of each wall tie. To measure localised section loss, the mass of each wall tie was recorded immediately before and after electrolysis. The localised section loss percentage was then calculated by multiplying the mass loss by the fraction of the mass of the mortar interface region (approximately 30 mm) to the mass of the full wall tie. The data in Table 1 indicates a linear rate of corrosion, typical of electrolysis; however, it is crucial to note that, as established by Melchers, the corrosion of steel is generally non-linear [13]. Therefore, the mass loss results from the electrolysis process used in this study may not accurately reflect natural corrosion behaviour and care should be taken when comparing the results of section loss shown in Table 1 and in-situ corrosion loss findings.

TABLE 1: LOCALISED SECTION LOSS AS A PERCENTAGE OF MASS LOSS MEASURED FOR EACH INDIVIDUAL WALL TIE SUBJECT TO ELECTROLYSIS AT THE MORTAR-INTERFACE ZONE.

Time in electrolysis cell (mins)	Sample ID	Localised section loss (%)	Average localised section loss (%)
0	1	0.00%	0.00%
	2	0.00%	
	3	0.00%	
5	4	14.81%	14.86%
	5	14.91%	
	6	14.86%	
10	7	28.51%	28.58%
	8	28.46%	
	9	28.76%	
12.5	10	35.44%	35.17%
	11	34.93%	
	12	35.54%	
	13	35.08%	
	14	34.88%	
15	15	41.41%	41.04%
	16	41.01%	
	17	41.46%	
	18	40.55%	
	19	40.76%	
17.5	20	48.59%	48.16%
	21	47.98%	
	22	47.93%	
	23	48.28%	
	24	48.03%	

COMPRESSION TESTING

Following electrolysis, each of the 24 wall ties were constructed into brick-tie-timber subassembly samples. The construction included embedding the wall tie into the 10 mm thick mortar layer between two standard clay bricks with dimensions 230 mm x 110 mm x 76 mm. The mortar used was M3 general purpose mortar specified in AS3700, which consisted of a ratio of 1 part cement to 1 part lime to 6 parts sand (by volume). With a 50 mm cavity achieved, the wall tie was then side fixed by nail into an MGP10 grade timber stud approximately 150 mm in length. The brick-tie-timber subassembly samples were constructed to match the subassemblies created for Muhit's study [2] and the more recent study on the tensile strength of wall ties subject to artificial corrosion [9].

After 7-days of curing, the subassemblies were subject to a standard compression test using a 100 kN MTS machine. The subassembly was placed into the machine and the timber stud was clamped to the fixture of the MTS load cell. The load cell was displaced at a rate of 1 mm/min. The force was increased steadily until the wall tie failed, in accordance with AS2699.1, and the resultant data was recorded for analysis. AS2699.1 specifies that the strength of individual wall ties can be taken as the force that induces the failure of the wall tie [14].

RESULTS

The results presented herein provide a direct correlation between the amount of localised section loss of the wall tie and the strength of the wall tie in compression. The results presented in this section highlight the change in strength due to corrosion, showcasing the change from the characteristic failure mechanism previously defined by Muhit [2]. Table 2 provides a summary of the peak compression load and corresponding displacement of each sample tested. For each of the sample sets, an average peak load and corresponding displacement is also shown.

TABLE 2: COMPRESSION TESTING RESULTS FOR ALL SAMPLES TESTED.

Time in electrolysis cell (mins)	Sample No.	Peak Load (kN)	Displacement at Peak Load (mm)	Average Peak Load (kN)	Average displacement (mm)
0	1	-1.07	-4.10	-1.12	-4.76
	2	-1.14	-1.82		
	3	-1.15	-8.35		
5	4	-1.16	-3.67	-1.13	-3.56
	5	-1.15	-3.18		
	6	-1.08	-3.82		
10	7	-0.88	-4.20	-1.16	-4.81
	8	-1.40	-4.95		
	9	-1.19	-5.27		
12.5	10	-1.14	-6.78	-1.10	-3.63
	11	-1.33	-4.48		
	12	-0.85	-2.28		
	13	-1.26	-2.58		
	14	-0.90	-2.03		
15	15	-1.11	-3.78	-1.12	-4.12
	16	-0.98	-1.37		
	17	-0.99	-3.05		
	18	-1.17	-7.32		
	19	-1.35	-5.08		
17.5	20	-1.28	-6.35	-0.82	-1.85
	21	-0.87	-2.20		
	22	-1.21	-5.73		
	23	-0.68	-1.70		
	24	-0.90	-1.65		

The compression testing completed for each of the sample sets were compared to understand and document any changes in the wall tie behaviour as the localised section loss due to corrosion increased. The 'ideal' curves shown in Figure 3 document the results presented in Table 2 and show the behaviour of the wall ties in compression with the load as a function of the displacement. Notably, both the load and displacement values in Table 2 represent the compression values (as negatives) showing how far the wall tie had displaced from its original cavity width of 50 mm. Figure 3 represents the results of each sample set with an ideal curve and highlights any difference between sets (based on the time in the electrolysis cell).

The ideal curves for each sample set shown in Figure 3 were established based on the average of the load-displacement relationship at specific points, in which the method Muhit (2021) used was replicated and adapted to accommodate the data presented in this report. This methodology included plotting five points on a curve with four distinct zones for all 24 samples. At each point, the data in the sample set was averaged to create the ideal curve. Point 0 was the origin point. The elastic linear region (zone 1) was between points 0 and 1, where Point 1 represents the yield point at a set value of 1 mm displacement. Zone 2 is a yielding stage between the yield point and the peak buckling load (Point 2). The peak buckling load (kN) indicates the maximum force the wall tie can withstand before the failure mechanism occurs. The third zone is a decreasing stage between the peak load (Point 2) and the load at 7 mm displacement (Point 3), indicating plastic deformation. The final zone is between 7 mm displacement (Point 3) and 10 mm displacement (Point 4), depicting the line continuously slightly decreasing, showing the full effect of failure. The ideal curves were created for all the sample sets and are shown in Figure 3. The ideal curve for the 17.5 minute sample set was created using a similar methodology to the other sample sets, however, only includes the yield point (Point 1), peak buckling load (Point 3) and decreasing load at 3 mm displacement (Point 3) for three out of the five samples. Two samples were categorised as outliers in this set. Figure 3 includes each sample set ideal curve and corresponding error bars to represent the +/- 1 standard deviation at each point.

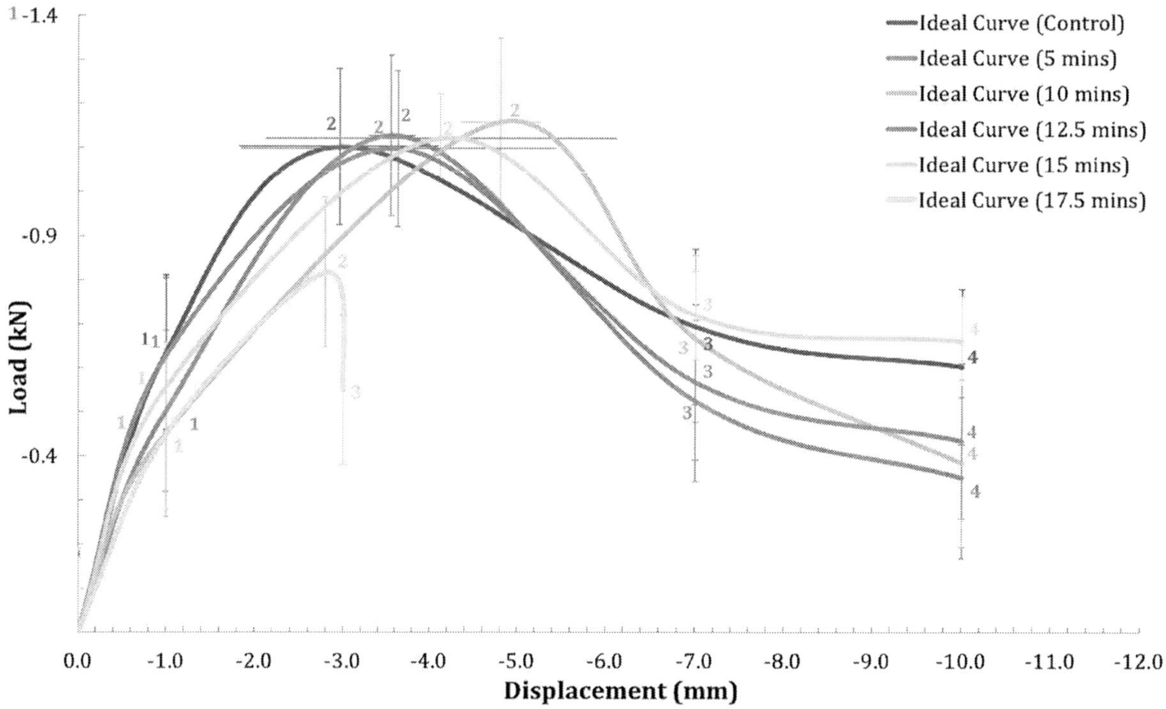

FIGURE 3: COMPARISON OF IDEAL CURVE RESULTS FOR EACH OF THE SAMPLE SETS BASED ON THE TIME IN ELECTROLYSIS CELL.

The primary failure modes of the 24 subassembly samples included axial buckling of the wall tie near the timber connection, pull out of the nail from the timber and buckling of the wall tie at the interface zone. The aim of the compression testing was to understand at what level of corrosion loss the failure mechanism changed to buckling of the wall tie at the interface zone and reduce the peak compression load. Figure 4 provides examples of the three failure modes observed in this study including failure by buckling or nail pull-out. Figure 4 (C) highlights the failure mode of buckling at the mortar interface zone of the wall tie, causing fracture of the wall tie. The wall tie samples that failed similarly to sample 23 in Figure 4 (C) had decreased peak compression load readings, this is highlighted in Figure 5. These included samples that had undergone 17.5 minutes in electrolysis. Four out of five samples that were artificially corroded for 15 minutes also failed by buckling at the mortar interface zone, however, did not significantly decrease in compression strength. Therefore, between approximately 41-48% and beyond 48% localised section loss, the wall tie is likely to decrease in peak buckling load, as illustrated in Figure 5.

FIGURE 4: EXAMPLE OF FAILURE MODES (A) BUCKLING NEAR TIMBER CONNECTION, (B) NAIL PULL-OUT AND (C) BUCKLING AT CORRODED REGION (MORTAR INTERFACE ZONE) RESULTING IN FRACTURE OF WALL TIE.

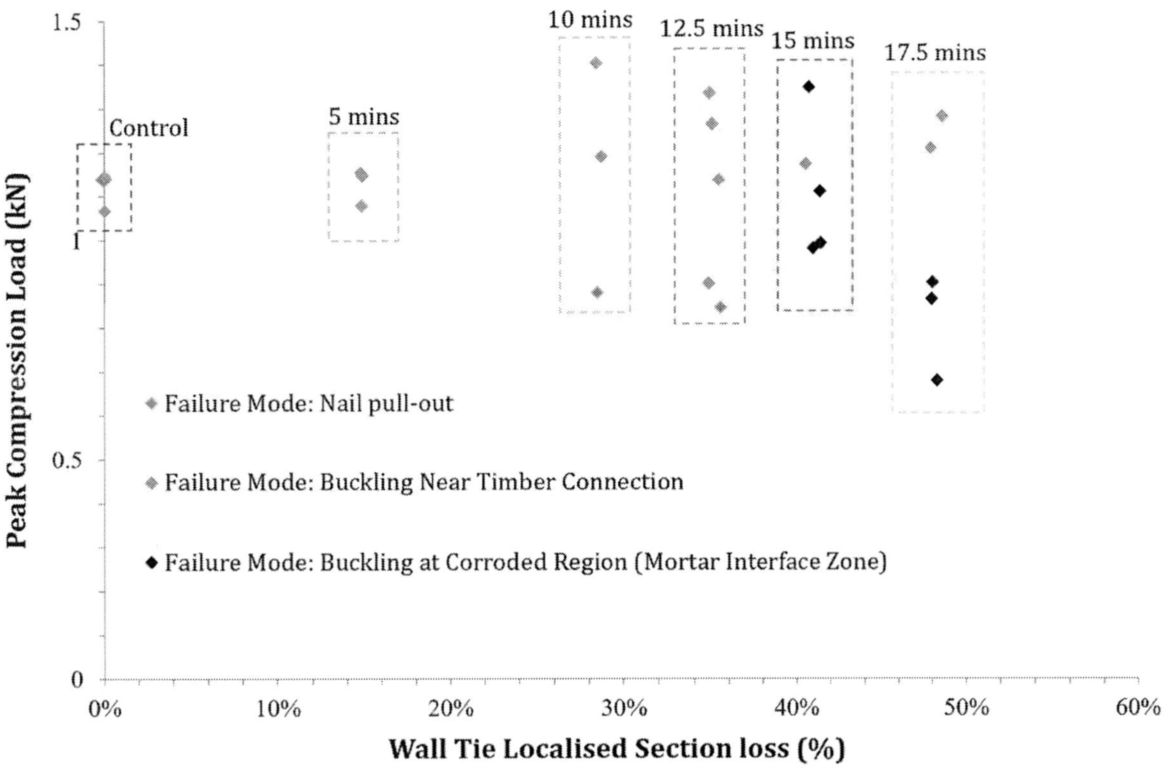

FIGURE 5: PEAK COMPRESSION LOAD VS WALL TIE LOCALISED SECTION LOSS BASED ON TIME (MINS) IN THE ELECTROLYSIS CELL.

The primary finding of this study is illustrated in Figure 5, which depicts the relationship between the peak load and localised wall tie section loss due to electrolysis. The results indicate that when the localised section loss at the mortar interface reaches approximately 41-48%, the peak load can decrease by up to 25%. Furthermore, this level of section loss alters the characteristic failure mode, shifting from the standard failure mechanisms to buckling at the mortar interface zone.

Discussion

The results presented in this study highlight that as the localised section loss of a wall tie increases, the strength in compression decreases. The decrease in peak buckling load occurs when failure by buckling of the wall tie is at the mortar interface zone, as shown in Figure 5 for several samples in the 15 minute and 17.5 minute sample sets. Figure 4 (C) shows the buckling failure occurring near the mortar embedded region of the wall tie, a segment which is shown to have points of complete section loss at the edges of the wall tie in Figure 2 (B). It is likely that the localised section loss applied through electrolysis at the mortar interface zone caused this section to be the weakest point of the wall tie and hence induced failure by buckling when subject to compression loading.

For the 15-minute samples, the dominant failure mechanism shifted from near the timber connection to the mortar interface zone. However, the ideal curve for these samples in Figure 3 shows slightly increased values compared to the control ideal curve, indicating that their compressive strength has not decreased. This observation is further supported by Figure 5, which demonstrates that the average peak load used to create the ideal curve is similar to that of other sample sets with lower localised section loss. Therefore, this study concludes that a localised section loss of 41% does not significantly affect the overall strength of the wall ties; however, it does lead to a change in the failure mechanism, causing it to shift to buckling at the mortar interface zone of the wall tie.

The ideal curve for 17.5 minute samples showed distinct behaviour compared to the other ideal curves (see Figure 3). This ideal curve was based only on the results of samples in the set that failed due to the buckling of the wall tie at the mortar interface region. The corrosion due to electrolysis is shown to alter the material properties of these three samples, making them less ductile. Figure 5 further illustrates a decrease in peak load for the 17.5 minute samples that failed by buckling at the mortar interface zone, with a reduction of approximately 25% compared to the control and 15 minute samples. This indicates that a localised section loss of approximately 48% leads to a change in the failure mechanism with the mortar interface becoming the critical zone influencing the strength of the subassembly. Additionally, Figure 5 shows that two of the five samples in the 17.5 minute sample set and one of the 15 minute sample sets failed by buckling near the timber connection. These three samples likely failed by buckling near the timber connection. This variation is likely due to material inconsistences, as masonry is known for its high variability. Overall, the samples in the 17.5 minute sample set that failed by buckling at the mortar interface zone exhibited reduced compressive strength. It is likely that as the localised section loss continues to increase beyond 48%, the peak compression load will continue to decrease.

Static compression loading in a brick veneer wall is unlikely to occur in the way that the compression test was conducted for this study. Instead, brick veneer wall ties may typically experience cyclic loading due to lateral forces from wind or earthquakes. For example, during an earthquake the wall tie may be subject to cyclic tension and compression loading as it shakes the structure, transferring loads between the external leaf of masonry and the internal framework. The samples tested in this study that experienced failure by buckling at the mortar interface zone saw the fracture of the wall ties (see Figure 4 (C)). If these samples were then subject to tension forces, the strength would be diminished and connection would deteriorate. Therefore, based on the results obtained in this study and existing literature [3, 8], it can be inferred that where a wall tie has experienced severe localised corrosion losses and is subject to cyclic compression and tension loading (depending on the load magnitude), the strength will be significantly decreased and may cause the structure to collapse.

CONCLUSIONS

This experimental study examined the impact of localised corrosion on the failure mechanism and strength of brick veneer wall ties under static compression loading. The analysis included 24 brick-tie-timber subassembly samples, featuring six sets of R2 light-duty galvanized wall ties exposed to varying timings of accelerated corrosion in an electrolysis cell, resulting in different recorded levels of localised section loss. The findings revealed that below a localised section loss of 35.2%, the subassembly predominantly failed by axial buckling of the wall tie near the timber connection. Between 35.2% and 41% localised section loss, the failure mechanism shifted to axial buckling at the corroded mortar interface (30 mm zone). Between 41% up to 48.2% localised section loss, the primary failure mode became buckling at the mortar interface, with peak buckling loads reduced by up to 25%.

ACKNOWLEDGMENTS

The authors acknowledge the continued financial support provided by The University of Newcastle School of Engineering and the Australian Research Council via Linkage Project LP220100028, as well as the technical guidance and advice from Prof. Robert Melchers and Dr Robert Petersen. The authors also acknowledge the unwavering technical guidance provided by the technical staff in the Civil Engineering Laboratory at the University of Newcastle.

References

1. Martins, A., G. Vasconcelos, and A. Campos Costa, Brick masonry veneer walls: An overview. Journal of Building Engineering 9 (2017) 29-41.
2. Bin Muhit, I., Stochastic assessment of unreinforced masonry veneer wall systems subjected to lateral out-of-plane loading, University of Newcastle, 2021.
3. Bin Muhit, I., M.J. Masia, and M.G. Stewart, Failure analysis and structural reliability of unreinforced masonry veneer walls: Influence of wall tie corrosion. Engineering Failure Analysis, 2023. 151(9).
4. Melchers, R.E. and A.W. Page, The Newcastle earthquake, in Proceedings of the Institution of Civil Engineers: Structures and Buildings. 1992. p. 143-156.
5. Page, A.W., et al., Structural aspects of the Newcastle earthquake. The Institution of Engineers Australia Structural Engineering Conference, 1990: p. 305-312.
6. Jardim do Nascimento, B., Predicting masonry brick-veneer and cavity brick wall-tie corrosion, University of Newcastle, 2021.
7. Chaves, I. A., Melchers, R. E., Jardim Do Nascimento, B., Philips, J., & Masia, M., Effects of inter-cavity corrosion on metallic wall ties in masonry structures. AIMS Materials Science 9(2) (2022) 311-324.
8. Choi, Y.-H. and J.M. LaFave, Performance of corrugated metal ties for brick veneer wall systems. Journal of Materials in Civil Engineering 16(3) (2004) 202-211.
9. Terry, L., Gillogly, G., Chaves, I. A., Masia, M., & Petersen, R., Impact of corroded galvanised wall ties on the structural integrity of masonry veneer walls, (In) Proc. 26th Australasian Conference on Mechanics of Structures and Materials (ACMSM26), Dec 3-6, 2023, Auckland, New Zealand.
10. ASTM G71-81, Standard guide for conducting and evaluating galvanic corrosion tests in electrolytes.
11. ASTM G31-21, Standard guide for laboratory immersion corrosion testing of metals.
12. LibreTexts, Standard Potentials see: https://chem.libretexts.org/Bookshelves/Analytical_Chemistry/Supplemental_Modules_(Analytical_Chemistry)/Electrochemistry/Redox_Potentials/Standard_Potentials. Last accessed 5 June 2024.
13. Melchers, R.E., Progress in developing realistic corrosion models. Structure and infrastructure engineering 14(7) (2018) 843-853.
14. AS2699.1:2020, Built-in components for masonry construction, Part 1: Wall ties. 2020.

AUTHOR DETAILS

	L. Terry is the corresponding author of this paper. She is a PhD student at the University of Newcastle since 2023. She is a Civil and Environmental engineer with 6 years experience in the industry as a student engineer. Her research topic is in the long-term corrosion of metal wall ties in brick cavity and veneer walls.
	S. Gillogly is a fourth-year student at the University of Newcastle studying a Bachelor of Civil Engineering and a Bachelor of Business. She has 3 years experience in the industry as a student engineer.
	Igor A Chaves is an Associate Professor of Civil Structural Material Science at The University of Newcastle, Australia. He is a civil engineer, holds a Masters degree in structural engineering and a PhD for his studies in corrosion analysis and prediction. His research interests also include structural reliability and risk assessment of infrastructure. Recent awards include Excellence in Student Experience (2021, 2022) and Industry Research Engagement (2021) from The University of Newcastle. He is the recipient of the 2017 & 2012 Marshall Fordham Award (Australasian Corrosion Association) and the 2016 Brian Cherry International Scholarship award.
	Mark Masia is a Professor in Structural Engineering, and Director of the Centre for Infrastructure Performance and Reliability, School of Engineering at The University of Newcastle. He has a special interest in the structural performance, assessment and design of masonry structures.

PROTECTION OF STEEL AT EDGES AND SCRATCHES BY ZINC COATINGS

R A Francis

Retired consultant, Victoria, 3147, Australia

ABSTRACT

Zinc coatings, whether metal coatings such as galvanizing or thermal spray, or zinc-rich paints, provide protection through a combination of galvanic and barrier protection. Barrier protection takes place where the coating is sound and intact, but galvanic protection generally only operates at cut edges, scratches or other regions where the underlying steel is exposed. The amount of galvanic protection can vary greatly depending on a number of factors reviewed in this paper. Analysing these factors in conjunction with research carried out has enabled a proposed mechanism of protection to be suggested that is somewhat more involved than simple cathodic protection typically presented. A topcoat further modifies the process of zinc protection and a mechanism is suggested.

Keywords: zinc coatings, zinc-rich paint, edge protection, galvanic protection

INTRODUCTION

Zinc coatings provide excellent barrier protection in many corrosive environments while the coating is intact. However, the unique property of zinc is that it provides sacrificial protection to steel in areas where the steel is exposed, such as in scratches or at cut edges. This is the property that makes zinc coatings, whether galvanized, thermal spray or zinc-rich paints (ZRPs), widely specified as a single coat or as part of a multi-coat protective system. However, the edge protection provided by zinc can range from excellent to poor (see Figure 1). This is because the efficiency and extent of this protection depends on many factors, some inter-related, and these are reviewed in this paper. The paper concentrates on practical issues and aspects of this protection, but does briefly look at some fundamental features. Background and other information on zinc coatings has been covered by the author in earlier papers, including a general review of protection by zinc-rich paints [1], laboratory and exposure test methods in [2] and the effect of fillers or extenders in [3].

This protection is often known as cut edge or scratch protection, but will be referred to as 'scribe protection' in this paper, as it is often investigated by cutting an intentional scribe on a coated test panel. It should be noted that some studies investigate the early stages of zinc protection by looking at the development of rust in the scribe, while most investigate later stages where the coating at the edges of the scribe lifts, and the extent of creep (also known as rust creepage, cutback or underfilm delamination) is measured. The amount of rust in the scribe or extent of creep enables quantification of scribe protection provided by a zinc coating. Undamaged or intact zinc coatings will rely on the barrier protection of the coating for durability and, in severe environments, this may limit coating life. The factors which determine the barrier protection offered by these coatings are not discussed in this paper, but are likely to be quite different from those influencing scribe protection.

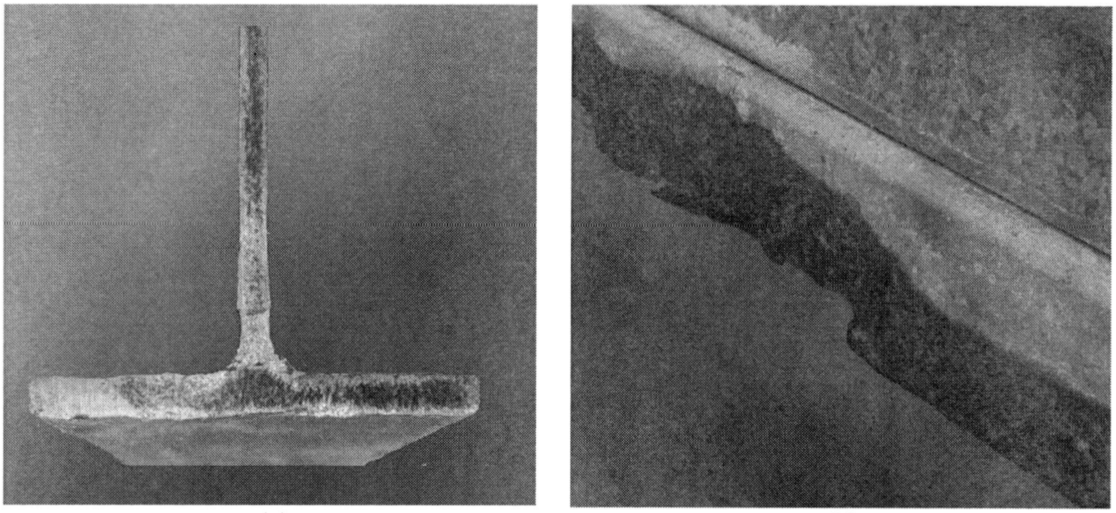

(a) (b)

FIGURE 1:EDGE PROTECTION OF BARE STEEL BY ZINC CAN RANGE FROM EXCELLENT TO POOR.

EFFECT OF ZINC COATING TYPE

While all zinc coatings will provide scribe protection, they function differently. An investigation was carried out by Lindqvist *et al* [4] who compared the time it took for rust to form in exposed steel edges with a range of zinc coatings in a sodium chloride solution. The coatings investigated were hot dip galvanizing, inorganic zinc and epoxy zinc (single coat and topcoated) paint coatings along with a conventional epoxy paint coating (without zinc) for comparison. Figure 2 shows the time required for first visible signs of steel corrosion to appear at the unprotected edges of the plates of varying thicknesses (note the logarithmic vertical scale to compress the time scale). Times ranged from a few hours for unpigmented epoxy to almost a year for hot dip galvanizing. Between these extremes, uncoated epoxy zinc coating lasts a week or two, while uncoated inorganic zinc provides 2 to 3 months in this environment. Topcoating these primers greatly reduces the time of protection and width that can be protected as there is much less available zinc for protection. The width of the steel edge to be protected is important for the topcoated ZRPs, but less important for the other coatings in this investigation.

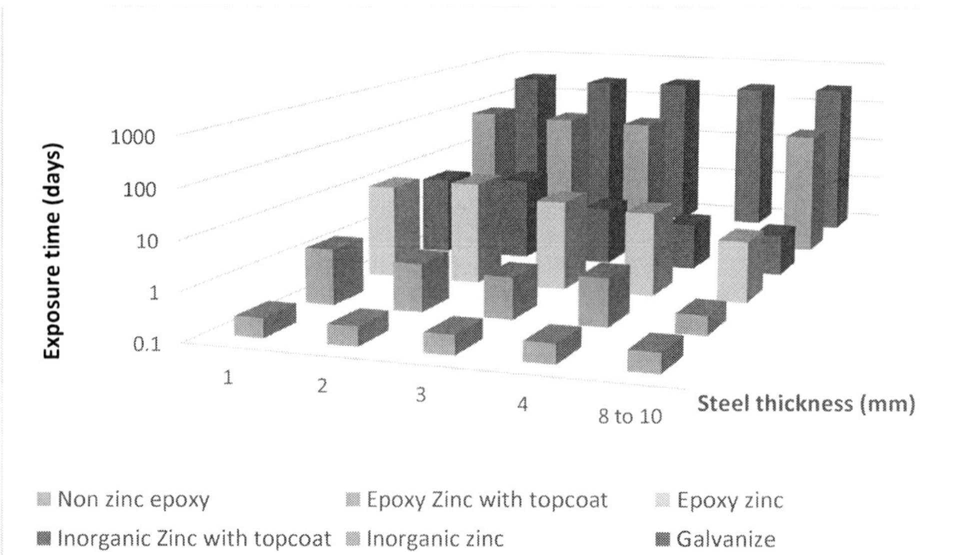

FIGURE 2: TIME TO FIRST RUST APPEARANCE ON EDGES OF STEEL OF DIFFERING THICKNESS IN A NaCl SOLUTION FOR DIFFERENT ZINC COATINGS AND A NON-ZINC EPOXY FOR COMPARISON.

Effect of the Environment

The above work showed the performance of zinc coatings in a highly-conductive salt solution, but different environments will influence the extent of sacrificial protection because of varying electrical resistance. Figure 3, based on the work of Zhang and Xing [5], shows schematically the 'protection distance' provided by a metallic zinc coating to an adjacent steel surface as a function of electrical resistance of the environment. Significant protection is achieved with a conducting environment such as full immersion which has a large volume of electrolyte even though the resistivity of the pure water may be high Conductive salt spray also shows a considerable protection. A cyclic environment, such as natural atmospheric exposure, has a limited protection distance as it is dry much of the time, although the drying-out period would influence the nature and protective properties of the corrosion products. Protection distance in a high resistance atmosphere such as 100% relative humidity with a the very thin layer of moisture is very limited. Conductivity of the environment would be expected to have a similar relative effect on ZRPs, although the absolute distance would be different from that provided by metallic zinc.

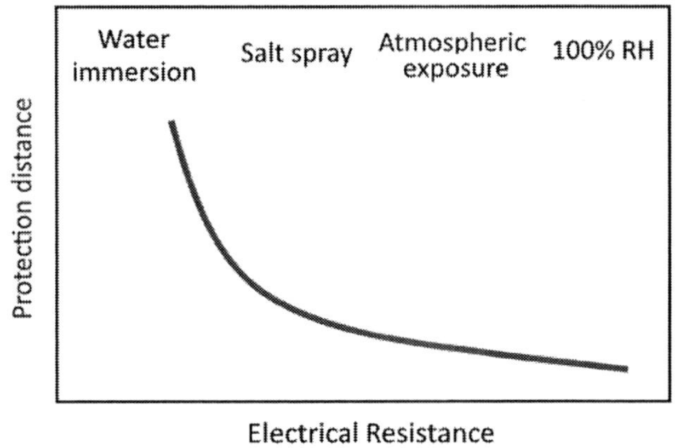

FIGURE 3: SCHEMATIC DIAGRAM OF THE PROTECTION DISTANCE PROVIDED BY A ZINC COATING IN ENVIRONMENTS OF VARYING RESISTANCE.

The environment along with zinc availability is the most likely cause of the difference in behaviour of the two galvanized items shown in Figure 1. Figure 1(a) is a hot dip galvanized section exposed to a mild atmospheric environment, with intermittent wetting and drying. The relatively thick zinc coating, with the wetting and drying, has allowed development of good edge protection, and an adherent, protective film of zinc corrosion products on the exposed steel with limited red iron rust. Figure 1(b), on the other hand, shows a continuously galvanized beam from the roof of a fresh water tank where it would be exposed to almost continuous wetting by condensation of highly corrosive oxygenated water. Protection to the cut edge would be minimal, and the thin zinc coating is consumed as it attempts to protect the exposed steel. Minimal protection by the zinc is provided, the remaining zinc is rapidly used up and the exposed steel immediately corrodes to red iron rust.

Zinc Content of ZRPs

We have seen that, with single-coat ZRPs, inorganic zinc shows superior scribe protection to epoxy zinc, even though the zinc content of these coatings is similar. It is worth investigating these differences further. Weaver *et al* [6] exposed scribed panels with a range of inorganic and organic zinc coatings to the severe marine environment at Kure Beach, North Carolina, and noted the time required (up to a maximum of 123 months) for rust to appear in a 10-centimetre-long scribe (the width was not given). The results, shown in Figure 4, clearly confirm the better performance of inorganic zinc over organic zincs. Furthermore, there appears to be a minimum zinc content for scribe protection of between 70% and 80% by weight for organic zincs, and somewhat less than 50% by weight for inorganic zincs. However, a number of results do not follow this trend and clearly factors other than zinc content and binder type are important. It should be noted that the coatings all had a minimum dry film thickness of at least 50 microns, many much more, and so coating thickness is unlikely to be a factor.

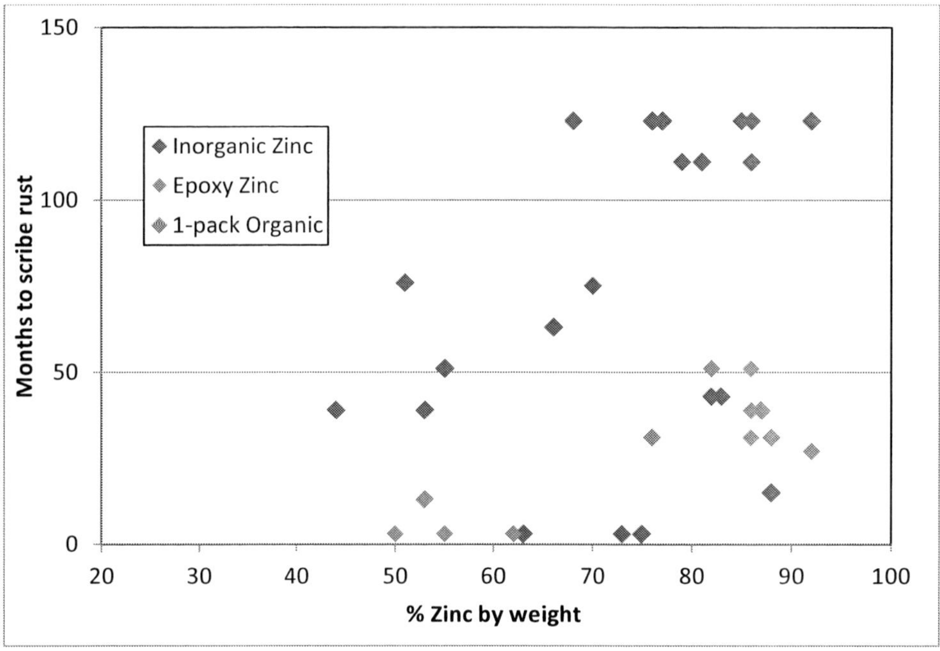

FIGURE 4: TIME TO RUST IN THE SCRIBE AS A FUNCTION OF ZRP TYPE AND ZINC CONTENT FOR ZRP-COATED PANELS EXPOSED AT KURE BEACH.

PERFORMANCE IN MULTI-COAT SYSTEMS

It was shown earlier that topcoating a ZRP significantly reduced the protection provided by the zinc, and it is worth looking further at this aspect. Ulkem and Winter [7] measured the amount of creep in a 2-millimetre-wide scribe on a number of single-coat and topcoated inorganic zinc coatings after 4200 hours of cyclic exposure testing according to ISO 20340, with results shown in Figure 5. The lower the creep distance, the better the performance. The single coat system shows far less rust creep than the multi-coat system. Unlike the results given in Figure 4, this work shows no correlation between creep after testing and the percentage zinc in the dry film, whether topcoated or not. It should be noted that rust creep is not the same as rust in the scribe, and accelerated testing may not degrade the coating in the same manner as natural exposure.

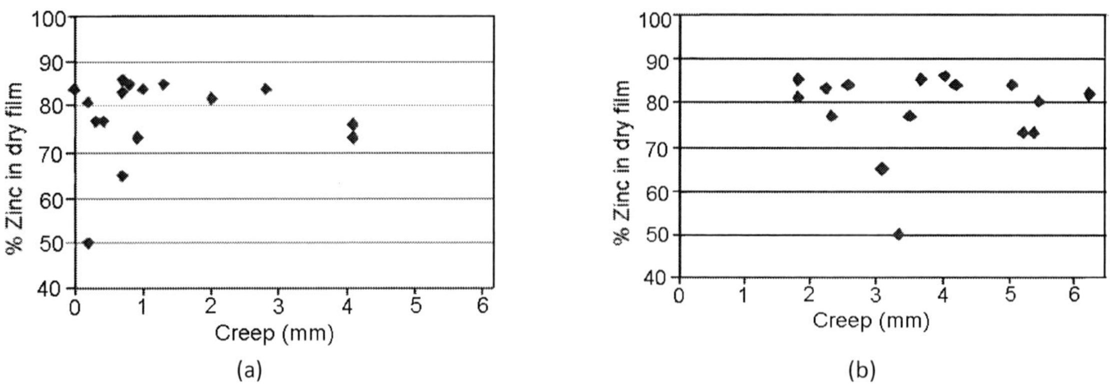

FIGURE 5: EFFECT OF ZINC CONTENT ON AMOUNT OF SCRIBE CREEP IN (A) SINGLE COAT AND (B) TOPCOATED INORGANIC ZINC SYSTEMS AFTER ISO 20340 CYCLIC TESTING.

POTENTIAL AND PH BEHAVIOUR IN SCRIBED ZINC COATINGS

The observations showing factors which influence exposure performance can be complemented by reviewing some fundamental studies. Many investigations into the fundamentals of zinc coatings look at the electrochemical behaviour in a bulk electrolyte solution, but local behaviour such as potential changes at a scribe is likely to be completely different to that which occurs in a bulk solution. Furthermore, reactions taking place on atmospheric exposure which allow build-up of protective corrosion products are likely to be different from those taking place in a bulk electrolyte. However, some investigations can provide a worthwhile insight into corrosion behaviour in a scribe.

A number of researchers, including Saeedikhani *et al* [8], measured the pH changes across a scribe in a high zinc content ZRP during weathering by neutral salt spray. The initial pH was a constant pH6 across the scribe, with the central region of the scribe rapidly increasing to pH9.5 within 2 hours of exposure, increasing to pH11 after 48 hours exposure. It then dropped back to approximately neutral after 168 hours exposure. This suggests fast production of hydroxyl ions in the cathodic region (exposed steel) from the classic cathodic oxygen reduction reaction between oxygen and water as a consequence of rapid anodic corrosion of the zinc. Hydroxyl production then drops away with extended exposure as the zinc corrosion rate falls away. This change in corrosion rate with time is typical of galvanic corrosion [9].

Knudsen *et al* [10] exposed 3-coat paint systems (270 to 320 microns total DFT) with various primers to an offshore environment and determined the average and maximum creep on a scribe with dimensions 3mm wide and 60mm length. The average and maximum scribe creep for each system are given in Figure 6. As expected, the zinc-free epoxy mastic (EM) primer system showed significant creep. The epoxy zinc (ZE) primer systems, however, showed wide variation in average and maximum creep even though they all had similar zinc contents (81 to 86 weight % zinc). The zinc silicate (ZS) primers performed well, with average scribe creep of the order of 1mm. The zinc silicate shop (SP) primers performed almost as well as the standard zinc silicates despite their low zinc content (although the level was not given, it is typically 65 weight % zinc for shop primers), a low thickness of 20 microns and a shot rather than grit blasted surface, all of which would be expected to diminish performance.

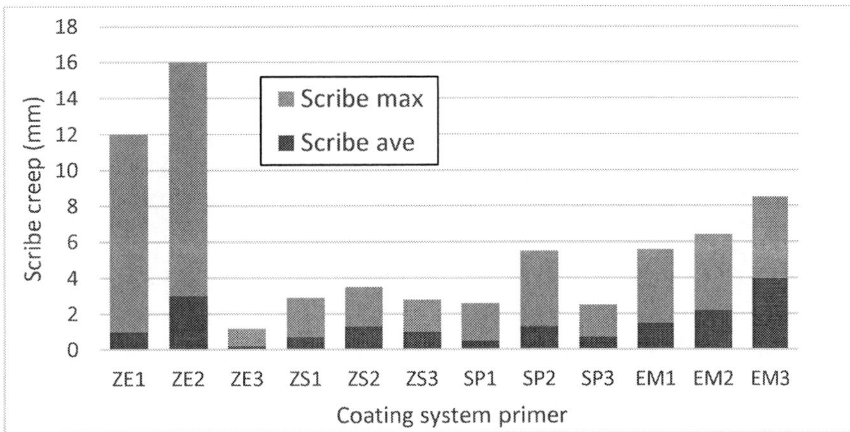

FIGURE 6: AVERAGE AND MAXIMUM SCRIBE CREEP FOR PRIMER SYSTEMS AFTER 5 YEARS OFFSHORE EXPOSURE (ZE: EPOXY ZINC, ZS: INORGANIC ZINC, SP: INORGANIC ZINC SHOP PRIMER, EM: EPOXY MASTIC).

These researchers also carried out electrochemical testing on the coatings. As expected, the three inorganic zincs showed protection potentials of -1.0 to -0.8 volts versus the Saturated Calomel Electrode, close to that of zinc metal, for the entire test period (100 days). The best performing zinc epoxy ZE3, showed similar performance to the inorganic zincs although zinc epoxy ZE2 lost protection with 10 to 20 days and ZE1 showed variable performance. At first glance, these results appear to correlate to the exposure testing. However, the shop primers, which performed similarly to the standard inorganic zincs in exposure testing, lost cathodic protection within a day or two. With such limited sacrificial protection, these coatings would be expected to show poor performance with atmospheric exposure. The authors noted "the zinc silicate shop primers all showed limited ability to provide cathodic protection, but in the field test the shop primer systems performed well. Evidently, the electrochemical properties of the primers alone cannot explain the performance of a coating system."

The above observations are consistent with the zinc coating corroding and providing sacrificial protection to the adjacent bare steel. However, this classic explanation would suggest a higher zinc content and thicker coating should provide longer protection than a low zinc content and thinner coating. A simple cathodic protection cell does not explain why an epoxy zinc of similar zinc content and thickness is inferior to an equivalent inorganic zinc. Nor does it explain why a low zinc, thin shop primer is as effective as a normal inorganic zinc.

Ross and Lingard [11] measured the change in potential with time of a steel panel coated with ZRPs of 85 to 89 weight % zinc in dilute sodium chloride solution, with a 3.2mm diameter hole in the coating exposing the steel substrate. As expected, the potential was initially close to that of freely corroding zinc, dropping to that of corroding steel after about 15 days, then remaining constant. The pH of the void rapidly rose to pH11 at the commencement of the experiment, and white zinc corrosion products formed at that site. What these researchers noted that is often ignored was that it was the rapid increase in pH from the oxygen reduction reaction as a result of rapid dissolution of zinc particles adjacent to the bare steel void that was key to providing protection, not the long-term potential. Rapid zinc dissolution and a rapid increase in alkalinity on the steel led to protection of the bare steel. The authors noted that cathodic protection is "short-lived but essential to the prevention of rusting at existing or newly-formed porosities in a zinc-rich paint film. Subsequent longer-term protection develops through the formation of zinc compounds".

Yadav *et al* [12] also studied potential and pH changes, this time in a scribe at the interface between a 20 micron zinc metal coating and steel substrate under a thin (0.1mm) layer of artificial seawater, using a device similar to a Scanning Kelvin Probe. The potential across both the zinc and steel initially was approximately −1100 mV versus the silver-silver chloride electrode, a value typically observed with zinc corrosion. With time the potential became more positive, but was still protective for up to 11 days exposure to this solution. Only then did the steel potential become more positive than the protection potential, then only at some distance from the interface. However, the surface did not show any red rust formation, but rather corrosion of zinc continued at the coating interface and a front of white zinc corrosion products moved across the steel surface away from the interface. These observations are consistent with the findings described above [11]. That is, while it is the environment and zinc coating which create conditions for the rapid development of a galvanic cell between the zinc and steel to generate zinc ions and hydroxyl ions, it is alkalinity and development of a protective film on the steel which continued to provide protection even when the galvanic cell is no longer in operation.

This mechanism can explain the unexpected performance of the epoxy zincs and the inorganic zinc shop primers discussed in [7]. In an epoxy zinc, the zinc particles are surrounded by insulating epoxy, and cannot activate or generate the rapid zinc dissolution and production of protective alkali on adjacent steel required for protection. By the time the electrolyte has penetrated to the zinc particles, steel rusting has already taken hold. However, it may be possible that some epoxy zinc formulations allow easier or more rapid contact between the electrolyte and the zinc particles, explaining their good performance. Extenders or fillers have been shown to be beneficial in ZRPs [3] and these may assist in the electrolyte accessing the zinc particles. For the shop primers, although they have a limited zinc content, the zinc particles within the permeable silicate binder can still activate rapidly and corrode creating a galvanic cell (albeit short-lived), providing the alkalinity and zinc corrosion products required for good protection. However, the low thickness and low zinc content means they may not be able to provide long-term protection in a severely corrosive environment or to a large area of bare steel. A zinc metal coating such as galvanizing, will be readily activated so provides rapid scribe protection and a thick zinc coating will provide long-term protection.

An important factor influencing both the short-term activation of the zinc and the long-term protection, will be the area of steel to be protected by the adjacent zinc coating. A relatively large area of bare steel will allow the cathodic reaction to rapidly commence, encouraging activation of the zinc. However, a large cathodic area will also cause more rapid consumption of the zinc than with a smaller cathodic area. On the other hand, a very small area of bare steel will not act as an efficient cathode and the zinc may not activate. Most testing these days requires a scribe with a width of 2 to 3 mm or so, which should provide sufficient cathodic area for this not to be an issue. In the past, scribing was done with range of tools giving great variation in width, which could be one reason for inconsistent results.

The stages in protection, or absence of protection, to the exposed steel in a scribe by a zinc coating can be summarised:

1. When the scribed zinc coating is exposed to moisture or other electrolyte, a potential will develop between the zinc coating and the exposed steel. With metallic zinc, or a ZRP with a porous binder, the zinc activates quickly and a protection potential will rapidly develop on the steel.
2. Zinc will corrode to produce zinc ions at the interface and oxygen and water will react to form hydroxyl ions on the exposed steel, raising the local pH considerably.
3. Hydroxyl ions react with zinc ions to form a protective inhibitive film and the pH will slowly decrease.
4. If the zinc cannot quickly activate, such as with an epoxy ZRP, the exposed steel will not develop a protection potential and an iron corrosion cell may form in the scribe. In an immersed environment, the

electrolyte will eventually reach the zinc particles allowing them to corrode and form a cell with the exposed steel, creating a protection potential. In an atmospheric environment, however, the electrolyte is likely to dry out before this cell can form and little protection of the steel will be obtained.

5. How long this protection will last will depend on many factors. The more zinc, the longer it will last, so thickness is important with any zinc coating, as is zinc content for a ZRP. An aggressive environment will consume the zinc faster, and it activates the zinc in a ZRP more quickly under cyclic exposure. Similarly, a large cathode or steel area will consume zinc quickly, but also allow more rapid activation of a ZRP. But in a mild environment, low thickness and low zinc content may provide acceptable protection.

Mechanism of Breakdown of a Multi-coat ZRP at a Scribe

Breakdown of a single coat ZRP at exposed steel was described in [1]. As noted, performance with a topcoat is somewhat different. A topcoat will slow down zinc consumption, but may not allow zinc to activate and corrode quickly enough to provide efficient protection to bare steel from the zinc corrosion products. Degradation at a flaw exposing the steel of a zinc-rich coating with a topcoat will probably follow the following mechanism, shown schematically in Figure 7.

a) The exposed steel becomes the local cathode as oxygen and the electrolyte contact the steel and hydroxide ions are formed. The exposed edge of zinc becomes the anode, but with such a small anodic area, its initial corrosion rate will be high.

b) The reaction between zinc ions and hydroxyl ions will produce an inhibitive zinc oxide/hydroxide precipitate on the steel adjacent to the cut edge. Zinc corrosion products may lift the topcoat by a wedging action. The zinc will corrode causing anodic undermining of the coating system, but form zinc ions which continue to provide inhibitive protection to any exposed steel. Poorly adherent, white zinc corrosion products may appear in the cathodic zone.

c) The rate of anodic undermining of the organic coating depends on the rate of oxygen reduction on the cathode and is inversely proportional to the amount of available zinc. With a large cathodic area or very thin zinc coating, low zinc content or protective binder, the cathode to anode area ratio is very large and galvanic protection is insufficient to provide protection for more than a small distance into the bare steel and more than for a short time. Iron corrosion products (red rust) will form on the exposed steel. A thick topcoat will result in a deeper scribe and may make electrolyte and oxygen access to the exposed steel more difficult than with a thinner topcoat, so paradoxically provide less protection.

d) Cathodic delamination of paint films with ZRPs is unlikely unless the scratch becomes filled with rust or zinc corrosion products. In this case, the barrier paint coating breaks down near the edge where it has been weakened, becomes porous and allows oxygen and water to form a cathode adjacent to, but some distance from, the scratch. Bubbling and blistering of the topcoat may be observed. A similar mechanism may arise if the scribe only penetrates the topcoat exposing the zinc coating but not the steel substrate.

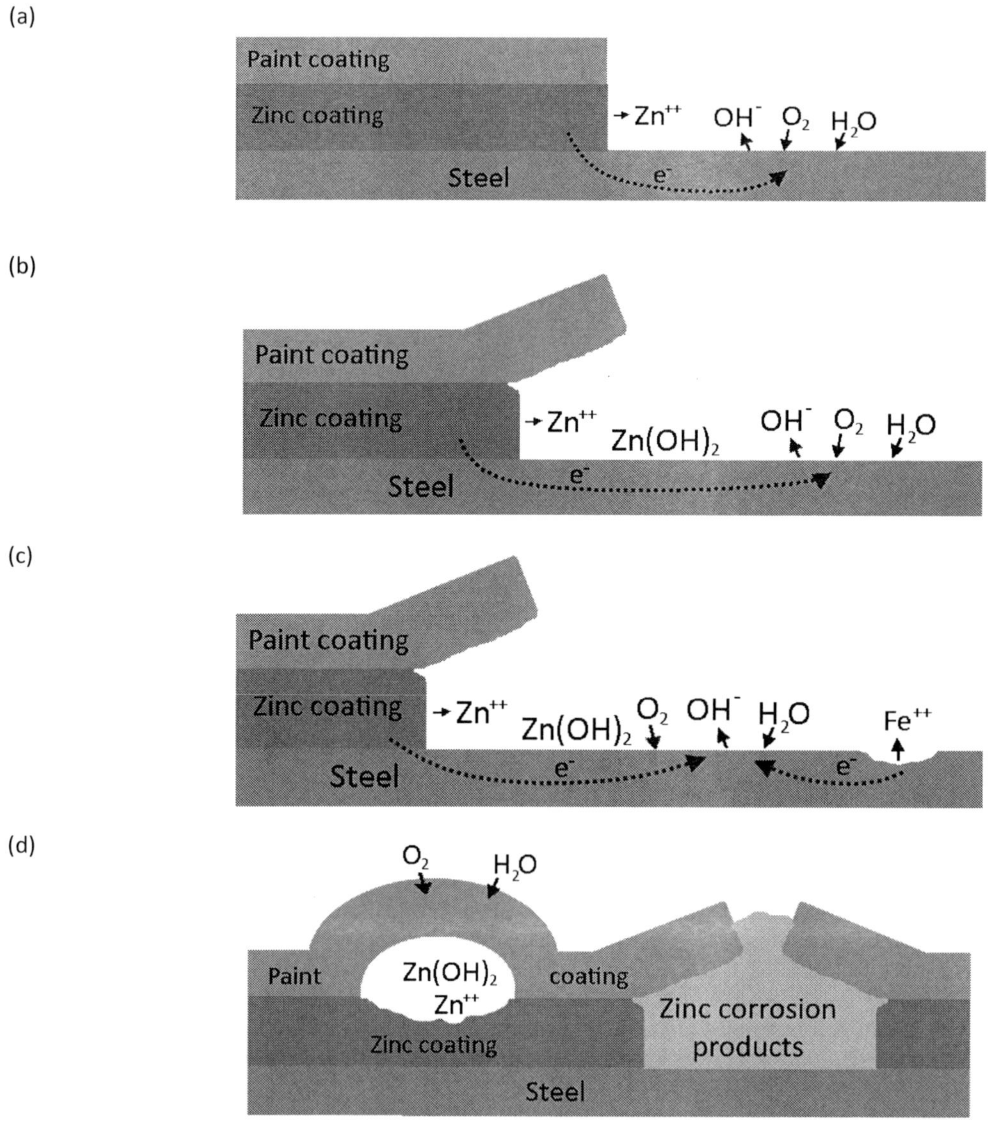

FIGURE 7: (A), (B), (C) DEVELOPMENT OF A CORROSION CELL UNDERNEATH A DAMAGED PAINT FILM. (D) WITH A NARROW GAP IN THE COATING, OR DAMAGE ONLY TO THE ZINC COATING.

Conclusions

1. The protection offered by zinc to bare steel at a scratch or similar defect in a zinc coating will depend on factors such as corrosivity of the environment, size of the defect and nature of the coating, including zinc content and binder type for a zinc-rich paint.
2. Topcoating a zinc coating will significantly reduce the amount of protection offered to bare steel in a defect.
3. Protection of exposed steel requires a rapid development of alkalinity as a result of the cathodic oxygen reduction reaction which in turn requires rapid formation of the galvanic zinc-iron cell. Metallic and zinc-rich paints with porous binders such as inorganic zinc allow this rapid activation so provide better protection than organic zincs which activate slowly.
4. Initial galvanic action by the zinc is essential to activate protection, but most protection to the exposed steel in the defect is provided by the alkalinity and zinc corrosion products which develop on the exposed steel.

ACKNOWLEDGMENTS

Thanks to Willie Mandeno for useful comments.

AUTHOR DETAILS

Rob Francis is a recently retired corrosion and coating consultant with over 40 years of academic, industrial and consulting experience in corrosion and protective coatings. Dr Francis has a B.Sc. in metallurgy and a Ph.D. in corrosion science. He has authored or co-authored over 40 technical papers or presentations on corrosion and coatings. He was given the Victor Nightingall award in 2014 and made an ACA Life Member in 2016.

REFERENCES

1. Francis, R A, "Zinc-rich Paints: Exploring Protection Mechanisms", Corrosion & Protection 2022, Newcastle, Australia, November 2022, Paper 326.
2. Francis, R A, "Laboratory Testing of Zinc-Rich Paints: What Can You Learn?", Corrosion & Protection 2022, Newcastle, Australia, November 2022, Paper 286.
3. Francis, R A, "Fillers or Extenders in Zinc-Rich Paints", Corrosion & Protection 2023, Perth, Australia, November 2023, Paper 0023-19838.
4. Lindqvist, S A, Meszarosi, L and Svensson, L, "Comments on the galvanic action of zinc-rich paints", JOCCA, 1985, (92), p34-40
5. Zhang, X G and Xing, A. "Galvanic action of zinc-coated steel - fundamentals and practical applications." Proceedings, Galvatech. Vol. 4. 2004.
6. Weaver, R E F, Boocock, S K and Appleman, B R, "Level of Zinc Required in Zinc-Rich Paints", ILZRO Project ZC-292, SSPC Project R-09, Final Report, January 1994.
7. Ulkem, I and Winter, M, "Understanding inorganic zinc-rich primers and specifications", Journal of Protective Coatings and Linings. 27, 40-48 (2010). Also, PACE 2010, Paper 78.
8. Saeedikhani, M, Wijesinghe, S and Blackwood, D J (2019), "Revisiting Corrosion Protection Mechanisms of a Steel Surface by Damaged Zinc-Rich Paints", Corrosion, 75(7), 756-770.
9. Francis, R A, "Who Rusts First: Revisiting Galvanic Corrosion", Corrosion & Materials, February 2016, 34-42.
10. Knudsen, O, Steinsmo, U and Bjordal, M, "Zinc-rich primers—Test performance and electrochemical properties", Progress in Organic Coatings, 54. 224-229 (2005).
11. Ross, T K and Lingard, J, "The Electrochemical Properties of Zinc-rich Paints", Transactions of the IMF, 40:1, 186-190 (1963)
12. Yadav, A P, Katayama H, Noda K, Masuda H, Nishikata A and Tsuru T, "Surface potential distribution over a zinc/steel galvanic couple corroding under thin layer of electrolyte," Electrochimica Acta 52.9 (2007): 3121-3129.

CORROSION & PREVENTION
CAIRNS 2024

THE AUSTRALASIAN CORROSION ASSOCIATION INC.

DURABILITY PLANS AND CORROSION IN MAJOR INFRASTRUCTURE PROJECTS

D.B. McDonald[1], A. Lu[*1], B. Eliasson[1]

[1]Vinsi Partners, NSW 2113, Australia

**Corresponding author email: Alu@vinsi.com.au*

ABSTRACT

Large infrastructure projects within Australia require durability plans to be developed that include consideration of deterioration mechanisms for all materials, primarily including concrete, steels, and polymers. There is also a growing trend for durability plans to be included for international projects. These require evaluation of environments as well as different mechanisms of corrosion and deterioration. This paper will provide examples of experiences with these obtained from several major Australian projects.

Keywords: durability plan, corrosion, seawater, adhesive, protective coating

INTRODUCTION

In 2009, Curtin University of Technology suggests that corrosion may be costing the Australian economy more than $30 billion each year, much of which is preventable [1]. This cost occurs throughout the world and has led to the development of requirements within major infrastructure within Australia to require the development of durability plans. These plans require formalised documentation of methods undertaken to ensure that the structure or facility has incorporated appropriate measures to ensure that the materials and construction methods utilised ensure that the required design life expected will occur.

During the past 10 years, significant work has been conducted by the authors developing these plans for projects totalling over $20 billion. These projects include tunnels, ports, railways and roadways that are in many different locations around Australia and in many different environments. While most of these projects are based upon total design lives of 100 years elements within each project may have different design lives, dependent on whether the elements are maintainable, replaceable or non-replaceable. Even with design lives as short as 10 years, some projects require evaluation of the products and systems to demonstrate that these will provide the appropriate durability in the specific environment.

This paper will outline how such planning occurs and some of the typical and atypical items that cause problems during these evaluations.

Requirements for Durability Plans

Large infrastructure projects within Australia have requirements for durability generally specified as part of the overall project requirements. This is generally included in Contract documents, such as Scope of Works and Technical Criteria (SWTC), Contract Scope and Contract Requirements (CSCR) or Project Scope & Technical Requirements (PSTR). These provide project specific requirements for various elements of the design, but many include requirements such as:

> *The Contractor must ensure and definitively demonstrate that the durability of all Assets comply with the Deed and the specified Design Life for each Asset.*

Within these documents criteria such as design life, exposure conditions and other specific durability requirements may be provided. These may or may not include the use of specific specifications and an order of precedence of documents, commonly requiring Australian standards to take precedence over ISO or other standards. Such requirements vary dependent on the contract or the agency.

Many of the Australian transportation agencies require the development of formal durability plans in a standardised method. This includes the Transport for New South Wales Durability Requirements for Civil Infrastructure [2]. As stated, *"The purpose of the standard is to specify requirements to increase the durability and reduce maintenance of new railway civil infrastructure;"* however, it is also used for roads and various other transport systems.

Within Queensland, the Department of Transport and Main Roads (TMR) Guideline for the Preparation of Road Structure Durability Plans [3] governs the durability planning. As stated in that document: *"This guideline sets out the requirements for the content of durability plan reports to be prepared by the designer and provides the format of summary tables to enable the standardised collation of information."* The methods presented in TMR require:

- *the identification of deterioration mechanisms;*
- *materials selection;*
- *the development of mitigation measures to ensure that the design intent is met;*
- *the identification of durability critical construction activities; and*
- *verification of constructed components to confirm compliance with durability requirements.*

Other Australian states do not have similar format requirements; however, many agencies have requirements for an overall durability plan that may be part of contractual documents. For example, in a recent bridge project in Tasmania, the Contract required:

> *The Contractor must ensure the durability of all New Assets. Durability must be addressed throughout the design, construction, refurbishment and maintenance of all Assets and must be reflected in the Project Plans and Maintenance Manuals.*

The Concrete Institute of Australia (CIA) has published documents relating to Durability Planning [4]. Within these series of documents, the CIA provides *"tools for managing durability though design, construction and maintenance."* These documents were key in the development of Concrete durability planning methods that are being developed worldwide.

The American Concrete Institute (ACI) has recently established a Technical Committee to develop a Concrete Durability Code. This Committee (321) has the mission to: *"Develop and maintain code requirements for the durability of concrete structures."* It has the following goal: *Develop "Code Requirements for Concrete Durability," a mandatory-language document addressing concrete durability...it will:*

- *Provide minimum prescriptive requirements and, where applicable,*
- *Provide alternative performance requirements for achieving concrete durability.*

The code will identify durability exposure categories and classes, and steps for durability planning, and provide requirements for durability design, construction/quality management, and maintenance.

British Standard BS 7543:2015[5] Guide to durability of buildings and building elements, products and components gives guidance on durability, design life and predicted service life of buildings, constructed assets and their parts. It applies to both new and existing buildings, defining the environment.

Experience with development of durability plans worldwide is limited; however, during the past 10 years, we have conducted a significant number of these plans ranging from small single span bridges to rail systems and stations, tunnels, and major infrastructure that required understanding of materials performance and durability along with modelling.

Exposure Environments

The initial phase of any durability plan is to accurately define the environment for each of the locations and structures within the project. These projects may consist of a single location or may extend over many kilometres, such as the Coffs Harbour Bypass, which covers some 14km with many different bridges and tunnels. Depending on local environmental conditions, various systems of "exposure classes" will typically be provided for different materials (**Figure 1**). It is generally impractical and not cost effective to specify different materials and concrete covers for each item and thus zones of exposure are generally allocated. This however is done on a case-by-case basis.

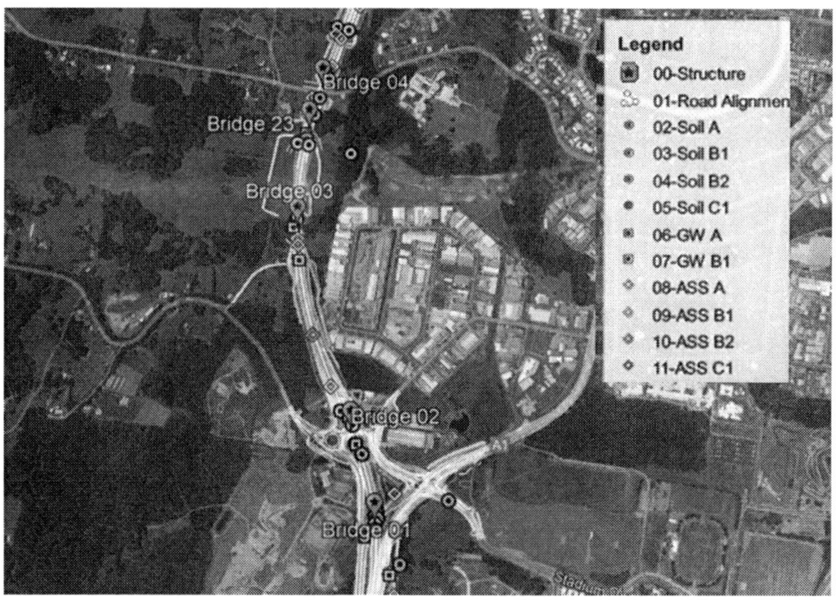

Figure 1 Typical project exposure map, showing locations of boreholes and related exposures.

Within each exposure, the exposures of concern are those associated with the soil or groundwater, with atmospheric exposures and those in marine environments. It is important to consider not only current usage, but also future usage. For example, if it is known that an airport or other industrial complex is to be constructed nearby, this should be addressed.

For marine exposure, the tide levels are required to be predicted. The National Oceanic and Atmospheric Administration (NOAA) [6] has predicted that there has been a 0.2m sea level change since 1880, and that if we are able to significantly reduce greenhouse gas emissions, United States sea level in 2100 is projected to be around 0.6m higher on average than it was in 2000. Under high emissions levels, this raises to almost 2m. Such changes would significantly affect exposures of coastal structures that are assumed. Similar data is being provided by the US Army Corp of Engineers as shown in **Figure 2**.

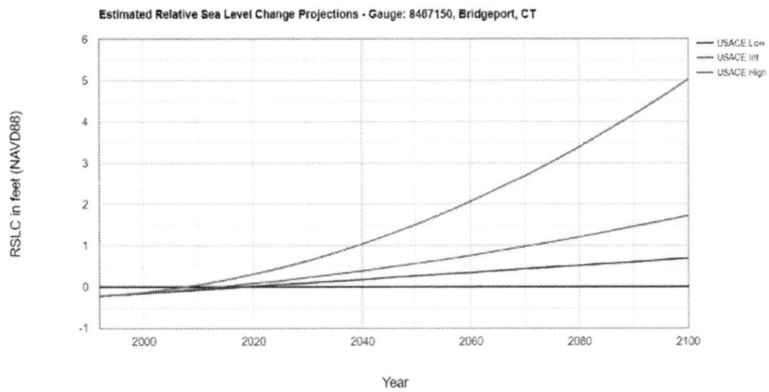

Figure 2 Predicted sea level changes at Bridgeport CT (From https://cwbi-app.sec.usace.army.mil/)

When the coastline is changing, either due to large changes in tides, beach conditions or where there are large tidal flats, significant differences in exposure classifications may result.

Atmospheric Exposure

In some durability plans it is relatively straightforward to define the atmospheric exposure using standards such as AS 4312[7] or AS 5100.5 [8]. This is based on a distance from a shoreline or coastline. As the distance of the structure to the shore decreases, the exposure becomes more problematic as the influence of additional factors such as microenvironments and minor changes in distances may result in changes to the assumed exposures.

It is not possible to predict the future development of lands near any construction site; however, the trends are that air and waters contain less particles and contaminants than have been in the past. Thus, concerns with acid rain, airborne sulfates and other deleterious materials within the air and water are less likely to be of concern within the future and design using today's values are most likely satisfactory for future exposure. The exception to this is future carbon dioxide (CO_2) in the atmosphere, and the associated temperature rises, discussed in detail below.

Effects of Increasing CO_2 and temperature

Many models have been proposed to estimate changes in CO_2 in the atmosphere over the next 100 years (**Figure 3**). The most severe of these Representative Concentration Pathways (RCP) is termed RCP 8.5, which estimates an atmospheric CO_2 of over 1400 ppm in 100 years, which would lead to a warming of over 4°C. Temperature range increases of 2 – 4°C are also being often considered in the design assessments, which affects the chloride and carbonation modelling for concrete structures and corrosion rates for metals. A more reasonable model is that of RCP 6.0 which is shown as leading to a CO_2 rise to 750 ppm in 100 years and a corresponding temperature rise of 2.8°C.

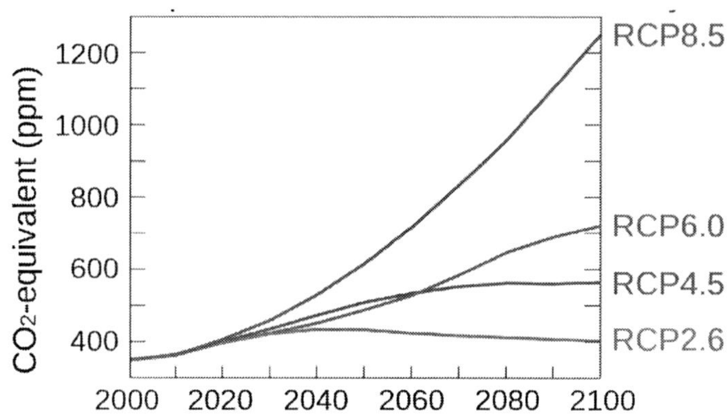

FIGURE 3 REPRESENTATIVE CO_2 CONCENTRATION PATHWAYS (FROM IPCC 2011)

The adoption of exposures to future carbon dioxide levels may significantly affect the choice of materials and the resultant concrete cover as outlined in Table 1. In this table, calculated carbonation rates for a particular location are shown for two mixtures, one containing 25% fly ash and the other with 50% granulated blast furnace slag (GGBS). For this example, the estimated difference between concrete covers required for the two scenarios (RCP 8.5 and RCP 6.0) may be significant, particularly if the concrete contains slag. These covers are frequently greater than those required by the AS 5100.5 for 100-year concrete design. This information becomes increasingly important as design chooses sustainable materials that require minimising total portland cement contents that reduce the overall carbon footprint of the concrete.

TABLE 1: CALCULATED CONCRETE COVER FOR RCP 8.5 AND 6.0 FOR 25% FLY ASH AND 50% SLAG FOR 100 YEARS.

Compressive Strength (MPa)	25% fly ash		50% slag	
	RCP8.5	RCP 6.0	RCP8.5	RCP 6.0
25	50	40	125	105
32	25	20	75	65
40	20	15	50	40
50	25	20	60	50

Atmospheric Chloride

Atmospheric chloride can result in the degradation of galvanized and stainless structural steels, as well as the reinforcement bars within concrete and is often the governing factor in standards and specifications for determining the exposure class of atmospherically exposed elements. AS 4312 shows that exposure to chloride is very highly correlated to the corrosion rate of metals.

The transport of chloride from seawater into the air is directly dependent on the bursting of bubbles at the water-atmosphere interface. Lewis and SE Schwartz [9] found that primary marine aerosol particles are produced mainly by bubble-bursting following wave breaking that is induced by the stress of wind on the sea surface. Varying surf conditions, wind speed, and sea foam all contribute to the presence of bubbles which form the salt aerosols. Based upon the research, the distance from breaking surf is more important than the distance from saltwater and that saltwater rivers, (such as the Paramatta River or Brisbane River) are often not significant sources of airborne chloride.

Standard documents

Standard documents such as AS 5100.5 and AS 4312 seek to provide guidance on risks due to deleterious substances such as chloride and sulfates, often grouped into different exposure classifications depending on distance from the coast or sites where aggressive pollutants are generated.

For atmospheric structures, distance to coast is conveniently estimated using GIS systems and public domain satellite photography. Generally, exposure zones further from the coast have wider bands, such that when structures are built further away, discrepancies in distance estimates are usually not of consequence; however, exposure zones closer to the coast are more sensitive to coastline distances and as discussed above such measurements may be subject to dispute between experts.

For example: in **Figure 4** the coastline at this particular location has changed over a distance of 80m during the period of 1989 and 2023 and there is no particular direction of change. Thus, if the design was conducted in 2023, a structure could be almost 80m closer to the Crowdy Head Road and be subject to an exposure of B1 per AS 5100.5, compared with if the design was conducted in 2013, where the same building may obtain an exposure of B2.

Figure 4 Change in Shoreline from 1989 to 2023 showing 80M change in location (From https://maps.dea.ga.gov.au/)

Metals

Difficulty arises when projects occur near boundaries of these defined exposure classification zones. For example, AS 4312 defines the C5 exposure classification as follows:

> *C5 category is normally found 50 m to 500 m inland from the shoreline of coasts exposed to surf or very rough seas.... On tropical coasts, the combined C4/C5 category extends 100 m to 500 m inland from the shore. Some of the damp or contaminated interior environments in category C4 may occasionally extend into this category.*

AS 4312 defines the C4 exposure classification as follows:

> *In areas with rough seas and surf, it extends from 0.5 km inland to about 1 km inland. Along semi-sheltered or less turbulent coasts, it extends inland from 50 m to 500 m from the shoreline. On quiet tropical coasts, it cannot be distinguished from the more severe C5 category, and a combined C4/C5 category is found in a narrow band approximately 100 m to 500 m inland from the coast.*

Standards and codes often specify exposure classification ranges based on distance, to address varying levels of risk resulting from chloride contamination on structures. However, distance from seawater alone does not entirely correspond to the availability of chloride in the air – there must also exist the appropriate mechanisms for the chloride to be released from the water. In a case of a project in Queensland located on a long tidal river, it was determined the lack of tidal activity, spray, and bubbles meant very little chloride was being transported from the water to atmospherically exposed structures.

It is noted that the ISO 12944-2 does not present information relating to distances but requires either the exposure to be specified or testing to be conducted [10].

Concrete

For concrete structures AS 5100.5 states:

> *...the coastal zone includes locations within 1 km of the shoreline of large expanses of saltwater. Where there are strong prevailing winds or vigorous surf, the distance should be increased beyond 1 km and higher levels of protection should be considered.*

In the context of Sydney, areas surrounding the sheltered Sydney Harbour or Botany Bay would typically be categorised in the same exposures (e.g B2 to AS 5100.5) as beaches directly exposed to surf such as Manly or Bondi Beach. However, our project experiences have found that the chloride content of concrete structures adjacent the sheltered waters of the Harbour are well below the concentrations expected at surface and subsurface concrete.

The issue of the exposure of concrete along long marine inlets requires the estimation of chloride exposure to determine how the salt changes within an environment. Within some analyses, estimates using linear functions within the exposures have been used, whereas others have used stepwise estimates for corrosion rates.

AS 5100.5 uses the terminology "large expanses of saltwater". In the 2004 version of AS 5100.5[11], the standard specifically indicated that Sydney Harbour only considered east of the harbour bridge and the spit bridge; however, this has been deleted in the 2017 version (**Figure 5**). The question of how bays are treated is left to the judgement of the durability design team.

Tests of concrete that were sampled between 400 and 1200m from the harbour shows that the surface chloride levels are very well below that which would suggest B2 exposures (**Figure 6**).

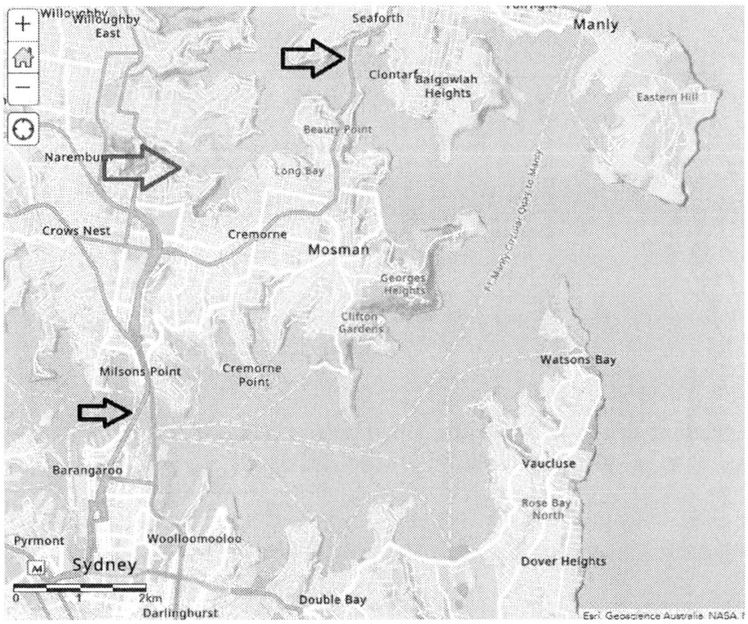

Figure 5: Sydney Harbour showing 2004 limits (black arrows) and areas that may or may not be considered in current code (red arrows).

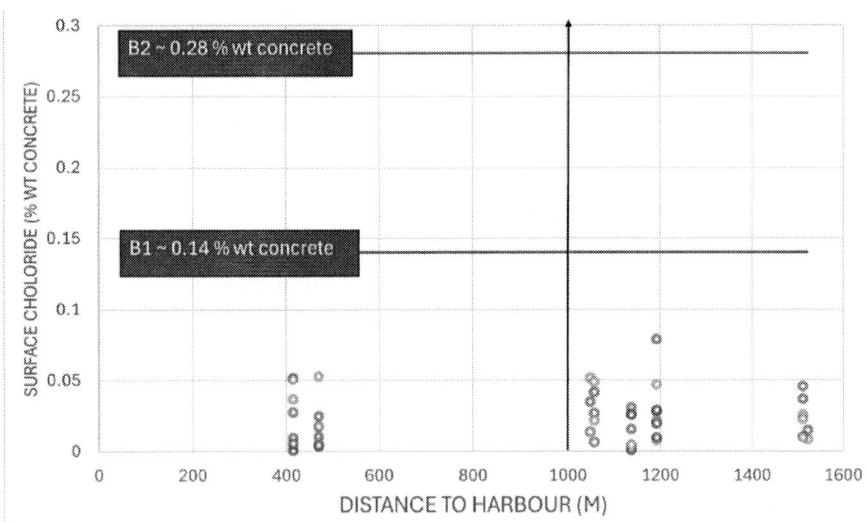

Figure 6: Concrete surface chloride contents for 40 to 60-year-old structures compared with distance from Sydney Harbour.

In the case of a new bridge in Tasmania, questions were raised as to the exposure relating to the large River inlet, shown in **Figure 7.**

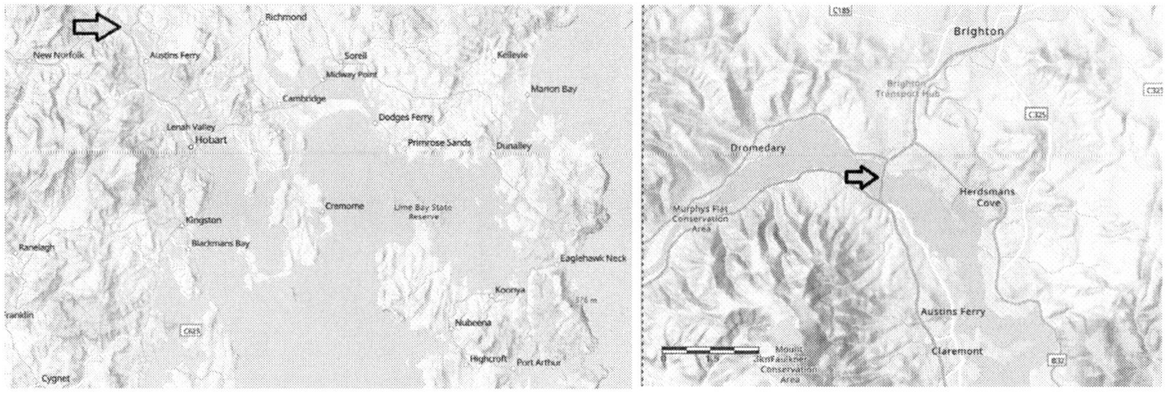

Figure 7: Location of bridge along the River Derwent. Local investigation was required to evaluate the exposure classification for the structure.

In both of these instances, the availability of local data assisted with the evaluation of the exposure environments; however, there are other cases where longer term chloride data are not available and such data cannot be rapidly obtained.

MICROENVIRONMENTS

In practice, exposure classifications due to environmental effects can also change depending on localised factors such as geography, built environment, and pollutants. There have been cases where the durability report adequately covers pre-existing environmental conditions, but the microenvironment created by the project itself upon completion, such as a building, tunnel or airport, has been overlooked. As an example, exposure classification for steels may increase from a C2 to a C3 according to the following excerpt from AS 4312:

> *Micro-environmental effects, such as proximity to airports and sewage treatment works, may also place a site within this category. Interior environments with category C3 corrosivity can occur in humid production rooms, such as food-processing plants, laundries, breweries, printing works and dairies.*

Exposure to rain or lack thereof due to buildings or shelter can constitute a significant change in microenvironment. Washing, whether from natural rainwater or as part of a maintenance plan, is crucial in washing off accumulated chlorides as well as preventing ingress of carbon dioxide gas through porous concrete. Counterintuitively, sheltering a structure can in some cases result in a more aggressive microenvironment. For this reason, although rural sites may be classed within the same exposure bands as inland cities purely based on distance from coast, actual observed corrosion rates can be an order of magnitude greater. For this reason, obtaining site specific data grows increasingly important.

OBTAINING DATA

Corrosion rates within various exposures have significant variance at lower and upper bounds. For example, in a C2 exposure classification, AS 4312 provides a mild steel corrosion rate of 1.3 μm/year to 25 μm/year, with the upper limit more than 19 times higher than the lower figure. Without finer resolution data, this can lead to structures experiencing corrosion significantly earlier than expected or being grossly overdesigned (i.e a total section loss from each surface of either 0.13mm or 2.5mm over 100 years). Fortunately, very few structures are designed using uncoated steel.

Although some locales have specific corrosion rate test data listed in AS 4312, structures not within areas covered by the standard may be required to conservatively assume the highest rate of corrosion. The need for location specific data becomes more relevant as development of large cities and infrastructure projects in Australia increasingly extends inland into the C2 exposure classification, such as in the case of the Western Sydney Airport and regional areas around Melbourne and Brisbane. As new cities are proposed, standardised exposures for that location need to be developed.

Measurements of atmospheric chloride or corrosion rates requires use of either relatively fragile chloride candles or corrosion plates that are measured over significant time. AS 4312 requires corrosion rate be measured over a minimum of 1 year to determine the exposure category, but this is not sufficient to determine the long-term corrosion rate. Methods to replace these long-term methods using electronic sensors are being reviewed; however, these newer methods cannot provide required exposure information within the typical design time periods. Methods using solid state electronics are being developed, such as that of Ahn et al. [12] and Yang et al. [13], who used ACM sensors composed of metal electrodes that are initially insulated from each other and measured galvanic current, Wade et al. [14], measured corrosion of industrial infrastructure using electrical resistance corrosion sensors, and Li et al. [15] who used an electrical resistance sensor and internet connectivity to automate corrosion monitoring.

GROUND AND GROUNDWATER EXPOSURE

For below ground structures, such as piles, determining exposure classifications are based on soil and groundwater data taken from borehole samples within the project bounds. However, the locations of these boreholes may not correspond to the location of structures, particularly if there is a change in alignment during design. From experience, soil and groundwater conditions can change significantly between boreholes as close as 50 to 100m, becoming more pronounced if data samples are also taken at different times. It is also important to understand if dewatering is to occur during operations or if adjacent facilities are to undergo construction that may change expected conditions.

SOILS AND ROCKS

Data obtained from borehole sampling may include various forms of chloride, sulfate, pH, and resistivity (or conductivity) information, as well as consider the likelihood of acid sulfate soils that may result in the conversion of neutral pH soils to acidic zones over time.

The location and type of groundwater is also required. In some projects this data may come from hundreds of locations and may still be insufficient for the analysis of the many structures that are along a project. In a recent project over 1200 separate borehole readings were reviewed and exposure classifications determined for both steel and concrete elements.

Certain soils and rocks may require additional evaluation. For example, pyrite (FeS_2) containing rocks may pose difficulty in the determination of their effect on exposed steel. Additional analysis may be required to determine whether these are appropriately buffered against forming later acidic conditions that may deteriorate embedded steel. Documents such as AMIRA P387A Test Handbook [16] may provide guidance.

The exposure classifications for the various codes may also not be consistent and that the exposure classifications between AS 2159 [17], AS 3600 and AS 5100.5 may not necessarily agree with each other; for example, sulfate exposure conditions do not align.

MATERIAL CHOICES

During the durability assessment process, the durability design team seeks to apply innovative methods to determine a range of materials applicable for the project that meet the project goals and client demands. Within a particular exposure, there are certain materials choices that tend to be the most cost effective while still meeting the required design life with minimal maintenance.

The most commonly used metal to use in civil construction is steel due to its high strength/weight ratio and its availability and flexibility. It is readily formed, cut, welded and coated. In most exposures it needs to be protected against corrosion using either an organic or galvanized coating.

Metals such as stainless steel are utilised close to the coast, while other materials, such as aluminium and copper are used for more decorative elements and plumbing. These tend to be used more selectively due to their inherent cost. Care also needs to be taken to understand risks related to galvanic corrosion.

Zinc Coatings - Galvanizing

Coating steel for civil structures with zinc can be done either with a batch process such as hot dipped galvanizing (HDG) typically meeting AS/NZS 4680 or an inline process such as used for continuously hot-dip metallic coated sheet steel & strip, typically meeting AS 1397.

Ranges of corrosion rates are provided for each exposure classification in AS/NZS 2312.2[18]. The International Standard ISO 9224[19], on which as AS/NZS 2312.2 was based, provides for reducing corrosion rates over time, but the Australian standards do not provide for this allowance. Thus, if Australian codes are utilised, lower design lives for galvanizing are determined compared with International Standards as outlined in Table 2.

TABLE 2: DESIGN LIFE OF GALVANIZING IN AS/NZS 2312.2 AND ISO 9224 FOR 85MM THICK COATING (ASSUMING MAXIMUM RATE FOR DESIGN).

Exposure	AS/NZS 2312.2		ISO 9224	
	Initial corrosion rate (μm/year)	Design life upper limit (years)	Design life average (years)	Design life conservative (years)
C3	0.7 < rcorr ≤ 2.1	>40	83	65
C4	2.1 < rcorr ≤ 4.2	>20	39	31
C5	4.2 < rcorr ≤ 8.4	>10	17	14

Further, there is little guidance provided in the standards as to whether the average or the greatest corrosion rates should be utilised in design. As the rates may vary by over 200%, the design lives calculated using the lowest and highest corrosion rate within an exposure vary considerably. Design should generally consider the highest rate for each exposure.

The composition of substrate steel has also been increasingly important in an Australian context. From recent experience, increasing quantities of foreign steel produced to different specifications and standards have been used domestically. Often the percentages of alloying elements vary to those permitted in Australian Standards. In particular, even small variances (0.01wt%) in silicon content can significantly alter the microstructure and thickness of zinc layers in hot dipped galvanized steel, leading to coating defects and reduced lifetime in aggressive Australian coastal conditions. While suitable in terms of structural performance, corrosion behaviour over the course of a 100-year design life for foreign specification steels can vary greatly.

Another recurring issue is the use of galvanized steels in direct contact with dissimilar metals. Recent experiences have shown that galvanic corrosion risks are frequently overlooked. Especially in higher corrosivity environments, designs may opt to substitute some HDG elements for stainless steels without ensuring elements in contact are appropriately isolated. An area where there appears to be little industry literature is in cases of three dissimilar metals in contact (e.g stainless steel fastener holding together an aluminium and galvanized plate) and the resulting corrosion behaviour.

There are a number of zinc alloys that are commonly used for coating steel, such as Zn-Al, that do not have commonly accepted corrosion rates. In such cases assumptions are made, based upon limited data. Sometimes the data may only be from salt spray tests, which is known to have a poor relationship to field performance. Additional work continues to find the appropriate correlations and include these in the standards.

ORGANIC COATINGS

Organic coatings are typically a polymer barrier applied to metallic substrates, either through a wet process (paint) or dry process (powder coatings). Coatings such as silanes can also be applied to concrete, in order to seal surface pores in the structure with a low permeability material.

Epoxy and polyurethane based systems are currently the primary coating systems used in construction. Durability information on these organic coating systems are covered in AS/NZS 2312.1[20] but are also in ISO 12944.5[21].

Powder coatings generally are more durable than paint coatings owing to a 100% solids content and the application of these within factory settings. These have most commonly been seen on aluminium architectural elements but are applicable to a range of substrates including stainless and galvanized steels. Powder coating Standards, AS 4506 [22] and AS 3715[23] are most commonly referenced.

As organic coatings typically do not degrade uniformly like galvanizing, it is not possible to provide a degradation rate. Instead, Australian Standards such as AS/NZS 2312.1 typically provide an expected time to first maintenance depending on exposure for various defined coating systems.

In highly corrosive environments, it may be necessary to utilise both galvanizing with an organic coating. Such "duplex systems" provide a "synergistic effect" where the total coating life is greater than the individual layers combined. AS/NZS 2312.2 provides guidance on coating systems compatible with galvanized substrates. The synergistic effect multiplier is noted to be 1.5x in a C5 exposure, whilst no guidance is given in this standard for the synergistic effects in other exposures. Often even a thick galvanizing layer is insufficient to provide a suitable design life in more aggressive exposures and a duplex coating is required. However, the American Galvanizers Association *Galvanized Steel Specification Guide* provides the following:

> *The service life of a duplex coating is 1.5 to 2.3 times the sum of the individual systems.*

There is also a disparity between the coating systems recommended by Australian Standards and those found in the ISO Standards. For example, as shown in Table 3, a coating is to be applied to steel lasting 25+ years to first maintenance in a C3 environment. According to AS/NZS 2312.1 an ACCC6, EHB6, or PUR5 coating system could be used. ISO 12944.5[24] provides for very high (VH) durability coatings with lives of more than 25 years. For corrosivity category C3, ISO System C3.04, C3.07 or C3.10 are provided. Comparison shows that the coatings provided by ISO are significantly lower in total coating thickness compared to the AS/NZS standards.

TABLE 3: SELECT COATINGS FROM AS/NZS 2312.1 AND ISO 12944-5 FOR EXPOSURE C3 WITH DESIGN LIVES GREATER THAN 25 YEARS.

Coating Layer	**AS/NZS 2312.1**			**ISO 12944-5**		
	ACC6	EHB6	PUR5	C3.04	C3.07	C3.10
Primer (μm)	75 zinc rich	75 zinc rich	75 zinc rich	60 – 80 misc	60 – 80 misc	60 – 80 misc
Intermediate Coat (μm)	200 HB epoxy	125 Epoxy MIO	200 HB epoxy			

Coating Layer	AS/NZS 2312.1			ISO 12944-5		
	ACC6	EHB6	PUR5	C3.04	C3.07	C3.10
Coat Final (µm)	50 Acrylic	125 Epoxy MIO	50 HB Polyurethane	2 to 4 coats Alkyd/ Acrylic	2 to 3 coats Alkyd/ Acrylic	2 to 3 coats Epoxy/ Polyurethane /Acrylic
Total Top Coats	2	2	2	2-4	2-3	2-3
Minimum Total Thickness (µm)	325	325	325	260	240	200

Newer coatings are difficult to handle within the durability assessment systems as adequate performance data is not yet available. Such materials include fluoropolymer coatings that have been applied extensively in Japan and elsewhere with good performance. Inclusion of these will depend on the ability of specifications to permit deviations from code or standard accepted products.

Stainless Steel

The grade of stainless steel is frequently critical to its performance. The most common stainless steels grades encountered in Australian projects have been Grade 304 and 316. Typically, Grade 304 is sufficient up to an exposure classification of C3 (AS 4312), but Grade 316 is recommended for more extreme exposures or longer design lives. Higher duplex grades may be necessary where additional aesthetic concerns are required or aggressive exposures are present.

Stainless steel may undergo deterioration particularly in high chloride environments, such as within water retaining or wastewater plants with pipe and plant equipment. Deterioration is primarily caused by localised pitting.

During fabrication, welding, and installation, contaminants such as grit can become embedded within the stainless steel surface serving as initiation sites for defects to develop. For these reasons, mechanical cleaning (blasting, abrasives) of stainless steels is not recommended. Treatment processes such as pickling are required, that remove surface layers of the protective oxide film and any roughness or embedded contaminants. The oxide film can then self-heal after passivation into a more uniform layer, providing better future protection.

Following installation, stainless steel may need to be maintained to ensure that the oxide film is kept intact. In high chloride environments, washing may be required, particularly if aesthetic concerns are present.

A common problem encountered is tea staining of stainless steels, that may be observed as a result of the surface profile of this material along with the exposure and type of steel grade. While this may be an aesthetic concern, it does not generally affect the long term durability.

Additional information on the performance of stainless steels is available from the Trade Associations including The Australian Stainless Steel Development Association (ASSDA) or the Nickel Institute.

Chloride Model

Modelling the effects of chloride ingress in concrete over time has proven more difficult as material selection and model choice is frequently left to the durability team. A general shift towards adoption of probabilistic rather than deterministic chloride models is being noted, particularly with the publication of the *fib* MC 2020 [25]. The choice of the model may significantly affect the results obtained from the analyses. Within the models, test data is frequently required, and often the choice of test methods may result in significant differences between models. For example, *fib* MC 2020 has chosen to use a non-steady-state migration coefficient from tests conducted according to Nordtest NT Build 492, which has been shown to have poor correlation with the more realistic chloride transport coefficient from the Nordtest NT Build 443 test.

Durability Limit State

For many materials there are no standardised methods for predicting the design life of the materials. For these materials, fundamental evaluation of degradation properties are reviewed. For example, for polymers, this may require consideration of all the various mechanisms that affect the life of a polymer, such as UV or the presence of various ionic species that may depolymerise the polymer and then this enables determination of the polymer life. Protection mechanisms for the life of the material may then be recommended, such as use of a UV stabiliser or a coating.

Sustainability Practices

The use of sustainable materials has been forefront in the construction of infrastructure throughout Australia. Replacement of portland cement has required the increased use of higher quantities of fly ash and slag within the concrete mixtures. While generally the modelling of durability, including corrosion factors, has been able to be conducted using standard methods, many of the existing databases do not have sufficient data on the properties of concrete at high replacement levels.

In addition, materials using recycled or novel additives are being recommended and each needs to be reviewed. By using a standardised method of evaluation that considers each of the degradation mechanisms and then considering the exposure, an appropriate design life or mitigation measure is then able to be assigned to the material.

Recommended Practices

Due to the variety of results that may occur relating to corrosion analyses from assumptions made, the following are some recommendations for uniform applications of corrosion and durability standards.

1. Project specifications should explicitly define where boundaries for basic exposure parameters are to be determined.
    - For example: the shoreline shall be taken from DEA Australia 2022 annual shoreline OR
    - Areas in the tunnel shall be designated Exposure Classification C3 OR
    - All steel east of Highway 1 shall be designated Exposure Classification C4.
2. Specific codes should be referenced in the design specifications to ensure that the owner, reviewer and designer obtain results that are consistent with owner expectations and are able to demonstrate appropriate design lives.
    - For example: Corrosion rates may be evaluated using ISO Standards OR
    - Decreases of corrosion rates over time shall not be considered OR
    - Coating life should be evaluated using AS/NZS 2312.1 and AS/NZS 2312.2 (vs ISO 12944).
3. Atmospheric assumptions should be provided within the specifications, particularly if long-term analyses are required.
    - For example: Future atmospheric assumptions shall utilise the RCP 6.0 Model for CO_2 increase over time.
4. Models for corrosion may be deemed to comply, deterministic or probabilistic. Appropriate models need to be defined. While the trend is towards probabilistic models, the use of these needs to be verified against actual structural performance.
5. Reference designs should be considered as a basis for durability design.

Summary

Large infrastructure projects within Australia require durability plans to be developed that include consideration of deterioration mechanisms for all materials, primarily including concrete, steels, and polymers.

The development of durability plans requires consideration of exposures, suitable materials and their deterioration mechanisms and methods to install and maintain these materials appropriately.

The exposure of environments may be viewed differently dependent on the consultant and to avoid disputes may be best defined specifically in contract documents. These disputes occur due to differences in definitions of tidal levels, future atmospheric conditions, and atmospheric chlorides. This requirement becomes more critical when structures are constructed around 1km from the coastline and around bays and harbours.

New methods for determining Atmospheric Exposure Classifications should be a goal of the corrosion community and the use of simple sensors appears to be a reasonable approach compared with the installation and cleaning of coupons or steel plates.

Below ground exposures will still depend on soil and rock sampling, and no simple method is viewed for replacement of this tedious process.

Steel and concrete will dominate the construction industry due to their unique properties of flexibility and ubiquity.

Steel will be generally galvanized and frequently coated, but new coatings with greater than 25 year design life need to be established within the standards. Additional data on the performance of zinc alloys should be pursued. The Australian standard and ISO standards for corrosion of galvanizing are significantly different and evaluation of the actual design life should be established.

In addition, the performance of organic coatings should be determined as the design lives in Australian and ISO standards are significantly different. Field tests on newer coatings indicate 35+ year design lives are possible, and these are not recognised within the current standards.

Modelling the effects of chloride ingress and carbonation of concrete is frequently left to the durability team. A general shift towards adoption of probabilistic rather than deterministic chloride models is being noted; however key values are still to be determined for some of these models to enable these to be more widely utilised.

Finally, sustainability practices are common, and these materials required additional consideration. By considering exposures, and potential degradation mechanisms, then appropriate design life or mitigation measure are able to be assigned to the material.

Finally, any durability plan requires working closely with the design team to develop innovative solutions that provide long-term durable designs that optimise material use.

References

1 Curtin University (2009) from:https://www.curtin.edu.au/news/media-release/research-shows-corrosion-costs-the-local-economy/

2 T HR CI 12002 ST Durability Requirements for Civil Infrastructure (2017), Transport for New South Wales.

3 Guideline for the Preparation of Road Structure Durability Plans, TMR, Queensland Government

4 Durability Planning, Concrete Institute of Australia (2014)

5 BS 7543:2015 Guide to durability of buildings and building elements, products and components, British Standards,

[6] Lindsey, R (2022) Climate Change: Global Sea Level, https://www.climate.gov/news-features/understanding-climate/climate-change-global-sea-level, April 19, 2022.

[7] AS 4312:2019 Atmospheric corrosivity zones in Australia, Standards Australia

[8] AS 5100.5:2017 Bridge Design – Part 5: Concrete, Standards Australia

[9] Lewis ER and SE Schwartz. 2004. Sea Salt Aerosol Production: Mechanisms, Methods, Measurements, and Models—A Critical Review. Washington DC: American Geophysical Union.

[10] BS EN ISO 12944-2: 2017, Paints and varnishes – Corrosion protection of steel structures by protective paint systems – Part 2: Classification of environments, BSI

[11] AS 5100.5:2014 Bridge Design – Part 5: Concrete, Standards Australia

[12] Ahn, J. H., Y. S. Jeong, I. T. Kim, S. H. Jeon and C. H. Park (2019). "A Method for Estimating Time-Dependent Corrosion Depth of Carbon and Weathering Steel Using an Atmospheric Corrosion Monitor Sensor." Sensors (Basel) 19(6).

[13] Yang, D., H. Mei and L. Wang (2019). "Corrosion Measurement of the Atmospheric Environment Using Galvanic Cell Sensors." Sensors (Basel) **19**(2).

[14] Wade, S. A., K. J. Begbie and A. Trueman (2007). Measuring atmospheric corrosion of industrial infrastructure using electrical resistance corrosion sensors. Corrosion Control 2007.

[15] Li, Z., D. Fu, Y. Li, G. Wang, J. Meng, D. Zhang, Z. Yang, G. Ding and J. Zhao (2019). "Application of An Electrical Resistance Sensor-Based Automated Corrosion Monitor in the Study of Atmospheric Corrosion." Materials (Basel) 12(7).

[16] AMIRA P387A Project Acid Rock Drainage (ARD) Test Handbook (May 2002)

[17] AS 2159:1995 Piling - Design and installation, Standards Australia

[18] AS/NZS 2312.2:2014 Guide to the protection of structural steel against atmospheric corrosion by the use of protective coatings – Part 2: Hot dip galvanizing, Standards Australia

[19] ISO 9224:2012 Corrosion of metals and alloys — Corrosivity of atmospheres — Guiding values for the corrosivity categories, International Standards Organization

[20] AS/NZS 2312.1:2014 Guide to the protection of structural steel against atmospheric corrosion by the use of protective coatings – Part 1: Paint coatings, Standards Australia

[21] ISO 12944-5:2019 Paints and varnishes — Corrosion protection of steel structures by protective paint systems Part 5: Protective paint systems, International Standards Organization

[22] AS 4506-2005 Metal finishing—Thermoset powder coatings, Standards Australia

[23] AS 3715-2002 Metal finishing—Thermoset powder coatings for architectural applications of aluminium and aluminium alloys, Standards Australia

[24] ISO 12944-5:2019 Paints and varnishes — Corrosion protection of steel structures by protective paint systems Part 5: Protective paint systems, International Standards Organization

[25] *fib* Model Code for Concrete Structures (2020), Lausanne Switzerland

AUTHOR DETAILS

	Dr. David McDonald is a Principal Civil Engineer at Vinsi Partners, a position he has held since 2019. He is frequently engaged to develop durability plans utilising his over 35-years of technical expertise in concrete, coating and corrosion. During his career he has been involved in the evaluation and recommendation of materials for bridges, roadways, tunnels and wharf structures throughout the world, including the USA, Canada, Australia and Pacific Islands. He has a Ph.D. in Civil Engineering, is a Fellow of ACI and Engineers Australia and is a Professional Engineer (NER) and Registered Engineer in Queensland (RPEQ). He has written many papers on concrete durability, corrosion and the use of construction materials.
	Andrew Lu is the corresponding author of this paper. He is a Graduate Materials Engineer from UNSW and is a subcommittee member of the NSW Branch of the Young Corrosion Group of the Australasian Corrosion Association (ACA). At Vinsi Partners he has conducted a number of durability plans for infrastructure projects including rail, architecture, bridge and tunnels.
	Brett Eliasson has 30+ years' experience in the consulting fields of Structural, Civil and Remedial Engineering coupled with Materials Technology. He has ensured the successful outcome of a wide range of investigation, refurbishment, and remedial works to buildings and civil structures. This has included the condition assessment and remedial strategy development of infrastructure assets. Having been involved with both new and existing structures he has developed a strong awareness of problems that can occur in the design and construction processes. He uses this knowledge to provide feedback and add value to new projects as well as identify faults in existing structures.

ACKNOWLEDGMENTS

The authors would like to thank the support of Vinsi Partners in preparation of this paper and in particular Warren Green, who reviewed and provided suggestions on the final version.

CORROSION & PREVENTION
CAIRNS 2024

THE AUSTRALASIAN CORROSION ASSOCIATION INC.

Identification & Control of a Persistent Sulphate Reducing Bacteria (SRB) & Tuberculate Microbial Induced Corrosion (MIC) Consortium from a Mild-Steel Closed-Loop Building Cooling System by a Novel Combination Strategy Relying on anti-Corrosion & anti-Microbial Additives.

Dr S. Nearhos[1,*]

[1]Baseline, Qld, 4014, Australia

**Corresponding author email: steven@baseline.com.au*

ABSTRACT

A Brisbane tower building with a closed loop air-conditioner cooling system was diagnosed in 2014 to have had corrosion in Victaulic joints between mild-steel pipe segments. This corrosion at the time was identified as from Sulphate Reducing Bacteria (SRB). No Iron Reducing Bacteria (IRB) or Iron Precipitating Bacteria (IPB) were identified and repairs were enacted. In 2021 when the joints again started leaking *Baseline* was engaged to investigate the causation. SRB were resolved but in addition both IRB and IPB were also found. The IPB were producing large fist sized polyps on the walls of the vertical pipes. The outside of these polyps was oxidized, and the inside was reduced and so able to actively dissolve and erode the pipe walls. It was observed that the historical management regime by a third-party company was focused on minimizing corrosion, however the chemicals used including Phosphorus, Nitrite and Sulphite directly or indirectly could be seen to have supported a persistent Microbial Induced Corrosion (MIC) consortium.

It was agreed by the building owners that a modified liquor based on recirculation of low mineral water might starve the problem microbes of many of their essential growth factors and so encourage a more sustainable system. *Baseline* also enacted routine testing for Heterotrophic Plate Counts (HPC) and plate based assays for IPB as these were more abundant and so more useful corrosion indicators than SRB. Minimal Inhibitory Concentration assays on isolated corrosion microbes indicated the Benzisothiazolinone (BIT) could be an effective biocide. A local manufacturer *Everything Water* was commissioned by Baseline to build a water treatment and dosing system, and in September 2022 the strategy was implemented.

Initially almost the entire population of HPC was also IPB, however over-time the proportion of IPB diminished and an adapted population of BIT degrading and Sulphate releasing microbes developed. This prompted the quest for alternative biocides and anti-corrosion agents without Nitrite, Sulphate or Phosphate and the microbial density by September 2023 had reduced by around 7 logs. Recirculation water turbidity as at July 2024 was less than 1 NTU and IPB, SRB and IRB were all less than 1CFU/100mL.

Keywords: Microbial Induced Corrosion, Minimal Inhibitory Concentration, (MIC) Sulphate Reducing Bacteria, SRB, Iron Reducing Bacteria, (IRB), Iron Precipitating Bacteria, (IPB) Closed Loop Cooling System, Mild Steel, Victaulic Joints, Anti-microbial, Bacteria Benzisothiazolinone BIT

INTRODUCTION

The Brisbane residential tower building where Microbial Induced Corrosion (MIC) occurred has a roof-based 30kL closed loop recirculation system feeding *ca 40cm* diameter mild steel pipes joined by Victaulic joints. At each level these steel pipes distribute water via subsidiary copper pipes to apartment air conditioners. Apart from the crevices between joints of the steel pipes the presence of these different metals further complicates corrosion minimization of recirculation water.

The building was constructed in the early 2000's but in 2014 leakage started to occur at the Victaulic joints. At that time Sulphate Reducing Bacteria (SRB) were identified and some pipes and faulty seals were replaced. When SRB are present the sulphide which is produced can interact with the Iron Oxide layer at the metal-water interface and cause pitting corrosion. The Iron Sulphide which is formed by SRB at the point of corrosion is chemically fragile and in the presence of oxygenated water the precipitate is easily oxidized so that sulphate is regenerated and so becomes available for a further iteration of corrosion by SRB.

Before Baseline's involvement the recirculation liquor contained Brisbane City council tap water which was supplemented with proprietary additives. Without necessarily disclosing the composition of these additives these included Sodium Sulphite which played a role as an oxygen scavenger and incidentally also increased the amount of recirculating sulphate,

In alkaline solution the spontaneous reaction between oxygen and sulphite is as below:

$$2SO_3^{2-} + O_2 \rightarrow 2SO_4^{2-} \quad +1.3v$$

Nitrite solution was another additive. The spontaneous protective alkaline reaction between Nitrite and Ferrous hydroxide is as below:

$$4Fe(OH)_2^{-} + 2NO_2^{-} + 3H_2O \rightarrow 4Fe(OH)_3^{-} \; N_2O \; + 2OH^{-} \quad + 0.71V$$

Bacteria can also utilize Nitrite and Sulphate, the starting and end points of the above reactions, both to supply respiratory Oxygen and also as sources of inorganic Nitrogen and Sulphur. Similarly, phosphates which are also often added to prevent corrosion are at the same time essential for microbial metabolism and so when phosphates are used to moderate corrosion, they also have potential to be diverted to sustain contaminant microbes recirculating in systems and so also to aggravate corrosion problems.

The process which was followed to tease out the issues was to make initial isolations of bacteria and evaluations of the chemistry of waters and corrosion deposits and then to develop an implementation and streamlined assessment strategy.

EQUIPMENT, PRODUCTS AND METHODS

With samples recovered from the recirculation system microbiological assays were undertaken to establish the presence or absence of SRB, IRB and IPB. Initially all these tests were done by the MPN methods using the media from Postgate (APHA, AWWA, WEF 1995) Gould *et al.* (2000) and UK Environment Agency (1998). The incubation period for these MPN procedures is quite long and when the infestations had been established other more streamlined tests were implemented for faster monitoring.

Baseline adapted a 96 well micro-titre plate assay for Minimal Inhibitory Concentration from Kowalska-Krochmal B, & Dudek-Wicher R (2021), Sub-contracted chemical analyses were enacted on both the waters and pipe deposits.

Results

In August 2021 samples were collected from various locations within the building and the water samples included sediment which was microscopically examined. Figure 1 is a photomicrograph of that sediment.

Figure 1.: Phase contrast image (x 1000) of sediment from recirculation water in August 2021

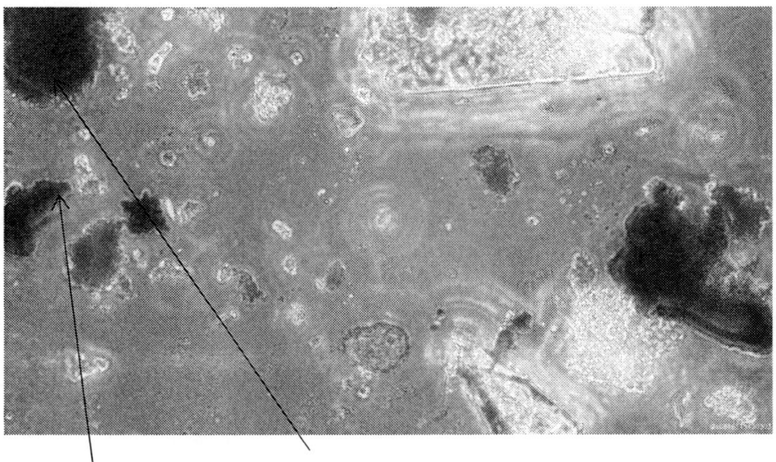

MPN assays were enacted on samples from 3 building levels for SRB, IRB, IPB and also for Heterotrophic Plate Counts (HPC) by Direct count. The MPN assays used 30 replicate tubes and had a sensitivity of less than one CFU/100mL to 23 000 CFU/100mL. These results are depicted in Table 1. In summary all three groups of corrosion bacteria were resolved however the SRB were least common and SRB and IRB seemed more prevalent in samples which also had sediment present. This is consistent with the recovery of these microbes from Victaulic seals in 2014. The Iron Precipitating Bacteria were most abundant including in the samples which had sediment where the population exceeded the capacity of the MPN method. The heterotrophic direct plate count data paralleled these population by tube methods.

At this point it was possible to inspect the pipe walls and these were covered by a blanket of tuberculate polyps (Figure 2) with downward pointing dendritic encrustations and a large encrustation was broken free. The fist sized polyps (Figure 3) were tightly fixed onto the walls of the mild steel pipe and bright orange on the outside but dark at the point of contact with the pipe (Figure 4). There were signs of pitting erosion on the under-surface of these polyps. The indicative analysis of the inner and outer polyp material (Figure 5) shown in Table 2 indicated similar total S but higher reduced sulphide in the inner material. This is consistent with the presence of a microbe consortium active of Sulphate Reducing Bacteria and Iron Reducing Bacteria in the inner core of the polyps and IPB actively precipitating the mobilized Iron on the growing outer dendritic tip of the polyps.

Table 1: The abundance of microbes in water from three building levels tested for SRB, IRB& IPB by MPN and incubated 14d @ 28°C and or 5d for HPC by Direct counts.

Sample Description	Sample Number	Baseline Sample Number	Sulphate Reducing Bacteria (SRB) Adapted from APHA (1995) in house Most Probable Number (MPN) 14d @ 28°C		Iron Reducing Bactera (IRB) Adapted from Gould *et al.* (2003) *in house* Most Probable Number (MPN) 14d @ 28°C		Iron Precipitating Bacteria (IPB) Adapted from UK UK Environment Agency (1998) Most Probable Number (MPN) 14d @ 28°C		Heterotrophic Plate Count (HPC) R_2A 5d @28°C
			MPN	95% Confidence Limits	MPN	95% Confidence Limits	MPN	95% Confidence Limits	CFU/mL
Level 68 Flushed Dosing Pot 07:30h 16/08/2021	1	26847	<1CFU SRB/100mL	<4 CFU* /100 mL	77 CFU IRB/100mL	24 to 250 CFU IRB/100mL	> 23 000 CFU IPB/100mL All tubes converted 48h	> 12 000 CFU IPB/200mL	8 000 CFU HPC/mL
Level 42 Flushed Deadleg 8:00h 16/08/2021	2	26848	2CFU SRB/100mL	<1 to 6 CFU /100 mL	220 CFU IRB/100mL	75 to 670 CFU IRB/100mL	> 23 000 CFU IPB/100mL All tubes converted 48h	> 12 000 CFU IPB/200mL	160 000 CFU HPC/mL
Level 25 Flushed Dead Leg 08:30h 16/08/2021	3	26849	<1CFU SRB/100mL	<4 CFU /100 mL	48 CFU IRB/100mL	15 to 160 CFU IRB/100mL	> 23 000 CFU IPB/100mL All tubes converted 48h	> 12 000 CFU IPB/200mL	5 600 CFU HPC/mL

* CFU Colony Forming Units

A dissecting microscope image of the tip of a polyp indicated a fine oxidized iron polyp (Figure 6) and when this was inoculated to an agar plate of WR medium a pure culture of IPB was recovered (Figure7). Similarly other microbes including IRB and SRB were isolated in our laboratory and these were tested against two biocides which were available through the previous water treatment supply company. These were a dialdehyde UN 2819 and Thiazolinone UN 3077. The Minimal Inhibitory Concentration assays were based on a standard protocol used for antibiotic susceptibility testing similar to Kowalska-Krochmal B, & Dudek-Wicher R (2021) but using resazurin as an indicator for bacterial growth and these are shown in Figure 8 and summarized in Table 4.

These initial assays indicated the Thiazolinone was more effective than the dialdehyde. The obligately anaerobic SRB were done in separate tubes and were more biocide-sensitive than all except the IRB which were also quite biocide-sensitive. IPB were most biocide-resistant.

The water chemical analyses indicated that high levels of elements which are essential for microbial growth present in the recirculating water. These elements included Nitrogen, Phosphorus, Sulphur, Potassium and Iron. Other elements including Calcium, Chlorine, Magnesium and Carbon were also present and these other elements might play a role in scale formation or copper corrosion. The carbonaceous material was assessed both for the presence of inorganic and organic carbon. It was unclear whether the initial assays indicated biomass or some other organic additive. The inorganic carbon was believed to have been respiratory carbon dioxide.

Based on these results, it was proposed by Baseline to refill the system as it stood in August 2021 with demineralized water to minimize Chloride, Nitrite, Phosphate, Potassium and Sulphate. To further minimize Sulphate in recirculating water Hydroquinone was used as an alternative oxygen scavenger to Sodium sulphite. Ciuba (1980) from Betz laboratories described this alternative oxygen scavenger application in a since expired patent.

Our client consented for the works to proceed and *Baseline's* Brisbane based partner *Everything Water* was engaged to design and build an adapted recirculating water treatment system. This new system had the capacity to demineralize water, filter and remove particulate debris and also to capture residues of the spent oxygen scavenger and to introduce measured doses of water-treatment chemicals.

The refill was enacted in September 2022 when the system was charged with around 30kL of demineralized water and treated with oxygen scavenger followed by adsorption of spent scavenger onto activated carbon. Thiazolinone (BIT) was added and also a solution of metal protectant Tolytriazole the pH was adjusted upwards with caustic soda. Measurements were made of pH, Conductivity and turbidity and regular direct counts were enacted for HPC and IPB. The data were recruited over six months and these are graphically presented in Figure 9. IPB counts were done direct onto WR medium by the drop count method Figure 10. This method had a sensitivity of 10 Colony Forming

Units (CFU)/mL after counts of HPC & IPB were reduced to less than detection resolution was improved by a thousand fold by using 100mL 0.2μM membrane filtered samples (Figure 11).

Figure2: Pipe walls showing carpet of polyps with downward facing Dendritic projections.

Figure 3: Fist sized Tuberculate Polyp recovered from Pipe wall. The image represents the dorsal surface.

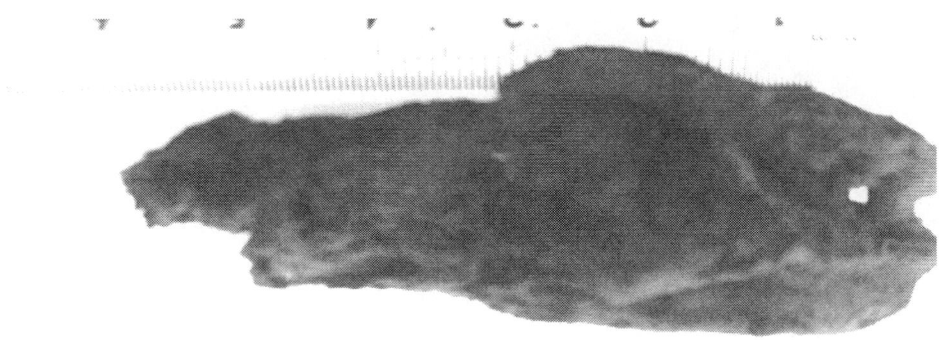

Figure 4: Pipe wall showing dark areas with pitting where polyps were attached.

Figure 5: Pulverized inner and outer material from Polyps

Table 2: Comparative Indicative Chemical analysis of internal and External Polyp Deposits

Metal deposits samples August 30th 2021.	Inner Pipe deposit	Outer Pipe Deposit
pH	6.4	4.8
Conductivity µS/cm	1520	1670
Moisture Content mg/kg	37400	34300
Aluminium mg/kg as Al_2O_3	2436	990
Calcium mg/kg as CaO	13626	9740
Carbon Inorganic mg/kg as CO_2	55733	1000
Carbon Organic mg/kg as CO_2	16133	320
Chloride mg/kg as Cl^-	4120	3660
Chromium mg/kg Cr_2O_3	287	50
Copper mg/kg as CuO	8380	4910
Iron mg/kg as Fe_2O_3	465000	463000
Magnesium mg/kg MgO	1800	1080
Manganese mg/kg MnO_2	2040	1360
Molybdenum mg/kg MbO_3	166	104
Nitrogen Kjeldahl as NO_2	1511	1511
Phosphorus mg/kg as P_2O_5	2270	2150
Potassium mg/kg K_2O	10	<50
Silicate SiO_2	ND	ND
Sodium mg/kg Na_2O	180	140
Sulphur as Sulphide mg/kg S_2^-	440	267
Sulphur as Sulphate mg/kg SO_3^-	2880	2800
Zinc mg/kg as ZnO	1420	1130
Imputed from the analysis	809g/kg	729g/kg

Figure 6: Filamentous dendrite from IPB ca 1cm in length from the tip of polyp (mm/cm scale).

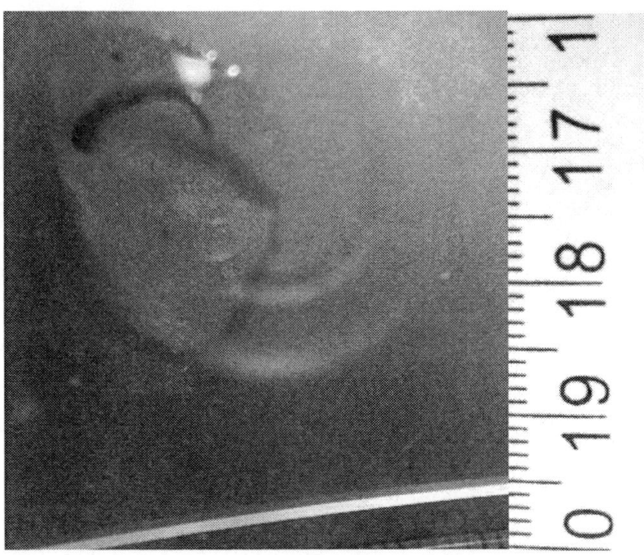

Figure 7: Pure culture of IPB from Dendrite on WR medium

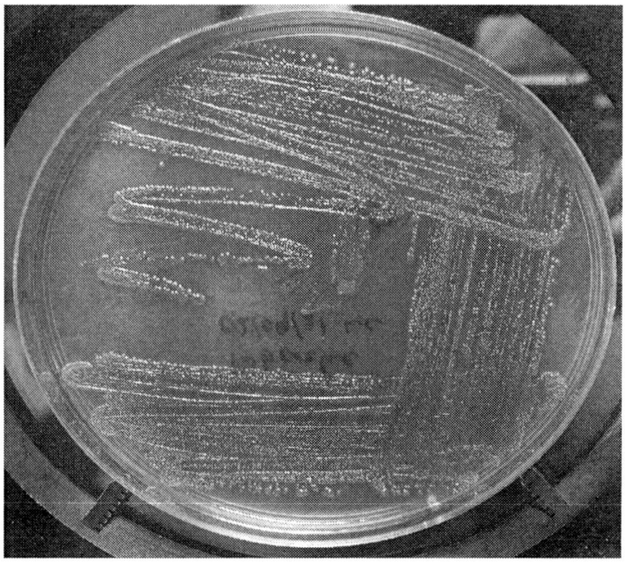

Table 3: Water Chemical analyses were undertaken on the same three samples assessed for microbial contamination in August 2021.

	Level 68	Level 42	Level 21
Sample Date	August 16th	August 16th	August 16th
pH	9.41	9.42	9.38
Conductivity µS/cm	962	962	964
Carbonate Alkalinity mg/L CaCO3	76	81	71
Bicarbonate Alkalinity mg/L Ca CO3	115	116	124
Total Alkalinity mg/L CaCO3	190	196	196
Sulphate as SO4 mg/L	36	36	36
Chloride mg/L	105	105	106
Total P as PO4 mg/L	15.4	16.7	17.3
Nitrite as NO2 mg/L	77.2	75.6	75.6
Calcium mg/L	12	12	12
Magnesium mg/L	5	5	5
Sodium mg/L	178	174	177
Potassium mg/L	4	4	4
Total Iron mg/L	0.09	1.92	0.12
Total Organic Carbon as C mg/L	12.8	11.9	4.6

Parameters highlighted in Red have significance for microbe growth.

Parameters highlighted in Green are relevant for copper corrosion.

Table 4: Comparative Efficacy of Two Proprietary Biocides

Strain	Origin	UN2810	UN3077
1 *Acidithiobacillus* sp IPB	Qld MBS 2009	1%	0.008%
2 NSW Bore IPB	NSW Bore 2021	0.5%	0.004%
3 Pipe Scale IPB	AT Level 65	0.5%	0.008%
4 Pipe Tubercle IPB	AT Level 65	1%	0.03%
5 IPB minor	AT Level 21	1%	0.06%
6 IRB	AT Level 42	1%	0.05%
7 IRB	AT Level 21	0.25%	0.016%
8 IPB	AT Level 42	1%	0.004%
9 *Shewanella* sp IRB	Queensland, IRB	0.004%	0.004%
10 IRB	Bris 2021	0.5%	0.008%
11 SRB	AT Level 42	<0.03%	<0.008%

UN3077 Thiazolinone

UN2819 Aldehyde

Figure 8: Six strains in duplicate. A quarter of the test panel. Red colour is from Resazurin a dye which becomes reduced when growth occurs. This result is from the di-aldehyde UN2819 biocide which was least successful. Iron Reducing Bacteria were entirely intolerant of this biocide.

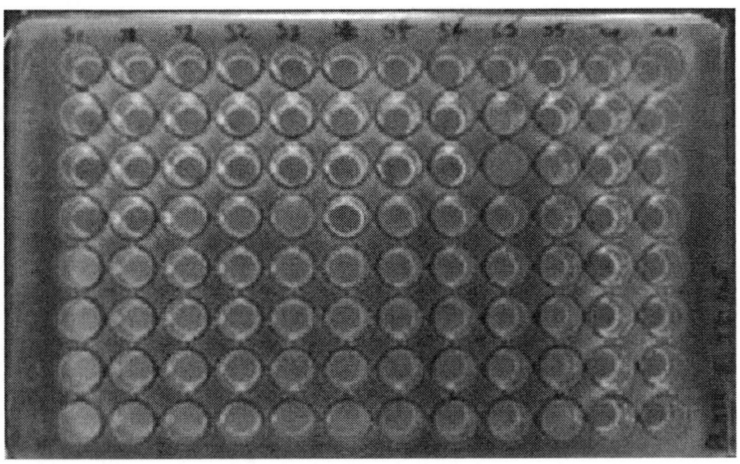

Figure 9:Plot of Recirculation Water Population of HPC and IPB Indicator Microbes (CFU/mL) over 6 months.

Baseline Ref 90301, March 2023

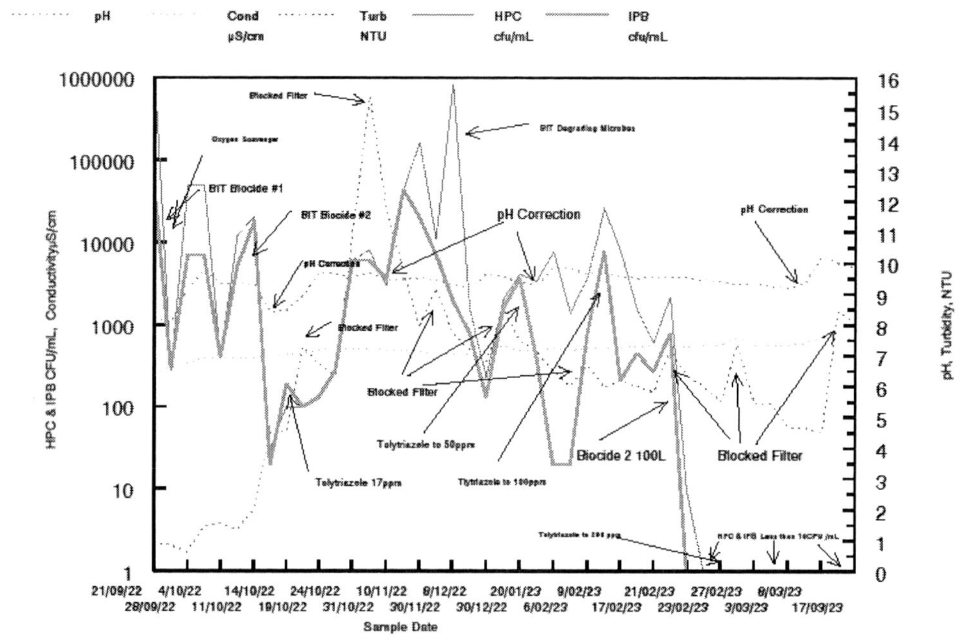

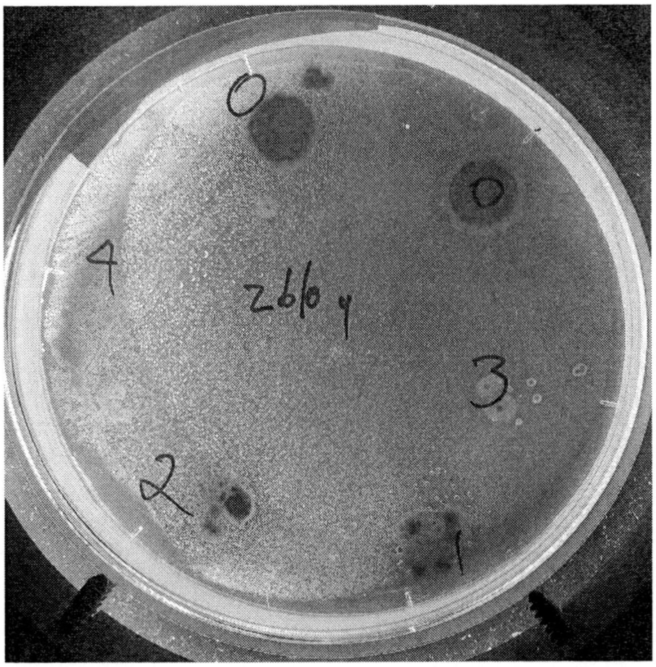

Figure 10: Plate count for IPB indicating around 300 000 CFU/mL

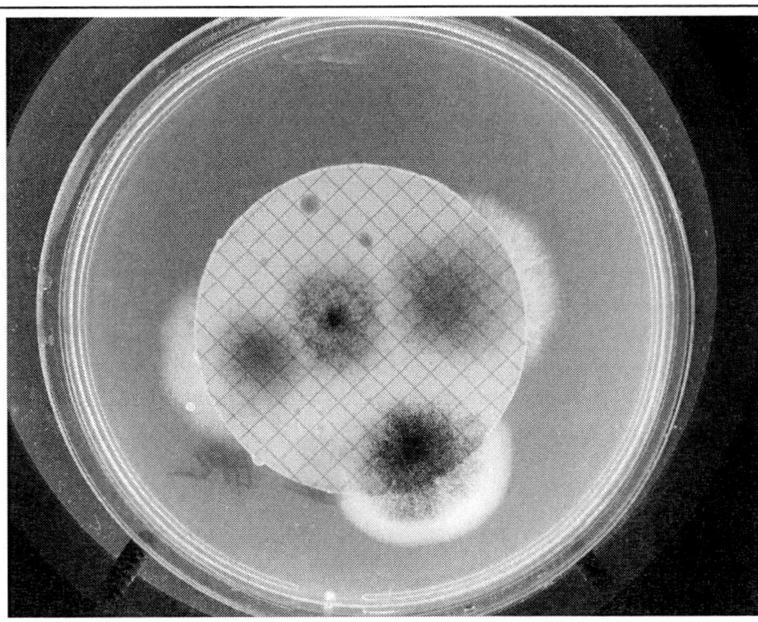

Figure 11: Membrane filter count for 100mL sample and the resultant

Heterotrophic Growth of non-IPB Bacteria and Fungi.

Discussion & Conclusion.

The initial conductivity after the system was refilled was less than 500µS/cm. The pH was maintained around pH 9.5 by periodic addition of caustic alkali prepared in distilled water. While the pH was less than 10 it appeared that pH dosing was followed by a spike of the bacterial population. The pH was not raised beyond 10 to ensure there was no secondary alkaline hydrolysis of the hydraulic services.

The Turbidity ranged between 1 and 15 NTU and was highest just before filter-socks were replaced. Before BIT was added the population of HPC in Figure 9 this corresponds to a purple line was almost entirely comprised by IPB (orange). The population count of these heterotrophs in September 2022 was around 300 000 CFU/mL.

Over time this correspondence between HPC and IPB fell away and it came apparent and was later chromatographically confirmed (Bruce Heath) that the BIT was being chemically attacked by an adapted population of BIT degrading non-IPB. Subsequent Chemical testing also indicated that sulphate rose as a result of this chemical degradation. Because BIT was found as ineffective a second biocide was selected and with this chemical microbe counts were reduced to less than 10 CFU/ml. At this microbe stage test methodology was revised and samples were based on membrane filter counts this improved the test sensitivity to 1 CFU/100mL.

The system was refilled again in September 2023 and a further metal preservative as introduced to the regime.

As at July 2024 three years after Baseline's initial involvement the persistent MIC problem is stable and under control.

Tests for MIC indicated less than 1/100mL for SRB IRB and IPB

Since 2022 when the new treatment system was installed the most abundant corrosion microbes, IPB reduced by more than seven orders of magnitude.

Since September 2023 Turbidity has been maintained at less than 1NTU and Total Iron in water has been less than 0.05mg/L and Total Copper has been less than 0.01mg/L

Acknowledgments

The author wishes to thank Rowan Dann and Everything Water for fabricating the water treatment and chemical dosing system which allowed the shift of the present water treatment from concept to reality.

The author wishes to thank Bruce Heath from Industrial Cleansers for the chromatographic assessment of the decomposition of the initially used BIT biocide.

The author is also grateful to our clients who open-mindedly provided the platform which allowed the implementation and evolution of this novel water treatment strategy.

References.

1. APHA,AWWA,WEF (1995) Standard Methods for the Examination of Water and Wastewater 9240D.3 Sulfur Bacteria. pp 9.82-3.

2. Ciuba S.J. (1980) Hydroquinone as and Oxygen Scavenger in an Aqueous Medium US Patent 4282,111 Awarded August 4th 1981 . Expired .

3. Cochran W.G. (2950) Estimation of Bacterial Densities by Means of the Most Probable Number. *Biometrics* **6**: pp 105-116.

4. Gould W.D. Stichbury M. Lortie. L. & Blowes D.W. (2000) An MPN Method for Enumeration for Iron Reducing Bacteria. https:/pdf.library.Laurentian.ca

5. El Ibrahimi & Guo L (2020) Azole-Based Compounds as Corrosion inhibitors for Metallic materials *Intertech Open* 28pp

6. Kowalska-Krochmal B, & Dudek-Wicher R (2021) The minimal Inhibitory Concentration of Antibiotics: Methods, Interpretation, Clinical Relevance . *PATHOGENS* (**10,** 165-186.

7. Moeller T (1952) *Inorganic Chemistry.* John Wiley & Sons. Canada. 922 pages.

8. UK Environment Agency (1998) The assessment of taste, odour and related aesthetic problems in drinking waters. Section 6. 6 Iron Precipitating Bacteria pp13-15.

Author Details

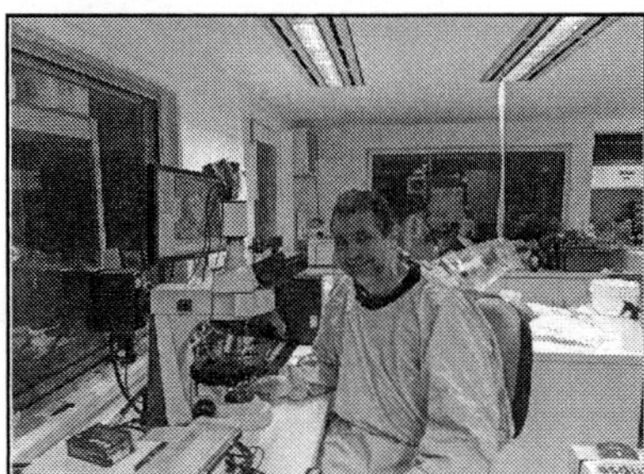

Dr Steven Nearhos is the founder of Baseline, a Queensland based Environmental Microbiology Services company with a large and dedicated private laboratory where since 2010 many thousands of assays for corrosion microbes including SRB, IRB and IPB have been enacted. These corrosion microbes have been isolated from ground and surface waters and from buildings in sprinklers or in cooling systems such as in the current case study.

Baseline also works in the area of microbial contamination and water-quality management and has adapted strains of rogue bacteria like the ones described here degrading BIT, for remediation of xenobiotic chemicals, and others including for nitrification and methane production. These days Steven has a lesser role in Baseline and focuses on R&D and mentoring staff.

CORROSION & PREVENTION
CAIRNS 2024

THE AUSTRALASIAN CORROSION ASSOCIATION INC.

Analysis Of Dust Levels Created By Centrifugal Abrasive Blasting

R. McHendrie[1], D. Fosdike[1], D. McDonald[1*], K. Sellars[1]

[1]BAE Systems Maritime Australia, SA, 5017, Australia

**Corresponding author email: darren.m.mcdonald@baesystems.au*

ABSTRACT

The prevalence of residual dust particles generated from industrial abrasive blasting processes has prompted extensive research to assess the risk of resultant premature coating failure on steel work pieces. This study presents an analysis of dust accumulation within a centrifugal abrasive blasting surface treatment line (STL), with a focus on pre- and post-cleaning assessments, tuning of processing parameters, blast media assessments, and dust chemical and morphological characterisation methods. In consultation with the Commonwealth of Australia and relevant paint manufacturers, adhesive tape testing shall be conducted in accordance with the industry standard ISO 8502-3. This paper seeks to engage industry to reach consensus of interpretation of densities and size of dust in relation to coating performance and ISO 8502-3, in addition to a comparative assessment of an alternate industry standard, AS 3894.6.

Keywords: steel, dust, shipbuilding, blasting, centrifugal abrasive blasting.

INTRODUCTION

Effective management of dust in industrial settings is critical for maintaining the performance and longevity of surface treatments and protective coatings. Reduction in adhesive performance is considered a primary casual factor in the consequent reduction of both chemical resistance and corrosion protection, where dust - defined as finely dispersed particles of matter poses challenges to the aforementioned properties. In industries such as shipbuilding where the preservation of high-grade steel is paramount, effective dust management is imperative to prevent premature coating failure and in support of fostering a strategy. Raw steel employed in the construction of maritime vessels undergoes a blast and primer painting process, followed by processing using specialised equipment to cut the steel into plates and profiles prior to fabrication. The abrasive blasting process, however, introduces significant dust levels onto steel substrates, potentially exposing steel to diminished coating adhesion and possible corrosion through moisture absorption.

To adequately assess the condition of blast-cleaned steel surfaces and the accumulation of dust, legislative documentation from the International Maritime Organisation (IMO) and literature produced by the relevant paint manufacturers have provided details of surface preparation by blast-cleaning, including the specification of surfaces free from contamination. In an effort to deliver a sustainable, sovereign naval shipbuilding capability that will serve the Government and Navy, cleanliness assessments of steel for the maritime defence industry have been undertaken through adherence to ISO 8502-3 [1]. This standard [1] dictates cleanliness assessments for steel surfaces prior to paint application, namely the determination of dust quantity and particle size using a pressure-sensitive tape method. This test involves the visual examination of dust adhered to tape, where a dust quantity R size C the maximum allowable requirement for steel in accordance with the IMO.

Despite efforts to address dust levels historically, challenges persist due to the subjective and qualitative nature of dust tape test results, highlighting the necessity for quantitative assessment. Furthermore, an alternate standard AS 3894.6 [2] has been used in industry which possesses dissimilar definitions of dust size classes and quantities compared to ISO 8502-3. This discrepancy underscores the need to achieve consensus through the establishment of semi-quantitative benchmarks adherent to each respective standard. In an effort to encourage objectivity, this paper therefore advocates for the adoption of semi-quantitative methods such as optical microscopy, multiphase particle analysis, scanning electron microscopy (SEM), and energy dispersive X-ray spectroscopy (EDS) to improve the reliability and precision of ISO 8502-3 evaluations. In doing so, this research aims to foster industry consensus on evaluating dust densities and sizes.

This case study is structured with sections covering the introduction, relevant methodologies, results, interpretation and discussions, and insights gained, all directed toward reducing subjectivity and ambiguity in ISO 8502-3 evaluations.

METHODS AND MATERIALS

SURFACE TREATMENT LINE FUNCTIONAL DESIGN

The STL used in the creation of the bulk of the data presented herein is an automated blasting and shop priming line, depicted in Figure 1. It is capable of processing both profiles and plates in the horizontal plane and can accommodate raw material up to 13,000 mm x 3,500 mm x 25 mm. The line comprises the following systems:

- Infeed conveying system
- Blow-off machine
- Pre-heating oven
- Blasting unit
- Painting unit
- Curing oven
- Marking station

The processing of raw materials through the STL is facilitated by transfer tables and conveyors. The blasting station is powered by eight turbines four positioned above and four below. Each turbine is equipped with six blades that centrifugally propel blast media onto the workpiece, effectively removing mill scale, rust, and surface contaminants. Post blasting, a brush and blow-off unit is used to remove any remaining blast media from the work pieces. The STL also incorporates a dust collection unit which is used to extract and exhaust dust from the blasting chamber and blow-off unit via a filter cartridge system. Following these stations, the work piece enters the inspection area where surface profile, cleanliness, and dust tests can be conducted and assessed. Post quality control inspections, the blasted work piece enters the spray booth for painting, followed by the curing oven for heat treatment. Finally, the marking station is employed for the addition of traceability data before the work piece enters the fabrication workshop for further processing.

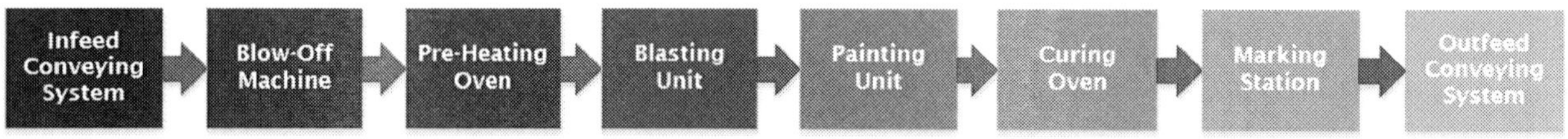

FIGURE 1: SURFACE TREATMENT LINE (STL) FUNCTIONAL DIAGRAM DEPICTING THE RELEVANT SYSTEMS (LEFT = BEGINNING OF LINE).

Dust Mitigation Strategies

High pressure water washing trials

Pre-blasted steel plates were subjected to high-pressure water washing to assess its impact on cleanliness levels and residual dust particles following blasting within the STL. The plates were treated with water jetting at 10,000 and 30,000 PSI, respectively. While improvements in pre-blast cleanliness and dust levels were observed, no significant reduction in post-blast dust levels was achieved after STL processing.

Blast media assessments

Testing of blast media, including shot and grit size and shape classification in accordance with ISO 3310-1, conductivity tests and water-soluble contamination in accordance with ASTM D4940, non-abrasive residue in accordance with SSPC-AB 2 4.1, and oil content assessment in accordance with SSPC-AB 2 4.4 was conducted, with nominal results adherent to each standard reported.

Dust Characterisation

Tape Tests

The assessment of the quantity and particle size of dust on steel surfaces prepared for painting is conducted using a pressure-sensitive tape method, as specified within ISO 8502-3 and AS 3894.6 [1] [2]. After processing steel material through the blasting unit of the STL, pressure-sensitive adhesive tape (length = approximately 150 mm, width = 25 mm, adhesion peel strength at least 190 N/m) are pressed onto the steel surface. Firm pressure is applied with a thumb, moving across the tape at constant speed. Alternatively, a spring-loaded roller may be used to apply a downward load of between 39.2 to 49.0 N while maintaining a constant speed along the tape. The method utilised is employed three times in each direction (each stroke 5-6 seconds). Following this, the tape is removed from the steel substrate and adhered to a paper display board, allowing the dust particles to be contrasted and examined visually.

The dust quantity and predominant particle size adhered to the tape was then estimated according to Figure 2 and Table 1 from ISO 8502-3. A hand lens, capable of 10× magnification, may be used to enhance visual aid.

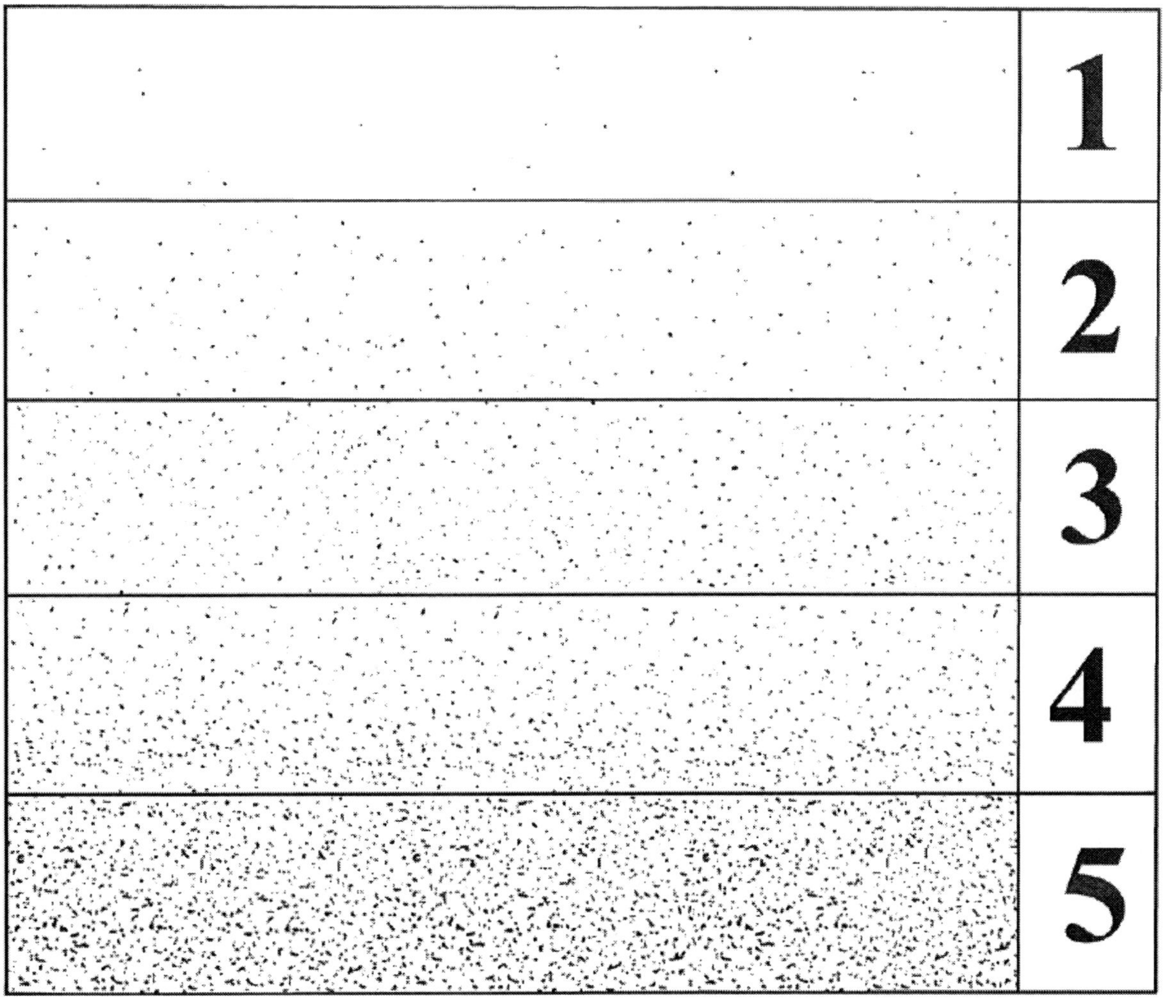

FIGURE 2: PICTORIAL REFERENCES CORRESPONDING TO DUST QUANTITY RATINGS 1, 2, 3, 4, AND 5, AS PER ISO 8502-3 [1].

TABLE 1: DUST SIZE CLASSES, DESIGNATED 0, 1, 2, 3, 4, AND 5, AS PER ISO 8502-3 [1].

Class	Description of dust particles
0	Particles not visible under 10× magnification
1	Particles visible under 10× magnification but not with normal or corrected vision (usually particles less than 50 μm in diameter)
2	Particles just visible with normal or corrected vision (usually particles between 50 μm and 100 μm in diameter)
3	Particles clearly visible with normal or corrected vision (particles up to 0,5 mm in diameter)
4	Particles between 0,5 mm and 2,5 mm in diameter
5	Particles larger than 2,5 mm in diameter

Similarly, within AS 3894.6, the test tape is compared to the visual reference standard in Figure 3, where a 0-5
excellent, less than Rating 1; Rating 5

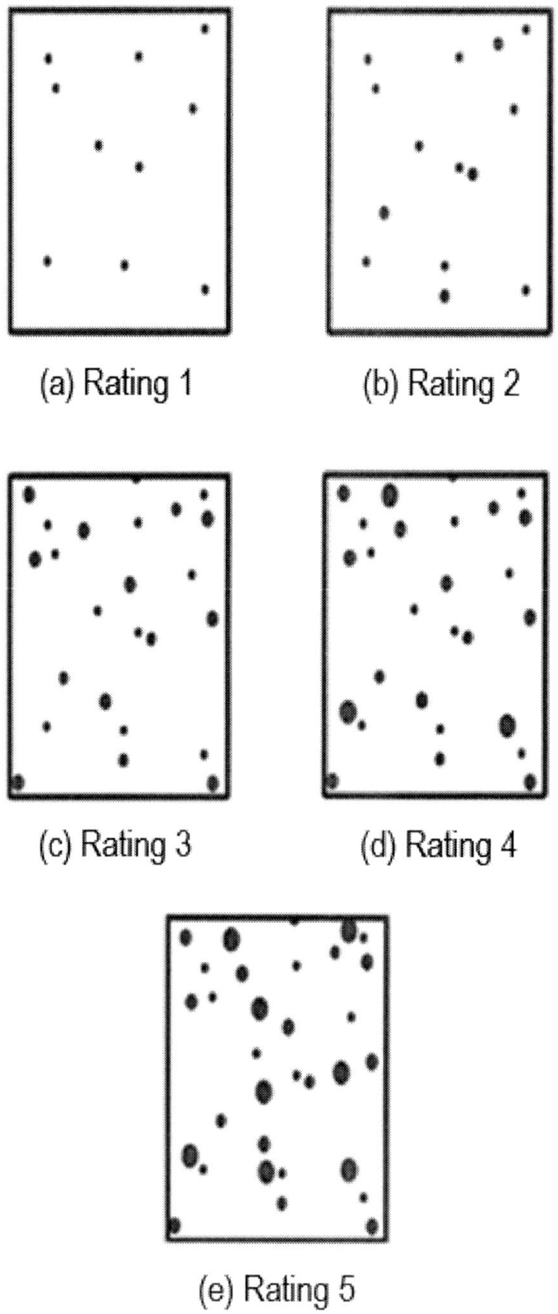

FIGURE 3: AS 3894.6 DUST DENSITY PICTOGRAMS [2].

An example tape test conducted on steel processed through the STL is depicted in Figure 4. Such sampling was conducted in accordance with ISO 8502-3, and initially compared to Figure 2 and Table 3 for dust characterisation.

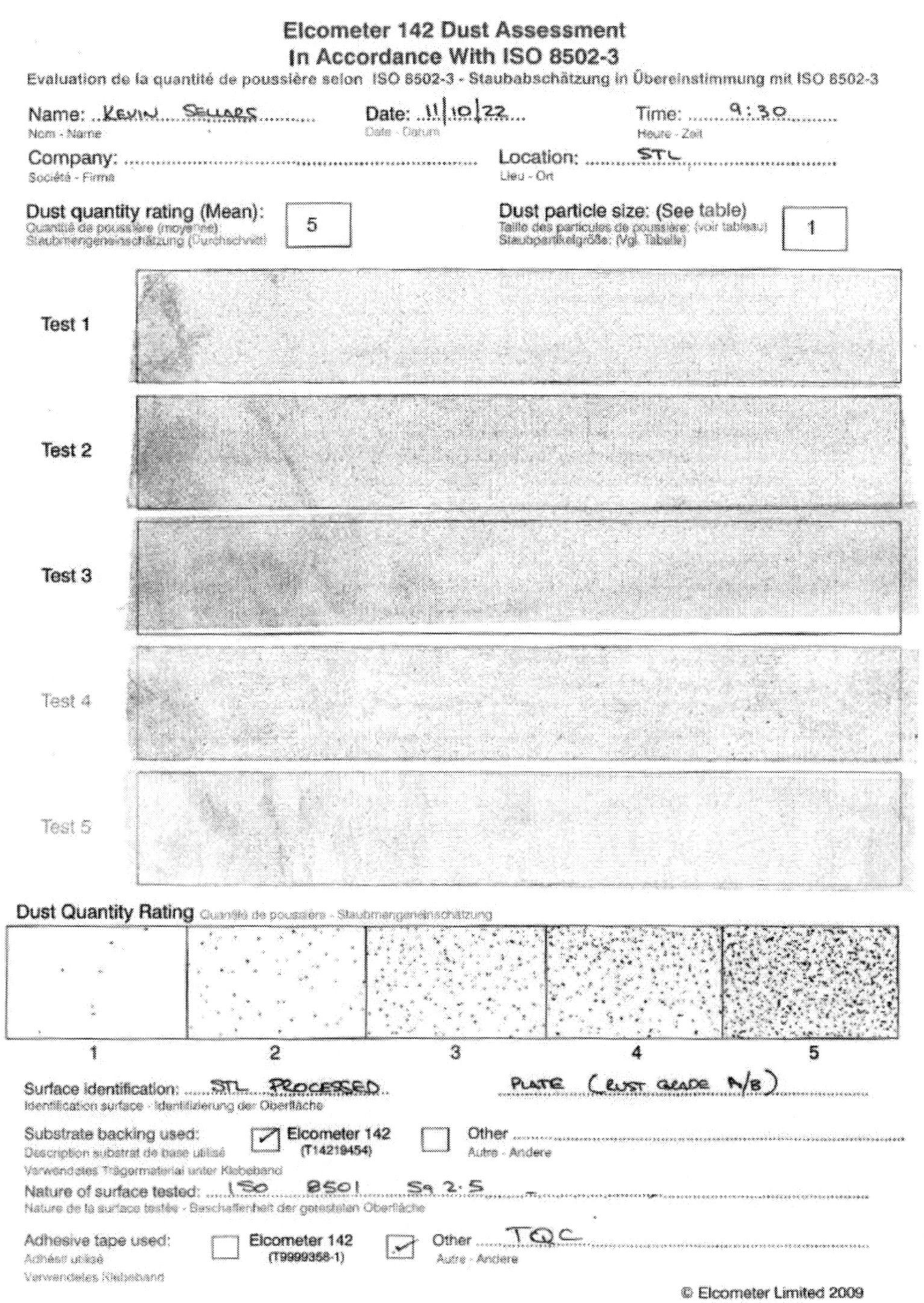

Elcometer 142 Dust Assessment
In Accordance With ISO 8502-3

Evaluation de la quantité de poussière selon ISO 8502-3 - Staubabschätzung in Übereinstimmung mit ISO 8502-3

Name: KEVIN SELLARS — Nom - Name
Date: 11/10/22 — Date - Datum
Time: 9:30 — Heure - Zeit
Company: — Société - Firma
Location: STL — Lieu - Ort

Dust quantity rating (Mean): 5 — Quantité de poussière (moyenne); Staubmengeneinschätzung (Durchschnitt)
Dust particle size: (See table) 1 — Taille des particules de poussière: (voir tableau) Staubpartikelgröße: (Vgl. Tabelle)

Test 1
Test 2
Test 3
Test 4
Test 5

Dust Quantity Rating Quantité de poussière - Staubmengeneinschätzung

1 2 3 4 5

Surface identification: STL PROCESSED PLATE (RUST GRADE A/B) — Identification surface - Identifizierung der Oberfläche
Substrate backing used: ☑ Elcometer 142 (T14219454) ☐ Other — Description substrat de base utilisé; Autre - Andere; Verwendetes Trägermaterial unter Klebeband
Nature of surface tested: ISO 8501 Sa 2.5 — Nature de la surface testée - Beschaffenheit der getesteten Oberfläche
Adhesive tape used: ☐ Elcometer 142 (T9999358-1) ☑ Other TQC — Adhésif utilisé; Autre - Andere; Verwendetes Klebeband

© Elcometer Limited 2009

FIGURE 4: TAPE TEST CONDUCTED ON STEEL MATERIAL PROCESSED THROUGH THE STL. TESTING WAS DONE IN ACCORDANCE WITH ISO 8502-3 [1]. NOTE: SCANNING OF TAPE TESTS DISTORTS TRUE DUST DEPICTIONS.

Optical Microscopy

Characterisation of dust density and sizes by multiphase porosity analysis was conducted using a Zeiss Axio Imager Z2 optical microscope paired with the ZenCore imaging software at Flinders University (Adelaide, Australia). Tape samples were imaged in an analysis area of 100 mm^2 under 10x objective lens magnification. Camera acquisition and multiphase porosity analysis settings were adjusted prior to data collection and remained constant throughout.

Scanning Electron Microscopy (SEM) and Energy Dispersive X-ray Spectroscopy (EDS)

The surface morphology and chemistry of dust collected from the abrasive blasting process within the STL was elucidated using a Scanning Electron Microscope by Flinders Microscopy and Microanalysis (Adelaide, Australia). Dust samples for SEM and EDS analysis were taken after single and double blasting from both the underside and topside of the material using adhesive carbon tape. All tape collection was completed in accordance with ISO 8502-3. Sample preparation for SEM analysis involved sectioning samples into $\sim 1\ \text{cm}^2$ pieces and mounting them to aluminium SEM stub holders. The microscope was equipped with an FEI Inspect F50 SEM paired with an EDT secondary electron detector, concentric backscatter electron detector and an Ametek EDAX EDS detector. Images were collected from a scanning height of 10 mm, a dwell time of 3 µs, and a voltage of 20 kV.

Results

Dust Characterisation

Optical Microscopy

To enhance the qualitative capabilities of the dust tape standards, OM analysis was performed on the quantity rating pictograms from Figure 2 for ISO 8502-3 and Figure 3 for AS 3894.6. The analysis shown in Figure 5 depicts the detection of dust particles for each quantity rating pictogram by the OM software, followed by the corresponding percentage area of dust detected. Evidently, significant differences in dust quantities exist between the standards, in addition to varied sequential increases in dust density between consecutive ratings. For example, a Rating 5 according to ISO 8502-3 corresponds to 14.9% dust, while a Rating 5 in AS 3894.6 corresponds to a percent area of 5.0% dust. This large variation between standards may result in poor quality control and regulatory compliance in industry. Moreover, visual inspection of the imaged pictogram ratings reveals significant density differences between the two standards, which are further corroborated by comparing the percent area of dust.

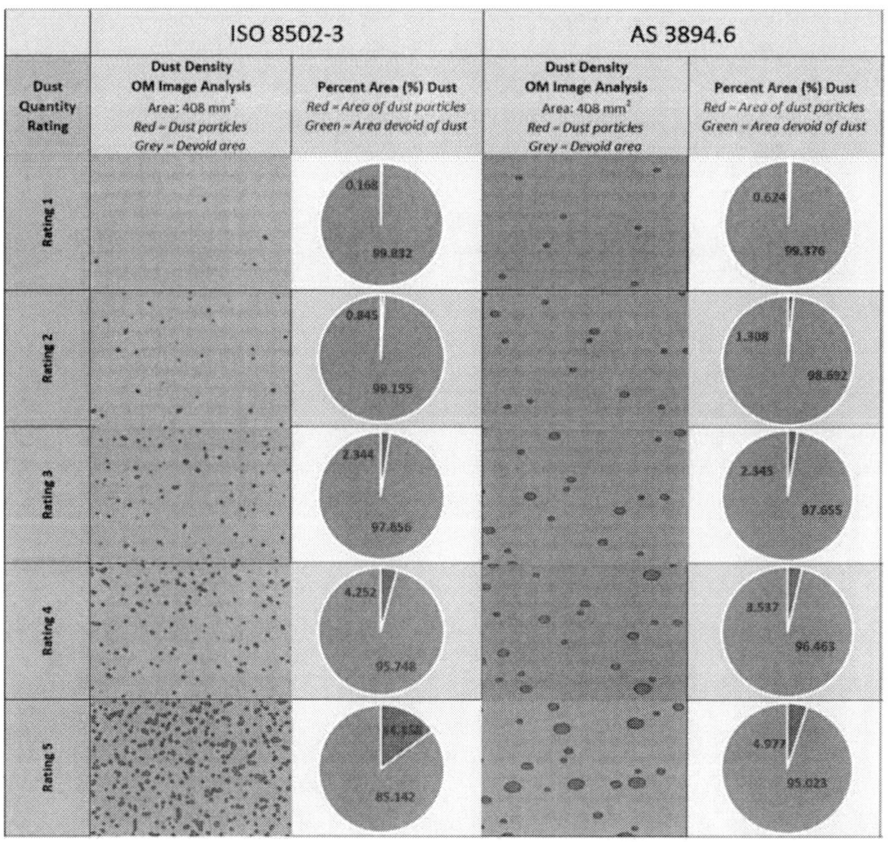

FIGURE 5: OPTICAL MICROSCOPY ANALYSIS OF QUANTITY RATINGS IN ISO 8502-3 AND AS 3894.6 (RED = DUST; AREA = 408 MM²).

Optical microscopy and multiphase porosity analysis was conducted on tape tests obtained from steel substrates subjected to various pre-cleaning and processing parameters through the STL, as shown in Figure 6. Dust distribution histograms depict the range of particle sizes within the analysis area, including feret maximum measurements for particles less than 10 μm through to those larger than 100 μm. All eleven samples primarily contain dust particles smaller than 30 μm, with the predominant size being less than 10 μm. Notably, the vast majority 91.38% of dust particles are less than 50 μm in size, which corresponds to a Class Size 1 in accordance with ISO 8502-3. The analysis in Figure 6 therefore reveals that the tape tests correspond to an ISO 8502-3 dust class size Rating Table 1), where the standard deviation is also provided to indicate the variability of data.

The assessment of dust particle quantity can be approached through initial qualitative analysis using dust density images that highlight dust particles over a 100 mm^2 sample area, in addition to semi-quantitative analysis via pie charts that depict the percentage area of dust. Comparing these results to those of the quantity ratings from each standard in Figure 5, ten of the eleven samples surpass a quantity Rating 5 (14.9% dust) from ISO 8502-3. The outlier, assessment area 5, still approaches a Rating 5 classification displaying 13.9% dust, as a Rating 4 is defined by a mere 4.3% dust (Figure 5) and is therefore not representative of this sample.

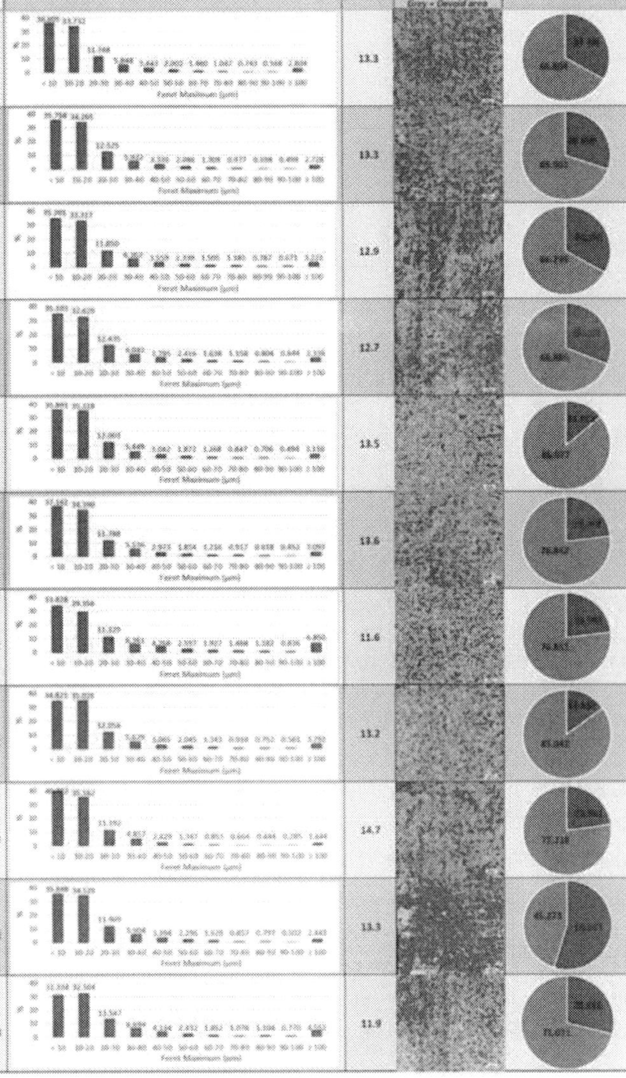

Assessment Area	Description	Date	Plate	Pre-Cleaning	Processing Parameters	Dust Size Distribution Histogram	Standard Deviation	Dust Density OM Image Analysis	Percent Area (%) Dust	Dust Quantity Rating ISO 8502-3	Dust Size Class ISO 8502-3
1	BAESAM STL Pre STL Clean Baseline	11/10/22	Top side	No	Normal		13.3			5	1
2	BAESAM STL Pre STL Clean Baseline	11/10/22	Underside	No	Normal		13.3			5	1
3	BAESAM STL Pre STL Clean	7/09/23	Top side	Pre-Cleaned HPWW	Normal		12.9			5	1
4	BAESAM STL Pre STL Clean	7/09/23	Underside	Pre-Cleaned HPWW	Normal		12.7			5	1
5	BAESAM STL Post STL Clean	13/12/23	Top side	Pre-Cleaned HPWW	Normal & Preheated 60°C		13.5			5	1
6	BAESAM STL Post STL Clean	13/12/23	Underside	Pre-Cleaned HPWW	Normal & Preheated 60°C		13.6			5	1
7	BAESAM STL Post STL Clean	13/12/23	Top side	Pre-Cleaned HPWW	Normal & No Preheat		11.6			5	1
8	BAESAM STL Post STL Clean	13/12/23	Underside	Pre-Cleaned HPWW	Normal & No Preheat		13.2			5	1
9	Dent UK STL	28/11/22	Top side	No; Double Processed	Normal & No Preheat		14.7			5	1
10	Samaras STL	6/04/22	Top side	No	Normal & No Preheat		13.3			5	1
11	Prokote STL	3/04/23	Top side	No	Normal & No Preheat		11.9			5	1

FIGURE 6: OPTICAL MICROSCOPY RESULTS FOR STL PLATES SUBJECT TO DIFFERENT PRE-CLEANING TREATMENTS AND/OR PROCESSING PARAMETERS.

Figure 7 presents a comparative analysis of tape data collection methods, specifically using a thumb versus a spring-loaded roller (note: the denser appearance of dust is due to variations in processing parameters inherent to the STL method and/or limitations of the OM software). Despite minor differences between the samples, all were consistently assigned a quantity Rating '5' with a Class Size '1', irrespective of the collection method used. While the use of a roller is advocated for improved consistency across data collection points, the findings indicate negligible variation between the methods.

Assessment Area	Description	Date	Plate	Pre-Cleaning	Processing Parameters	Dust Size Distribution Histogram	Standard Deviation	Dust Density OM Image Analysis Area: 100 mm² *Red = Dust particles* *Grey = Devoid area*	Percent Area (%) Dust *Red = Area of dust particles* *Green = Area devoid of dust*	Dust Quantity Rating *ISO 8502-3*	Dust Size Class *ISO 8502-3*
A	First blast, roller	28/03/2024	Top Side	No	Normal		14.2		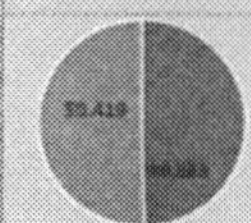	5	1
B	First blast, thumb	28/03/2024	Top side	No	Normal		14.6	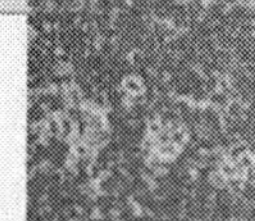	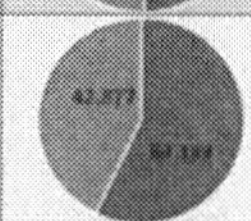	5	1
C	First blast, roller	28/03/2024	Underside	No	Normal	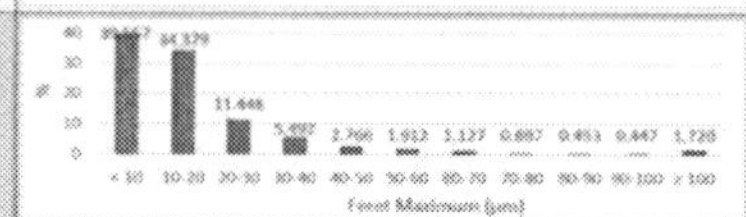	14.2	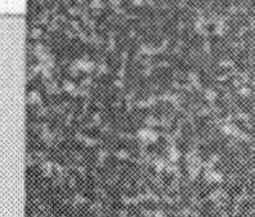		5	1
D	First blast, thumb	28/03/2024	Underside	No	Normal	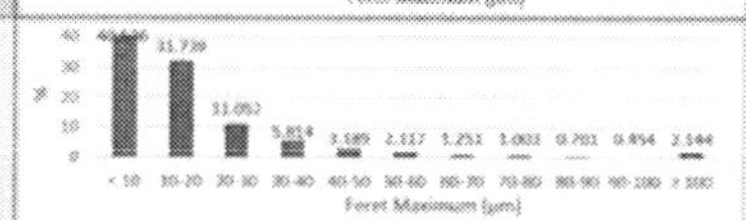	13.9		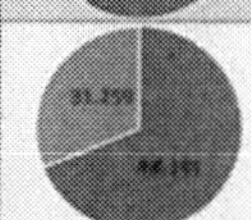	5	1
E	Second blast, roller	28/03/2024	Top side	No	Normal		13.8		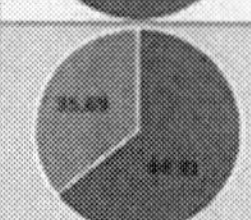	5	1
F	Second blast, thumb	28/03/2024	Top side	No	Normal	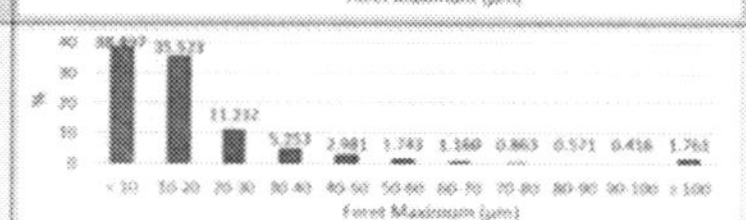	14.2	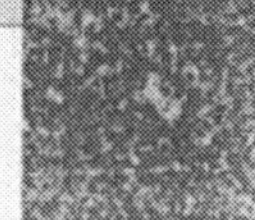	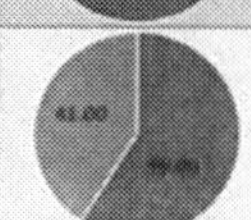	5	1
G	Second blast, roller	28/03/2024	Underside	No	Normal	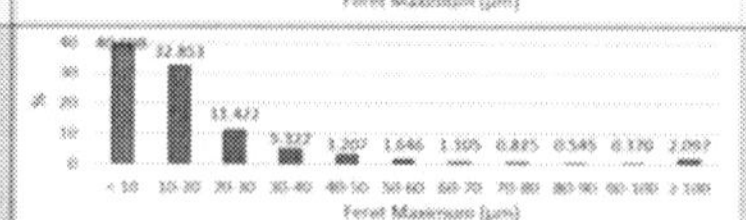	14.1		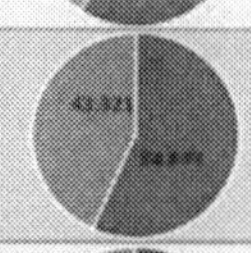	5	1
H	Second blast, thumb	28/03/2024	Underside	No	Normal	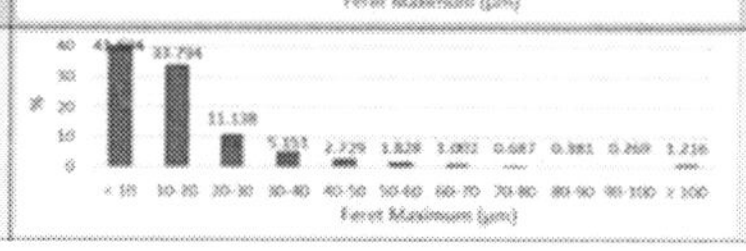	14.6		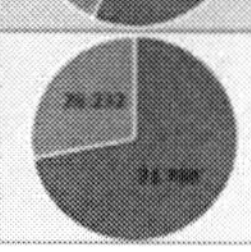	5	1

FIGURE 7: OPTICAL MICROSCOPY COMPARATIVE ANALYSIS OF TAPE SAMPLES COLLECTED USING A THUMB AND A SPRING-LOADED ROLLER ON THE TOP-SIDE AND UNDERSIDE OF STEEL PLATES PROCESSED THROUGH THE STL (RED = DUST).

SCANNING ELECTRON MICROSCOPY & ENERGY DISPERSIVE X-RAY SPECTROSCOPY

Secondary electron and backscattered electron (BSE) images for each sample, namely single blast underside (Sample 1), single blast topside (Sample 2), double blast underside (Sample 3), and double blast topside ample 4), are depicted in Figure 8a-b, Figure 8c-d, Figure 8e-f, and Figure 8g-h, respectively. Each of the dust tape samples depict iron and iron oxide rich particles, evidenced by the contrast provided by differences in average atomic number of the materials, with ing These features suggest that materials enriched in iron-based products are more prominent than those composed of carbonaceous materials or other contaminants. The BSE image for Sample 1 in Figure 8b depicts the vast majority of product to be iron enriched particles. Similarly, for Sample 2 shown in Figure 8d, several iron enriched particles are evident, with two examples circled. Figure 8f depicts the larger particles as consisting of non-oxidised iron, indicated by their poor brightness in the BSE image. As with the other samples, Figure 8h reveals numerous candidates for iron rich particles, where particles towards the bottom of the image are likely to be oxidised iron material.

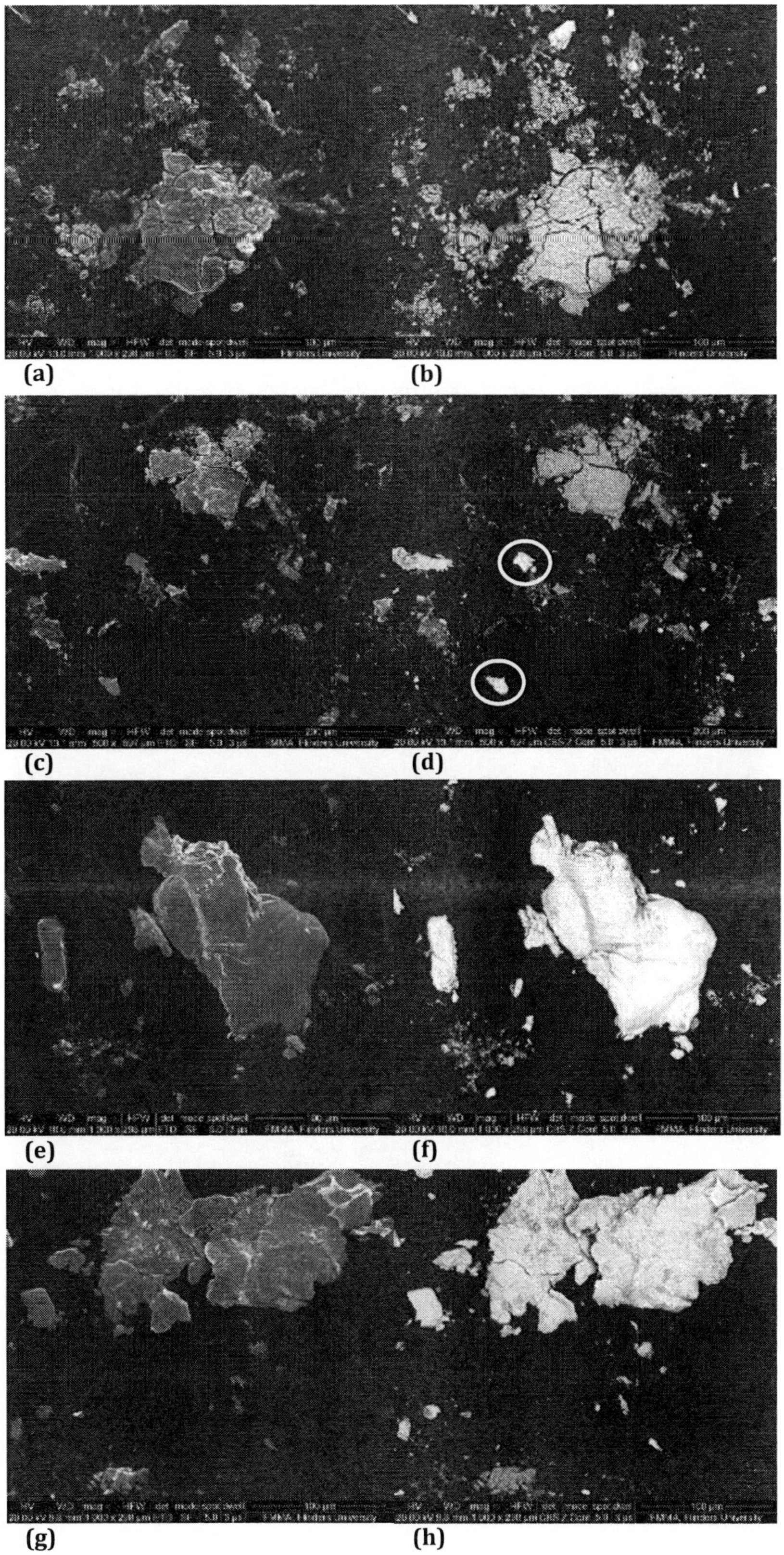

FIGURE 8: (A) SECONDARY ELECTRON IMAGE OF SAMPLE 1, 1000 × MAGNIFICATION, (B) BSE OF SAMPLE 1, (C) SECONDARY ELECTRON IMAGE OF SAMPLE 2, 500 × MAGNIFICATION, (D) BSE OF SAMPLE 2, (E) SECONDARY ELECTRON IMAGE OF SAMPLE 3, 1000 × MAGNIFICATION, (F) BSE OF SAMPLE 3, (G) SECONDARY ELECTRON IMAGE OF SAMPLE 4, 1000 × MAGNIFICATION, (H) BSE OF SAMPLE 4.

The results from EDS analysis aid in elucidating the chemical composition of the dust sourced from the STL. Elemental mapping of the imaged locations is shown in Figure 9 and confirms the elemental composition of the samples. Figure 9a-d provides the elemental maps overlaid on the BSE image, while Figure 9e-h provides lone overlaid maps. These overlays indicate that the majority of particles are iron or iron oxide, where the individual elemental maps shown from Figure 9i-a2 indicate trace alloying elements of aluminium, silicon, manganese, and copper for all samples. Notably, in Figure 9g and Figure 9f, a silicon and manganese rich particle is detectable (circled), which is absent within other samples.

EDS spot analysis was conducted to further reveal the semi-quantitative elemental concentrations of the four samples, as shown within Figure 10 and Appendix 1 (Table A 1, Table A 2, Table A 3, Table A 4). For Sample 1 (single blast, underside), the particle analysed in Spot 1 is primarily comprised of iron, with trace amounts of other elements (primarily carbon and oxygen, but likely contaminants). Spot 3 was expected to show a similar spectrum, however the results in Table A 1 indicate a primary composition of carbon and oxygen the composition of the carbon tape substrate. This observation is likely owed to the static charge introduced by the electron beam, leading it to move from the analysis area. Other analysis Spots (2, 4-6) indicate elemental concentrations rich in iron and oxygen. EDS analysis of Sample 2 (single blast, topside) indicate primary compositions of iron that is non-oxidised, evidenced by Spots 2-6. However, Spot 1 comprises iron oxide. Furthermore, trace elements or low concentrations of silicon, manganese, aluminium, and copper are present throughout. The elemental analysis and approximate weight% calculated from the spectra of Sample 3 indicates that Spots 1, 4 and 5 are enriched with non-oxidised iron particles, which is an expected result as per the elemental mapping in Figure 9. EDS spot analysis for Spot 3, which was previously identified to contain larger concentrations of silicon and manganese, confirms this observation, in addition to small weight% concentrations of chromium. Moreover, Sample 4 spot analysis indicated that Spots 1, 4 and 6 are oxidised iron, while Spots 2 and 3 are of non-oxidised iron composition. From the EDS mapping, Spot 5 was also anticipated to be non-oxidised, however, the electron beam has led to the displacement of the particle during analysis.

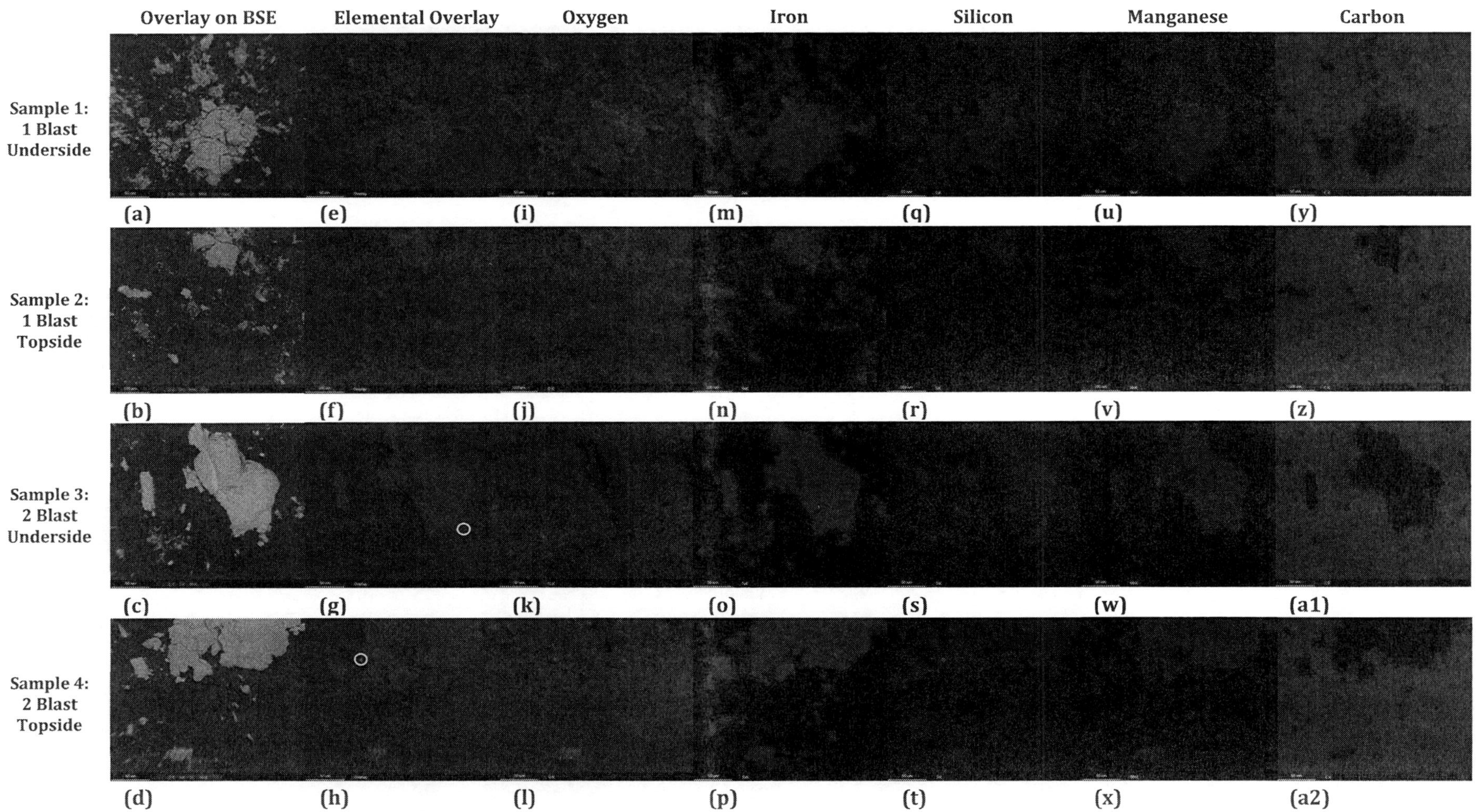

FIGURE 9: EDS ELEMENTAL MAPPING, (A-D) EDS ELEMENTAL MAPS OVERLAYED ON BSE IMAGE, (E-H) OVERLAYED EDS ELEMENTAL MAPS, (I-L) OXYGEN ELEMENTAL MAPS, (M-P) IRON ELEMENTAL MAPS, (Q-T) SILICON ELEMENTAL MAPS, (U-X) MANGANESE ELEMENTAL MAPS, (Y-A2) CARBON ELEMENTAL MAPS.

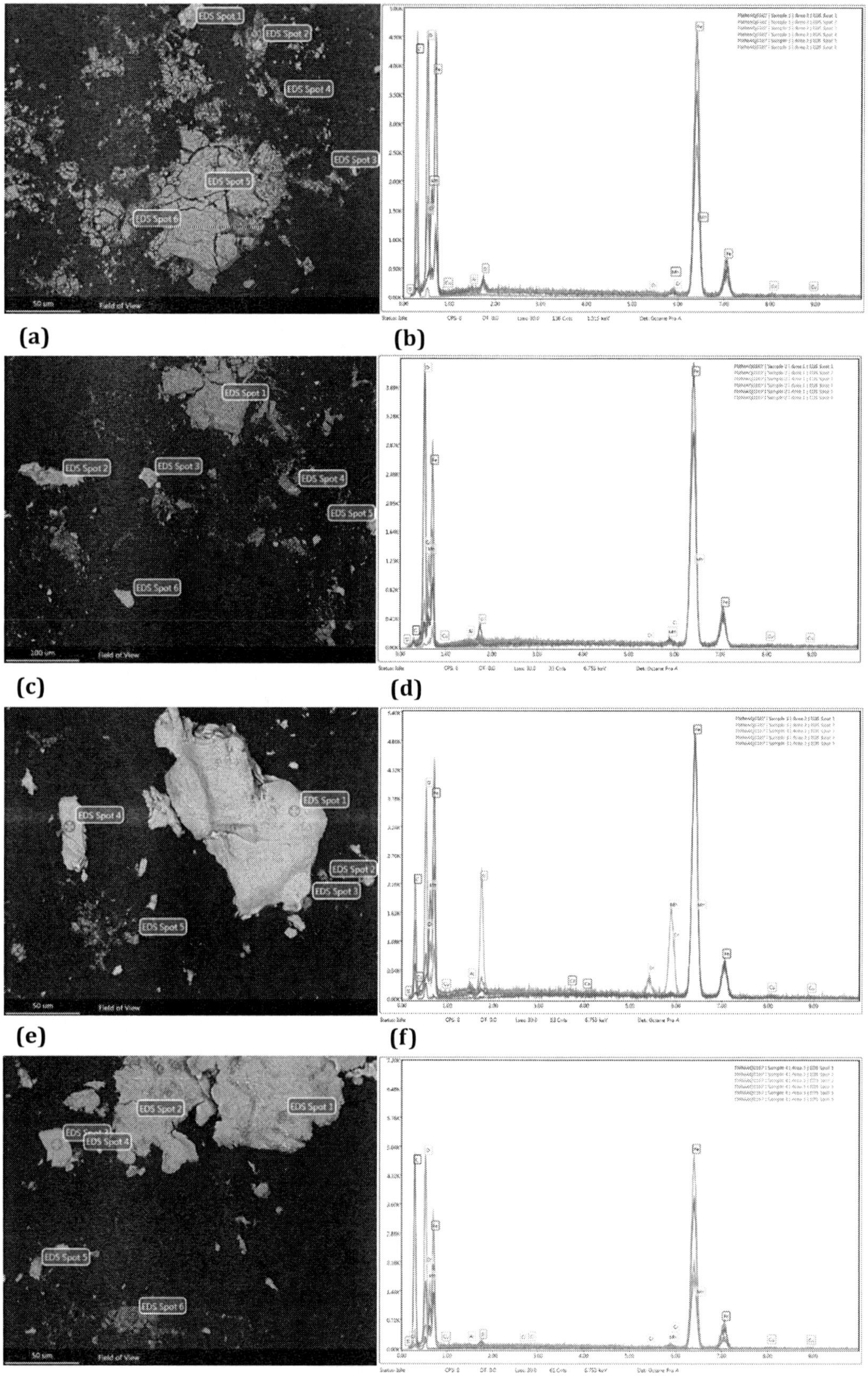

FIGURE 10: BSE WITH EDS SPOT ANALYSIS LOCATIONS MARKED FOR (A) SAMPLE 1 AT 1000 × MAGNIFICATION, (C) SAMPLE 2 AT 500 × MAGNIFICATION, (E) SAMPLE 3 AT 1000 × MAGNIFICATION, (G) SAMPLE 4 AT 1000 × MAGNIFICATION. EDS SPECTRA OF THE CHOSEN SPOTS, WITH ELEMENTAL LABELS ASSIGNED TO EACH PEAK, FOR (B) SAMPLE 1, (D) SAMPLE 2, (F) SAMPLE 3, AND (H) SAMPLE 4.

Discussion

Dust Characterisation

To address the subjective nature of traditional tape tests and dust assessments, microscopy methods with semi-quantitative capabilities were employed. Optical microscopy paired with multiphase analysis revealed the percent area of dust associated with each quantity rating in both ISO 8502-3 and AS 3894.6. Such results were then used to determine the quantity ratings of various tape samples from steel plates subject to different processing conditions through the STL, in addition to comparisons of using the roller versus the thumb. Analysis revealed that all samples correlated to a dust quantity Rating 5 (>14.9% dust) and a dust size Class 1 (predominant particle size <50 µm). As per the IMO and relevant paint manufacturers, the shipbuilding industry is required to meet a minimum requir
quantitative analysis techniques can confirm that all samples subject to processing within the STL meet a quantity Rating 5 for Class size 1, which is of higher cleanliness standard than the aforementioned requirement. Therefore, the results from multiphase analysis may be used within industry to foster improved consensus on dust ratings, as historically, assessment using ISO 8502-3 and AS 3894.6 was largely qualitative and therefore prone to subjective assessment. The results in Figure 5 therefore seek to reach consensus in industry regarding the interpretation of densities and sizes of dust in relation to coating performance.

Data ascertained from SEM and EDS analysis revealed that t
iron oxide enriched particles. These elements, in addition to trace concentrations of other elements, are as a result of the chemical composition of the parent material (DH36 steel, Table 2) and steel shot and grit (Table 3) from the centrifugal abrasive blasting process. The steel processed through the STL also comprises surface mill scale product, which is the metallurgical by-product produced by the rolling mill during the hot rolling process. Mill scale largely comprises elemental iron and iron oxides in various forms, -Fe_2O_3), and magnetite (Fe_3O_4) [3]. Mill scale from the parent steel therefore underpins the presence of iron oxides within the EDS analysis, where the existence of elemental iron with low concentrations of oxygen is likely owed to spent abrasive. The presence of minor alloying elements, such as silicon, manganese, copper, and aluminium, are also introduced by the parent steel and abrasive products, as depicted in Table 2 and Table 3. Silicon may be added to steel to increase the strength but can also be present as a residual product. Similarly, manganese is supplemented to provide enhanced strength and hardness properties, whereas aluminium is a deoxidising agent employed within the steel making process to act as a grain refiner.

TABLE 2: ELEMENTAL COMPOSITION OF DH36 STEEL (WT%) – SEE APPENDIX FOR FULL TABLE.

	Average Elemental Composition (wt%)									
	C	**P**	**Mn**	**Si**	**S**	**Ni**	**Cr**	**Mo**	**Cu**	**Al**
DH36 Steel (6.5 mm)	0.1502	0.0186	1.208	0.296	0.0092	0.0192	0.0366	0.0034	0.0514	0.0334

TABLE 3: ELEMENTAL COMPOSITION OF STEEL SHOT AND GRIT USED WITHIN THE STL (WT%).

	Average Elemental Composition (wt%)				
	C	**Si**	**Mn**	**P**	**S**
Steel Shot (S230)	0.87	0.86	0.83	0.025	0.025
Steel Grit (GL40/ GL25)	0.87	0.86	0.83	0.025	0.025

SUGGESTIONS FOR FUTURE WORK

Comparison between the data in Figure 5 further reveals the disparity between the dust density ratings in each standard, for example, a 'Rating 5' depicts 14.9% dust within the ISO 8502-3 standard compared to 4.98% within AS 3894.6, which has significant implications for industry practices and consensus. For shipbuilding and manufacturing industries where surface cleanliness is critical, these discrepancies can affect the performance and longevity of coatings, adhesives, and other surface treatments. Accurate and consistent dust measurements are essential for quality control, and differences in standards may require the adoption of multiple assessment methods, leading to increased complexities and operational costs. This variation therefore highlights the need to foster better awareness among coating inspectors that ISO and AS standards are different in their ratings. Moreover, future work should be directed towards adhesion testing to measure the effect of residual dust on the adhesion of coatings.

CONCLUSIONS

The implementation of advanced microscopy and compositional analysis techniques has significantly improved the precision and reliability of dust assessments in accordance with relevant standards. The combination of optical microscopy with multiphase analysis allowed for a more objective and quantitative evaluation of dust particle quantities and sizes, addressing the inherent subjectivity of traditional tape test analysis. SEM and EDS analyses provide detailed insights into the composition of dust, primarily consisting of iron and iron oxides, as well as minor alloying elements from both the parent DH36 steel and the abrasive materials used in processing. These comprehensive assessments may enhance quality control and compliance with regulatory standards throughout coating applications and wider industry. By facilitating better standardisation and understanding of dust particle characteristics, improved benefits to industries operating across diverse regulatory environments can be implemented. However, future endeavours following this work should focus on better teaching and awareness among coating inspectors highlighting the significant differences between the ISO 8502-3 and AS 3894.6 standards for coating and dust assessments.

ACKNOWLEDGMENTS

The authors wish to acknowledge the facilities and technical assistance of Microscopy Australia and the Australian National Fabrication Facility (ANFF) under the National Collaborative Research Infrastructure Strategy, at the South Australian Regional Facility, Flinders Microscopy and Microanalysis, Flinders University. Additionally, the authors extend their thanks to Flinders University for access to facilities at Tonsley, South Australia.

REFERENCES

[1] ISO 8502-3 - 2017, Preparation of steel substrates before application of paints and related products Tests for the assessment of surface cleanliness. Part 3: Assessment of dust on steel surfaces prepared for painting (pressure-sensitive tape method).

[2] AS 3894.6 - 2002, Site Testing of Protective Coatings. Method 6: Determination of Residual Contaminants.

[3] Jikar P and Dhokey N, Overview on production of reduced iron powder from mill scale waste, Materials Today: Proceedings, 44(4) (2020) 1-2.

AUTHOR DETAILS

R. McHendrie is the author of this paper. She is a Materials Engineer at BAE Systems Maritime Australia, a position she has held since 2024. She is currently involved in Manufacturing Engineering for the Hunter Class Frigate Program, and has worked on tasks associated with the Materials Technology team, including welding engineering, additive manufacturing, blasting and painting, materials, and build strategies.

	D. Fosdike is a joint author of this paper. He is the Materials Technology Lead at BAE Systems Maritime Australia, in this role he is responsible for the corrosion prevention across all platforms built and maintained at the Osborne Naval Shipyard as well as welding and additive manufacturing. He is also an AMPP Senior Certified Coating Inspector as well as a CIP Instructor. Daniel holds qualifications in Mechanical Engineering and Welding Engineering and is the Immediate Past President of the ACA South Australia Branch.
	D. McDonald is a joint author of this paper. He has extensive experience and is currently a Materials Technologist at BAE Systems Maritime Australia, a position he has held since 2021. He is responsible for the management and implementation of process strategies within the scope of Blast & Paint, Adhesives and Corrosion Mitigation. During this time, he has worked on Collins Class, Arafura Class, T26 (RN), and now concentrates on the Hunter Class Project.
	K. Sellars is a joint author of this paper. He is a Senior Production Engineering Officer (Materials Technologist) at BAE Systems Maritime Australia with responsibilities for the management and implementation of process strategies within the scope of Blast experience in the maritime industry having previously worked on Collins Class Submarine build and maintenance projects, build of the Hobart Class Air Warfare Destroyers and now the Hunter Class Frigate Project.

Appendix

TABLE A 1: APPROXIMATE WEIGHT% (NORMALISED TO 100%) OF ELEMENTS IN SAMPLE 1 CALCULATED FROM THE EDS SPOT ANALYSIS SPECTRA.

:	C	O	Si	Fe	Mn	Cu	Other Elements
Spot 1	3.0	1.9	0.9	93.3	0.7	0	Cr 0.3
Spot 2	11.4	11.3	0.6	74.4	0.9	0.6	Al 0.6
Spot 3	89.9	10.1	0	0	0	0	
Spot 4	24.6	31.3	0.9	42.0	0.6	0.4	Cr 0.2
Spot 5	2.0	24.5	1.3	70.9	1.1	0	Al 0.2
Spot 6	5.5	20.3	0.3	72.7	0.9	0.3	

TABLE A 2: APPROXIMATE WEIGHT% (NORMALISED TO 100%) OF ELEMENTS IN SAMPLE 2 CALCULATED FROM THE EDS SPOT ANALYSIS SPECTRA.

	C	O	Si	Fe	Mn	Cu	Other Elements
Spot 1	0.4	24.4	1.5	72.4	1.3	0	
Spot 2	1.8	1.6	0.2	94.4	1.4	0.5	Al 0.1
Spot 3	1.4	0.3	Trace	97.1	0.4	0.4	Al trace Cr 0.4
Spot 4	0.6	2.1	0.4	95.6	1.2	0	Al 0.2
Spot 5	2.5	2.3	0.2	93.4	1.6	0	
Spot 6	2.0	4.0	0.4	92.2	0.3	0.6	Al 0.2 Cr 0.2

TABLE A 3: APPROXIMATE WEIGHT% (NORMALISED TO 100%) OF ELEMENTS IN SAMPLE 3 CALCULATED FROM THE EDS SPOT ANALYSIS SPECTRA.

	C	O	Si	Fe	Mn	Cu	Other Elements
Spot 1	0.9	1.9	0.4	95.5	1.2	0	Al 0.2
Spot 2	25.2	18.9	0.7	53.9	0.4	0.4	Al 0.4 Cr 0.2
Spot 3	15	15.0	5.7	48.5	13.1	0	Al 0.5 Ca 0.3 Cr 2.0
Spot 4	1.3	4.3	0.5	92.8	0.4	0.5	Al 0.1 Cr 0.2
Spot 5	13.8	1.7	Trace	83.3	0.4	0.7	Al 0.1

TABLE A 4: APPROXIMATE WEIGHT% (NORMALISED TO 100%) OF ELEMENTS IN SAMPLE 4 CALCULATED FROM THE EDS SPOT ANALYSIS SPECTRA.

	C	O	Si	Fe	Mn	Cu	Other Elements
Spot 1	1.8	24.9	0.3	71.8	1.1	0	Al 0.2
Spot 2	1.2	0.8	0.2	98.6	1.5	0.7	Al 0.1
Spot 3	1.1	3.3	0.6	93.7	0.4	0.7	Cr 0.2
Spot 4	44.9	12.1	0.1	42.5	0.3	0.2	Al Trace
Spot 5	84.1	13.6	0	2.3	0	0	
Spot 6	12.6	40.9	0.6	44.4	0.8	0	Al 0.4 Cl 0.3

TABLE A 5: COMPREHENSIVE ELEMENTAL COMPOSITION OF DH36 STEEL (WT%).

Average Elemental Composition (wt%)															
C	P	Mn	Si	S	Ni	Cr	Mo	Cu	Al (total)	Ti	B (total)	N	Nb	Sn	V
0.150	0.019	1.21	0.29	0.0092	0.019	0.037	0.0034	0.051	0.033	0.017	0.0034	0.0021	0.001	0.0048	0.003

Superhydrophobic Coatings For Corrosion Control

David Morton[1,*], **Liliana Madaleno**[2],

[1] *Hempel A/S, Kongens Lyngby, 2800, Denmark*

[2]*Hempel (Wattyl) Australia Pty Ltd, WA, 6053, Australia*

**Corresponding author email: dmor@hempel.com*

ABSTRACT

During the last years, superhydrophobic coatings have been developed in order to prevent corrosion or reduce the rate of corrosion of steel. This papers describes the performance of water repellent polyurethane and polysiloxane topcoats containing a superhydrophobic technology that actively repels water from the coating thus greatly improving corrosion protection. The technology can be incorporated into various binder systems and when used in UV durable polyurethane or polysiloxane binders it can effectively eliminate the need for a primer or intermediate in many environments. The water repellent coating was tested in a single coating system and in a two coat system with a zinc rich primer for maximum protection in severe offshore environments. Several tests, including Salt Spray (ISO 9227), Blister Box (ISO 6270-12) and Ageing Cycling Resistance (ISO 12944-9) were conducted to compare the corrosion resistances of the different coating systems. The test results show that the incorporation of the superhydrophobic technology greatly enhances the ability of the coating systems to reduce corrosion of steel in an offshore environment.

Keywords: steel, seawater, water repellency, protective coating

Introduction

In the coating industry, a typical coating system exposed to atmospheric coastal and offshore environments would comprise a zinc rich primer applied on top of the steel to provide galvanic and barrier protection , an epoxy intermediate adding barrier properties to the coating system and a topcoat to provide aesthetic properties and UV light protection. Addition of each coating layer adds up in time, costs and environmental impact of the coating system application process. An alternative approach would be to develop a water repellent topcoat coating that in addition to UV protection could actively repel water from the coated surface, making the intermediate epoxy layer and the primer, redundant.

The interest in superhydrophobic or water repellent coatings has increased during the last years, driven by the potential industrial applications of these surfaces in different areas. Several techniques have been employed to create superhydrophobic coatings, and the majority rely on the generation of a coating surface profile or chemically modifying the coating´s surface. However, when these surfaces are damaged, the superhydrophobicity is usually lost and consequently their inherent water repellent properties. Therefore, a volumetric superhydrophobic coating would represent a breakthrough technology in the coatings industry.

We have previously developed a water repellent 2 coat system (zinc rich primer and water repellent polyurethane topcoat) that offers outstanding corrosion protection and mechanical performance for severe offshore environments[1]. Further internal studies have confirmed that the water repellent technology adopted can be incorporated into several binder systems and when used in UV durable polyurethane or polysiloxane binders it can effectively eliminate the need for an intermediate or primer in many environments.

In this paper, the testing results related to a water repellent polyurethane and polysiloxane (isocyanate free) topcoats, in terms of mechanical performance and corrosion protection, are reported. The water repellent topcoats (polyurethane and polysiloxane) were tested in a single coating system, applied directly to metal, and in a two coat system with a zinc rich primer for maximum protection in severe offshore environments.

Several tests, including Salt Spray (ISO 9227), Blister Box (ISO 6270-12) and Ageing Cycling Resistance (ISO 12944-9) tests were conducted to investigate the corrosion resistances of the coating systems tested.

Methods and Materials

Preparation of Materials

The different coatings were applied directly to metal and as part of two coating systems containing a zinc rich epoxy primer. The systems description are shown in **Table 1** and **Table 2**.

As a reference, a standard 3 coat system was used for ageing resistance testing as this is the typical system used by the industry today. This will be designated as System 7.

Table 1: One coat systems prepared for testing.

System 1	System 2	System 3
Conventional polysiloxane topcoat	Water repellent Polyurethane topcoat	Water repellent Polysiloxane topcoat

Table 2: Two coat systems prepared for testing.

System 4	System 5	System 6
Zinc primer 1 - Conventional polysiloxane	Zinc primer 1 - Water repellent Polyurethane topcoat	Zinc primer 1 - Water repellent Polysiloxane top coat

SUBSTRATE AND SURFACE PREPARATION

Mild steel panels of rust grade A (ISO 8501-1)[2] were abrasive blasted with iron grit (G070 acc. ISO 11124)[3] to a cleanliness degree equivalent to Sa 2.5 (ISO 8501-1)[2]. The surface roughness of the panels had a Rz between 40 and 55 µm (ISO 8503)[4].

TEST PROGRAM

Salt Spray ISO9227[5],Adhesion, Water condensation tests

Salt spray tests were conducted according to ISO9227 for systems 1-6. One coat, DTM systems were tested for 1440 hours and two coat systems were tested for period of 2200 hours. After exposure, the degree of blistering, rust, cracking and flaking were assessed on both panel and around the scribe. Pull-off adhesion[6] was evaluated 24h after the end of the exposure. The corrosion creep from the scribe was measured at least at nine points and the average width, M, was calculated according to Equation 1, where C is the average of the measurements and W is the original width of the scribe.

$$M = \frac{C-W}{2} \qquad (1)$$

Blister box tests were conducted according to ISO 6270-1[7] for systems 1-3 for 1440h. After exposure, the evaluation of the panels was performed as described above.

Ageing resistance ISO12944-9[8]

Ageing resistance test, according to ISO 12944-9, was performed for systems 5, 6 and 7 (standard 3 coating system as reference). Tests were performed for 4200 hours and after exposure, the evaluation of the panels was performed as described above.

Crack Resistance TM0404-2004[9]

Thermal cycling crack resistance was conducted on systems 1-3 under the same conditions as NACE standard TM0404-2004, Section 9, but using internal test Hempel method, that was empirically proven to replicate NACE standard TM0404-2004 test results. The coatings were post cured at 60°C for one week and then subjected to upper and lower temperatures of 60°C to -30°C, within a two hour period of time, for three weeks. After the test period, all the samples were examined for cracks under a microscope.

Flexibility Test ASTM D 522-2008 (Method B)[10]

Flexibility tests were performed on systems 1-3 to determine the flexibility of the top coats. Panels were tested after room temperature curing for 2 weeks. Respective flexure strain was measured using a fixed-radii mandrel bending machine. The flexure strain, F_{strain}, of the coating systems material were calculated using the equation:

$$F_{strain} = \frac{\frac{t}{2}+c}{R+\frac{t}{2}} \qquad (2)$$

t = steel panel thickness; c = coating system thickness; R = mandrel radius (mm). The deformed coatings were evaluated visually and microscopically for signs of cracking.

Results

Salt Spray Resistance ISO9227

DTM panels (Systems 1-3) were exposed to salt spray testing for a period of 1440 hours. Results are shown in table 3. After 1440 hours no blistering, flaking or cracking were observed. When we examine the scribe line area, it is notable that the creep line width is larger for the Systems 1 and 2 compared to system 3 as shown in figure 1. Pull off adhesion strength of the system 2 and 3 did not change more than 10 % of the initial values. However, system 1 pull off adhesion average value drop was higher, up to 30%, and failure mode changed from cohesive to adhesive particularly close to the scribe line and panel edges indicating signs of water ingress under the coating.

Table 3: Results from ISO9227 after 1440 hours

	System 1 (100 µm)	System 2 (190 µm)	System 3 (95 µm)
Creep line (mm), after SST exposure for 1440 hours	1.8	1.30	0.25
Pull-off Adhesion Average before / after SST (MPa)	11.70/ 7.0	7.60/7.30	7.70/7.0

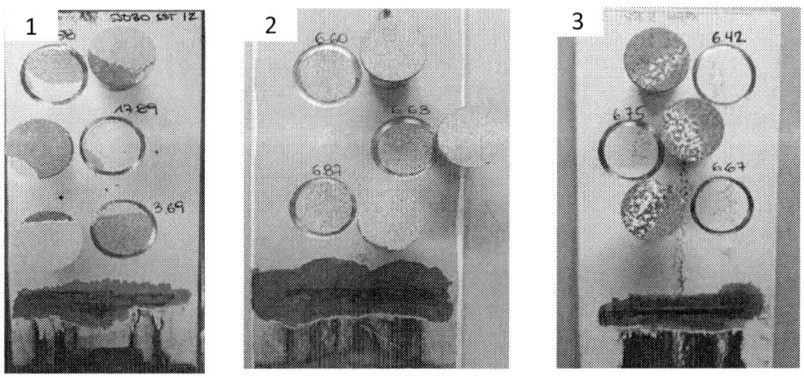

Figure 1: System1,2 &3 after salt spray test for 1440 hours.

Two coat systems (Systems 4, 5 and 6)

SST results for the systems 4, 5 and 6, exposed for 2200 hours, are shown on Table 4 below. The panels were observed after different times of exposure in order to evaluate the degree of blistering, cracking, rusting and/or flaking. No visible defects were observed after the exposure times. For all of the tested coating systems, after 2200 hours of exposure, the corrosion creep line is below 0,5 mm. It was observed that pull off adhesion values did not decrease for system 5 and increased for system 6. Increase of the values can be explained by further curing of the coating system when exposed to the temperature conditions of SST. Meanwhile, the pull off adhesion values of the system 4 have decreased up to 40% of the initial value.

Table 4: Results from ISO9227 after 2200 hours

System	System 4	System 5	System 6
DFT (µm)	Primer: 105 Top coat: 195	Primer: 100 Top coat: 290	Primer: 100 Top coat: 220
Observations	Blistering:0(S0) Rusting:Ri0 Flaking:0(S0) Cracking:0(S0)	Blistering: 0(S0) Rusting: Ri0 Flaking: 0(S0) Cracking: 0(S0)	Blistering: 0(S0) Rusting: Ri0 Flaking: 0(S0) Cracking: 0(S0)
Corrosion creep line, M, mm Average	M = 0,31	M = 0,01	M = 0,10
Pull Off Adhesion (MPa), Average	5,90	6,90	9,40
Adhesion Failure	100% adhesion primer/steel	100% cohesive within the top coat	100% cohesive within the top coat
	Reference (non- exposed)	*Reference (non-exposed)*	*Reference (non-exposed)*
Pull-off Adhesion Average (MPa)	9,50	7,10	7,80
Adhesion Failure Mode)	100% cohesion within the top coat	100% cohesion within the top coat	100% cohesion within the top coat

Blister box test – ISO 6270-1 DTM System 1-3

All systems showed no defects after 1440 hours exposure and the results are documented in Table 5. Pull off adhesion values of all systems are quite stable during the test.

Table 5: DTM Systems 1-3 after blister box test for 1440h.

System	System 1	System 2	System 3
DFT (µm)	90	190	90
Observations	Blistering:0(S0) Rusting: Ri0 Flaking: 0(S0) Cracking: 0(S0)	Blistering: 0(S0) Rusting: Ri0 Flaking: 0(S0) Cracking: 0(S0)	Blistering: 0(S0) Rusting: Ri0 Flaking: 0(S0) Cracking: 0(S0)
Pull-off Adhesion Average before / after BBT (MPa)	11.70 / 12.20	7.30 / 6.70	7.80 / 8.80
Adhesion Failure Mode	100% cohesion within the coating	100% cohesion within the coating	100% cohesion within the coating

Ageing resistance ISO12944-9

System 5, 6 and reference 3 coat scheme, System 7, were tested according to ISO12944-9. The results obtained for the cycling ageing resistance test after 4200 hours of exposure are shown in table 6. Corrosion creep from the scribe is significantly larger for the refence 3 coat scheme (system 7) compared to the two coat water repellent systems (System 5 and 6). All meet the requirements of the standard. Again we see very stable pull off adhesion for systems 5 and 6 while adhesion is greatly reduce in system 7 after the test. We see a change in the failure mechanism on system 7 indicating early signs of system failure even though the system meets the requirement of the standard (retain at least 50% of the initial adhesion and be above 5 MPa).

Table 6 Results after cycling ageing according to ISO12944-9

System	System 5	System 6	System 7
DFT (μm)	Primer: 95 Top coat: 180	Primer: 70 Top coat: 100	Primer: 60 Intermediate: 120 Top coat: 80
Observations	Blistering: 0(S0) Rusting: Ri0 Flaking: 0(S0) Cracking: 0(S0)	Blistering: 0(S0) Rusting: Ri0 Flaking: 0(S0) Cracking: 0(S0)	Blistering: 0(S0) Rusting: Ri0 Flaking: 0(S0) Cracking: 0(S0)
Corrosion line, M, mm Average	M = 0.43	M = 0.41	M = 2.53
Pull off Adhesion Average (MPa)	5.90	6.90	7.50
Adhesion Failure	90% cohesive within the top coat and 10% adhesive between the primer and steel surface	90% cohesive within the top coat and 10% adhesive between the primer and steel surface	100 % adhesive between the primer and steel surface
	Reference (non-exposed)	*Reference (non-exposed)*	*Reference (non-exposed)*
Pull-off Adhesion Average (MPa)	7.0	8.7	12.9
Adhesion Failure Mode)	100% cohesive within the top coat	100% cohesive within the top coat	100% cohesive within the top coat

Crack Resistance TM0404-2004

The DTM systems were examined for cracking of the coating, at different ranges of dry film thicknesses (DFT) using the methodology described in TM-404. System 2 did not show any cracking even at a DFT above 900 microns. System 3 cracked in the range of 500-600 microns, which is at least three times above the recommended application DFT. System 1 cracked in the range of 200-250 microns, which is two times higher the recommended DFT. The results are summarized in table 7 and pictures are shown in figure 2.

Table 7: Cracking DFTs for tested systems after 3 weeks of testing

System	System 1	System 3	System 2
Cracking DFT (μm)	200-250	500-600	900

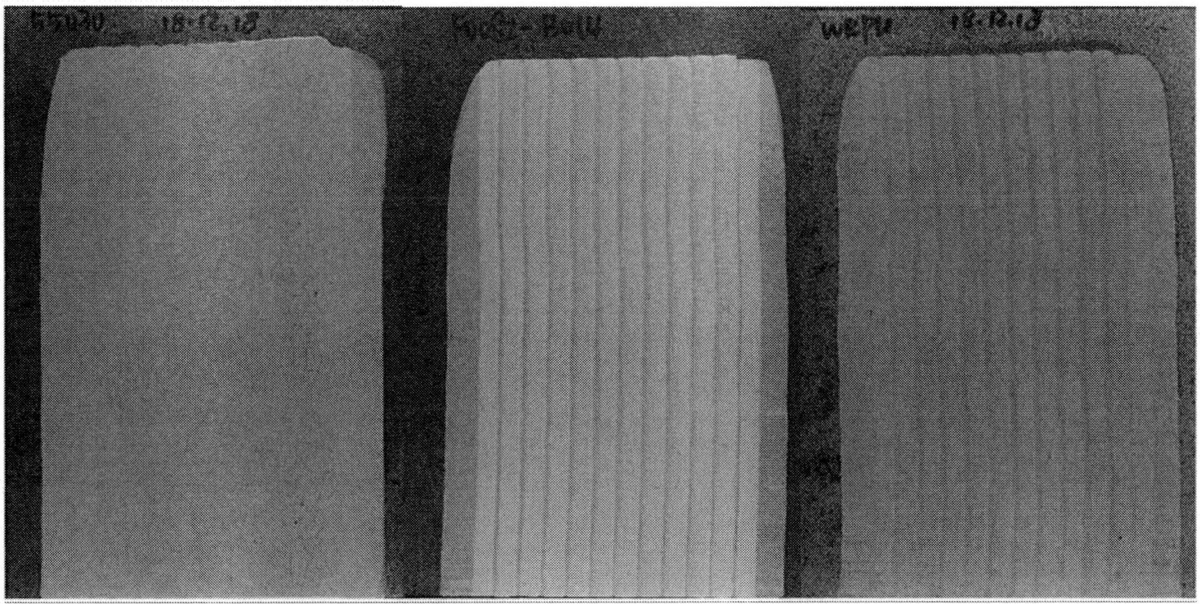

Figure 2: Test panels after TM-404. From left to right: System 1, System 3 and System 2

Flexibility ASTM D522

Systems 1, 2 and 3 were bent over different radii mandrels . System 2 had no observable cracks while system 1 and 3 cracked at different mandrel radii. It can be concluded that system 2 is highly flexible while system 3 shows good flexibility.
Flexural strain and DFT values are presented as averages of six measurements. The results are shown in table 7.

Table 8: Flexural strains of the tested systems.

System	System 1	System 2	System 3
DFT (μm)	100	275	120
Flexural strain (%)	1.50	7.82	3.82

Discussion

In this work, we evaluated the potential of water repellent topcoats (polyurethane and polysiloxane) for improved corrosion resistance, especially in severe coastal or offshore environments. The salt spray test results of the one coat (DTM systems) indicated that the water repellent topcoats can reduce the corrosion creep and preserve the adhesion to the substrate. Salt spray test results of the two coat systems showed the benefit of the water repellency with both very low rust creep and very high adhesion retention.
Blister Box test results of one coat, DTM systems showed good resistance to fresh water. However, the tests should run for longer periods as it is not possible to discern any notable difference between the systems.

Ageing Cycling Resistance tests have indicated that two coat systems with water repellent topcoats, both PU and polysiloxane, perform better than a conventional three coat system. The two coat system with zinc primer and the isocyanate free water repellent top coat have passed Ageing Cycling Resistance test without any visual degradation, with low rust creep and retained pull off adhesion values, which indicates robust anticorrosive properties of the coating system.
Regarding the mechanical performance of the topcoats, the water repellent coatings showed excellent flexibility and crack resistance at much higher thicknesses when compared to a standard coating.

CONCLUSIONS

In conclusion, the results of this study indicate that water repellent topcoats made using both polyurethane and polysiloxane resins have a strong potential to improve anticorrosive properties as compared to standard coatings, both when applied directly to metal and in the two coat systems with the zinc rich primer. The excellent flexibility of the resulting coatings can overcome the main challenges related to the development of cracks when in service, which are one of the main reasons for the coating systems failure. The observed outstanding anti-corrosive and mechanical properties of the two coat system indicate that these coatings can be considered as viable candidates for areas exposed to aggressive coastal and offshore environments.

References

1. Corrosion 2018 (C2018-11144): An innovative, long life, superhydrophobic/water repellent coating system for offshore application.
2. ISO 8501-1:2007 Preparation of steel substrates before application of paints and related products — Visual assessment of surface cleanliness — Part 1: Rust grades and preparation grades of uncoated steel substrates and of steel substrates after overall removal of previous coatings
3. ISO 11124 -1:2018 Preparation of steel substrates before application of paints and related products — Specifications for metallic blast-cleaning abrasives.
4. ISO 8503-1:2012 Preparation of steel substrates before application of paints and related products — Surface roughness characteristics of blast-cleaned steel substrates — Part 1: Specifications and definitions for ISO surface profile comparators for the assessment of abrasive blast-cleaned surfaces
5. ISO 9227:2022 Corrosion tests in artificial atmospheres — Salt spray tests
6. ISO 4624:2023 Paints and varnishes — Pull-off test for adhesion
7. ISO 6270-1:2017 Paints and varnishes — Determination of resistance to humidity — Part 1: Condensation (single-sided exposure)
8. ISO 12944-9:2018 Paints and varnishes — Corrosion protection of steel structures by protective paint systems — Part 9: Protective paint systems and laboratory performance test methods for offshore and related structures
9. NACE TM0404-2004, Offshore Platform Atmospheric and Splash Zone New Construction Coating System Evaluation
10. ASTM D522-2008 (method B). Standard test methods for Mandrel Bend Test of attached organic coatings.

AUTHOR DETAILS

	David Morton is a Chief Scientist at Hempel A/S based in Copenhagen, Denmark He has been with Hempel since 2013 and has nearly 40 years' experience in the coatings industry. Originally from the UK with a PhD from the University of Saint Andrews he has worked previously at AkzoNobel and Barloworld coatings.
	Liliana Madaleno is a Materials Engineer and has been with Hempel for more than 10 years. Before moving to Australia in 2019, Liliana was part of the Technology and Innovation department at Hempel HQ in Copenhagen, working on research and development of novel coating technologies. She holds a PhD in Materials Science from Aalborg University, Denmark.

CORROSION & PREVENTION
CAIRNS 2024

THE AUSTRALASIAN CORROSION ASSOCIATION INC.

Hot water pitting and corrosion loss trends for phosphorous deoxidised (DHP) and phosphorous deoxidised arsenical (DPA) copper alloys

P. Richardson[1,*], R. Melchers[1], D. Nicholas[2]

[1]Centre for Infrastructure Performance and Reliability, The University of Newcastle, NSW, 2308, Australia

[2]Nicholas Corrosion Pty Ltd, NSW, 2299, Australia

**Corresponding author email: peter.j.richardson@newcastle.edu.au*

ABSTRACT

Phosphorous deoxidised (DHP) copper alloy is the usual material of choice for piping in recirculating hot water systems in Australia due to its resistance to corrosion. However, there have been industrial cases noted where severe corrosion, including perforation pitting, have occurred in these systems. In this work, the corrosion behaviour of DHP copper coupons and corrosion-resistant phosphorous deoxidised arsenical (DPA) copper coupons immersed in a 70 °C hot water tank was investigated over 2 years. The presented data include corrosion loss (including a bimodal corrosion trend) and SEM analysis with respect to observed pitting behaviour. Some pitting of the DHP coupons was observed early in the experiments which did not translate into long-term, stable pitting. This was attributed to the presence of metallic contaminants on the surface of the coupons. The arsenical copper showed no susceptibility to pitting corrosion. A brief hypothesis is presented to explain this behaviour.

Keywords: copper alloys, pitting, hot water corrosion.

Introduction

Copper and its alloys are used extensively in situations exposed to mild corrosion conditions due to its natural passivation abilities, formability and in some cases thermal properties. One of its primary applications is as piping in recirculating hot water systems, where hot water is delivered from a heated water source, such as a boiler, to various outlets where hot water is required. The relevant Australian standard, AS/NZS 3500.4:2017 [1] dictates that stored hot water must be kept at least 60 °C and no more than 65 °C. This allows water exiting an outlet to be limited to <50 °C as required by the standard, and at the same time the growth of Salmonella and Legionella bacteria can be prevented in the system. However, reports from contractors and consultants in the field have revealed that, in practice, stored hot water is often observed at 70 °C or higher. Further to this, perforation pitting has been observed on occasion, particularly in some commercial settings where the temperature has been noted in excess of 70 °C. The cause for such pitting is currently debated but it is speculated that the increased temperature could play a role. Additionally, the link between the cases where pitting occurs, compared to where it does not occur, has not yet been established.

Pitting of copper alloys has been observed in some cases to be due to water chemistry (i.e. increased chloride and sulfate concentrations) [2], in others due to under-deposit corrosion [3-5] and in others due to increased temperature facilitating cathodic deposit regions and subsequent anodic copper regions [6]. These are aside to the initiation of corrosion due to poor design, materials, workmanship and handling of the copper pipe [7].

The present study aimed to investigate and understand the pitting behaviour of two grades of copper in a purpose-built hot water system under elevated temperature (70 °C). The first grade, phosphorus deoxidised (DHP) copper was selected as it is commonly used for a variety of general-purpose applications, including in recirculating hot water systems. The second grade, phosphorus deoxidised arsenical (DPA) copper was selected for the investigation due to its known corrosion resistance properties, endowed by the small amount of arsenic acting as a corrosion inhibitor.

Methods and Materials

Two sets of coupons were created for the DHP and DPA copper alloys. The DHP copper (grade C12200 - UNS) was sourced from Australian Metals Pty Ltd in the form of 1.2mm thick sheet and was cut into 100×50×1.2 mm rectangular coupons by guillotine. The DPA copper (grade C14200 - UNS) was sourced from Columbia Metals in the form of 2" diameter round bar and processed into 50.9×2 mm circular disc-shaped coupons by parting and facing on a metal lathe. Both types of coupons had holes drilled in them, to allow them to be suspended by overhanging bars in the hot water immersion tank via cable ties. Photographs of the two types of coupons are provided in Figure 1. Prior to placing in the immersion tank, each coupon was photographed, weighed, and tagged with an identification number. All coupons were weighed before being submerged in the immersion tank to record their initial mass in g (3 d.p.). The chemical composition, in weight %, of the two copper alloys is as follows: C12200 (DHP) – Cu 99.96%, P 0.0298%, Pb 0.0001%; C14200 (DPA) – Cu 99.47%, As 0.45%, P 0.034%, Ni 0.013%, Fe 0.009%, Pb 0.002%, Se+Te 0.002%, Bi 0.001%, Sn 0.001%, Sb 0.001%.

(A)

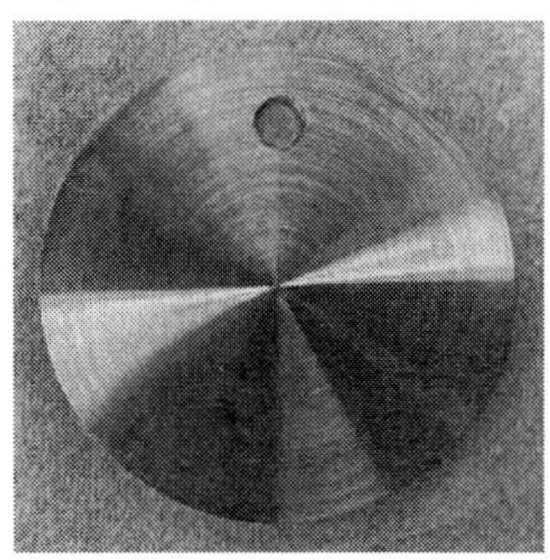

(B)

FIGURE 1: PHOTOGRAPHS OF COUPONS PRIOR TO IMMERSION IN THE HOT WATER TANK: (A) C12200 - DHP GRADE, 50×100×1.2 mm, (B) C14200 - DPA GRADE, 50.9×2 mm.

The hot water immersion tank, shown in Figure 2, consisted of a 205 L mild steel drum, coated internally and externally with epoxy-based paint, which was filled with water. Ambient temperature tap water entered the bottom of the tank via a hose at a rate of ~10 L/h and water exited the tank through an overflow tube at the top of the tank. The water composition was tested and is provided in Table 1. A compressed air line was used to bubble air continuously through the bottom of the tank at a rate of ~200 L/h to avoid anoxic conditions. Two 750 W drum heaters were integrated with a purpose-built temperature controller to maintain the water at a constant temperature of 70±1 °C. The water temperature was continuously measured and feedback provided to the controller, using a T-type thermocouple which was positioned at the top of the tank where the coupons were located. The tank was covered with 25 mm thick thermal insulation blanket, including on the lid, to maintain the temperature.

FIGURE 2: HOT WATER IMMERSION TANK SHOWING AIR AND WATER INLET, OVERFLOW AND SAMPLES SUSPENDED BY CABLE TIES FROM PVC HANGING RACKS. TANK LID REMOVED FOR VISUALISATION PURPOSES.

TABLE 1: MEASURED CHEMICAL AND PHYSICAL CHARACTERISTICS OF FRESH TAP WATER. UNITS ARE IN mg/L EXCEPT FOR pH.

pH	Dissolved Solids	Hardness ($CaCO_3$)	Chloride	Sulfate
7.45	119	45	27	19
Manganese	**Iron**	**Magnesium**	**Sodium**	**Aluminium**
0.117	0.58	3	16	0.34

Five coupons of each grade were recovered every 3 months and, after removal of the thin layer of corrosion products, weighed for determination of the mass loss. The corrosion products were removed from the coupons using the process outlined in ASTM G1-03 [8]. In brief, coupons were submerged in 2M sulfuric acid solution for 3 min. The coupons were then rinsed thoroughly with water followed by ethanol, dried with hot air, and weighed to record mass loss. Non-immersed samples were also exposed to the acid cleaning process to evaluate the impact of cleaning on mass loss and changes to surface morphology. The corrosion loss induced by the cleaning process was found to be negligible (~0.03 μm surface thickness loss) compared to the loss due to corrosion after immersion for 3 months (~1 μm surface thickness loss). Nevertheless, the loss due to cleaning was used to compensate for the actual thickness losses for the coupons presented in the results. The error presented for the corrosion loss was calculated by adding the corrosion loss due to cleaning to the standard error obtained for the population of 5 replicate samples for each exposure period.

Optical microscopy was performed for imaging of the coupons before removal of the corrosion products, and for pit depth analysis of larger pits. The depth was measured by focusing on the top of the coupon and then focusing on the bottom of the pit. A home-built device employing a rotary position sensor attached to the focus knob was used to measure the distance travelled by the sample stage during focusing between the two locations and therefore acted as a direct measure of the pit depth. Scanning electron microscopy (SEM) was conducted to obtain images of the surface morphology for coupons exposed to the hot water environment, using a Zeiss Sigma VP FESEM operating at 2 kV with an in-lens secondary electron detector. Analysis of the larger, individual pits was conducted using a Thermo Scientific Phenom XL benchtop SEM operating at 15 kV, equipped with a four-quadrant backscattered electron (BSE) detector, energy dispersive spectroscopy (EDS) detector and 3D Roughness Reconstruction software. Coupons were analysed by SEM following the acid cleaning process and were uncoated.

RESULTS

GENERAL CORROSION

The morphological impacts of the acid cleaning process, used to remove corrosion products from the metallic Cu coupons, are shown in Figure 3. Some minor attack of the surface was observed leading to a slightly different surface texture of the cleaned specimens. This material loss translated to a uniform corrosion thickness loss of 0.03 μm, or 3% of the loss experienced by the coupons exposed for the shortest amount of time (3 months).

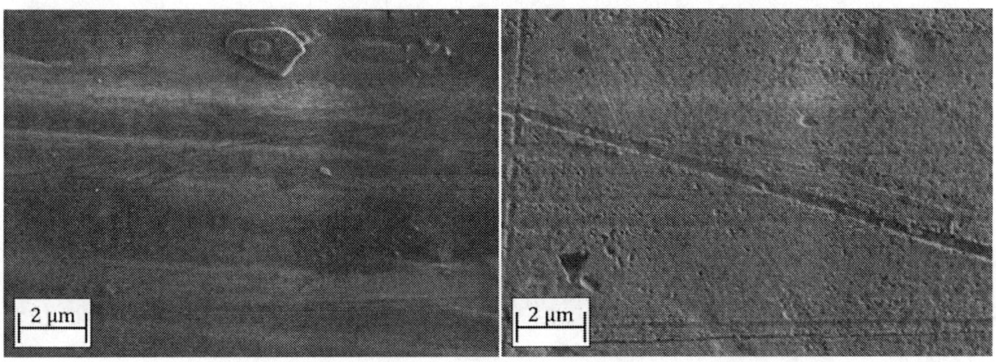

(A) (B)

FIGURE 3: EFFECTS OF ACID CLEANING - SEM IMAGES (IN-LENS SECONDARY ELECTRON DETECTOR, 2 KV, 20K X MAGNIFICATION) OF C12200 COUPONS: (A) AS-SUPPLIED, (B) AFTER ACID CLEANING.

SEM images show that the extent and morphology of the corrosion loss experienced by the coupons due to the immersion corrosion conditions, seen in Figure 4, was substantially greater than for the cleaning process. For both the C12200 and C14200 copper grades, the duration of exposure up to 15 months did not impact the morphology of the corroded surface. However, the C12200 coupons corroded to expose coarse features, while the C14200 coupons corroded to expose finer features. In both cases, dark spots are associated with deeper penetration of the corrosion attack, which is taken to be a form of micro-pitting. The scale of these pits is on the order of 1 μm or less for both coupon grades, with C14200 coupons having a higher proportion of smaller pits.

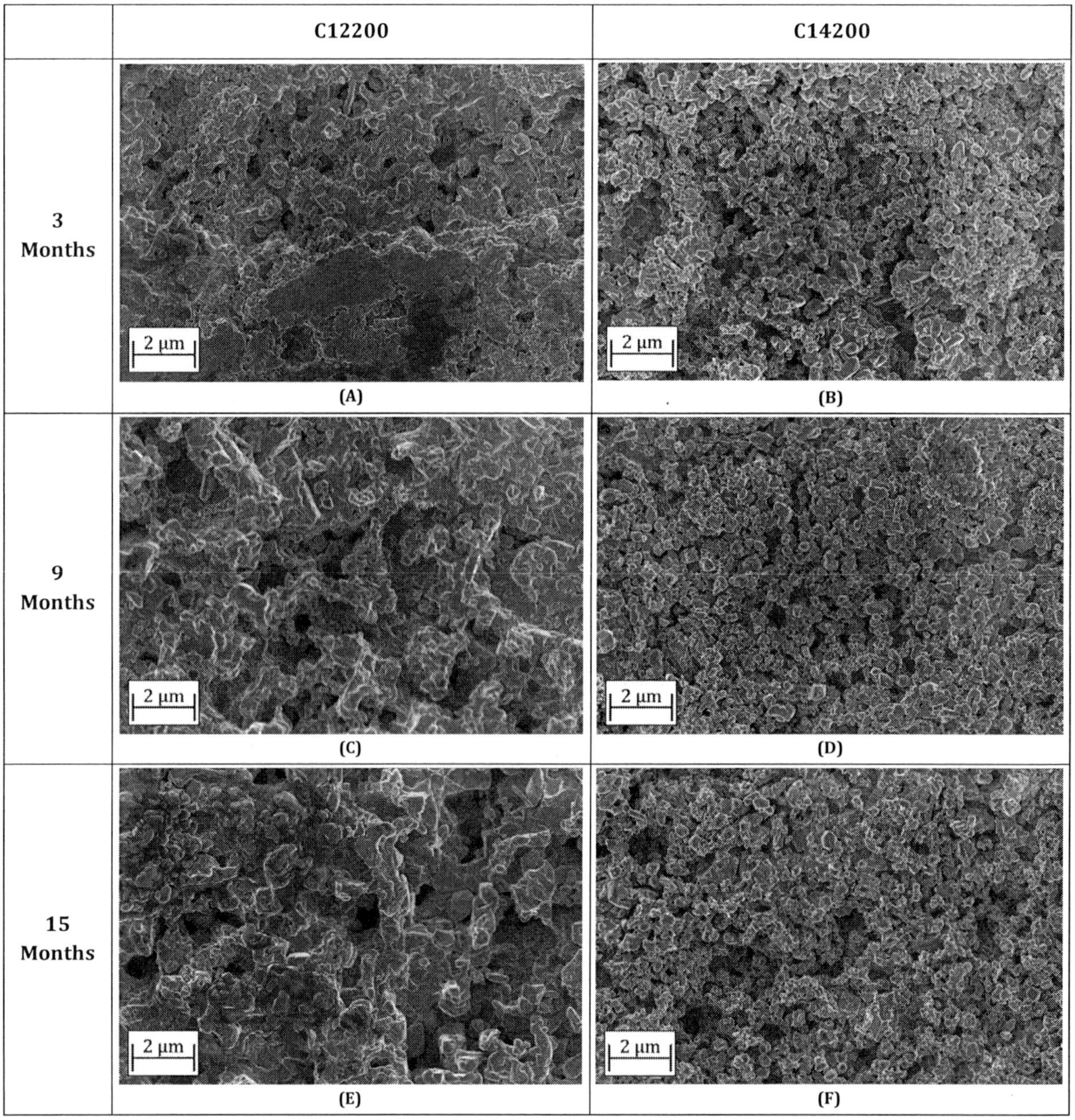

FIGURE 4: SEM IMAGES (IN-LENS SECONDARY ELECTRON DETECTOR, 2 KV, 20K X MAGNIFICATION) OF COPPER ALLOY COUPONS CORRODED BY HOT WATER IMMERSION FOR DIFFERENT EXPOSURE PERIODS: (A) C12200 - 3 MONTHS, (B) C14200 - 3 MONTHS, (C) C12200 - 9 MONTHS, (D) C14200 - 9 MONTHS, (E) C12200 - 15 MONTHS, (F) C14200 - 15 MONTHS.

Pitting Observations

While the general corrosion appeared to be reasonably uniform and entangled micro-pitting for both copper grades, there were a few instances of larger, isolated, more noticeable pits exclusively for the C12200 grade. These pits first appeared for samples recovered after 9 months of immersion and could be observed by naked eye, due to the presence of bright spots distributed within the patina of the corroded coupons. The bright spots represented cathodic regions where no corrosion was observed, usually surrounding a pit in the centre of the region. An optical microscope image of a typical example of one of these regions is shown in Figure 5. The pit depth was estimated to be 80 μm.

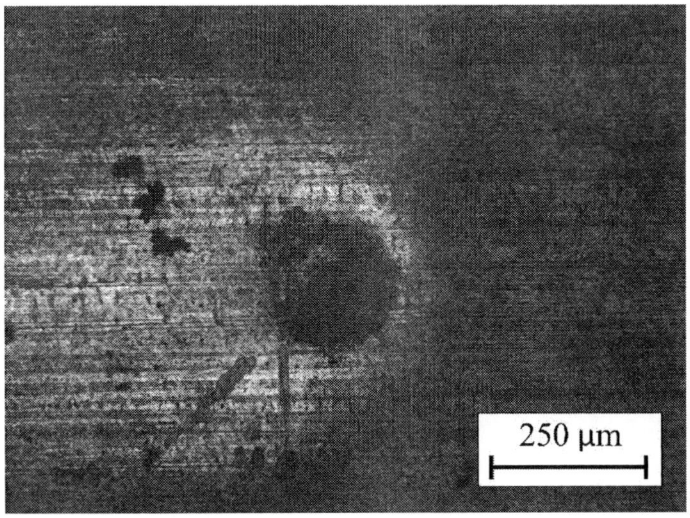

FIGURE 5: LARGE PIT OBSERVED IN C12200 COUPON AFTER 9 MONTHS EXPOSURE.

It is noted that the number and size of such pits did not increase with increasing exposure period, up to 15 months. After 15 months, these pits could no longer be identified.

Chemical investigations via EDS in and around the pitted regions during SEM analysis revealed that the pits appeared to be correlated with the presence of metallic impurities. For example, as seen in Figure 6, a large Fe signal was measured within a pit (~40% wt.% of total metal signal).

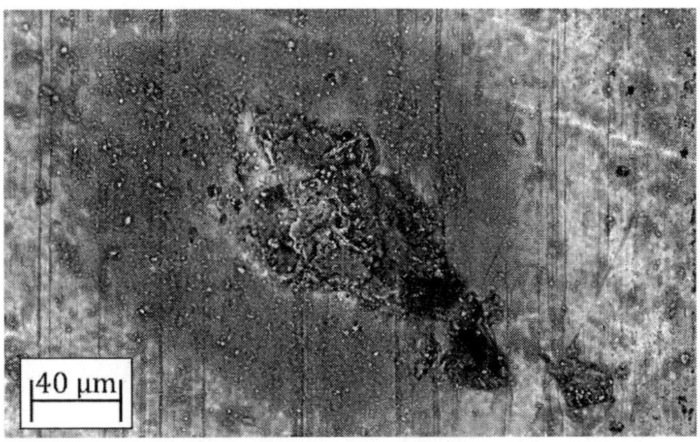

FIGURE 6: PIT IN C12200 COUPON WITH PROTECTED CATHODIC REGION SURROUNDING IT. RED CROSS REPRESENTS LOCATION FOR EDS ANALYSIS, MEASING HIGH LEVELS OF IRON.

Chemical analysis of the as-supplied C12200 copper returned a 100% Cu signal as the low concentration of the alloying elements (mainly P), were too low to be detected by EDS. No impurity regions, such as that seen in Figure 6, were observed during the screening of several coupons cut from the as-supplied material (i.e. non-immersed).

Corrosion Loss

Corrosion loss, measured as surface thickness loss from each face of the coupon, showed that both copper grades corrode at a similar rate in the investigated environment (see Figure 5). The main difference between the two is in the variability of the data as a function of exposure period, with the C12200 grade experiencing more deviation from linearity than the C14200 grade. The deviation is not explained by systematic error sources which are respectively small for each recovery period, indicating a high level of confidence in the measured data.

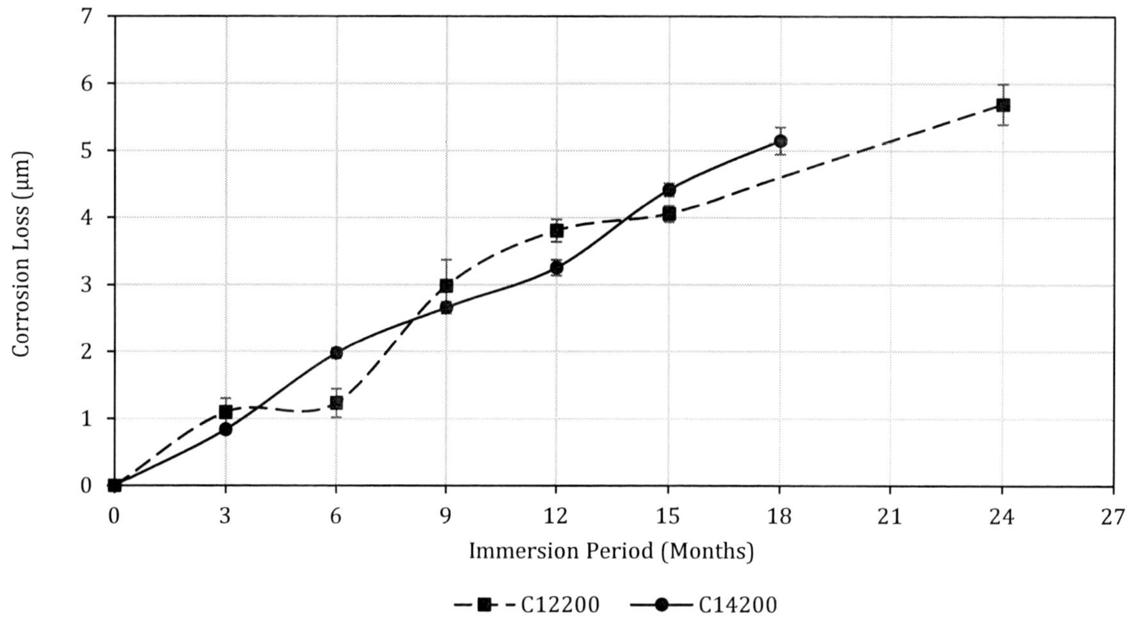

FIGURE 7: CORROSION LOSS FOR C12200 AND C14200 COPPER ALLOY COUPONS IMMERSED IN 70°C HOT WATER TANK.

DISCUSSION

As seen in Figure 4, there are some conclusions that can be drawn regarding the general corrosion behaviour of the C12200 and C14200 copper grades. Firstly, what is considered to be 'general corrosion' or sometimes 'uniform corrosion' is, in this case, actually a fine distribution of many closely-packed regions of localised corrosion, i.e. micro-pitting. This micro-pitting was found to be uniform in nature at the microscopic level and due to the dispersion of the pitting, stable growth of individual pits was not observed. This is further evidenced by the fact that during the 1 year of exposure between 3 months and 15 months (see Figure 4), there were essentially no differences in the microstructure of the corrosion morphology. However, the corrosion loss data (Figure 5) indicates that within this time period, a loss of ~3µm from the surface of both grades was experienced. This corrosion loss is greater than the size of the typical corrosion features for both grades and therefore the microscopy of each new recovery period represents images of a freshly corroded surface.

Interestingly, the small number of isolated, larger pits that appeared on the C12200 copper coupons (e.g. Figure 5 and Figure 6) were associated with impurity elements such as Fe, Si, Mn and elevated amounts of C and O. These pits did not increase in size with exposure period, suggesting that they were metastable. This is reminiscent of under-deposit corrosion caused by metal oxide deposits, for example Fe and Mn oxides, which have been noted in the case of pitting of copper pipes in the literature [5]. The as-supplied copper did not contain evidence of the impurity elements observed in and around the metastable pits. However, over the course of the experimentation, significant corrosion of the internal surface of the steel immersion tank was noted, with obvious presence of iron corrosion products suspended in the water. This may have led to some iron corrosion products being transplanted onto the copper surface. It is possible that these products may have acted as pit initiation sites. Further investigation into this phenomenon is required to fully understand and validate the mechanisms involved.

Aside from the pitting associated with the impurity elements, no other pitting was observed for the duration of experiments. Certainly, no stable pitting was observed which could lead to perforation, as has been identified by contract workers in recirculating hot water systems. Based on the current results, there is no evidence to suggest that such perforation pitting is due to excessively high temperatures, and that this pitting originates from some other cause. Two remaining possibilities for this behaviour are candidates for future investigation. The first possibility is related to the presence of impurities and other up-stream corrosion products within the water source entering the recirculating hot water systems. If the tubing is exposed to contaminants, such as Fe and Mn-based oxides, these impurities could lead to under-deposit corrosion on the under-side of the interior surface of the pipes, or cathodic protection of the pipes (for temperatures >71 °C) leading to pinhole pitting located at the tops of the inside of the pipes [5, 6]. The second possibility is on the immersive nature of the copper piping in operation, where it has been stated that water is on occasion left in a non-circulating state. This suggests the possibility that the water in the pipes sits up to a certain height, rather than filling the pipe and completely immersing its surfaces, with the rest of the pipe filled with air. In this case, an oxygen differential mixed with wet-dry cycling would be established, which can cause significant pitting corrosion; even for copper.

The corrosion loss data seen in Figure 5 shows clear non-linear behaviour, with a greater extent of non-linearity observed for the C12200 coupons compared with that of the C14200 coupons. An explanation for this is the presentation of a bi-modal corrosion trend, which has been extensively reported for steels [9-12], aluminium [13] and copper (in seawater environments) [14]. Details surrounding this corrosion model have been published extensively elsewhere [10]. The bi-modal trends are seen very clearly in Figure 7, with the transition from Mode 1 (oxygen reduction reaction) to Mode 2 (hydrogen evolution reaction) occurring at approximately 6 months and 9-12 months respectively for the C12200 and C14200 grades. However, longer-term corrosion rates could not be accurately determined as there was insufficient data collected to confirm a steady-state corrosion rate. Longer term (>2 years) data is required to confirm these values and is the subject of continued experimental work on this project, with additional samples yet to be recovered.

CONCLUSIONS

Copper coupons of two grades (phosphorus deoxidised – DHP – C12200, and phosphorus deoxidised arsenical – DPA – C14200) were immersed in an aerated hot water tank (70 °C) for up to 2 years to assess their pitting and general corrosion trends. The following conclusions were drawn from the work:

- No pitting of the C14200 coupons was observed for any of the recovered coupons, while some evidence of minor pitting was observed for the C12200 coupons, particularly for early exposure periods of ~6-12 months. This pitting was always accompanied by impurity elements as determined by EDS analysis, namely Fe and Mn and/or their oxides. The pits were deemed to be metastable, as they did not grow larger with increasing exposure period. Nor did they increase in number with increasing exposure period.
- The general corrosion for both grades of copper showed signs of extensive micro-pitting which was uniformly distributed across the sample surfaces. The features of this micro-pitting were finer in nature for the C14200 coupons than for the C12200 coupons.
- Both grades of copper showed signs of non-linear (bi-modal) corrosion trends, with the non-linearity more pronounced for the C12200 coupons than for the C14200 coupons. However, the overall rates of corrosion were quite similar for the two grades.

ACKNOWLEDGMENTS

The authors acknowledge the assistance of the workshop staff within the University of Newcastle's School of Engineering in preparing the copper coupons prior to starting the experiment. The authors also acknowledge the Electron Microscopy and X-ray (EMX) unit of the University of Newcastle's Central Analytical Facilities (CAF) division for the use of their SEM equipment.

REFERENCES

[1] AS/NZS 350.4, Plumbing and drainage, Part 4: Heated water services.

[2] Sato S, Minamoto T, Seki K, Yamamoto H, Takizawa Y, Okada S, et. al. Case studies on pitting corrosion failures of copper tubes in hot water, Corros. Eng. 31(1) (1982) 3-11.

[3] Cong H and Scully JR, Effects of aluminum solids on the under deposit corrosion of copper in synthetic potable water: The arguments for and against a semi-permeable membrane, J. Electrochem. Soc. 160(9) (2013) C403.

[4] Lee SH, Kim JG and Koo JY, Investigation of pitting corrosion of a copper tube in a heating system, Eng. Failure Anal. 17(6) (2010) 1424-1435.

[5] Wait K, Roy S, and Edwards MA, Rapid pitting of copper tube from deposits of cathodic manganese in hot water systems: Mechanism and effects of water chemistry, Corrosion 79(4) (2023) 459-468.

[6] Oliphant RJ, Causes of Copper Corrosion in Plumbing Systems, (In) Review of Current Knowledge, Foundation for Water Research, Allen House, The Listons, Marlow, UK, 2010.

[7] Charlton RS, Failures of copper potable water piping due to design, materials and poor workmanship, Corrosion (2001) NACE-01486.

[8] ASTM G1-03, Standard Practice for Preparing, Cleaning and Evaluating Corrosion Test Specimens

[9] Petersen RB and Melchers RE, Bi-modal trending for corrosion loss of steels buried in soils, Corros. Sci. 137 (2018) 194-203.

[10] Melchers RE, Modeling of marine immersion corrosion for mild and low-alloy steels—part 1: Phenomenological model, Corrosion, 59(4) (2003) 319-334.

[11] Melchers RE, A review of trends for corrosion loss and pit depth in longer-term exposures, Corros. Mater. Degrad. 1(l) (2018) 42-58.

[12] Melchers RE, Long-term immersion corrosion of steels in seawaters with elevated nutrient concentration, Corros. Sci. 81 (2014) 110-116.

[13] Melchers RE, Bi-modal trend in the long-term corrosion of aluminium alloys, Corros. Sci. 82 (2014) 239-247.

[14] Melchers RE, Bi-modal trends in the long-term corrosion of copper and high copper alloys, Corros. Sci. 95 (2015) 51-61.

AUTHOR DETAILS

	P. Richardson, BE, BSc, PhD, is the corresponding author of this paper. He is a post-doc research associate in the Centre for Infrastructure Performance and Reliability (CIPAR) at the University of Newcastle, Australia. With a strong background in materials science, he has a particular interest in characterising new materials formed during the interactions between known materials and their environment and understanding their reaction mechanisms. His current research focus is on the immersion corrosion of metals in different environments.
	Robert E. Melchers, Professor of Civil Engineering (Emeritus), The University of Newcastle, Australia, BE, MEngSC (Monash), PhD (Cambridge, UK), Fellow, Australian Academy of Technology and Engineering. Honorary Fellow, The Institution of Engineers Australia. Awards include 2009 ACA Corrosion Medal, 2012 Jin S Chung Award (ISOPE), 2013 John Connell Gold Medal (The Institution of Engineers, Australia), 2014 Eminent Speaker, Structural College, The Institution of Engineers, Australia, 2006 and 2018 Guy Bengough Award (IOM3).
	David Nicholas has been for 17 years the Principal of Nicholas Corrosion Pty Ltd, a consultancy firm specialising in corrosion and materials engineering issues in the Water Industry. David has over 40 years' experience in this Industry working for both Hunter Water Corporation and Hunter Water Australia for much of this time. He has authored or co-authored over 50 papers and is a former president of the ACA, holder of the Corrosion Medal, and a Life member of the Association.

DETERIORATION OF CONCRETE PILES UNDERWATER

P. Karajayli[1,*], S. Odlin[1], D. Cukierski[1]

[1]MEnD Consulting / Duratec Ltd, New South Wales, 2000, Australia

**Corresponding author email: phil.karajayli@duratec.com.au*

ABSTRACT

Reinforced and prestressed concrete piles submerged in marine or brackish waters are known to be subject to highly aggressive conditions. While it is well established that significant chloride contamination of the concrete cover, combined with intermittent wetting and high oxygen in the tidal/splash zone of the pile can often result in severe steel reinforcement corrosion and spalling, it is sometimes wrongly assumed that the submerged zone of the pile is in good condition without adequate confirmation. It has been reported that, in general, lower oxygen levels underwater typically limit the uniform corrosion rate of steel reinforcement in sound concrete, and this would appear to be the experience in most cases, however there are also limited reports of anaerobic corrosion and other deterioration mechanisms which can occur underwater, some of which can result in severe deterioration of the concrete and subsequently allow accelerated corrosion of the steel reinforcement. This paper explores typical risks of underwater deterioration and corrosion, requirements for underwater investigations and presents some case studies where underwater diving inspections revealed unexpected concrete deterioration underwater.

Keywords: piles, concrete, submerged, deterioration, sulfate

INTRODUCTION

Deterioration of reinforced concrete piles in the tidal and splash exposure zones of marine and brackish waters due to chloride-induced corrosion of steel reinforcement is a common and reasonably well understood mechanism that limits the service life of structures and often requires major maintenance intervention. Durability planning for new structures often involves understanding the different exposure zones of the structure, applying exposure classifications and designing accordingly using modern codes and/or durability modelling (which is typically based on time to corrosion) in order to achieve the required design life. While modern structure designs attempt to address potential durability issues to ensure the required design life is achieved, older structures were often not designed and constructed with this in mind. Many older structures are constructed in aggressive environments with poor quality concrete and poor quality construction that is susceptible to deterioration due to a variety of potential mechanisms.

The challenge for the asset manager is to understand the durability risks of existing structures, and this includes parts of the structure that are often hidden and cannot be easily investigated. There is often a trade-off between carrying out minimalist routine inspections and the cost to carry out out more detailed investigations which may uncover hidden deterioration.

The permanently submerged zone of concrete piles in marine and brackish waters presents such a challenge. While it is often assumed that the submerged zone of concrete piles is at low risk of steel reinforcement corrosion due to low oxygen levels and the risk of concrete deterioration appears to be a rare occurrence, there are examples of where this has been found to not be the case. This paper provides a discussion on the primary mechanisms encountered that may cause corrosion and deterioration in underwater concrete piles, with some relevant case-studies, and appropriate investigation methodologies that can be used to address some of these risks.

LITERATURE REVIEW

A primary outcome of the literature review confirmed there is very limited published technical literature and real-world case studies on this topic, particularly for piles in brackish waters. This is likely due to the difficulty in comprehensively inspecting piles underwater, the uncertainty on the condition underground which cannot be confirmed without excavation or exhumation of the pile, the potential corrosion mechanism (if present) being anaerobic corrosion which makes corrosion difficult to detect, and it would appear severe pile deterioration underwater remains a relatively rare occurrence. Given the lack of published information, further large-scale studies would be required to confirm this.

CORROSION RISKS

When assessing the durability risk of concrete piles in marine and brackish water based on exposure zone, the primary consideration is justifiably most often given to the tidal and splash zones, where factors such as chloride contamination and intermittent wetting often result in severely corroded reinforcement causing concrete cracking and spalling. Much of the literature states or infers that there is generally a lower risk of steel reinforcement in the submerged zone.

In concrete structures in seawater or brackish water, chlorides diffuse into the concrete until eventually the depth of the steel reinforcement is reached. Once a threshold level of chloride concentration is met, chloride ions attacking the passive layer protecting the steel may allow the corrosion process to proceed, subject to several variables [1]. The exposure condition of the concrete, primarily the height relative to the mean water level, has an effect on the depth of penetration of chloride ions. The highest surface concentrations occur at relatively high position (atmospheric zone), where the accumulating effect of drying out will be strong. At lower positions (tidal and splash zones), the concrete is wetter and transport towards the inner parts will be faster and consequently the penetration will be deepest. In the submerged part, accumulation due to wetting and drying is absent; here, equilibrium between bound and free chloride governs the total chloride content, together with relatively fast transport to the inner parts [2].

The submerged zone of the piles may not be adequately accounted for during inspections, due to assumptions that the condition of the concrete in the submerged zone is good [3], or to the difficulty inherent in accessing, cleaning, and inspecting concrete underwater [4]. As the concrete is fully saturated with water, the diffusivity of oxygen is reduced and thus also the availability of oxygen at the surface of the embedded steel which lowers the steel potential (reducing likelihood of corrosion initiation) and, should initiation occur, supresses the corrosion rate [2], [3], [5]. This can result in instances where high concentrations of chlorides (generally considered to be above 1.0% concentration by weight of cement [1]) are present in concrete and yet limited uniform reinforcement corrosion occurs.

The rate of corrosion of steel within good quality, well-compacted, dense, and uncracked concrete with good cover to the steel and low permeability, even with chloride contamination, can also remain low due to the protection provided by the high pH environment of the concrete forming a passive layer on the steel and continuing to provide alkaline buffering [6]. For older structures with poorer quality concrete this is often not the case, and in instances where reinforcement becomes exposed underwater, such as in a location where concrete is cracked, or the concrete is deteriorated, macrocell corrosion effects may occur [2].

Macrocell corrosion is typically seen as localised corrosion (rather than uniform corrosion) with a separation of the anodic reaction and the cathodic reaction between two separate areas [1]. When a macrocell forms, anodic areas experience accelerated corrosion rates while the passivity of cathodic areas is stabilised. For submerged structures, this may occur where reinforcement is not entirely embedded in concrete, caused by a difference in potential between steel embedded in concrete and the exposed steel, and may be exacerbated by direct exposure to chloride-rich water [2].

Corrosion of steel reinforcement in oxygen-deprived areas (which is the case for submerged areas of piles) is known as anaerobic corrosion and often referred to as "black rust" (magnetite, Fe_3O_4). The identification of anaerobic corrosion or "black rust" in structures is typically very difficult, as unlike "red rust" resulting from aerobic corrosion (ferric oxide, Fe_2O_3), it does not cause expansive forces to be applied to the concrete and therefore usually does not cause concrete defects such as cracking, spalling or delamination. Studies have postulated that black rust occurs under the conditions that the anode and cathode are separated (by at least a few hundred millimeters) and in low oxygen conditions. When these conditions are satisfied, the iron at the anode may stay in solution, rather than forming Fe_2O_3 [1].

While not commonly observed, black rust has been found in areas with limited oxygen concentration, such as under damaged waterproofing membranes [1], in the submerged portion of concrete piles [3], in culverts [7], and in reinforced concrete elements where corrosion was at an advanced stage, cutting oxygen off from the uncorroded central portion of the rebar [8].

A study of 3 out-of-service bridges in Florida in a marine environment aged between 51 to 66 years old examined sample piles that were exhumed [3]. Interestingly, the study found that for the most part the steel reinforcement was in good condition as expected, however there were areas of localised severe corrosion of between 1-12% of the length of steel examined that had appreciable section-loss of steel of 20-29%. It is noted that the concrete used for these older bridges would likely have been highly permeable and much lower quality than modern concrete mixes. The study hypothesises that that cathodic reaction rates under oxygen diffusional limitation that are negligible in cases of uniform corrosion can nevertheless support substantial corrosion rates if the corrosion becomes localised. In addition to oxygen limitation, the study also notes that corrosion in the submerged zone can be somewhat suppressed by galvanic coupling with reinforcing steel in the tidal and splash/evaporation zones which started corroding earlier in the life of the structure. This coupling tends to make the potential of the submerged steel more negative, and this effectively elevates the corrosion threshold and thereby retards the initiation of corrosion below water (a form of cathodic prevention). However, the effect of the small but finite cathodic reaction taking place on the entire submerged region (plus any additional coupling to cathodic regions above water) can be enough to sustain appreciable anodic activity over small regions. The study proposes a steady-state predictive model for this scenario and projects that eliminating corrosion in the splash/evaporation zone could in some cases increase corrosion vulnerability of steel in the submerged region, as the above-water region could then act as a supplemental cathode to the underwater area.

While the Florida study presents some potentially serious concerns and challenges to the asset manager, it is only based on a very small sample size. This issue of anaerobic corrosion on piles is largely unreported in Australia, but this may be because there is limited opportunity to perform such destructive investigations unless a pile is being removed from service. In addition, the risk-based structural implications of localised corrosion would need to be assessed on a case-by-case basis. Further investigation into this issue is required to understand the real-world risks.

Concrete Durability Risks

There are various mechanisms by which concrete can deteriorate underwater in marine/brackish environments. Concrete deterioration may have structural implications such as loss of strength and loss of concrete section. While concrete deterioration underwater may not be typically caused by steel reinforcement corrosion, degradation of the concrete cover can eventually directly expose steel reinforcement/strands to the chloride-rich water and allow corrosion to occur, further weakening the pile.

Publications such as ACI 201.2R-16 *Guide to Durable Concrete*, ACI 546.2R *Guide to Underwater Repair of Concrete*, and BRE Special Digest 1 *Concrete in Aggressive Ground* provide a good summary of the primary risks to durability of concrete in water and soil. In addition to steel reinforcement corrosion risks, the primary durability risks to piles underwater in natural waterways typically include:

- Deficient construction practices and errors
- Sulfate attack – conventional and thaumasite
- Alkali Silica Reaction (ASR)
- Delayed Ettringite Formation (DEF)
- Acid attack
- Leaching
- Industrial chemical attack
- Mechanical damage – impact, abrasion and cavitation
- Damage due to loads
- Freezing and thawing – although not typically an issue in most regions of Australia

As this paper focuses on piles in natural waterways, other potential concrete deterioration mechanisms are not discussed, such as acid attack from acid-producing bacteria as found in sewers.

Sulfate attack is a mechanism which may result in deterioration of the concrete causing both chemical and physical attack of the concrete. Concrete is typically exposed to sulfates from exposure to soil and water with high sulfate concentrations. The mechanism by which conventional sulfate attack occurs is that the sulfate ions react with calcium hydroxide within the cement to form gypsum, and with calcium aluminates to form ettringite, the growth of which causes expansive pressure on the concrete resulting in cracking/failure of the concrete [9]. Other types of sulfate attack (such as thaumasite sulfate attack) may occur under specific environmental conditions.

The aggressivity of the sulfate attack also depends on which cation accompanies the sulfates. Magnesium sulfate is a particularly aggressive species. In chloride containing water (i.e. seawater or brackish water), the risk of sulfate attack can be reduced by the presence of chloride ions. AS 5100.5 section 4.8.2 [10] states that "sulfate ions become aggressive at levels of 600 to 1000 ppm when combined with magnesium or ammonium ions. In the presence of chloride ions, however, attack by sulfate ions generally exhibits little disruptive expansion" and "if magnesium ions exceed 1000 mg/L together with sulfate ions (SO_4) more than 1000 ppm in soil or 400 ppm in ground water, an aggressivity or exposure classification of one higher class should be adopted, additional protective measures should be considered and expert advice should be sought".

The combined attack of magnesium and sulfate in seawater or brackish water occurs when the magnesium and sulfate ions react with portlandite and aluminate phases in hardened concrete to form gypsum ($CaSO_4 \cdot H_2O$) and brucite ($Mg[OH]_2$), the combination of which can be diagnostic of its occurrence.

The formation of gypsum can be catastrophic for hardened concrete due to its two-fold deterioration potential:

- Gypsum growth is an expansive reaction which can result in physical cracking of the paste and aggregate, exposing unreacted paste to the system which will allow for further deterioration.

- Gypsum is soft and non-durable, such that cement paste that has been replaced by gypsum will have a greatly reduced compressive strength and can be easily worn away/removed by even the slightest physical abrasion.

Unlike gypsum, the formation of brucite (often coupled with aragonite ($Ca[CO]_3$)) can form a somewhat protective barrier to prevent further deterioration. However, this quasi-protective layer can be removed by physical abrasion or deterioration of the underlying concrete [2].

Alkali-Silica reaction (ASR) is a mechanism which occurs as a result of a reaction with certain types of aggregates that are susceptible to ASR within the concrete mix. The ASR Field Identification Handbook defines ASR as a chemical reaction between the alkali hydroxides in the pore solution of concrete and certain forms of reactive silica minerals occurring in some aggregates. The reaction product, an alkali-silica gel, is hygroscopic, and will absorb water and swell if the concrete is in a moist environment [11]. ASR requires high levels of moisture to occur, and thus is susceptible to form in submerged structures which contain reactive aggregates. The significance is that deleterious expansion can cause cracking (exposing the reinforcement/strands directly to chloride-rich water and allowing corrosion to occur) and reducing tensile strength of the concrete.

Delayed Ettringite Formation (DEF) can occur when under certain conditions, such as concrete elements exposed to high temperatures during curing, and later exposed to moisture, which can result in expansion and cracking. Modern codes have limitations on the curing temperature, which is particularly important when steam-curing is used. Submerged areas of structures may be susceptible to this form of deterioration, and therefore signs of cracking should be investigated petrographically to confirm whether DEF or ASR is the cause. In two of the case studies detailed below, deteriorated concrete attributed to ASR and DEF was identified in prestressed concrete piles, primarily in the tidal zone and below [12], [13].

Acid attack of concrete (which may be combined with sulfate attack) may occur in certain conditions such as natural waters with low pH and in acid-sulfate soils (ASS). While durability planning for the design of new structures typically attempts to address this issue through compliance with codes and guidelines (e.g. AS5100.5, TfNSW TS 02014 *Provisions for Concrete Structures in Acid Sulfate Soils*), it is largely unknown what effect ASS is having on older structures with potentially poorer quality concrete that have been exposed to this environment over the long term.

Other less common types of deterioration can occur in certain conditions such as leaching of cement compounds in poor quality concrete with high water flow resulting in a weakening the concrete, and attack by aggressive industrial agents (such as ammonia) in contaminated soils and water.

INVESTIGATIONS

It is prudent that a comprehensive condition investigation includes an assessment of potential deterioration mechanisms based on exposure conditions.

It is typical practice to carry out periodic routine underwater visual inspections using divers or remote-operated vehicles (ROVs) to check for any obvious defects. This inspection should first involve cleaning of a representative sample of piles as marine growth could mask potential hidden issues. Road authorities in Australia generally have underwater inspection procedures which outline the frequency, level of detail, and proportion of piles to be examined on bridge piles. The Ports Australia Wharf Structures Condition Assessment manual (WSCAM) also provides some guidance on underwater inspections.

Piles should be cleaned to remove marine growth, which is typically completed using a combination of hand scraping and high-pressure water blasting, where high pressure water blasting will achieve better results and is best used for detailed pile examination. After cleaning, visual inspection should be completed by divers under supervision of an engineer with sufficient experience to interpret the results of the inspections and direct the diver should any unusual features be encountered. The results of the inspection should be recorded in a logical manner so that the location (pile face/cardinal direction and depth), dimensions, and type of defects are recorded.

If any unexpected or significant deterioration has occurred, consideration should be given to further testing. Petrographic examination of core samples in accordance with ASTM C856 is an ideal technique for identifying potential deterioration mechanisms. Concrete petrography is a technique which involves examination of polished sections and thin sections of concrete using microscopes to provide insights into the mineralogical, chemical and microstructural features of concrete. Petrography may be used to identify and confirm the presence of a number of concrete deterioration mechanisms including ASR, DEF, leaching, sulfate attack and acid attack. This may also be complemented by other laboratory techniques such as X-ray Diffraction (XRD) or X-ray Fluorescence (XRF).

Core sampling underwater can be difficult in practice however, as there is a risk of cutting through steel reinforcement / pre-stressed strands, and it may not be appropriate to carry out (particularly for pre-stressed piles).

The condition of the buried part of piles is an unknown factor and is also difficult to examine. A risk-based approach should be used in making this assessment. If safe to do so, consideration may be given to various investigative methods. Testing of the soil and water aggressivity can provide an indirect indication of potential issues (noting that aggressivity can change over time), combined with understanding the age of the structure. Localised shallow excavation around the pile (if structurally safe to do so) may provide an indication of potential deterioration in the buried zone. In addition, testing the concrete quality above the buried area may provide an indication of the likely durability below-ground.

Case Studies

Four case studies where unexpected levels of concrete deterioration in the submerged zone were identified during underwater dive inspections are discussed in this section.

Bridge in New South Wales

The first case study covers a small reinforced concrete bridge in NSW. The bridge was constructed approximately 60 years ago, with piles constructed from pre-cast prestressed concrete, driven into a tidal creek with brackish water.

Investigation Methodology

Reinforced concrete piles were cleaned of marine growth and visually inspected for defects below and above-water. While the piles were in good condition above-water, including the tidal/splash zone, the inspection revealed loss of concrete section underwater. Materials testing was then completed on two concrete samples: the first collected from the tidal zone in a sound concrete area (as a reference), and the second sample collected from a deteriorated pile in the submerged zone. Testing was carried out for chloride content profiles, sulfate content profiles, petrographic examination, X-ray Diffraction (XRD), X-ray Fluorescence (XRF), and compressive strength testing.

Preliminary water and soil aggressivity testing were also conducted on a water sample taken at the turn of an outgoing tide, and of the topsoil layer near deteriorated piles.

Findings

The dive inspections highlighted the presence of concrete section loss on the underwater portions of some piles, including the identification of large areas of concrete loss and cracking. The condition of the piles was obscured by thick marine growth which is likely why the deterioration had gone unnoticed. The extent of concrete section loss was not uniform across all piles. Some of the piles had reduction in concrete cross-sectional area at the most deteriorated portion of the pile (Figure 1), while no section loss was observed in the pile in the best condition, however softening of the cement paste causing aggregate to become exposed was observed (Figure 2). Spiral reinforcement had been exposed due to the deteriorated concrete on the pile in the poorest condition, allowing corrosion of the reinforcement to occur.

FIGURE 1 CONCRETE DETERIORATION ON UNDERWATER PILES.

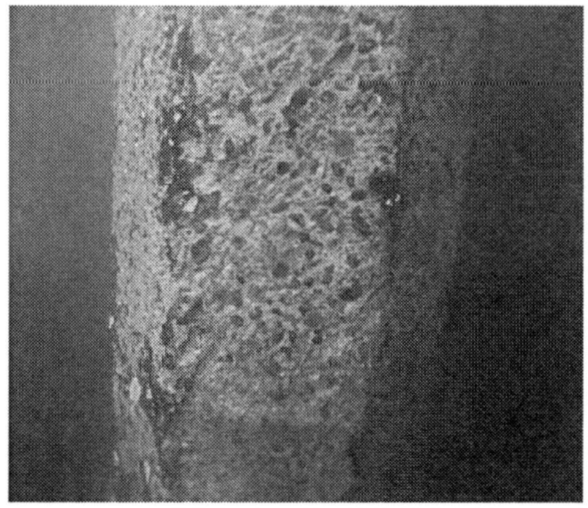

FIGURE 2 CONCRETE PILE AT SAME BRIDGE WITH NO SECTION LOSS.

A concrete sample was retrieved from the pile in the poorest condition. The concrete was weak and friable, allowing the sample to be retrieved by hand. Figure 3 shows an image of a cut section of the sample taken in the laboratory after drying, including cracking through a piece of aggregate.

FIGURE 3 CROSS SECTION OF SAMPLE COLLECTED FROM DETERIORATED PILE.

Petrographic analysis indicated that the entire thickness of the sample, around 40 mm, was found to have been replaced by gypsum (Figure 4). In addition, voids and microcracks were filled with gypsum and the paste was soft and powdery throughout. Another petrographic texture identified in the above water sample was popcorn-calcite deposition (PCD). PCD is deposit which forms in cement paste that is depleted of residual cement. It typically occurs in concrete with a reduced pH, exposure to moisture ingress and can be (but not always) associated with mechanisms like ASR and specific types of sulfate attack. Concrete with this deposit has typically been decalcified and exhibits weaker, more friable characteristics compared to non-impacted concrete.

The samples were also petrographically examined for signs of ASR. The aggregate fractions of the samples had ASR potential per HB 79. However, evidence of ASR was not identified in either of the samples, and the condition of the piles in the tidal and atmospheric regions showed no evidence of ASR cracking. Therefore, ASR was considered unlikely as a potential cause of the concrete deterioration.

Analysis by X-ray Diffraction (XRD) confirmed the presence of brucite in the sample. Evidence of similar attack was not identified on the sample collected in the tidal zone of the piles. Both gypsum and brucite were identified via XRD.

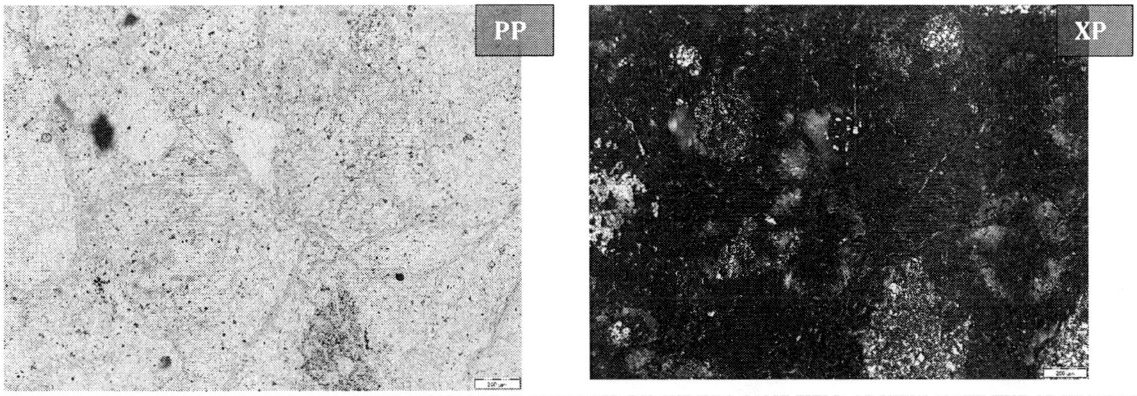

FIGURE 4 AREA OF SB-P1P5 THAT HAD BEEN COMPLETELY REPLACED BY GYPSUM, SAME FIELD OF VIEW, NOTE THE ABSENCE OF GP.

Petrography also confirmed that the concrete only contained Portland cement, and no supplementary cementitious materials (SCMs) were identified. The design strength was nominated as 38MPa. This could not be confirmed by core sampling due to the very high congestion of embedded steel. For new bridges, AS5100.5 requires SCMs to be used in the concrete mix to enhance the durability of concrete to achieve the required design life.

Water testing of a sample collected from the creek did not highlight any particularly high concentrations of sulfates (1,700 mg/L), chlorides (13,000 mg/L), or magnesium (820 mg/L), the levels of which were not found to be greater than that typically found in seawater. However, ion concentrations are expected to vary over time, and this was a single test at a point in time. It is also possible that historic uses of the land around the creek may have contributed to high sulfate levels in the past. The water flow of the creek is relatively fast due to the tidal effect and it was noted that the pile in worst condition is facing downstream. Based on the limited test samples carried out and the petrographic and XRD findings, the cause of deterioration was most likely caused by the combined attack of sulfates and magnesium from exposure to the creek water.

Rawdon Island Bridge, New South Wales

In 2021 flooding events prompted dive inspections to be conducted on Rawdon Island Bridge, which spans the Hastings River in Sancrox, NSW. The Bridge was constructed around 63 years ago. The bridge piles are reinforced concrete which were anchored into bed rock and cast in-situ and are exposed to tidal brackish river water.

The dive inspection uncovered significant loss of section of the reinforced concrete piles, in one case a 90% reduction in section was observed, as seen in Figure 5 [14]. Generally, the piles in the deepest part of the river were the most affected. Where steel reinforcement was exposed, generally section loss of the rebar was also observed.

A soft white residue was found near the deteriorated areas of the piles, and concrete core samples were examined petrographically and via XRD techniques. Figure 6 shows a cross section of one of the core samples taken from the piles and the change in the cement paste morphology as sample depth increased. The white residue was found to be an amorphous material, thought to have formed due to excessive leaching of the calcium bearing phases in the concrete.

Directly beneath the amorphous residue, the cement paste in the cores was usually completely leached of portlandite and lacking in residual cement. In some cases, the cement paste had minor amounts of another phase present (likely ettringite or gypsum). The section shown in Figure 6 had cracks that were filled with ettringite that passed through aggregate particles, indicative of sulfate attack, however attack by sulfates was deemed unlikely to be the cause of deterioration, due to the low sulfate content measured in the concrete.

The type of cement or presence of SCMs could not be identified as the portlandite had largely leached out. The compressive strength of the concrete (as tested in conformance to AS1012.14) was found to be an average of 42MPa.

FIGURE 5 PILE WITH 90% SECTION LOSS AT RAWDON ISLAND BRIDGE.

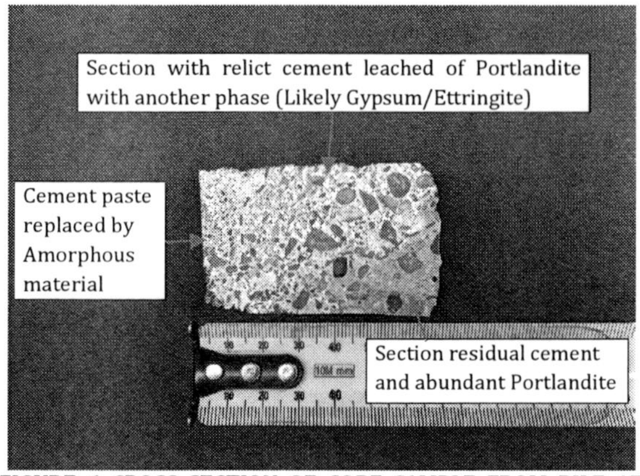

FIGURE 6 CROSS SECTION OF CORE SAMPLE FROM PILE AT RAWDON ISLAND BRIDGE.

Concrete that is exposed to water migration while in service is at risk of being leached. Leached concrete can be depleted of calcium-bearing phases that are critical to its integrity, leaving behind a husk of weak, softer material. The degree of leaching (or more severe degradation and material loss) is dependent on numerous factors including pH, dissolved salts content/chemistry, suspended load and more. Excessive leaching of the portlandite and Portland cement phases was determined to be the cause of concrete deterioration.

Pile repairs were carried out by encasement of the original pile with micropiles drilled into the surrounding riverbed.

TEMPE BRIDGE, SYDNEY

A major bridge over the Cooks River in Tempe (Tempe Bridge) which is part of the Princes Highway, built in 1960, was found to have severely deteriorated piles during an underwater inspection in 2002 [13]. The precast octagonal piles were visually obstructed by concrete skirts, which prevented inspection during low tide, therefore preventing the section loss from being flagged earlier during routine visual inspections.

Some of the pile concrete was found to be completed disintegrated in the outer annulus, up to the depth of the pile reinforcement (approximately 60 mm depth), shown in Figure 7. Beyond this, the concrete was found to be deteriorated, cracked, and weak, increasing in strength towards the centre of the pile. The cause of the deterioration was believed to be AAR or DEF. Urgent actions including restricted load limits and strengthening works were required to avoid potential catastrophic failure of the bridge.

FIGURE 7 PILE CONCRETE DETERIORATION IN BRACKISH WATER ATTRIBUTED TO ASR OR DEF.

The bridge was rehabilitated in two stages. In Stage 1, additional cast-in-place bored piles were constructed adjacent to the existing piers. A steel underpinning system was used to jack up the pier substructure and transfer the loads to the new piles. In the Stage 2, a new prestressed concrete pile cap was constructed around the underpinning system and existing piers, which was used to both provide durability corrosion protection to the exposed steel, and to integrate the existing bridge substructure with the Stage 1 remedial works.

Two Bridges in the South Island of New Zealand

In 2000 in the South Island of New Zealand, piles from two bridges constructed in 1969 were found to have cracking, splitting, and spalling on a quarter to a third of bridge piles [12]. The piles were of prestressed concrete and were submerged in brackish water with tidal movement. The piles were generally in good condition above the high tide level, and cracking was observed to extend from the tidal zone to below low tide level and sometimes below bed level, with softened concrete. The paper notes that the same aggregate was used in both bridges, though the piles were manufactured by different suppliers. The design compressive strength was 45 MPa for one of the bridges, and cement content was measured at around 320 kg/m^3.

In total, thirteen concrete samples were collected across both bridges. ASR was conclusively identified in 2 samples and was identified as a possible deterioration mechanism in cores from all other piles (6 samples). DEF was identified in 2 samples, and those 2 samples were identified in cores showing signs of ASR, both of which were also located within the tidal zone. Crack patterns observed by microscope showed that the concrete was subjected to expansive stress, consistent with ASR-induced DEF occurring in permanently wet concrete. There was some difficulty in diagnosing the cause of the deterioration observed on the piles, as evidence of ASR and DEF were not conclusively identified in all samples, despite targeting concrete sampling in areas showing deterioration.

Conclusions

This paper has discussed the various typical mechanisms by which concrete piles underwater in marine and brackish water may be subject to corrosion of steel reinforcement and deterioration of the concrete. While corrosion of steel reinforcement in concrete underwater is mostly inhibited by limited oxygen, there have been limited reports overseas that anaerobic corrosion can occur at localised locations. With respect to concrete deterioration, depending on the specific site location, exposure and concrete quality, this can include such effects as external attack (such as from sulfates in combination with magnesium ions and acidic water/soil), or internal attack such as from ASR and DEF. Deterioration of the concrete if left unchecked can then result in exposure of the steel reinforcement/strands to the environment and allow corrosion to occur.

To address this largely hidden risk, it is prudent to implement regular routine inspections which should include representative marine growth cleaning and underwater visual inspection, followed by more detailed testing and sampling if deterioration is observed. Techniques such as petrographic examination and XRD are ideally suited for identifying potential causes of deterioration combined with understanding the exposure condition and aggressivity of the soil and water.

While it is outside the scope of this paper to discuss remedial options for deteriorated underwater piles in detail, there are various techniques which have been employed. The type of repair selected will need to address the root cause of the issue to ensure adequate extension of service life and that the problem is not simply being masked. Typical repair details may include patch repair and pile encasements (reinforced or unreinforced) combined with cathodic protection to arrest corrosion if adequate electrical continuity is present. In extreme cases of deterioration, new piles may need to be constructed.

While concrete deterioration can usually be visually observed, the hidden risk of anaerobic corrosion of steel reinforcement in piles underwater is not well understood and there is very limited published data on this topic, particularly in Australia. Further investigation into this topic, including a risk-based assessment of potential structural implications is required for asset managers to make informed decisions. This would ideally involve detailed destructive investigations of piles of older bridges that are demolished in the future when the opportunity arises.

REFERENCES

[1] J. P. Broomfield, Corrosion of Steel in Concrete, 2nd ed., Oxon: Taylor & Francis, 2007, pp. 9-10.

[2] L. Bertolini, B. Elsener, P. Pedeferri and R. P. Polder, Corrosion of Steel in Concrete, Weinheim: WILEY-VCH, 2004.

[3] M. T. Walsh, *Corrosion of Steel in Submerged Concrete,* Florida, 2015.

[4] ACI Committee 546, "Guide to Underwater Repair of Concrete," American Concrete Institute, 1998.

[5] Standards Australia, HB84, Guide to Concrete Repair and Protection, 2018.

[6] R. E. Melchers, "Field Experience and the Long Term Durability of Reinforced Concrete Structures," in *International Conference on Performance-based and Life-cycle Structural Engineering*, Brisbane, QLD, 2015.

[7] H. Fatemi, M. Moore and R. Khatri, "Importance of Detection of Black Rust Formation," in *Austroads Bridge Conference, 8th, 2011*, Sydney, NSW, 2011.

[8] R. O'Donovan, B. D. O'Rourke, K. D. Ruane and J. J. Murphy, "Anaerobic Corrosion of Reinforcement," *Key Engineering Materials,* Vols. 569-570, 2013.

[9] BRE Construction Division, BRE Special Digest 1:2005, BRE Bookshop, 2005.

[10] Standards Australia, *AS 5100.5 Bridge Design Part 5: Concrete,* 2017.

[11] M. Thomas, B. Fournier, K. Folliard and Y. Resendez, "Alkali-Silica Reactivity Field Identification Handbook," U.S. Department of Transportation, 2011.

[12] S. M. Bruce, S. A. Freitag, A. Shayan, R. Slaughter and R. Pearson, "Identifying the cause of concrete degradation in prestressed bridge piles," in *Corrosion & Prevention 2008*, 2008.

[13] Y. C. Chong, L. Prasad, S. Deshpande, H. Madrio and M. Ali, "Bridge over Cooks River at Tempe," in *AustRoads Bridge Conference*, 2022.

[14] J. Duffy, G. Deutsch and L. Holloway, "Case Study - Rawdon Island Bridge Remediation Project," in *Concrete Institute of Australia, Concrete 2023*, 2023.

[15] A. B. Poole and I. Sims, Concrete Petrography: A Handbook of Investigative Techniques, Second Edition ed., Boca Raton, Florida: CRC Press, 2016.

AUTHOR DETAILS

	P. Karajayli is the corresponding author of this paper. He is a Technical Director at MEnD Consulting/Duratec Ltd and a chartered engineer with more than 23 years' experience specialising in infrastructure condition assessments, durability, repair and protection, corrosion management, cathodic protection design and life-cycle maintenance planning. His industry experience spans across a wide range of infrastructure including bridges, tunnels, marine, rail, water and wastewater assets with experience in both consulting and remedial contracting.
	S. Odlin is a Consulting Engineer at MEnD Consulting/Duratec Limited, a position she has held since 2022. She is responsible for conducting inspection and condition assessment of structures and providing durability advice to extend their life.
	D. Cukierski is the Laboratory Manager at MEnD Consulting/Duratec Limited. With over 10 years of experience, he specialises in the forensic and compositional analysis of concrete and other materials utilising optical petrography, X-ray diffraction, scanning electron microscopy, wet chemistry and other methodologies.

CATHODIC PROTECTION SYSTEM DESIGN – ACTUAL VERSUS DESIGN OUTCOMES FOR CP DESIGNS FROM AEM (AIRBORNE ELECTROMAGNETIC) SURVEYS

A. Sterling[1]

[1]APA Group, Queensland, Australia

ABSTRACT

Accurate soil resistivity data is critical to the design of high current output land based cathodic protection systems. To obtain this data APA undertook an airborne electromagnetic conductivity (EMC) survey. This paper discusses real world outcomes, comparing actual results to desktop design predictions. According to the mathematical modelling, EMC data should resolve low resistivity (1-5ohm.m) layers in the near surface (0-50m), however struggle in areas with higher average resistivities (10 ohm.m +). In higher resistivity areas EMC should only point to areas with lower bulk average resistivity, but not resolve thin 10 & 20 ohm.m layers near surface. To test this theory, the difference between design and actual system loop resistance was compared to determine if there was any correlation. In general, where layer soil resistivity was low the EMC data reflected that and a successful construction outcome was achieved. As soil resistivity increased the likelihood of success reduced.

Keywords: soil resistivity, cathodic protection, design

INTRODUCTION

Metal pipelines are frequently protected from external corrosion through a combination of coatings and impressed current cathodic protection. In the case of many aging pipelines, CP current demand is high and increasing, leading to CP site spaced at 2 to 20km along the pipeline. Individual sites are optimised to provide protection to the local pipeline segment with anode beds being electrically remote, generally between 300 and 1200m away from the pipeline. A critical element in anode bed design is ground resistivity profile. It determines both the type of anode construction that can be used and the separation distance. Importantly, lower ground resistivity is sought because it means a more efficient design through lower driving voltages and generally more easily constructed anodes.

Typically, soil resistivity profiles are determined by carrying out ground resistivity soundings (4-Pin Wenner test method or similar) in areas that appear favourable from the aerial imagery (Google Earth). That is, areas with trees / shrubs possibly indicating good underground water, or clay pans or flood plains or cropping areas. To complete this across many sites (~180 sites over 1300km) is expensive, labour intensive and time consuming, not least because Google Earth is not a particularly reliable indicator of ground resistivity. Wenner surveys themselves cover only a single point, and more importantly, ground access can be difficult particularly in cropping land. Furthermore, the final anode bed location is often 100's of meters from the test site, potentially resulting in ineffective CP anode beds. Given a typical CP site can cost more than $500,000, accurate ground resistivity data is critical.

Airborne electromagnetic conductivity (airborne EMC) surveys rely on *inductive* coupling with the ground, and thus are capable of surveying large areas in single day. More importantly, airborne surveys do not require ground access and thus remove a significant obstacle to data acquisition.

Mathematical modelling indicated EMC would resolve the conductive layers (<5ohm.m) better than the ground resistivity and would be adequate for the CP design purposes (Sterling & Hine [1]). Ground resistivity surveys were more effective in areas where the 'best' conductor layer was >10ohm.m, or when the conductive layer was right at surface. However, in most cases, given that the ideal layers for anodes are <5ohm.m and >5m depth, EMC was sufficient and, importantly, significantly faster and covered a much larger area.

After preliminary concept proofing, APA undertook an extensive airborne EMC survey to collect soil resistivity data, from which CP design were completed. This approach had many benefits for CP design including:

- Simplified the process of site selection
- Provided target location for anode beds
- Allowed quick validation of relocated anode bed
- Provided better resolved soil data for design purposes
- Reduced the risk of poor outcomes.

The purpose of this paper is to present the results of designs completed using EMC soil profile data.

The Project

The Moomba Wilton Pipeline CP Augmentation project has been tasked with upgrade / construction of new CP sites.

APA undertook an airbourne EMC survey and backcalculated the soil resistivity profile at approximately 1.2 million locations along the pipeline to facilitate the design process. Figure 1 shows a pipe section and the points for which soil resistivity profiles were derived and a typical soil profile.

(a) **(b)**

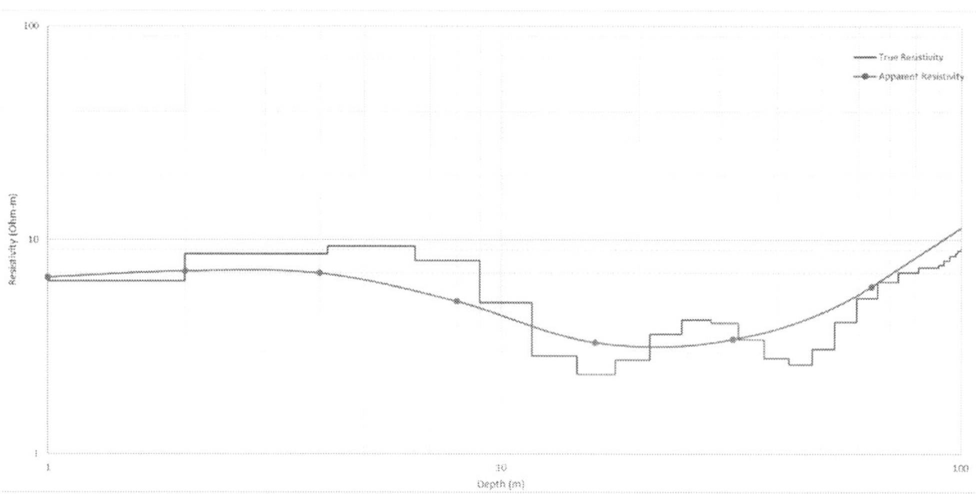

(C)

FIGURE 1: FIGURES (A) AND (B) SHOW A PIPELINE SECTION WITH POINTS (DOTS) WHERE SOIL PROFILES HAVE BEEN DERIVED. FIGURE (C) SHOWS A TYPICAL SOIL RESISTIVITY PROFILE WHICH IS AVAILABLE AT EACH POINT.

The project identified about 180 CP sites (new and existing) which are likely to be needed to ensure long term integrity. About 140 of these sites need to be constructed in the short term (3 to 5 years) to set the pipeline up for long term integrity and allow CP to move into a routine maintenance phase. Approximately two thirds will be solar powered.

All CP sites required new or extended anode bed easements as the CP current has increased due to coating deterioration. From a long term pipeline operation perspective, designs considered the worst case scenario for current demand (pipeline with no coating).

Due to the scale of the project and for efficiency, a modular design approach was adopted to simplify construction as much as practical. Standardisation of materials also allowed the procurement of long lead time items. Bulk materials procurement was undertaken for anode bed kits, carbon backfill, header cables, CP units and solar power modules.

THE DESIGN PROCESS

For maximum CP coverage, anode beds should be electrically remote, this required pipe to anode bed separation of up to 1200m. The process for acquiring an easement is quite complex and time-consuming, requiring cultural heritage (CH) and environmental considerations as well as negotiation with land holders. In most cases the easement process took 12 months or longer.

As a result, designs required a multi-step approach. Initially, a preliminary design is developed which is passed to the Lands & Acquisition team. The Lands team would undertake CH and Enviro assessments and negotiate with land holders. Often, CP sites must be moved to satisfy the requirements of all parties. Once all assessments were completed and the LH agreements in place, the easement is surveyed and submitted to government for final gazettal which can also take 6 months. The total process process can take years, particularly if the system is to be installed in a National Park. Once the easement is gazetted, the final design can be completed, documents issued and construction can commence, pending weather and land use.

DESIGN BASIS AND ASSUMPTIONS

The CP Site design basis was as follows:

- Shallow vertical (SV) design preferred (reasons explained below)
- Target Loop Resistance ~0.1 ohms (reduce power demand for solar sites)
- Anode bed remoteness (voltage gradient < 0.3V at the pipe for maximum "throw")
- Design Current of 120Amps (KP002 to KP1060) (or 80Amps if soil resistivity not ideal)
- Design Current of 40Amps (KP1060 to KP1299) (coating condition significantly better)
- Easement length < 1200m (for practical purposes only)

Generally design current was derived from worst case scenario (pipeline essentially bare steel) and utilised predominantly to ensure acquired easements would be suitable for end of effective service life.

In addition to completing CP designs to Australian Standards (AS2832.1 [2]), APA engaged an external CP Specialist [3] to provide an independent review of the design process. A number of recommendations were made and implemented in the design process. In particular, a "rule-of-thumb" was adopted for shallow horizontal (SH) design calculations. Shallow horizontal (SH) design calculations applied the apparent resistivity at a depth equivalent to anode bed length. For example; for an anode bed consisting of 2 kits in a single continuous column (80m) the apparent resistivity at 80m deep would be used in the calculations for separation distance and resistance to earth.

For design loop resistance calculations, the following assumptions were employed:

- Pipe to earth resistance assumed to be 0.05 ohms.
- Resistance between anode and carbon backfill ignored

- Header cable resistance based on easement length

Figure 2 below shows a typical design calculation sheet for a SV anode bed. The designer would update the sheet with the following data:

- target layer resistivity from the soil profile
- separation distance based on the acquired easement, and
- header cable size (generally 500mm2 aluminium) based on available materials.

The design would be reviewed for power demand and driving voltage, with consideration of additional anode bed kits for power reduction for solar powered sites.

Ground Bed Calculations

Description:

The following calculations use soil, anode and T/R data to calculate anode resistance. The calculation uses AS2832.1 Equation G6.2 and assumes a number of vertical anodes installed to a depth greater than the length. The anode to carbon resistance is ignored.

		Vertical Anode Arrangement Drawing 530-DWG-CP-0021		
	Site:	Shallow (Auger)	Shallow (Auger)	Shallow (Auger)
	Anode Bed Kits (10 anodes / kit)	1 Kit	2 Kits	3 Kits
B.1	Soil Resistivity (SkyTEM Data) ohm-cm	500	500	500
	ohm-m	5	5	5
B.2 Anode Specification & Details				
Material:	**MMO coated Titanium**			
	Length (mm)	1500	1500	1500
	Anode diameter (mm)	25	25	25
	Anode rating (A for 20 year life) (min)	4.0	4.0	4.0
B.3 Anode Bed Details				
Backfill:	**Loresco SC-3 (or equiv.)**			
	L - Column length (m)	5	5	5
	d - Column diameter (m)	0.30	0.30	0.30
	N - Number of anode wells (n)	10	20	30
	S - Anode Spacing (m)	5	5	5
	Active Zone Depth (m)	6 - 11 m	6 - 11 m	6 - 11 m
	Anode bed service life (Years)	25	25	25
	Anode Bed Current Rating (A)	**40.0**	**80.0**	**120.0**
	CP Unit Voltage (V)	50.0	50.0	50.0
	Target loop resistance (ohms)	1.200	0.600	0.400
	Anode bed - Structure separation (m)	430	430	430
	Anode bed - Foreign structure separation (m)	450	450	450
	Anode bed - Cable run length (m)	480	530	580
	Anode bed - Header cable size (mm2)	140	500	500
	Design Structure separation (AS2832.1 Eqn G3.1) (m)	106	212	318
	Pass/Fail at V < 0.3 @ design output (desirable)	Pass	Pass	Pass
	Voltage gradient @ Primary Structure (V/m)	0.07	0.15	0.22
	Voltage gradient @ primary structure allowing for depth based dissipation (V/m)			
	Voltage gradient @ Foreign Structure (V/m)	0.07	0.14	0.21
B.4 Anode Bed Calculations				
	Soil Resistivity (ohm-m)	5	5	5
	Resistance of backfill to soil (ohms)	0.123	0.072	0.052
	Current Density (A/m^2) (Criteria <1.6A/m^2)	0.849	0.849	0.849
	Carbon backfill density (kg/cu m)	1185	1185	1185
	Active zone coke volume (cu. m)	3.534	7.069	10.603
	Carbon backfill to be installed (kg)	4188	8376	12564
	Carbon backfill service life (years) [1kg/A/year]	105	105	105
	Total loop Resistance (ohms)	0.24	0.16	0.14
	Driving Voltage (V)	**12.0**	**15.04**	**19.3**
	Power Demand (W)	**479**	**1203**	**2318**
	Power Cost Over Service Life ($)	$29,123.86	$73,203.49	$141,003.93
	Ground Bed - Recommended Option	Shallow	Shallow	Shallow

FIGURE 2: TYPICAL DESIGN CALCULATION SHEET FOR SHALLOW VERTICAL CONSTRUCTION.

Shallow Horizontal Designs

Shallow horizontal (SH) designs were completed utilising Equation G6.4 from AS2832.1 [2]. As mentioned earlier, the apparent resistivity at a depth equivalent to anode bed length was utilised for design calculations.

Shallow horizontal construction was the least preferred construction option due to a number of reasons:

- Construction required substantial earthworks due to the need for battering (1m batter for 1m of depth)
- Minimum trench width at depth ~ 600mm leading to excess carbon use (~10 pallets per anode bed kit compared to 5 pallets / kit for vertical construction = ~$15,000 per kit)

Generally, SH construction was only adopted when low resistivity target layers were more than 18m deep. The depth limit was driven by the economics of drill equipment. Figure 3 shows the standard SH anode bed configuration.

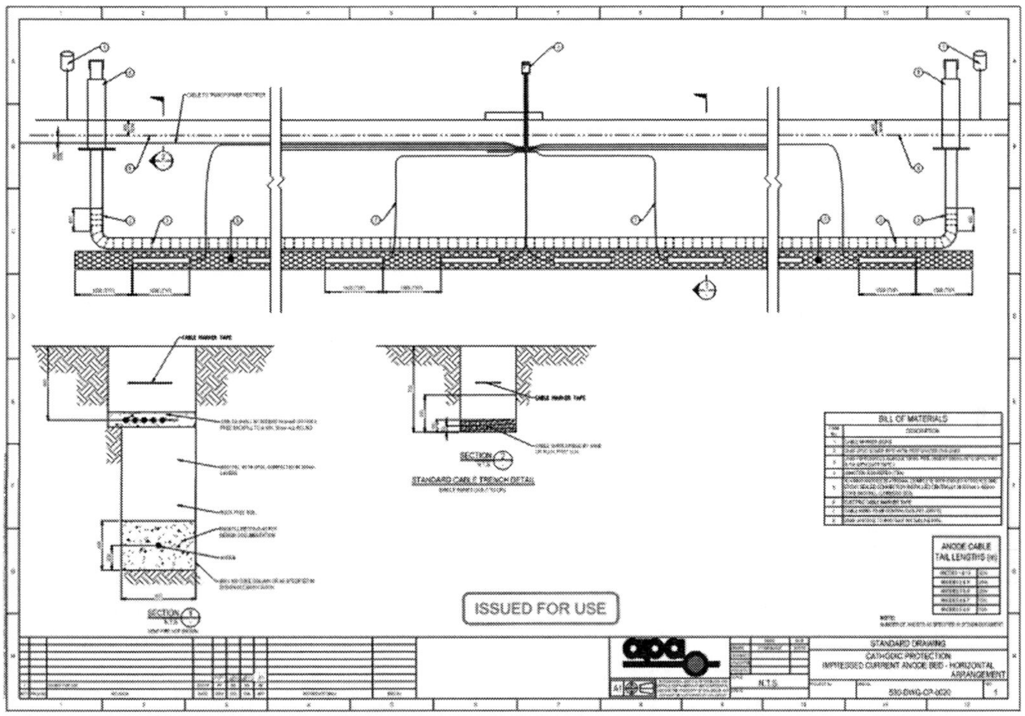

FIGURE 3: TYPICAL SHALLOW HORIZONTAL CONSTRUCTION.

Shallow Vertical Designs

Shallow vertical (SV) anode bed type is the preferred construction option as it requires much less civil works and uses less carbon backfill (as discussed above). Additionally, the electrode resistance calculations are somewhat better understood and more straightforward. Additional advantages over SH construction include:

- Easier to construct (SH construction requires excavation battering)
- Easier to target low resistivity layers.
- Ability to measure soil resistivity to confirm target zone reached

Shallow vertical design calculations utilised the Modified Dwight Equation G6.2 (AS2832.1 [2]) and the actual target layer resistivity. Generally, SV construction was adopted when low resistivity target layers were less than 18m deep. This depth limit was driven by the economics of drill equipment. Deeper drilling required much bigger equipment (air drilling versus auger) at significant additional cost. Imposing a depth "limit" also allowed adoption of standard anode header cable lengths. Figure 4 shown the standard SV anode bed configuration.

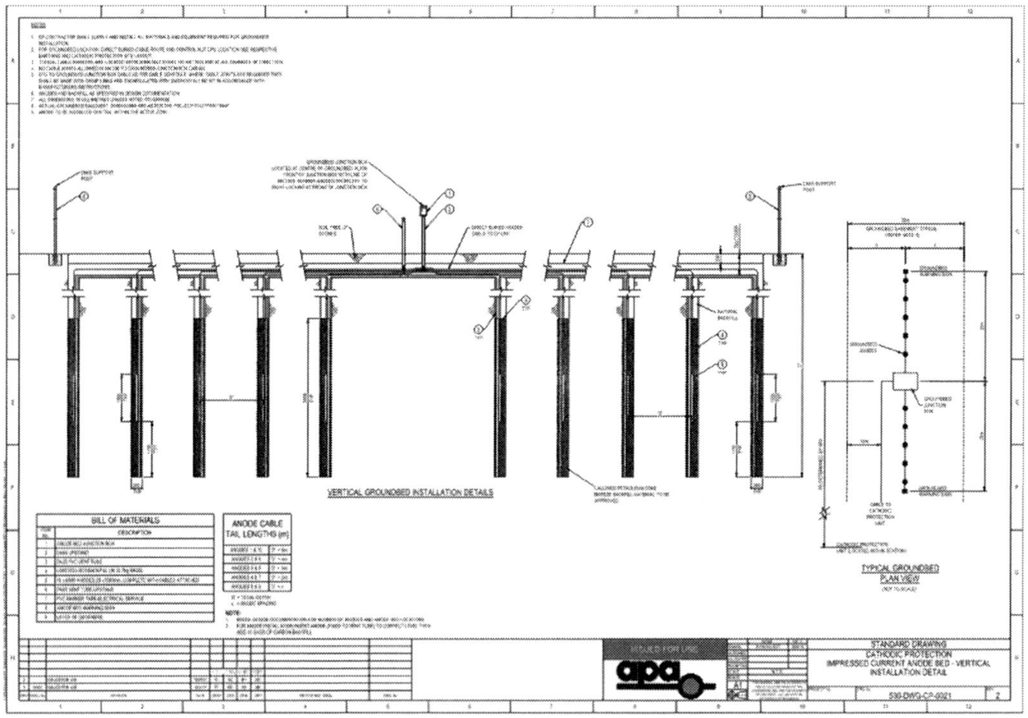

FIGURE 4: TYPICAL SHALLOW VERTICAL ANODE BED CONFIGURATION.

Outcomes

Success has been defined as the actual loop resistance being +/- 0.1 ohm of the design loop resistance. Noting the target loop resistance was 0.1 ohm, this does not mean that constructed sites outside this criteria do not work; it means the driving voltage and hence power demand are higher which may limit output current particularly for solar sites which have limited power (typically 850W total load).

Actual results indicate EMC data can be very effective and accurate for cathodic protection designs. Overall, EMC data lead to a "successful" outcome for 80% of the sites. Success improved to 90% if the soil profile was suitable for SV construction at a depth less than 18m.

The results indicate EMC data was accurate for identifying very low soil resistivity with 96% success rate when the indicated layer resisitivty was less than 5 ohm m.

The results for SH construction are likely skewed due to the small sample size and only being considered when high resistivity was indicated at the surface. High resistivity layers are not readily resolved by the EMC data.

The following sections provide a breakdown of the results for SH and SV constructed sites.

Shallow Horizontal Design Outcomes

To date, SH construction has been undertaken at 11 sites with a success rate of 55%. Figure 5 and Table 1 show a plot of the difference between design and actual loop resistance versus soil resistivity and data specifics. As soil resistivity increases the chance of success reduces. As mentioned above, the results for SH construction may well be skewed due to being only considered in "problem" soil areas where SV construction was not an option. The sample size for shallow horizontal construction was particularly small and the results may have been similar if shallow horizontal had been the preferred construction method.

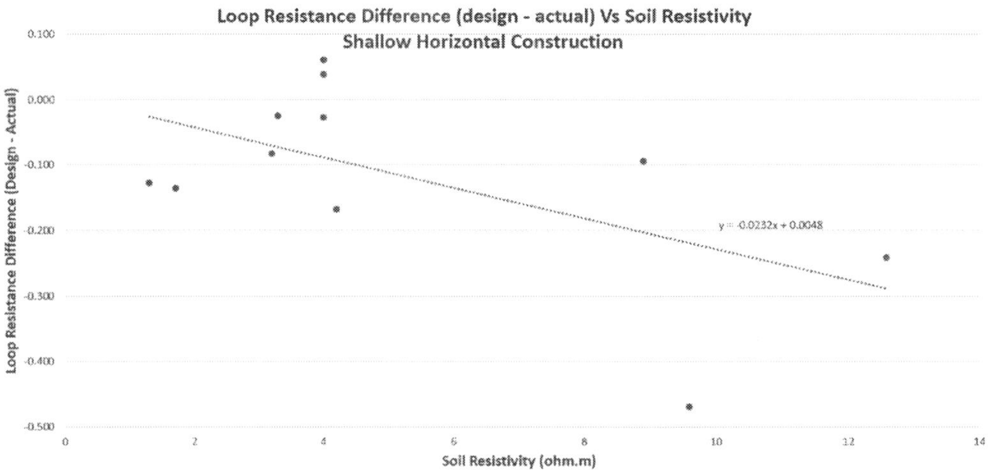

FIGURE 5: TYPICAL SHALLOW VERTICAL ANODE BED CONFIGURATION.

TABLE 1: OUTCOMES FOR SHALLOW HORIZONTAL CONSTRUCTION.

Shallow Horizontal Designs	
Soil Resistivity (x)	**Loop Resistance within 0.1 of Design**
All	55% (6 of 11)
x < 5 Ohm m	63 % (5 of 8)
5 < x < 10 Ohm m	50 % (1 of 2)
x > 10 Ohm m	0 % (0 of 1)

Shallow Vertical Design Outcomes

To date, SV construction has been undertaken at 55 sites with a success rate of 80%. Achieved success hits a high of 96% if the target layer resistivity is less than 5 ohm m. Figure 6 and Table 2 show a plot of the difference between design and actual loop resistance versus soil resistivity and data specifics.

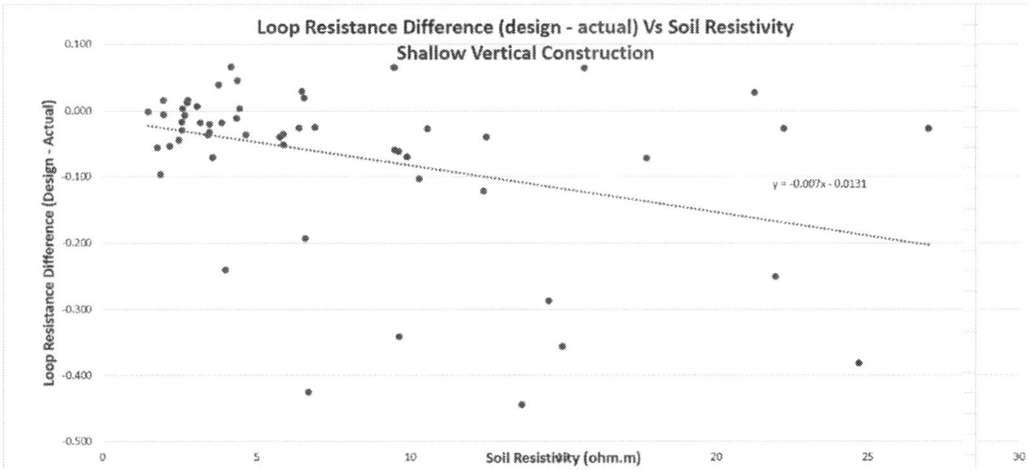

FIGURE 6: TYPICAL SHALLOW VERTICAL ANODE BED CONFIGURATION.

TABLE 2: OUTCOMES FOR SHALLOW VERTICAL CONSTRUCTION.

Shallow Vertical Designs	
Soil Resistivity (x)	**Loop Resistance within 0.1 of Design**
All	80% (44 of 55)
x < 5 Ohm m	96 % (26 of 27)
5 < x < 10 Ohm m	80 % (12 of 15)
x > 10 Ohm m	46 % (6 of 13)

Summary

Real world outcomes show that EMC data is effective and accurate for cathodic protection designs, particularly for shallow vertical construction. The sample size for shallow horizontal construction was particularly small and results may have been similar if horizontal had it been the preferred construction method.

The simplification of the design process and the ability to validate anode bed relocations using EMC data easily justifies the initial expense of data collection and processing.

Acknowledgments

The author acknowledges the permission and support of APA and the CP Augmentation Project team members over the last 6 years to make this project the professional production it now is. The random meeting that lead to APA & Mitre Geophysics testing and adopting the use of EMC data has been instrumental in effectively automating CP design to a 30 minute task. Special mention should also go to our construction contractor, Acacia Electrical who have done all the hard work under some very trying conditions (heat, rain, rock & Covid).

References

1. Sterling A., Airborne EM Surveys for Large Scale Pipeline Cathodic Protection Design, Proc. Corrosion & Prevention 2022, Newcastle, Australia, Paper 310.
2. AS 2832.1 Cathodic protection of metals Part1: Pipes and cables, 2015
3. Marilyn Lewis PE, CP Specialist. , Private communications 2021.

Author Details

Allan Sterling is a Senior Project Engineer (Cathodic Protection) at APA Group, a position he has held since 2018. Allan is responsible for cathodic protection design and specification for the CP Augmentation works currently being undertaken on a major APA gas pipeline asset. This project includes the construction of very large on-shore cathodiuc protection systems using a variety of power supplies in a wide range of environments.

CORROSION & PREVENTION
CAIRNS 2024

THE AUSTRALASIAN CORROSION ASSOCIATION INC.

IN-SITU METAL WALL TIE CORROSION INSPECTION OF PRE-1990 BRICK BUILDINGS

S. de Prazer[1], I. Chaves[2]*, C. Middleton[1], B. Smith[1], L. van Woensel[1], M. A. Hossain[2], M. J. Masia[2]

[1] *Dept. of Energy, Mines, Industry Regulation and Safety – Western Australia (WA) Government, WA, 6107, Australia*

[2] *Centre for Infrastructure Performance and Reliability – The University of Newcastle, NSW, 2308, Australia*

**Corresponding author email: igor.chaves@newcastle.edu.au*

ABSTRACT

The collapse of external brickwork due to corrosion of its supporting wall ties, is a danger facing millions of buildings the world over. In Australia, medium rise buildings with external brick (cavity masonry) leaves, constructed before the 1990's are of particular concern. Without rectification works, they face weaknesses due to inferior durability of the built-in masonry ties and an absent design requirement to resist earthquake loads. Preliminary tests indicates that the corrosion and strength of wall ties inferred by current Australian standards may overestimate the ability of ties to resist design action effects. This paper explores how microclimatic effects within the cavity leaf brickwork (particularly severe corrosion near the cavity mortar interfaces) and variances across wall faces relative to prevailing winds carrying salt laden air and rainfall; building heights and ages, and geographic locations, may cause variance in corrosion losses from those predicted by Australian Standards. It also explores whether the predicted capacity of wall ties is directly related to cross-sectional loss of the wall ties as is the premise of masonry design standards, or whether corrosion can promote a premature pull-out failure. This document, and others referenced herein, report on several in-situ wall tie corrosion measurements and elaborates on the implications for public safety.

Keywords: Masonry Structures, Wall-Ties, Durability, In-Situ Corrosion, Wall Collapse

INTRODUCTION

There is a stock of buildings in Australia and indeed throughout the world that contain a hidden danger. This danger includes corrosion of concealed wall ties [1,3], weakened mortar [2,3] and inherent design defects [2,3], that can cause bricks, or entire brick wall panels, to dislodge and fall. There have been several related wall collapses investigated by the Western Australia Government Department of Energy, Mines, Industry Regulation and Safety (the Department) since 2017, one investigated by The University of Newcastle in 2015 and several in the wake of the 1989 Newcastle earthquake [3-7]. Based on these investigations, detailed literature review, interviews with industry stakeholders, inspections of existing buildings and a wet candle test regime; pre-1990's buildings of medium to high rise and/or proximate to the ocean are of particular concern. The danger of collapse for external cavity walls with brickwork facing due to deterioration of concealed ties, is considered of high importance. This danger is currently being further investigated in detail by a research group led by The University of Newcastle and supported by the Department, Curtin University, consultant EngAnalysis and key stakeholders under the financial support of the Australian Research Council.

Cavity brick wall stability relies on wall ties (traditionally steel fitments) cast into the bedding joints of external brick walls. The ties allow the stabilised internal masonry leaf or framing structure to support the outer brick leaf against horizontal forces (e.g. wind and earthquake actions). Around the world, issues of corrosion in brick wall ties have been prevalent for some time, with millions of buildings considered at risk in the UK alone some decades ago [1]. Forensic analysis of a multitude of buildings severely damaged by the 1989 Newcastle earthquake [3], led to the development of Australian Standards that helped reduce risk of tie corrosion for subsequent buildings such as AS3700 - Masonry Structures [8] and the AS/NZS 2699 series – Built-in components for masonry construction [9,10]. Many buildings approaching or exceeding their design life, will have deteriorated ties and reduced capacity to resist brick dislodgement.

Durability requirements for cast-in masonry components across Australia are currently based on in-situ tests in exposed areas of select cities. These tests inform the predicted material loss of the cast-in wall ties. Wall failure investigations have suggested buildings near or past their design life have had greatly reduced wall-tie capacity. Observations of wall tie condition in existing buildings have prompted concern for the in-situ capacity of corroded ties. This paper summarises observations from numerous building investigations, in combination with wall tie field trials and wet candle testing in Perth, WA and Newcastle, New South Wales (NSW).

METHODS AND MATERIALS

Forensic analysis of external cavity brickwork failures in Australian buildings; visual and laboratory observations of wall ties from existing buildings and custom-built test couplets; and wet candle testing, has informed this study.

After a wall failure, the main action effect that led to the collapse is determined. When wind is the primary effect, it is compared to the event that applicable building standards require the building to resist. The wind speeds are determined to suitable levels of accuracy through interpolation between weather station readings and deduction from localised damage in the built and natural environment, using established methods. Wind loading was found to have caused brickwork collapse in Bar Beach Newcastle in 2015 [4]; Fremantle (2017, 2020) [5], Northbridge (2019) [6] and Bedford (2020) [7] in Perth, WA. For all such cases, the wind loading that caused the wall failure was significantly less than the design wind loading. Of these five buildings, three were well beyond, one marginally beyond and one marginally before, their fifty-year assumed working life. In all cases, moderate to severe deterioration of wall ties was observed; however surprisingly, the condition of ties varied significantly in different areas of the wall and along the length of the tie crossing over from the mortar joints to open cavity in between.

(a) (b) (c) (d)

FIGURE 1: BRICKWORK COLLAPSE IN BUILDING OF WALL BUILT 55-60 YEARS PRIOR WITH SEVERED CAVITY TIE (A, B) & CIRCA 48-YEAR-OLD WALL WITH HEAVY CORROSION STAINING REMAINING ON BED JOINTS (C,D) COURTESY G. SIMUNDIC [4].

For a building marginally beyond its design life, the cavity portion of a wall tie appeared uncorroded at a location where the external leaf had been removed. The part of the wall-tie remaining in the internal leaf pulled out of the wall through a pinching action, revealing the rust coated portion from the bedding joint. The wall-tie capacity was significantly less than the 0.6kN load capacity required of a medium duty cavity tie (AS 3700: 2018; Table 3.6 [8]). The masonry standard in effect at the time of failure (still current) assumes the tie durability will allow it to continue to provide the required capacity throughout the building's design life. The ties in this elevation of wall appeared in much better condition than those in the ocean-facing elevation. A similar difference in tie condition between wall elevations was observed in two buildings inspected in Perth, WA (in 2024) through the removal of bricks.

Building inspections were performed by technical officers of the Department with observations made on the location of the building relative to saline water (durability classification as per AS/NZS 2699.2 [10]); the condition of the brickwork; building age; ventilation and shielding of the inspected walls relative to wind, rain and irrigation. The time and date of inspection was also recorded for comparison against seasonal effects, location and building age. Where possible, representative samples of both mortar and wall ties were collected. When required, samples were posted to The University of Newcastle in airtight bags for further analysis. In-situ moisture readings, scanning electron microscopy (SEM) and x-ray diffraction (XRD) chemical compositional analysis of mortar samples, and measurements of corrosion to the ties were also recorded where applicable. In-situ moisture readings were recorded with a certified and calibrated pad type moisture meter, with vernier callipers used as initial check of the wire tie diameter.

Alongside building investigations, the Department has ongoing wet candle apparatuses set-up in several locations across the Perth metropolitan area, as per standardised methodology [11]. The suitably exposed wet candle consists of a hermetically sealed glass flask with a protruding test tube candle approximately 150mm in height and 25mm in diameter when wrapped in medical gauze. The soaked gauze in contact with the distilled water is then allowed to absorb airborne salts over a period of typically one month. A conductivity meter was then used to determine the retrieved solution's conductivity, and in accordance with ISO 9223 [12], ISO 9225 [13] and ASTM G91-97 [14], calculate and convert the conductivity values to chloride deposition rates. These wet candles exposed in Western Australia enable comparison of salt deposition rates across various proximities to saline water bodies, as well as to the test results reported for The University of Newcastle, Belmont, NSW test site [15].

RESULTS

FORENSIC ANALYSIS OF BUILDINGS WITH WALL FAILURES

Three buildings inspected in Perth, WA following partial wall failures revealed that intuitively the condition of ties worsened from the youngest (circa 60-year-old) building to the eldest building (circa 100 years old). The building that was middle in age (circa 80 years) was however found to have ties that were in relatively better condition than the 60-year-old building. This building was furthest from the ocean (circa 12.5 km), whereas the youngest was closest (circa 2.5 km) and the eldest approximately 10 km from breaking surf and 1.5 km from part of the Swan River considered saline (**Figure 2**).

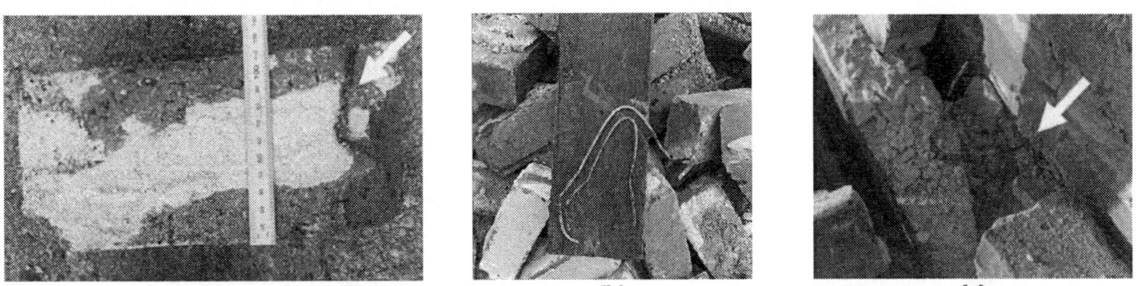

(a) (b) (c)

FIGURE 2: REMAINING WALL TIE CONDITION, OFTEN COMPLETELY DETERIORATED, FROM COLLAPSED WALLS FROM ELDEST CIRCA 100-YEAR-OLD BUILDING (A), TO CIRCA 80 -YEAR-OLD BUILDING (B), AND YOUNGEST CIRCA 60-YEAR-OLD BUILDING (C) SHOWING THAT AGE ALONE CANNOT PREDICT CONDITION

Investigation of such buildings was limited to the parts of the walls that had fallen/were exposed following failure (southwest, northwest and western faces for oldest to youngest buildings respectively). The northern face of the youngest building was later inspected during rectification works and found to have some ties appearing in similar condition to those in the western face and other ties in better condition (similar to the central photo of **Figure 2**).

A circa 50-year-old building within 200 m of breaking surf in Newcastle, NSW was found to have severely corroded wall ties after a collapse (refer photos on right side of **Figure 1**). This is the youngest of the buildings for which wall failure is reported herein. Of the subject buildings, the youngest has an R4 classification, least corroded but second to oldest building is R2 and other two buildings rated R3 as per AS/NZS 2699.1 [9]. All four buildings inspected contained galvanized wire ties. Considering that the amount (g/m^2) of sacrificial zinc galvanizing increases from R2 to R4 rating [8-10], it appears that corrosion rates are most greatly influenced by exposure conditions rather than age.

INSPECTION OF BUILDINGS WITH NO APPARENT EXTERNAL WALL DISTRESS

Inspection of two single-storey buildings and one two-storey double brick cavity building in the metropolitan area of Perth, WA occurred when opportunity presented around demolition schedules. These inspections revealed significant variation in condition of ties between elevations, with severely corroded ties found in all buildings despite no walls displaying visual distress. Variation in corrosion severity was noted in the two buildings where different elevations were able to be inspected.

In the Perth area, the prevailing winds are north-westerly to south-westerly sea breezes. These winds also accompany most rainfall (in the winter months of June to August). In the warmer months easterly breezes typically blow from inland areas in the mornings. Two buildings had northeastern walls that were relatively shielded from the wind and exhibiting only slight corrosion, whereas partially or unshielded walls facing the southeast through to southwest directions suffered severe corrosion (**Table 1**).

TABLE 1: OBSERVATIONS OF WALL TIE CONDITION FROM BUILDINGS DEMOLISHED FOR RE-DEVELOPMENT

Approximate building age [years]	Dist. to breaking surf [kms]	Elevation 1	Exposure as per AS4055 [16]	Corrosion	Elevation 2	Exposure as per AS4055 [16]	Corrosion
60	0.1	south	part shielded	severe	--	--	--
75	12.5	southwest	part shielded	severe	northeast	fully shielded	slight
50	1	southeast	not shielded	severe	northeast	fully shielded	slight

The circa 65-year-old building was inspected prior to demolition, where full removal of ties from only the southwest elevation was permitted with outer brick removal aiding inspection of the northeast elevation. Ties collected from the other two buildings were able to be completely detached from both leaves as they lay in rubble, however in the circa 60-year-old building only ties from one elevation could be collected. Ties from different elevations of a building approximately 1 km from the 60-year-old shows the significant difference in wall tie condition along both the length of the tie and between fully shielded and un-shielded elevations facing away from breaking surf (**Figure 3**).

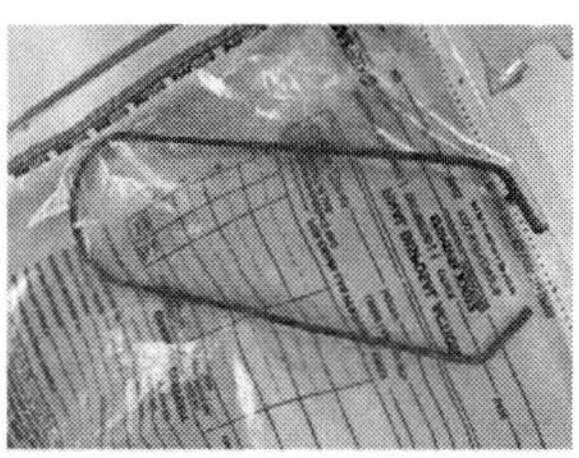

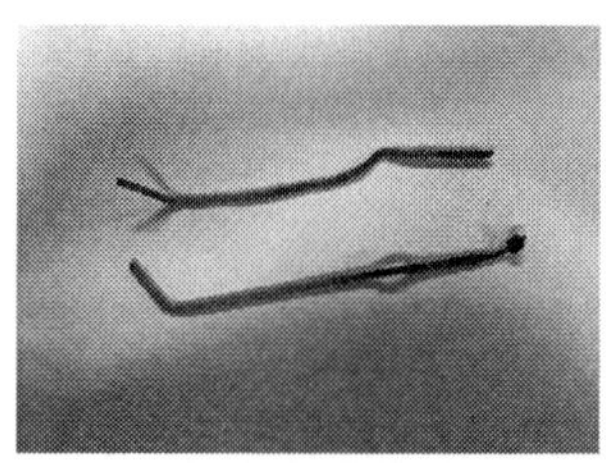

FIGURE 3: WALL TIES FROM NORTHEAST (LEFT) AND SOUTHEAST (RIGHT) DEMOLISHED 50-YEAR-OLD BUILDING

WALL TIE FIELD TRIALS

Brick couplets with embedded galvanized masonry veneer plate type ties were placed similar distances from breaking surf in Trigg, WA and Belmont, NSW. A couplet each of M1, M2 and M3 mortar mix and with R3 and R2 galvanized ties were exposed over a period of fourteen months on roofed test racks. No discernible difference in corrosion rates was noted between the two sites or between the seaward or opposite facing cast-in ties. The Trigg and Belmont couplets indicated higher corrosion losses for the galvanized ties than those predicted by published models as noted by Chaves et al. [17].

WET CANDLE TEST RESULTS

Wet candle tests were conducted to provide an indication on how chloride deposition rates vary geographically across Perth, WA and as compared to Belmont, Newcastle (NSW) [15]. Corrosion rates are estimated for Perth and based on corrosion mapping (in 1992) for Newcastle in AS 4312: 2019 - Atmospheric corrosivity zones in Australia [18]. Results are provided below for the Perth sites. Noting that wet candle results are known to be affected by seasonal effects and wind shielding [19] readings were recorded monthly, with some months not being recorded due to logistical challenges. Figure 4 plots average monthly deposition per season, with minimums and maximums shown for comparison. The Trigg and Belmont wet candle site recordings are provided from preceding studies, with the Trigg reading from the single month of July 2020 and Belmont averaged across April to June [15]. An outlier in the January reading at the Champions Lake may be due to a new inspector taking over that test, or as the site is located between two artificial lakes that have salt added to the brackish water from which it is filled (particularly in the summer month of January). Results for the wet candle sites are presented below, with the Champions Lake data average included with and without the January spike for completeness (**Table 2**). The general variability in readings and lack of clear trend between sites seasonally, casts doubt over the accuracy of this method. It is provided as a general comparator, with average readings following the general order of minimum to maximum deposition rates expected as compared to distance from saline water bodies. It is noted that various topography, terrain and shielding from the natural and built environment between test sites may provide greater variation than distance from saline water, as well as the nature of the saline bodies to which the chloride deposits are attributed.

TABLE 2: WET CANDLE TEST COMPARISONS ACROSS PERTH, WA AND BELMONT, NSW WITH AS/NZS 2699.2 [10]

Site	Annual Average deposition [kg/ha/yr]	Distance from coast [km]	Distance from saline lake [km]	Classification as per AS/NZS 2699 [10]	AS/NZS 2699 Prediction [kg/ha/yr]	% Difference AS/NZS2699 vs wet candle
Armadale, Champion Lakes, WA *	804	--	<0.1	R4	1095	27
Armadale, Champion Lakes, WA (w/out outlier)	737	22	<0.1	R4	1095	33
East Rockingham, WA	502	0.3	--	R4	1095	54
Serpentine, WA	250	22	--	R2	73	-242
South Lake, WA	460	6.8	--	R3	219	-110
Wattleup, WA	779	2.3	--	R3	219	-255
Trigg, WA	1073	0.3	--	R4	1095	2
Belmont, NSW	1095	0.2	--	R4	1095	0

Wet candle results indicate a strong correlation between Perth and Newcastle sites near to surf beaches and the AS/NZS 2699.2 prediction [10]. A Perth beach with calmer waters (East Rockingham) but a similar distance to the ocean and in an industrial area was found to have chloride deposition rates of approximately 50% lower. This is not unexpected considering it is partially shielded by Garden Island approximately 6 km to the west. All three sites are classified as R4 as per AS/NZS 2699.2 [10], with the Champions Lake inland site also considered R4 owing to the salt in the lake. These results drew reasonable correlations with the AS/NZS 2699.2 predictions.

Variance between the AS/NZS 2699.2 [10] predictions and wet candle results are prominent for the Perth sites classified as R2 and R3. The deposition rates are approximately 100 – 250% greater for these sites. These findings suggest a more gradual decline in sea salt deposition rates from the coastline than values given in AS/NZS 2699.2 [10]. Serving as a second reference over time, the CSIRO 1976 sea salt chart [20] shown below (**Figure 5**) also suggests a more gradual decline. The CSRIO chart has a maximum value of less than 300 kg/ha/yr which is considerably less than the correlated R4 wet candle and AS/NZS 2699.2 values [10].

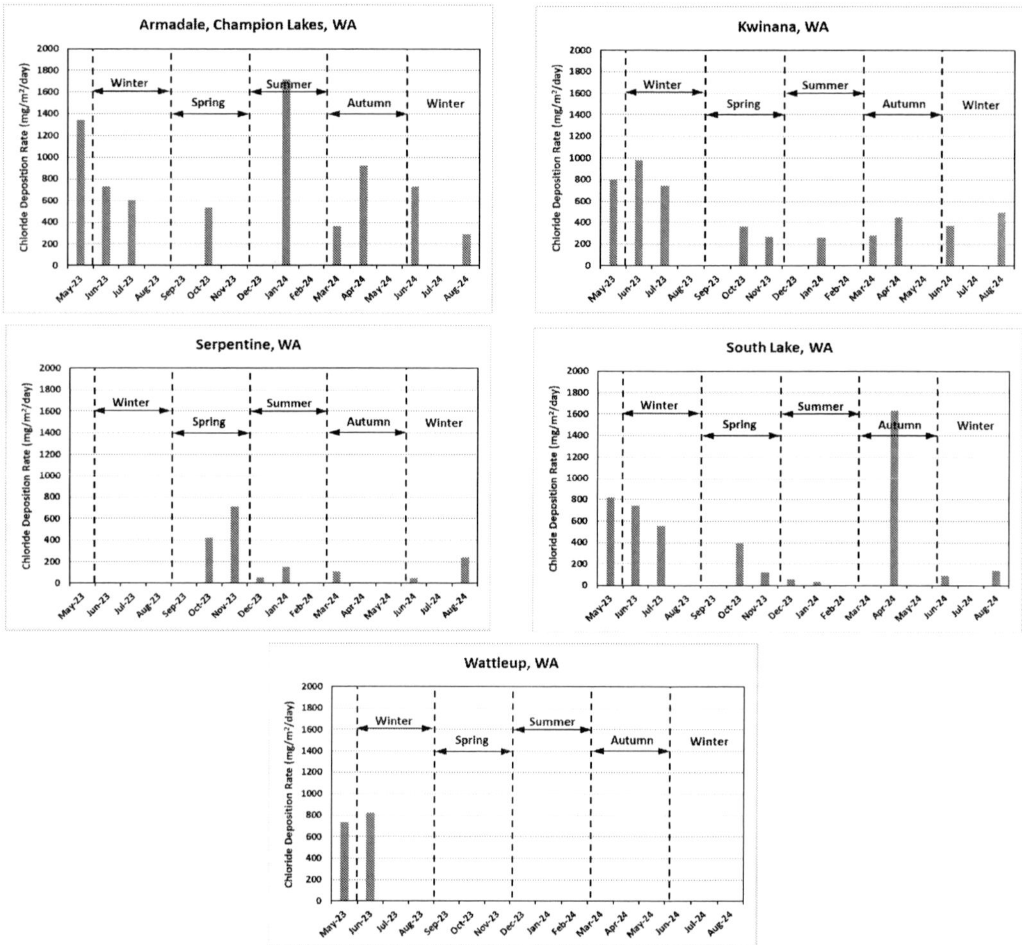

FIGURE 4: RECORDED WET CANDLE SEASONAL COMPARISONS ACROSS PERTH, WA.

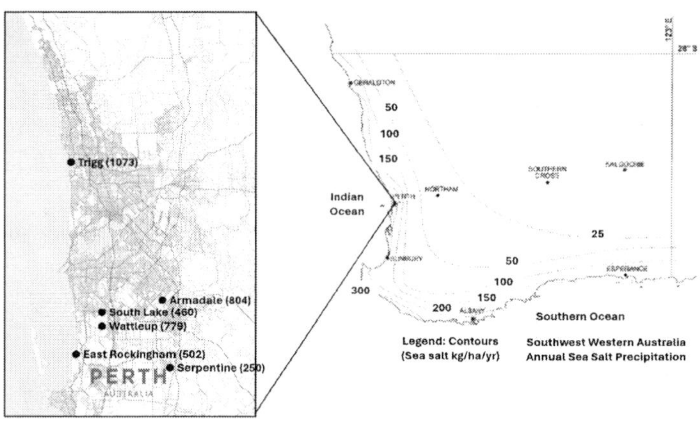

FIGURE 5: 1976 CSIRO SEA SALT CHART FOR SOUTHWEST WA [20]. ANNUAL AVERAGE DEPOSITION ACROSS SITES IN PERTH SHOWN FOR COMPARISON (KG/HA/YR)

DISCUSSION

Through the study of brick wall tie condition versus building age, elevation relative to the nearest saline water bodies and wet candle results, concern on the stability of external masonry cavity walls approaching or beyond their design life and reliant on galvanized wall ties is reinforced. Wet candle tests have confirmed that the AS/NZS 2699.2 [10] predictions for chloride deposition rates near to the ocean (R4 classification) are reasonably accurate, however for R3 and R2 classifications deposition rates may be underestimated. As such, further research is recommended into whether corrosion protection requirements for R2 and R3 durability classifications will result in wall ties achieving their required capacity for their entire working life.

As is the case for the Bar Beach bowling club [4] that experienced wall failure, there remains a stock of pre-1990's buildings where galvanized wall ties were employed prior to improvement in durability requirements. A comprehensive inspection regime is highly recommended for such buildings.

It is important to note that corrosion mapping applied to built-in masonry components links back to exposed test conditions [9,10,17]. For reference, table 3 compares AS/NZS 2699.2 classifications with the more commonly used ISO 9223 classifications. Reinforced by recent publications [15,17], the corresponding ingress of chloride deposits through masonry and resulting effect on cavity wall tie corrosion is not currently considered by standards applicable to masonry construction such as AS/NZS 2699.2. Testing of tie corrosion rates within a controlled test wall environment where the internal wall can be sealed by a built structure as is the case in real buildings is further recommended.

Results of destructive testing of buildings has demonstrated a difference in amounts of corrosion between various elevations. Couplet testing where the same salt laden air can pass over the couplet to ties embedded in either side of it will not be able to replicate the effect of varying building elevations. It is also suggested by Chaves et al that a microclimate occurs within a building's external cavity [17] which can be explored through a sealed cavity wall environment, where moisture readings of the internal and external leaf are compared to long-term corrosion and checked as like an in-service building. A study from BRANZ [21] reinforces this argument where their study reports on different material degradation on the building envelope by characterising micro-environments, i.e. sheltered and exposed, on seven residential buildings across New Zealand. This will further allow the observed variation in corrosion along the length of the wire ties to be more closely studied.

Table 3: Wall tie exposure classification comparison between AS/NZS 2699 [9,10] and ISO 9223 [12].

Wall Tie Material	Classification as per AS 2699.2	Classification as per ISO 9223
Galvanized steel bar (coating thickness ≥ 42 μm)	R1-R2	C2
Lightly galvanized steel sheet (coating thickness ≥ 42 μm)	R1-R2	C2
Heavily galvanized steel sheet (coating thickness ≥ 65 μm)	R3	C3
Lightly galvanized steel wire (coating thickness ≥ 42 μm)	R1-R2	C2
Heavily galvanized steel wire (coating thickness ≥ 65 μm)	R1-R2	C2
Stainless steel sheet/wire (SS304)	R3	C3
Stainless steel sheet/wire (SS316)	R4	C4-C5

A key finding reported on above is that a wire ties which looked relatively sound had little pull-out capacity from the mortar joint. This suggests that corrosion can lead to premature failure in a mechanism not contemplated by the masonry design standards which bases capacity on tensile failure of the tie, rather than withdrawal from the bedding joint. It is hypothesized that the outer layer of rusted steel forms an expanded sheet from which the inner wire can withdraw. Further testing utilising advanced corrosion techniques should be used to test the capacity of ties with various degrees of corrosion, on full-scale wall samples. A link between expected rates of corrosion and expected wall moisture content, time of wetness, external chloride deposition rates and effect of rain washing on removal of deposits and the migration of corrodents through bed joints and across the tie length also requires study.

CONCLUSIONS

- Investigation of failed external walls of double-brick cavity construction suggest that the numerous buildings across Australia nearing or surpassing their design life may be at risk of brick dislodgment and wall collapse.
- Corrosion of galvanized ties may lead to failure through pull-out of the mortar joint, a mechanism not contemplated by test regimes in building standards currently referenced in Australia's National Construction Code.
- The corrosion of ties in a building can be highly variable based on the exposure of different walls, direction the wall is facing, and along the length of wall ties (worst in external bed joint near to air cavity)
- Wet candle testing suggests that the chloride deposition standardised predicted rates for buildings with an R2 and R3 classification may be underestimated
- As durability classifications relate to deposition rates on exposed surfaces, it is recommended that the migration of chlorides through a cavity leaf construction is explored and compared to rates of observed corrosion for a simulated building environment (with internal wall sealed)
- For all pre-1990's external brick cavity buildings in Australia, it is recommended that appropriate inspection, maintenance and repair regimes are chosen to reduce the risk of wall collapse on the lives and livelihoods of those to which the buildings relate.

ACKNOWLEDGMENTS

The Authors would like to acknowledge the pivotal financial support from the Australian Research Council and the Western Australian Government Department of Energy, Mines, Building Regulation and Safety (DEMIRS) via Linkage Project LP220100028. Technical assistance by Mr G. Flowers is also greatly appreciated. The authors also would like to thank the various industry partners who have formally committed their involvement in supporting the project, particularly those who have facilitated inspections and supported project participants and who will carry findings forward to improve safety relating to the built environment.

References

1. de Vekey RC, Corrosion of steel wall ties: history of occurrence, background and treatment. BRE Information Paper, IP12/90 (1990).
2. Yardim Y and Lalaj O, Shear strengthening of unreinforced masonry wall with different fiber reinforced mortar jacketing, Construction and Building Materials 102 (2016) 149-154.
3. Melchers RE and Page AW, The Newcastle (New South Wales) Earthquake, Proceedings of the Institution of Civil Engineers-Structures and Buildings, 94(2), 1992, 143-156.
4. Email exchange with Goran Simundic, former technical officer at the University of Newcastle, Australia, 2015.
5. ABC news, Matilda Bay Brewery façade collapse, crushing car in Fremantle as strong winds hit WA, 2017 see https://www.abc.net.au/news/2017-08-14/8805008/ Last accessed 28 July 2024.
6. ABC News, Perth nightclub evacuated after wall crumbles during party, 2018 see https://www.abc.net.au/news/2019-06-30/11265134/ Last accessed 28 July 2024.
7. ABC News, Storm hits Perth as ex-Tropical Cyclone Mangga delivers wild weather across WA, 2022 see https://www.abc.net.au/news/2020-05-25/12281886/ Last accessed on 28 July 2024.
8. AS 3700:2018, Masonry Structure, Fifth Edition, Standards Australia, 2018, Sydney, Australia.
9. AS/NZS 2699.1:2020, Built-in components for masonry construction, Part 1: Wall ties, Standards Australia, 2020, Sydney, Australia.
10. AS/NZS 2699.2:2000, Built-in components for masonry construction, Part 2: Connectors and accessories, Standards Australia, 2000, Sydney, Australia

11. ASTM G140-02, Standard test method for determining atmospheric chloride deposition rate by wet candle method, ASTM International, 2019.
12. ISO 9223, Corrosion of Metals and Alloys, Corrosivity of Atmospheres: Classification, Determination and Estimation, ISO/TC 156, ICS: 77.060, 2020.
13. ISO 9225, Corrosion of metals and alloys: Corrosivity of atmospheres; Measurement of environmental parameters affecting corrosivity of atmospheres, ISO/TC 156, ICS: 77.060, 2022.
14. ASTM G91-97 (2018) Standard Practice for Monitoring Atmospheric SO2 Deposition rate of Atmospheric Corrosivity Evaluation, ASTM International, 6 pages.
15. Jardim do Nascimento B, Chaves IA, Masia MJ, Melchers RE, Long-Term Atmospheric Corrosion of Mortar Embedded Wall Ties, in Proc. of Corrosion & Prevention, November 11-18, 2018, Paper 33.
16. AS 4055, Wind Loads for Housing, Standards Australia, 2021, Sydney, Australia.
17. Chaves IA, de Prazer S, Jardim do Nascimento B and Flowers G, Empirical Coastal Atmospheric Corrosion of Masonry Metal Wall Ties. Corros. Mater. Degrad 2 (2021) 657-665.
18. AS 4312, Atmospheric Corrosivity Zones in Australia, Standards Australia, 2019, Sydney, Australia.
19. Haberecht P, Pollution deposition rates on insulator (HV) surfaces for use in atmospheric corrosivity estimation, PhD Thesis, The University of Newcastle, Australia, 2008.
20. Hingston FJ. and Gailitis V, The geographic variation of salt precipitated over Western Australia, Aust. J. Soil Res. 14(1976) 319-335.
21. Li Z and Marston N, Positional material deterioration over the building envelope, BRANZ Study Report SR457, New Zealand, ISSN: 1179-6197, 2021.

AUTHOR DETAILS

	S. de Prazer (S.deP.) is the (inaugural) Principal Building Engineer (Structural) for Western Australia's building regulator. He is experienced in structural design and asset management in several industries across the UK and Australia. He gained Chartered status (2008), and Fellowship (2018) with Engineers Australia, who he has volunteered with for 20 years, including as Chair of the Western Australia Structural Engineering Group and Standards Australia review group member for the Structural College Board.
	Associate Professor I. Chaves (I.C.) is the corresponding author of this paper. He is the Deputy-Director in the Centre for Infrastructure Performance and Reliability at The University of Newcastle, Australia. He is a chartered materials professional academic specialised in Civil/Structural/Material Science, holds a MEngSci degree in structural engineering and a PhD for his studies in corrosion analysis and prediction. His research interests also include structural reliability and risk assessment of energy infrastructure.
	L. van Woensel (L.vW) is a Structural Engineer who worked for the Department of Energy, Mines, Industry Regulation and Safety, Government of Western Australia, in the role of Inspector Engineer. He set up the wet candle test regime in Western Australia and performed most testing before handing the role on.

	B. Smith (B.S.) - Department of Energy, Mines, Industry Regulation and Safety, Government of Western Australia Responsible for ongoing wet candle testing and providing engineering support to building inspections. She is a Chartered Engineer and Principal Building Engineer (Structural) for Western Australia's building regulator, and a panel member of Engineers Australia's Western Australia Structural Engineering Group. She brings experience from the private sector in design, construction support, temporary works, and large iconic infrastructure projects.
	C. Middleton (C.M.) - Department of Energy, Mines, Industry Regulation and Safety, Government of Western Australia Responsible for co-ordinating and performing building inspections and providing technical support. He is a qualified bricklayer and registered Level 2 building surveyor. He brings decades of experience in the building industry to the project.
	M. A. Hossain (M.A.H) is a Research Associate at Centre for Infrastructure Performance and Reliability, The University of Newcastle, Australia. He assists with all aspects of this project's experimental and field research. Dr Hossain has over 12 years of applied structural engineering research and development experience. He specializes in experimental and numerical modelling analysis of structural (masonry) response.
	Professor M. Masia (M.M.) is the Director of the Centre for Infrastructure Performance and Reliability at The University of Newcastle, Australia. He specialises in the structural performance, assessment, and design of masonry structures. He has an established and longstanding reputation with Standards Australia, the Australian Research Council, and both national and international professional masonry societies.

THE REASON FOR CONTRASTING COLOURS

D. McKeown[1] , B. Lichtendonk[2]

[1]*Dulux Protective Coatings, Victoria, 3156, Australia*

[2]*Dulux Protective Coatings, Queensland, 4106, Australia*

**Corresponding author email:* Daniel.mckeown@dulux.com.au, Brendan.lichtendonk@dulux.com.au

ABSTRACT

Most specifications contain a requirement that the coating sequence is made up of contrasting colours, including the topcoat. This is sometimes challenged within the project hierarchy, typically to reduce cost or program time. Alternately when there is no specification issued by the asset owner, many different proposals will be put forward, particularly in the repaint or maintenance space. This paper will examine the reasons behind why contrasting colours are stipulated in a specification and look at what it is aiming to achieve, and potentially what can go wrong if the statement is ignored.

Keywords: specification, colour, intermediate, topcoat, opacity

INTRODUCTION

Some specifications contain written direction that states contrasting colours be used in sequence for the protection of steel assets. This is well established practice within the protective coatings world, particularly the infrastructure space, and most roads authorities have this as part of their standard requirements. For the most part, when work is being carried out in a factory prior to site installation, this requirement is understood and followed. Experienced protective coatings applicators are also used to this type of requirement in a specification.

But when a painting project is being carried out on site, some painting contractors wish to use the same coloured primer as topcoat, particularly in a refurbishment or repainting situation. This is commonly requested by most of the larger commercial painting companies who don't specialise in corrosion protection. The reason behind this request is typically cost (reducing the number of coats to be applied), and thus making the program faster. Many of these repainting projects may not have a formal specification prepared by the asset owner or their representative.

One of the other challenges with repainting projects is the application technique. Brushing and rolling will always deliver lower coating thicknesses when compared to spray application. Depending on the choice of colour, it is highly likely that two or more coats of the topcoat in the nominated colour are required to achieve opacity.

This paper will review some common specifications and explore reasons for these specifications to nominate contrasting colours for each layer. It will also look at real life case histories where this practice has and has not been bypassed, the outcomes and provide some insight into how to approach potential solutions, particularly for in-situ repainting projects.

Case Histories

As outlined, many well written specifications contain a requirement to use different coloured coatings in the sequence. Some examples of this are presented in the following few pages.

Example 1: DOT (Vic Roads) Specification

Coating Systems

Dulux	Product	DFT / WFT (microns)		Mix Ratio	Recoat (Min / Max) @ 15°C		Solvent
Primer	Zincanode 402	75	160	4A:1B	10 hrs	30 days	Epoxy
Intermediate	Durebild STE N35 Light Grey	150	178	4A:1B	24 hrs	14 days	Epoxy
Stripe Coat	Durebild STE Buff	100	120	4A:1B	24 hrs	14 days	Epoxy
Intermediate	Durebild STE MIO Natural Grey	150	178	4A:1B	24 hrs	5 days	Epoxy
Finish Coat	Weathermax HBR MIO Mid Grey	75	115	4A:1B	24 hrs	4 weeks	DuThin 040
Finish Coat (Props)	Weathermax HBR RAL 9002 Grey White	75	115	4A:1B	24 hrs	4 weeks	DuThin 040

AkzoNobel	Product	DFT / WFT (microns)		Mix Ratio	Recoat (Min / Max) @ 15°C		Solvent
Primer	Interzinc 315	75	108	4A:1B	3 hrs	30 days	GTA 220
Intermediate	Interplus 1180 N35 Light Grey	150	183	4A:1B	11 hrs	14 days	GTA 220
Stripe Coat	Interplus 1180 Buff	100	122	4A:1B	11 hrs	14 days	GTA 220
Intermediate	Interplus 1180 MIO Natural Grey	150	183	4A:1B	11 hrs	5 days	GTA 220
Finish Coat	Interthane 870 MIO Mid Grey	75	150	4A:1B	16 hrs	4 weeks	GTA 713
Finish Coat (Props)	Interthane 870 RAL 9002 Grey White	75	150	4A:1B	16 hrs	4 weeks	GTA 713

Notes

- There is no suggestion, implied or otherwise, that the above coating systems or individual products are equal.
- Consecutive coats shall be different colours.
- The finish coat colour shall be MIO Mid Grey on the main bridge steelwork and RAL 9002 on the cantilever props.
- More than one application of the intermediate or finish coats may be required to achieve the nominated DFT.
- The specified DFT shall prevail, not the number of coats.
- Apply a stripe coat of the high build epoxy to all welds, edges and corners after the first HB coat.
- The total system DFT shall be 450 microns minimum measured in accordance with AS 3894.3.
- This Coating Schedule shall be read in conjunction with the latest Product Data Sheets and MSDSs.
- The use of thinners other than those specified by the manufacturer is not permitted.
- Following award of the contract no alternative system will be considered.

Figure 1: Coating System sequence from Department of Transport Contract 9536

The above specification (Refer to Figure 1) is an excerpt from the Westgate Bridge repaint project from 2019-2021 and it requires a clear contrast in each coat. This project was fully inspected after application of each coat (bar the stripe coat, which was just a visual inspection). The contrast in colour made every layer easily identifiable.

As an aside, the stripe coat colour was changed to X42 Biscuit from X41 Buff to use a colour that contained less tinter (which can prolong curing times). The colour was a similar shade and still gave a significant contrast in colour but gave advantages of price and lower product tinting levels.

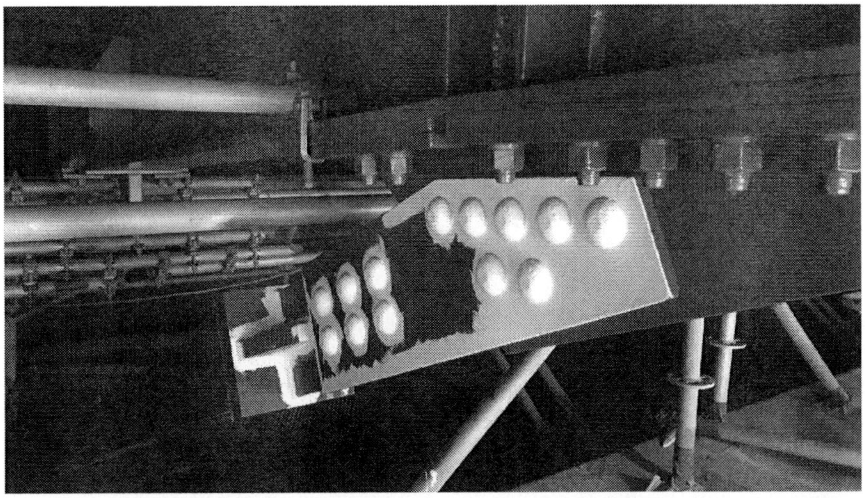

Picture 1: Process of overcoating White over MIO Mid Grey

Picture 2: Stripe Coating X42 Biscuit over N35 Light Grey

Example 2: TfNSW standard specifications

1	2	3	4	5
Item	**Description**	**All surfaces excluding Column 4 (DFT Range)**	**Internal surfaces latticed members (DFT Range)**	**Reference**
Preliminary Cleaning	Remove mud, dirt and other loose contamination. Wash using low pressure water jetting or spot clean using buckets, brushes and water.			B220 (3.7.2)
Oil and Grease Removal	Remove oil and grease in accordance with AS 1627.1.			B220 (3.7.3)
Removal of Mill-scale and Rust	Abrasive blast clean in accordance with AS 1627.4 to achieve a Sa 2½ cleanliness as described in AS/NZS 2312 and as depicted pictorially as Sa 2 ½ in AS 1627.9. The surface profile of bare metal must be 40 to 70 µm measured in accordance with AS 3894.5.			B220 (3.7.5)
Zinc Primer	One full coat of inorganic zinc silicate primer. Applied to all surfaces. Colour: Green	75 µm – 125 µm	75 µm – 125 µm	B220 (3.8) B220 (3.8.3) 3651 (5.1)
Stripe Coat	One stripe coat of a high build, two-pack epoxy paint. Applied to all crevices, edges, bolts, rivets, and other nominated surfaces. Colour: R63 Red Oxide to AS 2700	75 µm – 100 µm	75 µm – 100 µm (applied to all surfaces)	B220 (3.8) B220 (3.8.5) 3651 (5.5)
Build coat	One coat of a high build, high solids, MIO pigmented, two-pack epoxy coat. Applied to all surfaces. Colour: Natural Steel Grey	200 µm – 400 µm	125 µm – 250 µm	B220 (3.8) 3651 (5.6)
Topcoat	One coat of a high build, acrylic modified, MIO pigmented, two-component polyurethane. Applied to all surfaces. Colour: RTA Bridge Grey	75 µm – 125 µm	75 µm – 125 µm	B220 (3.8) 3651 (5.7)
Total Dry Film Thickness	Without stripe coat With stripe coat	350 µm – 650 µm 450 µm – 750 µm	NA 375 µm – 650 µm	B220 (4.1)

Figure 2: Coating System sequence from TfNSW B220 document System E1

The above specification (Refer to Figure 2) again has clear guidelines on colour for each coat. TfNSW typically have high levels of inspection on their assets and each layer will be visible under magnification with an obvious contrast (if inspection using a Tooke Gauge is carried out).

Other roads bodies around the country also require contrasting colours in their coating sequence for the protection of steel assets.

EXAMPLE 3: SPORTS STADIUM REPAINT (EXAMPLE 1)

The roof steelwork on a sports stadium was repainted between 2014 and 2018. There were a number of corrosion affected areas that required remediation, and the existing topcoat had faded. The specification issued by the asset manager required spot remediation of any corroded areas, a full primer on all surfaces and 2 applications of topcoat in the required colour. A contrasting primer colour was used for this repaint and is shown in the photo below (Refer to Picture 3). The contractor on this project followed through with the specification and was working directly for the facility manager.

Picture 3: Contrasting primer colour used on steel repaint.

Example 4: Sports Stadium Repaint (Example 2)

The roof steel on another stadium was repainted in 2018-2019. The condition of the structure was excellent, there was virtually no spot corrosion, and the existing dry film thickness was typically around the 500 micron mark. The repaint specification contained one full primer coat and one full topcoat. It is important to note that this specification was driven by the construction team, not the asset owner. It should be noted specific white colours (N14 White, RAL 9003 Signal White and other shades of white) have low opacity when they are tinted from a strength aligned base. Some topcoats are available in factory produced batches of white, but they are not aligned to any particular colour standard. They are just labelled as "white", but they contain a higher loading of pigment, specifically designed to give better opacity.

A high opacity factory package white was used as the topcoat for this repaint, - there was no requirement to match a particular white. (Refer to Pictures 4 & 5)

Pictures 4&5: Same colour primer / topcoat used on this steel repaint. DFT survey of coated surfaces provided limited information.

Example 5: Sports Stadium Repaint (Example 3)

Sports stadiums provide many examples of steel structures that require ongoing maintenance. This particular structure was first repainted in 2005. An inspection was carried out in 2015 prior to it being repainted in 2016. There was minimal corrosion and the existing coating thickness was in excess of 500 microns. This steel was to be repainted primarily for cosmetic purposes. The repaint specification was offered as a 2 coat system by the facility maintenance contractor – one primer and one topcoat. The required colour was AS2700 T63 Teal, which was supplied from a clear base. In the 2005 repaint, the painting contractor requested the primer to be the same colour as the topcoat. The same was proposed for the 2016 repaint. All the application was by roller as the steel has sufficient surface area to make this practical. The steel was in reasonable condition upon being inspected, there were some areas of flaking coating, but a significant loss in gloss across all areas was apparent on all surfaces exposed to UV light. (Refer to Pictures 6 & 7).

No measurements were taken to determine the film thickness of the two coats that had been applied, but in some locations the topcoat was assessed as very thin and most likely under the required minimum dry film thickness.

The use of a contrasting primer colour may result in the topcoat requiring application in 2 coats (given we are talking roller application). It is highly likely that this would result in a dry film thickness of the topcoat being in the range recommended by the technical data sheet or specification. It would certainly result in a thicker layer of topcoat being applied to the asset, as the same colour primer will give a false opacity. Most coatings, and particularly topcoats will offer different performance at different dry film thicknesses. It is expected that for applications under the minimum recommended thickness, the performance in gloss retention and film integrity will be poorer than for an application within the recommended film thickness range. However, this can impact on projects, as the cost of another coat can be quantified and is an up front cost, but the value of the third coat, in terms of extended time between repaints is not known.

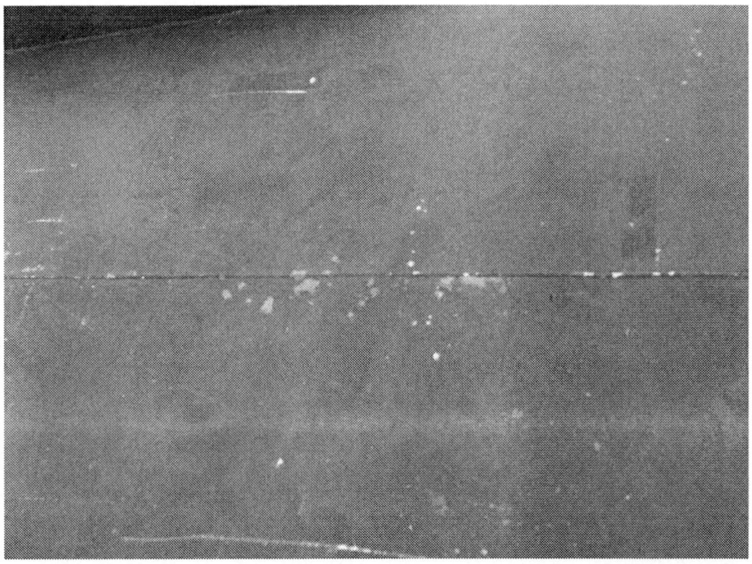

Picture 6: 10 Years on from the last repaint, the finish is chalky and faded.

Picture 7: Areas subject to more UV exposure were showing topcoat that was able to be removed easily with a knife.

EXAMPLE 6: HIGH RISE BUILDING HANDRAILS

A repaint project was carried out on handrails on a high rise accommodation block in Queensland. The requested final colour was Aubergine, a purple colour. This colour is usually delivered out of a clear tint base, and will have low opacity. Paint application was mainly by brush due to the size of the steel items being painted, predominantly thin diameter handrails and balustrade. The project was carried out using a surface tolerant epoxy primer and a high build polyurethane topcoat but both were tinted to the same colour at the request of the coating applicator. After 3 months, the applied coating system was showing peeling, chalking and coating breakdown across most areas of the structure. There were a couple of major contributing factors to this; firstly it was hard for the contractor to know what had been top coated after the primer coat had been applied, so misses of top coat were apparent after 3 months, showing severely chalked epoxy coating. It should be noted that these misses were not as obvious at the time of application but became readily apparent after 3 months of UV exposure, causing the epoxy to chalk.

The second mode of failure was the lack of opacity in the topcoat. While it looked like coverage had been achieved over the primer, this was not the case, and UV light penetrated the topcoat and reached the epoxy coating. It should be noted that epoxy coatings are not UV stable and will chalk (surface breakdown of the epoxy resin) when exposed to UV light. The result of the epoxy coating chalking underneath the applied polyurethane lead to topcoat delamination.

Had the epoxy primer been applied in a contrasting colour, this short term coating breakdown would have most likely been avoided, as the topcoat would have been applied at the correct film build, preventing the UV light reaching the epoxy primer. And any missed areas would have been obvious at the time of application.

The rectification for this project required the entire structure to be repainted. The coloured topcoat for the repaint was delivered by a factory milled batch. This process uses ground pigments in the factory to deliver the colour, instead of using in store tinting. Factory milled colour will usually have better opacity than colour delivered by in store tinting. (Refer to Picture 8 for a comparison between a factory made batch of colour vs colour tinted in a store from a tint base). The application contractor agreed to use a contrasting colour for the primer to eliminate the issues of misses in the painting process..

Picture 8: Aubergine; left panel is factory made batch, right panel is clear base tinted in store, showing variation in opacity.

Example 7: Site Applied MIO Coatings

The last of the case history examples is another in-situ repaint. The application contractor proposed a 2 coat system of a surface tolerant epoxy primer and a high build polyurethane topcoat. Again, there was no corrosion issue, and the existing coating system was typically measured at 400 microns, so the repaint was mainly for aesthetic purposes.

The topcoat colour was requested as MIO mid grey in the high build polyurethane. When MIO coatings are applied by brush and roller they can be challenging to get full opacity due to the flake pigment being aligned in one direction, in contrast to the random scattering of particles that is achieved by spray application. The application using a brush and roller can also push the particles around, resulting in uneven distribution of the MIO pigment and uneven opacity.

The photos below (Refer to Pictures 9 & 10) show a short-term delamination occurring 2 years after the coating was applied. One of the photos (Refer to Picture 9), shows the result of a cross-cut adhesion test, which resulted in topcoat delamination from well outside the test area, suggesting that the adhesion of the topcoat was completely compromised.

Pictures 9 & 10: Delamination due to low thickness (transparent) topcoat applied over epoxy primer

Flakes of topcoat that were removed on site were transparent and under the required minimum thickness for the topcoat that was used. The recoating interval was also a potential contributor to the poor adhesion. A more robust repainting option would have been to use a system that had 2 coats of the topcoat, but also to use a solid colour (eg N44 Bridge Grey) in lieu of the MIO colour.

EXAMPLE 8: EASY SITE IDENTIFICATION OF COATING SEQUENCE

Contrasting colours in a coating sequence allows for easy identification of all layers by forensic analysis. Not all projects require this type of investigation, but if they do, the use of contrasting colours makes for easy detection in the field. The following example was steelwork that was specified as a 3-coat system, comprising of an epoxy zinc primer (dark grey), intermediate epoxy (light grey) and MIO polyurethane topcoat (charcoal). From the site inspection using a Tooke Gauge, the coating layers are easy to identify, and the absence of the zinc primer was easy to observe. (Refer to Picture 11)

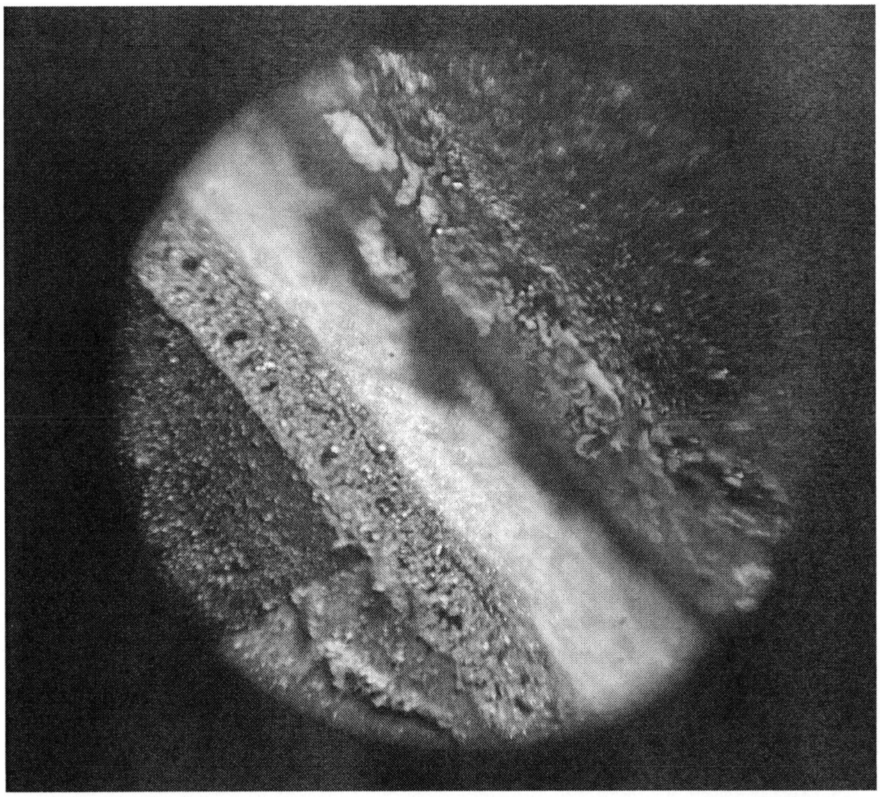

Picture 11: Photo from Tooke Gauge site analysis of coating sequence, showing 2 coat system. From the centre cut to the left, a light grey layer is shown, followed by the MIO polyurethane topcoat. No zinc primer is observed.

DISCUSSION, OBSERVATIONS AND SOLUTIONS

Well written specifications contain a requirement for contrasting colours being used in a system. The benefits of contrasting colours being applied in a system are as follows;

1. Allows for easy identification of all layers through a visual assessment (Tooke Gauge). This provides for a simple technique to verify that each coating is present.

2. It makes for easy visual assessment at application stage that coatings are applied to full opacity. Opacity can be determined visually with the contrast in preceeding coat, and it allows for easy identification of any missed areas. If missed areas are obvious, they are generally detected and fixed upon inspection immediately after the final coat is applied. Alternately, when the undercoat is the same colour as the topcoat, these misses may not become obvious until the steel has been exposed to the weather for several months, meaning rectification work takes longer, are more costly, and can have more requirements for surface preparation as more time has elapsed. Tank lining applications is one area where the use of contrasting colours will improve the "right first time" percentage, as reworking coatings intended for immersion service may result in delays in returning that asset to service along with impacting on the quality of the installed lining, in a higher risk service environment.

3. It provides more confidence that a coating is applied at its nominated dry film thickness. To achieve opacity, coatings typically have to be applied at the dry film thickness nominated on their data sheet. If a topcoat is applied over the same coloured intermediate coat, it is easy to apply that topcoat at a thickness lower than the recommended minimum on the product data sheet due to the perception of opacity, rather than actual opacity, with the primer coating being the same colour. Coating suppliers provide all their resistance testing at film thicknesses nominated on the technical data sheet. A reduction in film thickness will typically result in a reduction in performance of that coating. When coatings are applied below their recommended thickness and are transparent to UV light, it is likely that breakdown and delamination of the topcoat will occur in 2-5 years depending on the elevation and sunlight exposure. When the breakdown starts, it will likely spread and will require removal of the compromised topcoat, and full recoating on site. This is often, as has been shown in this paper, long before the corrosion issues present, which is driving premature and costly, maintenance being undertaken. The application of all coatings at their recommended thicknesses will impact on the system performance and durability.

4. The choice of colour will impact on the opacity of the coating. Pigments are classed as either organic or inorganic. Inorganic pigments tend to be higher opacity, earthy and muted colours. The Colourbond range (refer to Picture 12) typically highlights inorganic pigments, which are usually high opacity colours.

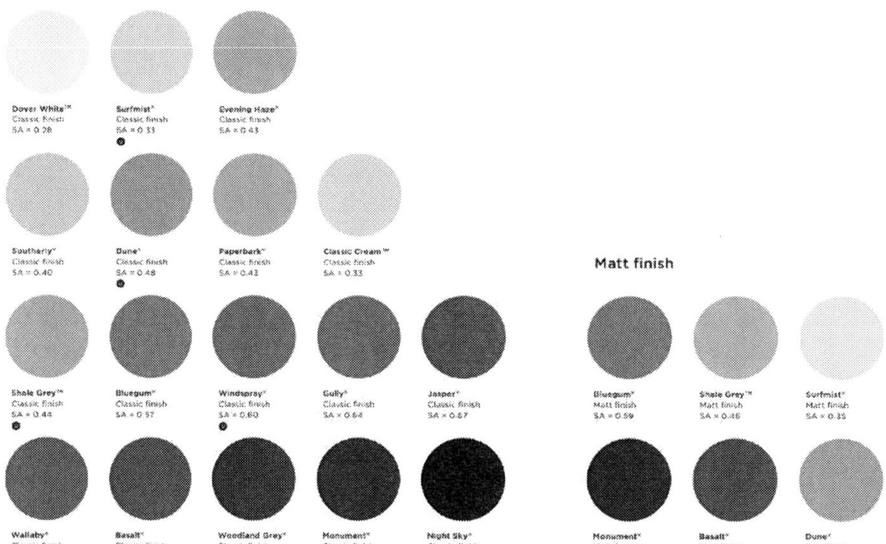

Picture 12: The Colorbond Steel Core Colour Chart showing a range of colours from inorganic pigments.

The colours on the façade at the Cairns Convention Centre are mostly derived from inorganic pigments (Refer to Picture 13). It would be expected that this choice of colour will result in excellent long term colour retention, even in the tropical exposure environment.

Picture 13: Cairns Convention Centre façade showing the use of inorganic pigments in the colour selection.

Bright colours are delivered by organic pigments. These types of pigments are known for delivering bright yellows, reds, oranges etc, but have typically much lower opacity than inorganic pigments due to their chemical structure. (Refer to Pictures 14 & 15 for examples of low opacity colours).

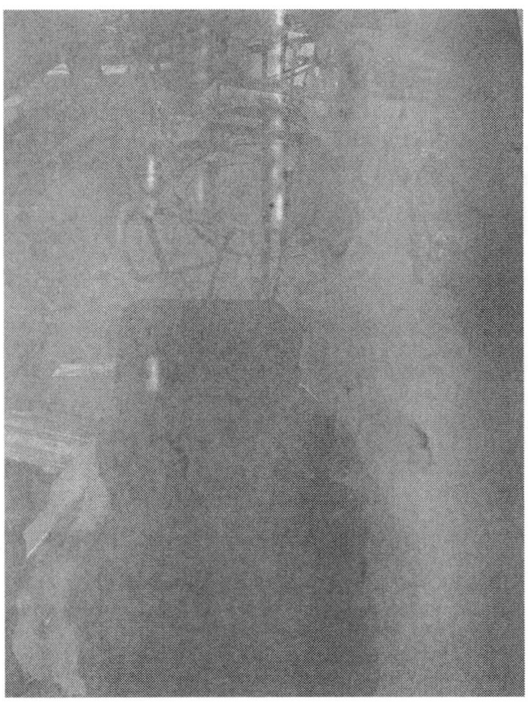

Pictures 14 & 15: A bright green, based on organic pigments, showing the transparency (or lack of opacity) at recommended product dry film thickness.

Some of these bright colours may come with a warning – for example; "low opacity – multiple coats may be required". In these cases, additional coats may be required to achieve opacity for particularly bright colours. This can certainly raise complications on projects where a specification is issued and priced before a colour is selected. Where there is doubt, it is strongly recommended that a sample be carried out which allows for verification on the opacity of the colour selected, along with setting a standard for the project.

5. It is important to understand how colour is delivered in a topcoat. Paint is either made in the factory in that colour (factory made/batched), or tinted in a store. Typically, in store tinting is easier, faster, requires little lead time and has no real minimum order quantity. In store tint systems are largely fixed in terms of colour set and lack the flexibility to customise the colourants with respect to durability and opacity to the requirement of the project.

   While the organic colourants tend to deliver lower opacity than the inorganic colourants, it is possible to find organic colourants with better opacity than those used for in-store tinting. Refer to the Aubergine example outlined earlier for an example of achieving higher opacity with a factory made batch of colour.

   If there is sufficient time allowed within the project, it can be worth pursuing if a factory batch of a particular colour in the nominated topcoat can be supplied. This option will most likely also provide a better long term durability result.

Lessons Learned

For the most part, critical infrastructure projects will continue to use a series of contrasting colours for the coating sequence. This is not done to make the applicators job harder, but help to safeguard the durability of the asset. For the most part, factory application contractors accept and understand the requirement for contrasting colours.

If a site refurbishment/repaint project is to be undertaken, and the coating system is to be applied by brush and roller, deviation from the recommended practice of a contrasting colours may cause a significant reduction in the expected durability of the coating system, despite it offering a cost reduction in the overall project. This type of practice can result in steel structures requiring a repaint prior to the expected time to first maintenance with the lack of the topcoat film build leading to poor system performance. If a second opinion is required, there are many coating consultants in the industry who can provide advice around higher durability outcomes.

Acknowledgments

1. Jason Farrugia – Dulux Protective Coatings for the proof reading and review process.
2. Brendan Lichtendonk – Dulux Protective Coatings for the ideas, proof reading and review process.
3. Will Burge - Dulux Protective Coatings for the proof reading and review process.
4. ACA Reviewers – you know who you are.

References

5. Department of Transport; Contract 9536; Westgate Bridge Repaint; Page 190.
6. Transport for New South Wales; B220 Standard; System E1; Page 28.
7. Colorbond Steel Core Colour Chart.

AUTHOR DETAILS

	A. D. McKeown is a Business Development Executive at Dulux Protective Coatings, a position he has held since 2009. He started at Dulux in 1996 as a chemist in Research and Development and has worked in several areas before moving into Protective Coatings in an account management role. He interfaces with architects, engineers, coatings applicators, coatings consultants in the course of his role.
	B. B. Lichtendonk is the Queensland State Manager at Dulux Protective Coatings a position he has held since 2015. He started in Dulux in 2009 in an account management role before moving to Queensland to manage a team. He interfaces with architects, engineers, coatings applicators, coatings consultants in the course of his role.

CORROSION & PREVENTION
CAIRNS 2024

THE AUSTRALASIAN CORROSION ASSOCIATION INC.

CAIRNS MAIN WHARF 1-5 REHABILITATION

M.Panafieu[1], N.Berends[1]

[1]Marine and Civil Maintenance Pty. Ltd NSW, 2111, Australia

**Corresponding author email: maxp@marineandcivil.com.au*

ABSTRACT

As a heritage-listed asset of regional and national importance, the Cairns Main Wharves required specialized remediation to extend their service life and restore their structural capacity. Ports North engaged Marine and Civil Maintenance (MCM) to undertake this complex project with a minimum design life of 30 years. The scope of work focused on addressing severe delamination, spalling, and advanced corrosion in key structural elements, particularly in the columns and soffit regions.

The project was guided by a comprehensive engineering assessment that identified significant section loss, with some areas requiring full reinforcement bar replacements. Hydro demolition was chosen for concrete removal due to its precision and ability to prevent micro-cracking, ensuring a superior bond between old and new materials. Repairs included structural concrete replacement, gunite application, and the welding of new reinforcement bars in critical areas.

Throughout the repair process, strict environmental controls were maintained to protect the sensitive marine ecosystem, while operational constraints were managed to minimize disruptions to the port's activities, including cruise ship operations. Staging restrictions were applied, preventing simultaneous repairs on adjacent piles but allowing flexibility in soffit work.

In addition to the primary repairs, the project uncovered further deterioration in soffit panels, beams, and piles that had not been previously addressed. These areas were also remediated to ensure the overall integrity of the structure. The extensive survey and collaborative approach ensured that the wharves remained operational during the repair process while preserving their heritage value.

This successful remediation project not only addressed the critical issues faced by the Cairns Main Wharves but also demonstrated the effectiveness of advanced repair techniques in prolonging the service life of essential marine infrastructure in harsh environments.

Keywords: Concrete Repair, Reinforcing Corrosion, Heritage Structure

INTRODUCTION

Corrosion of reinforcement within concrete structures has long been a significant concern and a primary cause of asset deterioration, particularly in structures over 100 years old.

If left unchecked, this corrosion can lead to the degradation of critical structural elements, such as beams and piles, posing considerable risks to safety and functionality. Unaddressed corrosion can negatively affect the operational capacity of the wharf by reducing load-bearing capabilities and creating exclusion zones. In extreme cases, prolonged neglect may ultimately result in structural failure.

Given the advanced age of the Cairns Main Wharves, Berths 1 to 5 have been undergoing continuous repairs for the past decade to mitigate these risks and preserve their integrity.

Wharves 1 to 5 are utilised by several vessels, including:

- Cruise ships
- Cargo ships
- Dive boats
- Boutique cruise operators
- Commercial fishing vessels
- Navy vessels

This is also a publicly accessible wharf, so it was critical to ensure that the works executed were done in a way to minimise their effects to the surrounding environment and wharf users.

FIGURE 1: GENERAL VIEW OF THE BERTHS

FIGURE 2: GENERAL VIEW UNDER THE BERTHS

CORROSION MECHANISM

The deterioration of concrete is primarily due to the corrosion of embedded reinforcing steel. During corrosion, the steel reacts with oxygen and moisture to form iron oxides and hydroxides, which have a significantly greater volume than the original steel. This volumetric expansion exerts tensile stresses on the surrounding concrete. When these stresses exceed the concrete's tensile capacity, they lead to cracking, delamination, and spalling [1].

Concrete typically has a high pH (above 12) due to the presence of hydroxides, which creates an alkaline environment that passivates the steel surface by forming a thin, stable oxide layer. This passive layer is protective, as it prevents further corrosion of the steel under normal conditions [1].

However, in marine environments, chloride ions (Cl^-) can diffuse into the concrete through its pores. Chloride ions do not chemically attack the concrete itself but instead migrate toward the steel reinforcement. Once enough Chloride ions reach the steel and exceed the critical threshold, the rebar passive layer is removed, leading to corrosion initiation. This corrosion can propagate, leading to significant damage, including cracking and spalling of the concrete [2].

THE STUCTURE

The Cairns Main Wharves consists in 6 berths, each of the structures being approximately 90m long and 27m wide and spans for 595m long. Its height is around 4.9m above port datum.

The structure is made of supporting piles coupled with primary beams and secondary beams. Piles are 450mm square column with longitudinal R40 bars and R10 ligatures at 100mm centres.

The primary beams are approximately 10m long by 2m deep and 0.8m wide while the secondary beams were 6m long, 0.5m wide and varies from 0.5m deep to 0.4m.

Constructed at the beginning of the 20th century, the Cairns Main Wharves now serve as the main regional multi-purpose port, supporting the communities of the Northen region, the Great Barrier Reef as well as the Torres Strait Islands and the Gulf of Carpentaria. The wharves handle a variety of goods including petroleum, fertiliser, general cargo and fishing resources. Additionally, reef passenger ferries and international cruise ships use the wharves as a terminal for their voyages.

The scope of work focused on addressing defects in the piles, primary beams, and soffit panels, which were identified as the primary areas of concern. A detailed survey was conducted to prioritize the repair locations based on the extent of deterioration in both the concrete and reinforcement.

FIGURE 3: AERIAL VIEW OF THE BERTHS

Ports North, the asset owner and operator, has conducted various inspection and remediation campaigns to keep the Cairns Wharves operational despite the defects.

The following list outlines the main remediation campaigns and specific structural elements that were prioritized:

- 1997 = Concrete repairs to primary beams, columns and soffits
- 1999 = Concrete repairs to primary beams, columns and soffits
- 2001 = Concrete repairs to primary beams, columns
- 2002 = Concrete repairs to primary beams, columns (Marine and Civil Maintenance was contractor)
- 2004 = Concrete repairs to primary beams, columns (Marine and Civil Maintenance was contractor)
- 2005 = Concrete repairs to primary beams, columns
- 2006 = Concrete repairs to primary beams, columns and soffits (Marine and Civil Maintenance was contractor)
- 2007 = Concrete repairs to primary beams, columns and soffits
- 2010 = Concrete repairs to primary beams, columns
- 2016 = Concrete repairs to primary beams, columns (Marine and Civil Maintenance was contractor)
- 2022 = Concrete repairs to primary beams, columns and soffits (Marine and Civil Maintenance was contractor)

FIGURE 4: DEFECT SELECTED FOR REPAIR - EXAMPLE 1

FIGURE 5: DEFECT SELECTED FOR REPAIR - EXAMPLE 2

FIGURE 6: DEFECTS SELECTED FOR REPAIR - EXAMPLE 3

Figures 4, 5, and 6 above illustrate the widespread corrosion observed throughout the structure, affecting all berths and structural elements along the wharf. In certain areas, section loss in the reinforcement bars exceeded 50%, prompting the design team to recommend the full replacement of the bars within the patch repair zones.

The drawings in Figure 7 below depict the various structural elements that were repaired during earlier remediation campaigns, while Figure 8 provides an overview of the repairs carried out by MCM during the most recent campaign in 2023.

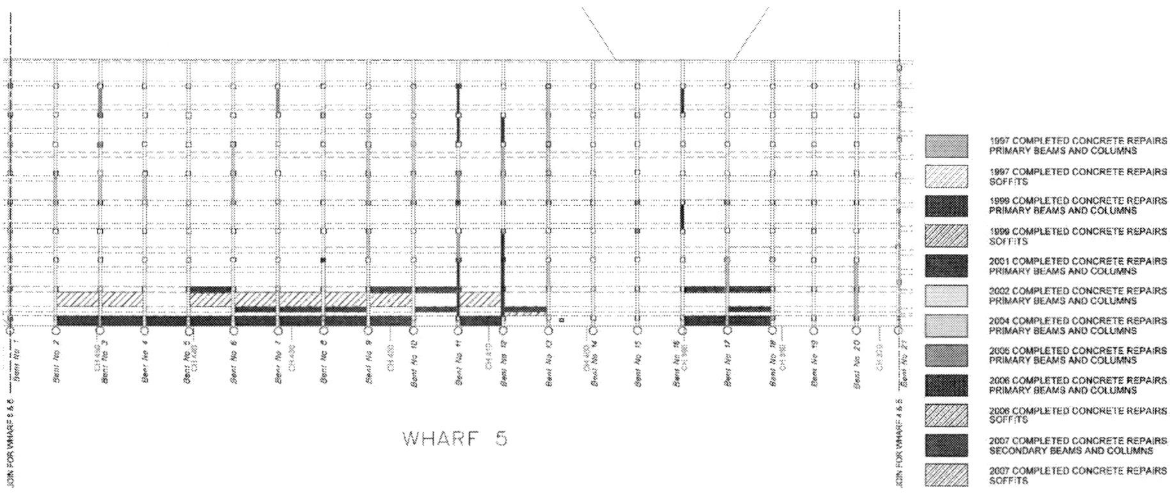

FIGURE 7: REPAIRS FROM PREVIOUS CAMPAINS

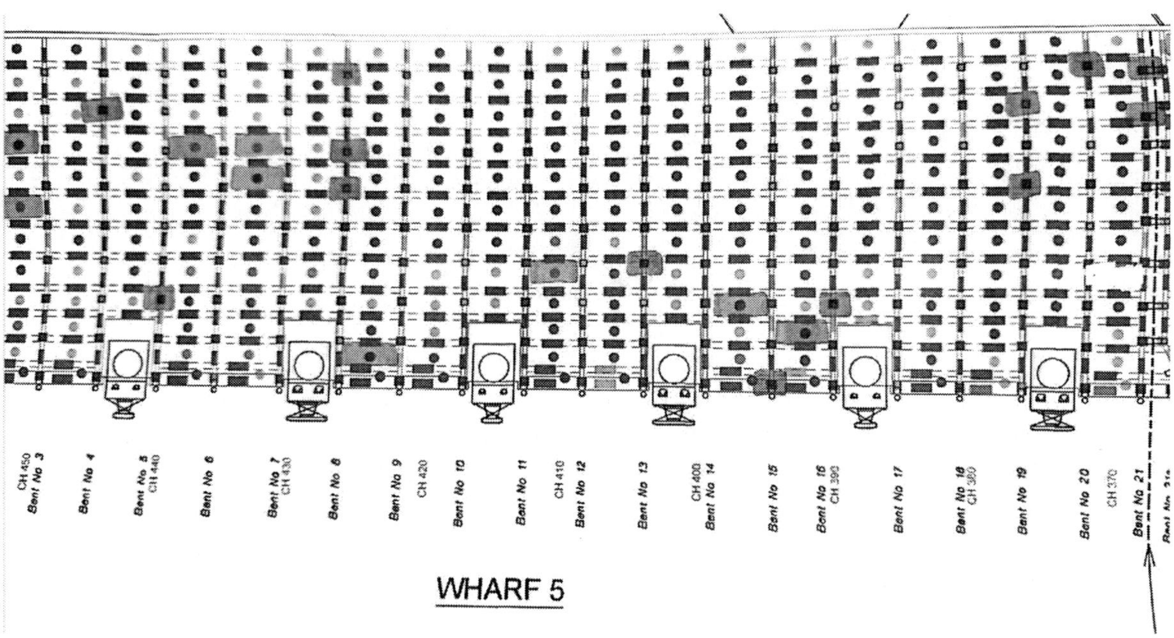

FIGURE 8: MCM REPAIRS CAMPAINGN IN 2023

From the drawings in Figures 7 and 8 above, it is clear that each remediation campaign targets different structural elements. These may include piles, soffits, or primary and secondary beams, with the prioritization determined by the consultant in collaboration with the asset owner. The contractor is then provided with a survey to guide the execution of repairs.

The photos also show that, on rare occasions, rework is necessary on previously completed repairs. Although uncommon, this can occur due to various factors, such as impact damage or overloading of the structure, among other potential causes.

REMEDIATION CONCEPT

During the design phases, varying levels of chloride contamination were detected, prompting Ports North to invite tenders for a new phase of remediation work, primarily focused on Berths 3 to 5 due to the high shipping activity at Berths 1 and 2. The asset owner's main priority was selecting a contractor capable of delivering the project without impacting port activities or causing damage to the existing structure or services. The project involved concrete patch repairs on the piles and soffits.

Regarding repair staging, limitations were imposed on the proximity of repairs to prevent compromising the structural integrity during the remediation process. Specifically, no adjacent piles could be repaired simultaneously to avoid potential structural issues. However, no such restrictions were placed on soffit repairs, allowing for greater flexibility and progress in these areas. This approach ensured that the repairs could be carried out efficiently and safely, while maintaining the wharf's operational capacity.

The markup in Figure 9 below highlights the repair locations:

- Piles: 19 locations (Black on Figure 9)
- Soffit 45 locations (Red on Figure 9)

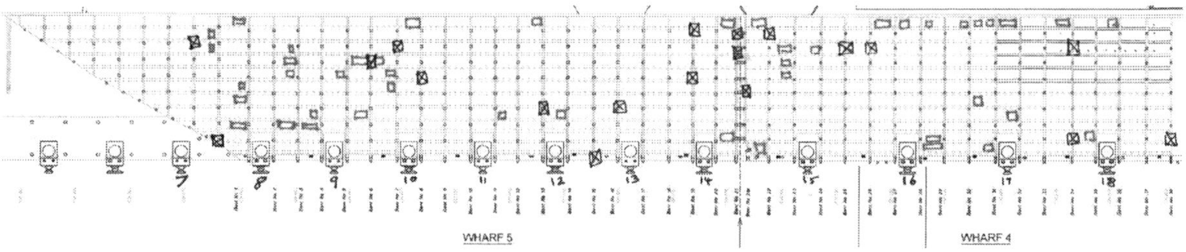

FIGURE 9: LOCATION OF REPAIRS

The general layout of the underside of the wharves is depicted in Figure 10 below. In the scope of work, the term "pile" refers to the length of the pile that extends through the primary beam and possibly through a secondary beam up to the underside of the slab soffit. The piles do not necessarily align with the secondary beams.

FIGURE 10: GENERAL ARRANGEMENT UNDER WHARVES

Two types of repairs were specified for the piles.

1) <u>Type A repairs</u>

Breakout done on one full face of the pile from 1000 below the bottom of primary beam to the 1000mm above the bottom of the primary beam (2000 vertical height) as shown in Figure 10 above.

Breakout was done to the required depths to provide clearance to the vertical reinforcing and ligatures in the piles.

Reinforcement bars were welded to existing reinforcement bars (x2 units). N16 deformed bars to be used, 1800mm long. Bars to be added in a manner which does not reduce cover

2) <u>Type B repairs</u>

Breakout was extended from 1.70m Chart Datum (Mean Sea Level) to the underside of the slab.

All deteriorated and drummy concrete was removed for the extent of the pile repair. If necessary, further concrete was removed to a point where sound reinforcement (zero loss of section) was exposed, and l removed from around the reinforcing bars to ensure that no chloride laden concrete remained in contact with the pile reinforcement. The detail for pile rehabilitation is shown below.

One type of repair has been specified for the soffits.

The rebar cover across the different wharves varied, reflecting the age of the structures. While the records provided during the tender process were helpful, there were some discrepancies in accuracy. The table below in Figure 11 presents the range of cover values encountered during the project.

LOCATION	COVER (mm)
	Columns
Wharf 1 Pier 96	63
Wharf 3 Pier 55	80
Wharf 5 Pier 20	90
Wharf 5 Pier 4	100
	Primary Beam
Wharf 1 Pier 96	65
Wharf 3 Pier 55	60
Wharf 5 Pier 20	125, 88, 165, 98
Wharf 5 Pier 4	170
	Secondary Beam
Wharf 1 Pier 96	85
Wharf 3 Pier 55	80
Wharf 5 Pier 20	75
Wharf 5 Pier 4	120
	Slab
Wharf 1 Pier 96	35, 40
Wharf 3 Pier 55	35, 50, 44
Wharf 5 Pier 20	50, 35, 65
Wharf 5 Pier 4	55

FIGURE 11: APPROXIMATE COVER TO REINFORCEMENT FOR EACH WHARF

Therefore, the consultant opted for s full replacement of the reinforcement bars within the soffit patches repairs, at every 150mm centres each-way. The bars had to be N12, deformed.

Please refer to the list below for numbers of bars per patch (indicative only):

- Soffit patch size 1m2: 12 bars
- Soffit patch size 2m2: 18 bars
- Soffit patch 3m2: 22bars
- Soffit patch 4m2: 24bars
- Soffit patch 5m2: 30 bars
- Soffit patch 6m2: 32 bars
- Soffit patch 7m2: 35 bars
- Soffit patch 8m2: 37 bars

ACCESS SYSTEM FOR THE WORKS

The proposed access system is based on MCM designed and prefabricated steel frames and mesh decking supported by chains from the superstructure. On completion, all bolts were removed 50mm below the surface and disposed of, and the holes mortared up with a suitable repair mortar to avoid any further chlorides egress to the structure.

For this project the decking was fixed at a height in consideration of the optimum work height and the movement of tides. It utilised 3.0m x 1.5m deck units, complemented by other size deck units as required.

Handrails or fence panels were installed on the perimeter to prevent falls and act as barriers for the safety of the site team, the consultant or client. Handrails also acted as a barrier and fence against possible crocodile presence on site. Additionally, ladders were used to access the decks from above and boats were used when the ladders weren't sufficient due to tidal restrictions.

Geofabric was also laid down on the platforms to contain demolition debris during the hydro demolition process, but also during the duration of the other works to minimise the impact onto the waterway.

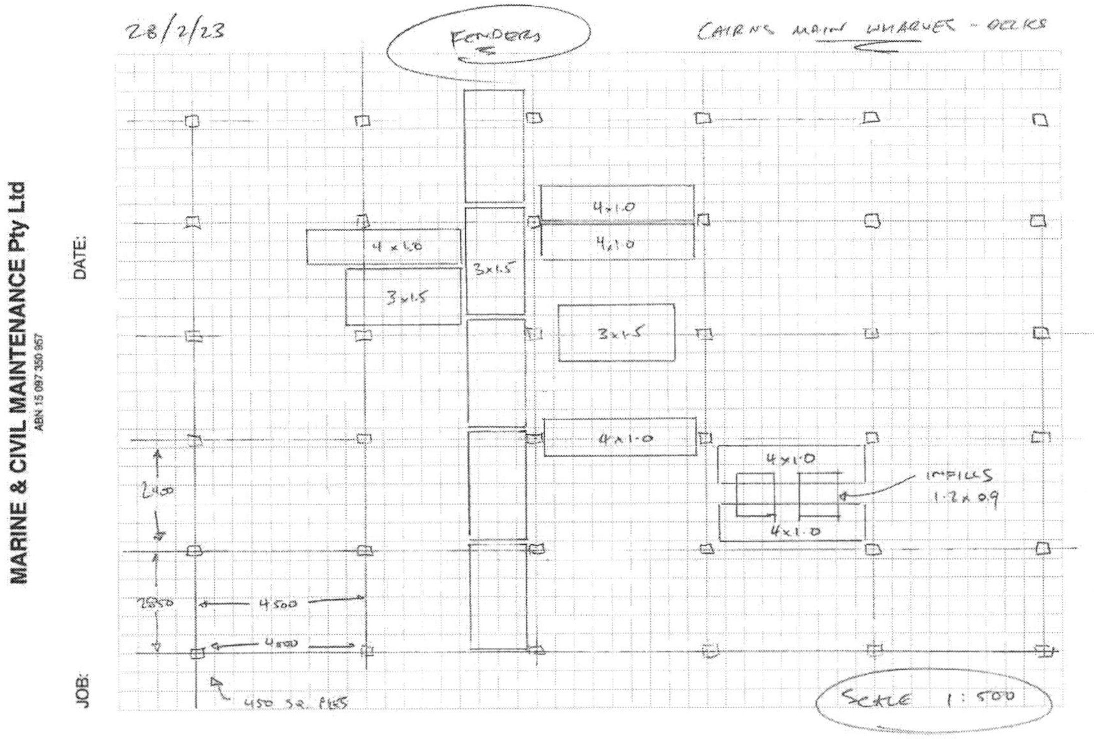

FIGURE 12: EXAMPLE OF DECKS INSTALLATION PLAN

EXECUTION OF THE WORKS

From the original scope (19 patch repairs on piles and 45 patches repairs on soffits), there was no major variance at the end of the project which demonstrated effectiveness of the survey done by the consultant and the client.

15 patches were delivered on piles and 50 soffits areas were repairs between Berths 3 to 5.

Overall close to 200m^2 of concrete repairs were completed as part of the project with an average depth of 150mm. The breakout extended to a depth of 25mm behind the layer of steel reinforcement.

Rubble from the breakout was caught on the access platforms and transferred daily to skips for disposal to a recycling facility. When breaking out concrete from the work punt, the boat was securely moored adjacent to the repair area so that the rubble was collected within the gunnel of the boat.

STEEL REINFORCEMENT CLEANING AND REPAIR

After breakout of defective concrete, the cleaning of the steel was performed, and the additional reinforcement bars were installed as per the consultant's instructions to replace all bars.

The below drawing show the details of the reinforcement repairs.

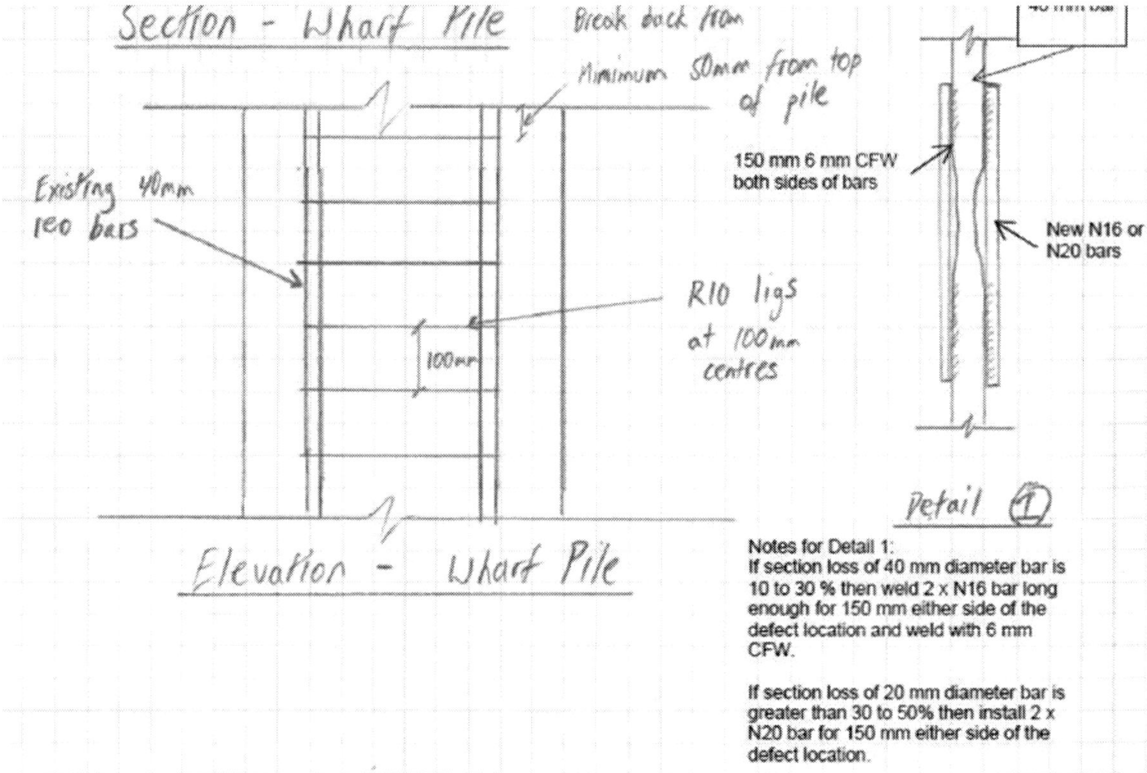

FIGURE 13: DETAIL OF REBAR INSTALLATION

The threshold was about the section loss of the bars to know if additional reinforcement had to be welded. Details as per the above drawing.

All welding was carried out with a portable electric-arc welder and complied with the AS1302, AS1304 standard.

CONCRETE REINSTATEMENT

The repairs were reinstated with cementitious dry-sprayed gunite. A pre-bagged proprietary product with specifications including low shrinkage, high compressive strength (50MPa) and moderate to high resistivity.

The specifications nominated a compressive strength of 50MPa at 28days which was achieved with values above 53 and 54MPa. Test boxes were done on site for these tests.

The substrate was thoroughly wetted until it was in a saturated, surface dry (SSD) condition before the gunite was pumped from the deck level, with a compressor and gunite machine.

The repairs were cured using a water-based curing compound to avoid affecting the application of silane.

FIGURE 14: EXAMPLE OF PILES REINFORCEMENT BARS PREPARATION

FIGURE 15: EXAMPLE OF SOFFITS REPAIRS

All repairs areas were broken out and steel reinforcement was prepped as per the illustrations above. Particular care was given to the application of zincrich products to ensure all steel surfaces were adequately covered and to avoid contamination of the concrete surfaces and have full coverage of the steel reinforcement bars. A proper application of the zincrich product is a key step for conventional repairs which is the case on this project.

FIGURE 16: FINAL REPAIRS PHOTOS

CHALLENGES ENCOUNTERED DURING THE WORKS

Various challenges have presented themselves during the duration of the project. All of them due to different reasons.

SHIPPING CONSTRAINTS

As stated within the tender documentation, flexibility during the works will have to always been given to the Port Operations due to the variety of the vessels using the berths.

As mentioned before, cruise ships, commercial vessels or navy vessels are using these Berths and all of them have different requirements in term of safety (Maritime Security Identification Card, MSIC required or not), exclusion zones, stopping works before and after the mooring of a vessel of total bans on work if a Navy vessel was at berth.

MCM brought its expertise when working in such environment and the key factor which helped delivering such a complex project was a constant communication between the site manager and team and the Ports North representative from the shipping point of view.

WORKS IN TIDAL AREAS

The range of the tides at the Cairns Port offered an extra challenge to the project due to large amplitude.

During some periods of an amplitude of 4m of the tides could be noticed and the access system proposed for the works had to be suitable to allow for the continuity of the works, no matter the level of the tides.

HERITAGE LISTED STRUCTURE

The wharf is a heritage listed asset and therefore a General Exemption Certificate from Queensland Heritages Places was obtained (dated 01/01/2020), by the Client in order to start the works.

Such exemption allows for works to be done on various structures as long as they follow the conditions of works:

- Removal and replacement of items can only be undertaken when the existing item is deteriorated extensively and can no longer be conserved
- All introduced materials or elements much match the existing one as much as possible.
- In any 3-year period replacement must not exceed 20% of an existing element
- Structure can be re-levelled but not raised or levelled.

It is also mentioned that all services need to be protected if not replaced and not to be disturb during the works.

CONCLUSION

Working on a heritage structure presents unique challenges, and the scope of work often differs from what was originally anticipated. Executing the repairs within a fully operational port facility added further complexities, which were successfully navigated through continuous communication between MCM, the asset owner, and various stakeholders.

The Cairns Main Wharves exemplify the corrosion mechanisms typically encountered on maritime structures, illustrating the extent of deterioration that can results from advanced corrosion if not addressed promptly. The client's priority repairs revealed significant surface loss and spalling, necessitating full replacement of all the reinforcement cage (both longitudinal and transverse). In some instances, the spalling due to high corrosion levels accounted for up to 20% of the total surface area.

Over the past 30 years, multiple remediation campaigns have been organized, generally occurring every two years or more frequently depending on the structural elements targeted for repair. The asset owner and the consultant have adapted the maintenance regime to accommodate the heritage nature of the wharf, conducting repairs in select areas at a time. This targeted approach is influenced by heritage restrictions, budgetary considerations, and the need to maintain operational functionality for all stakeholders in the region.

Given that most repairs are focused on the existing original concrete of the wharf, this targeted approach to conventional concrete repairs has proven effective in this context.

ACKNOWLEDGMENTS

The author would like to thank Ports North for permission to submit this paper as well as all the stakeholders using the wharf which we work with closely to deliver a successful project.

References

1. Chess P.M (2005) Cathodic Protection of Steel in Concrete

2. Tuutti, K. (1982). Corrosion of Steel in Concrete. Swedish Cement and Concrete Research Institute.

AUTHOR DETAILS

	***Max Panafieu*** Maxime is a project Manager with over 10years experience within the remediation industry Focused on various remediations projects across the country, he has worked in numerous wharves and jetties involving concrete repairs, installation of cathodic protection systems and coating works.
	***Nate Berends, BEng (CivInfra)(Hons)*** Nate is a Senior Technical Engineer at Marine and Civil Maintenance, a position he has held since the 2023. As part of the Engineering Services Team Nate supports MCM's engineers in the delivery of projects to ensure compliance with contract technical specifications, Australian and International Standards.

CORROSION & PREVENTION
CAIRNS 2024

THE AUSTRALASIAN CORROSION ASSOCIATION INC.

PORT OF HASTINGS MULTIFACTED DESIGN

N.Berends[1], D.Branca[1], B. McGuiness[1]

[1]Marine and Civil Maintenance Pty. Ltd VIC, 3071, Australia

**Corresponding author email: NateB@marineandcivil.com.au*

ABSTRACT

Port of Hastings Corporation (PoHC) engaged Marine and Civil Maintenance (MCM) to undertake specialized concrete remediation for three different structures: Long Island Point, Crib Point Dolphin 4A, and Stony Point Tug Berth/Approach Jetty. Pre-design findings revealed that a multifaceted strategy was necessary to customize the remedial works for each structure.

At Long Island Point, a crude oil wharf, a dual-phase hybrid anode system was employed to address high chloride content while meeting the safety requirements of its intrinsically safe exclusion zone. For the Stony Point Tug Berth, Approach Jetty and Crib Point Dolphin 4A, Electrochemical Chloride Extraction (ECE) was utilized to reduce chloride concentrations in the concrete, mitigating corrosion risk without the extensive monitoring typically associated with traditional impressed current cathodic protection (ICCP) systems.

Additionally, MCM mitigated the chloride-induced corrosion in the tidal zone of the Tug Berth piles by installing a string array of galvanic anodes, providing passive corrosion protection in the highly corrosive marine environment. The project faced challenges such as unforeseen repair volumes, shipping delays, and restrictive access conditions. However, adaptive planning and flexible methodologies enabled the successful execution of the works.

Preliminary results suggest the ECE treatment is effectively reducing chloride ion concentrations, with further monitoring needed to evaluate the long-term impact. This case study underscores the importance of site-specific approaches in concrete remediation, emphasizing that a one-size-fits-all solution is inadequate for addressing the complexities of corrosion in marine environments. Ultimately, the successful application of these strategies contributes to the long-term sustainability of the Port of Hastings infrastructure.

Keywords: Concrete Repair, Electrochemical Chloride Extraction, Cathodic Protection, Galvanic Anodes, Hybrid Anodes

INTRODUCTION

Concrete structures in marine environments are particularly susceptible to corrosion due to exposure to chloride ions and tidal conditions. Previous studies have highlighted the importance of customized approaches in mitigating corrosion in these settings, demonstrating that one-size-fits-all solutions are often inadequate. Effective remediation strategies must consider specific environmental conditions, such as varying chloride concentrations and moisture levels, which can significantly impact the longevity of structural integrity [1]. This case study presents tailored remediation efforts for three structures at the Port of Hastings, showcasing the effectiveness of these customized strategies in enhancing durability and performance.

The Port of Hastings, a critical infrastructure hub in Victoria, Australia, encompasses a range of marine structures, each with unique exposure conditions and operational constraints. The Port of Hastings places high priority on the environmental aspect works conducted within the facility, and in recent years this has been at the forefront of considerations when remediation works were being packaged. As such, Marine & Civil Maintenance were able to evaluate green alternatives to traditional cathodic protection systems when developing the design for the project. Marine and Civil Maintenance Pty. Ltd. (MCM) was commissioned by the Port of Hastings Corporation (PoHC) to undertake specialized concrete remediation for three distinct structures: Long Island Point, Crib Point Dolphin 4A, and Stony Point Tug Berth/Approach Jetty. This paper explores the multifaceted design strategies implemented to address the diverse needs of these structures, focusing on tailored approaches for concrete repair and corrosion mitigation.

The complexity of the project was underscored by the varying degrees of chloride exposure and environmental conditions at each site. At Long Island Point, a crude oil wharf with high chloride content and stringent safety requirements, a dual-phase anode system was employed. In contrast, Crib Point Dolphin 4A and Stony Point Tug Berth/Approach Jetty, which faced mid-level chloride exposure, benefited from Electrochemical Chloride Extraction (ECE) systems and galvanic anodes. These customized solutions aimed not only to enhance the durability and longevity of the structures but also to accommodate site-specific constraints such as accessibility, power availability, and operational safety.

This paper details the design philosophy and remedial strategies employed, including system selection criteria, design assumptions, and the challenges encountered during implementation. By examining these elements, the paper highlights the importance of a tailored, site-specific approach in achieving effective and sustainable concrete remediation in marine environments.

FIGURE 1 - VIEW OF THE STONY POINT TUG BERTH

Corrosion Mechanism

The deterioration of concrete is most commonly caused by corrosion of the reinforcing steel embedded within. Most of the time when steel corrodes the iron oxide (Fe2O3) produced has a substantially greater volume than the steel. This expansive force produces by this process create tensile forces within the concrete and when the tensile capacity of the concrete is exceeded these forces cause cracking, delamination and spalling [2].

Typical concrete is an alkaline material and generally has a pH over 12 [2]. This alkalinity acts as a protective barrier for the steel through the formation of a passive layer on the reinforcement surface. This layer is a dense and impermeable film that, when fully established and maintained, prevents further corrosion of the steel.

Over time structures in marine environments the migration of negatively charged chloride ions through the surface cover of the steel can cause corrosion even if the concrete remains highly alkaline (greater than 12). This is because the chloride ions do not directly attack the concrete but in fact diffuse through the pores within the concrete until they eventually reach the steel surface. These chloride ions then react with the oxide passive film and once this is disrupted it initiates to pitting corrosion that propagates and causes spalling of the concrete.

Remediation Contract Scope

Scope of Work

Due to the unique requirements of each location, a bespoke scope of work was designed for each individual structure. The design and construction were undertaken in accordance with AS 2832.5-2008, Cathodic Protection of metals, Part 5: Steel in concrete structures [3] and the Infracorr Performance Specification [4].

Stony Point – Approach Jetty & Tug Berth

1. Localised full depth conventional concrete repairs to beams, soffits, and approach jetty piles, including reinstatement of corroded reinforcement and replacement of concrete using high performance concrete Additionally, Includes installation of galvanic anodes around perimeter of repairs.
2. Localised part depth removal of concrete across the tug berth piles. Reinstatement of corroded reinforcement and replacement of concrete using high performance concrete repair mortar.
3. Installation of galvanic anodes within the tidal zone 1.9m – 3.2m as shown below in Figure 2.
4. Installation of an electrochemical chloride extraction on all tug berth piles and select piles on the Approach Jetty above 3.2m to 4.7m.

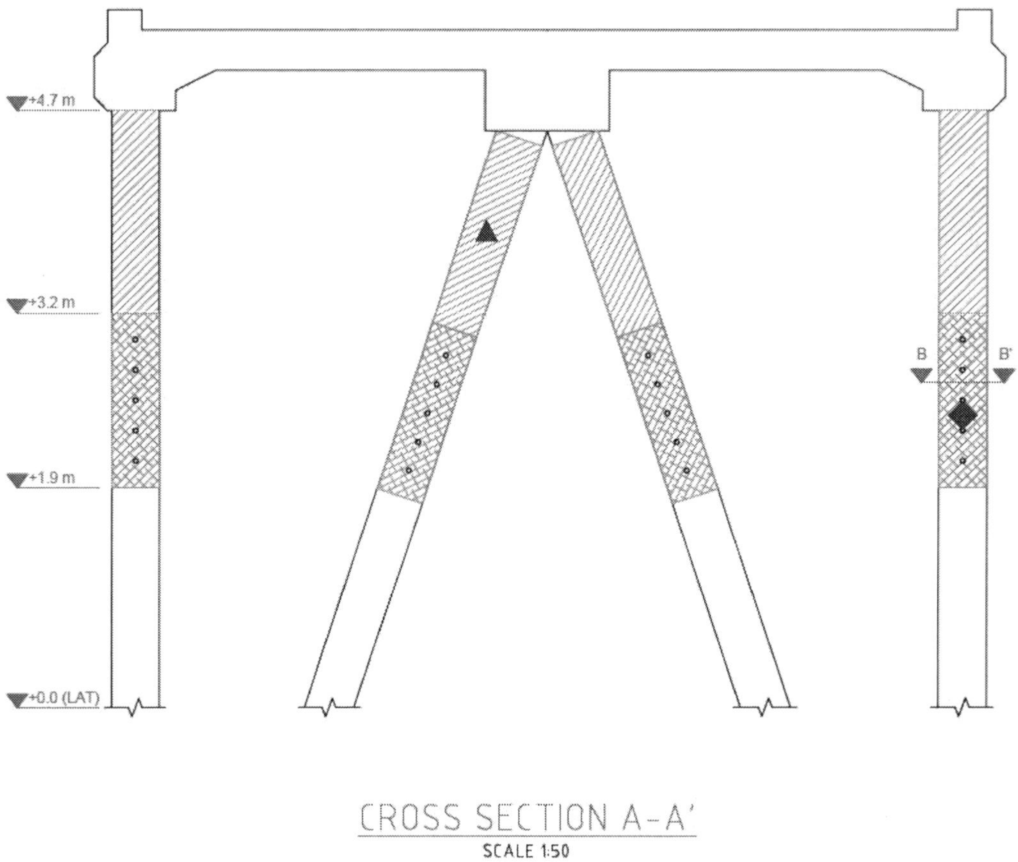

FIGURE 2 – CROSS SECTION OF STONY POINT TUG BERTH

CRIB Point – BERTHING DOLPHIN 4A

1. Localised full depth conventional concrete repairs to were undertaken to the soffit and face of the dolphin, this included reinstatement of corroded reinforcement and replacement of concrete using high performance concrete repair mortar.
2. Hydro-blasting and HDPE wrapping to steel tubular piles from LAT to underside of soffit.
3. Installation of an Electrochemical chloride extraction system to the Dolphin soffit, Dolphin faces, and Beam faces as shown below in Figure 3.

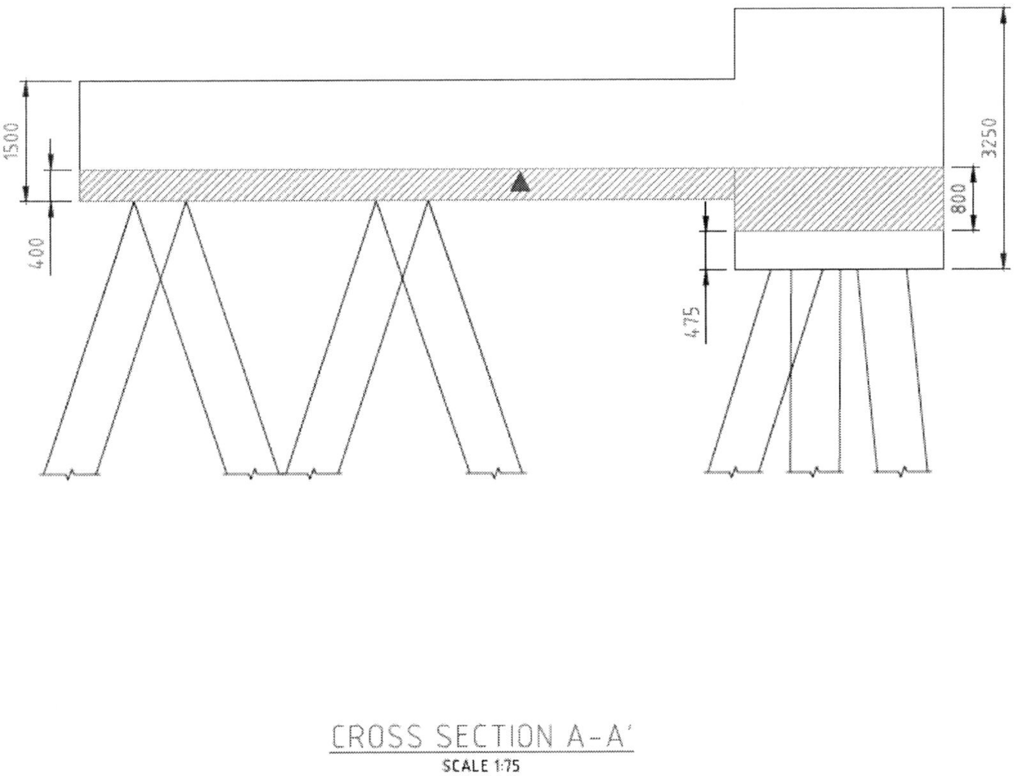

FIGURE 3 – CROSS SECTION OF CRIB POINT BERTHING DOLPHIN 4A

Mooring Dolphin Nos. 5 & 6 at Long Island Point

1. Full depth removal of concrete across the full soffit area to behind reinforcement depth. Replacement of corroded reinforcement and reinstatement of concrete using high performance concrete repair mortar.
2. Localised full depth removal of concrete across the dolphin faces to behind reinforcement depth. Replacement of corroded reinforcement and reinstatement of concrete using high performance concrete repair mortar. Includes installation of galvanic anodes around perimeter of repairs.
3. Localised part depth conventional concrete repairs to concrete piles. Reinstatement of corroded reinforcement and replacement of concrete using high performance concrete repair mortar.
4. Installation of a dual phase anode system from Mean High-Water Neaps (MHWN) to soffit level (From +1.9m to +3.66m) as shown below in Figure 4.

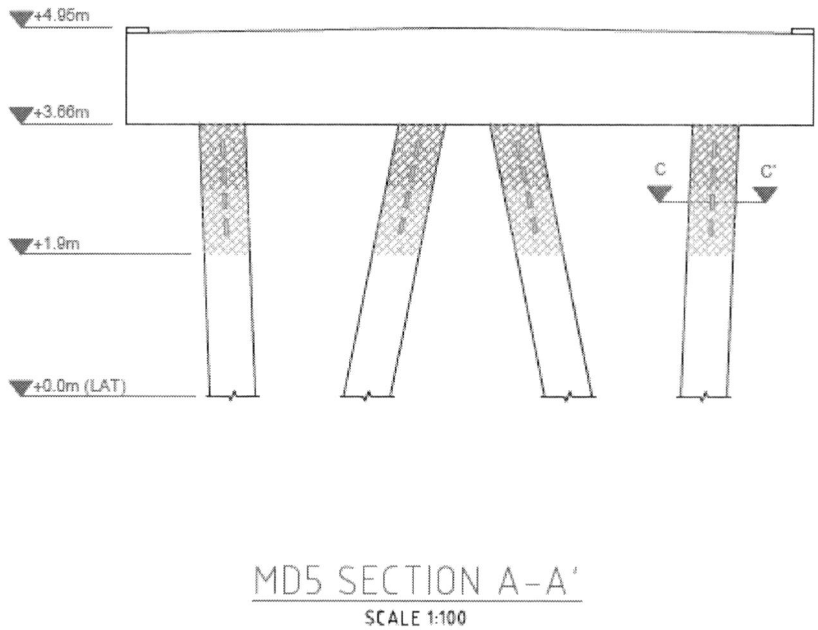

FIGURE 4 – CROSS SECTION OF LONG ISLAND POINT MOORING DOLPHINS

Design Philosophy

In general terms the design philosophy that was adopted can be summarised by 4 main points:

- Achievement of a 30-year design life
- System suitability based on site conditions and restrictions
- Ongoing performance
- Ease of system monitoring and maintenance

Furthermore, with regards to each unique project site (Stony Point, Crib Point, and Long Island Point), each of these points listed above will have a varying degree of influence. For example, accessibility and the availability of mains power is greatly different when you compare Stony Point, and Long Island Point. Additionally, concrete condition characteristics, and exposure conditions also differ not only from site to site, but also from element to element (including variations within single elements).

As such, the following design principles applied:

Stony Point Tug Berth:

- **Exposure Conditions:** Lesser degree of exposure, partially sheltered.
- **Site Accessibility (for installation and ongoing maintenance):** Simpler access due to locality of boat ramp, and lower restriction site access via land.
- **Availability of Power:** Readily available and accessible.
- **Proposed System(s):** In the tidal zone (+1.9m to +3.2m) where chlorides are present, but to a lesser degree than say Long Island Point, the installation of a string array of pre-encapsulated galvanic anodes.
  For the upper splash/atmospheric zone (+3.2m to +4.7m), where chlorides are present, however showing less visible deterioration, it is proposed to treat these areas with a Chloride Extraction System, followed by an application of silane to restrict reoccurring diffusion of chlorides back through the cover concrete.

Crib Point Dolphin 4A:

- **Exposure Conditions:** Mid exposure conditions.
- **Site Accessibility (for installation and ongoing maintenance):** Access has a higher degree of difficulty due to distance from boat ramps, distance from shore, high degree of restricted access, and swell influences.
- **Availability of Power:** Power available, with some pre-work requirements.
- **Proposed System(s):** For the dolphin sides and soffit, it is proposed to treat these areas with a Chloride Extraction System, followed by an application of silane to restrict reoccurring diffusion of chlorides back through the cover concrete.
  Due to mid-level chlorides and isolated identified areas of repair, a chloride extraction system is proposed to provide increased durability of the structure, whilst minimising the presence of exposed temporary cabling (in a restricted fuel loading facility) and minimising the requirement for site-based monitoring and maintenance.

Long Island Point:

- **Exposure Conditions:** Higher Degree of Exposure/aggressive conditions.
- **Site Accessibility (for installation and ongoing maintenance):** Access has a higher degree of difficulty due to distance from boat ramps, distance from shore, high degree of restricted access, and swell influences.
- **Availability of Power:** Limited/not available, also within an intrinsically safe exclusion zone.
- **Proposed System(s):** Internally powered dual phase hybrid anodes, due to increased performance in chloride laden concrete, eliminating the requirement for AC mains power, and exposed temporary cabling during the installation process within a restricted fuel loading facility, and minimising the requirement for site-based monitoring and maintenance.

System Design Assumptions

Throughout the design process, the following assumptions were made:

- Concrete repairs and associated electrochemical protection systems to be designed to achieve a 30-year design life
  - Galvanic and Hybrid anode systems design life calculated via mass of zinc and charge capacity (A.hrs) of individual units
- Based on the observed steel condition and associated pre-design investigation results, the following parameters were assumed for the design of the Dual Phase Hybrid Anode System(s):
  - 50kC/m2 of steel during the initial impressed current phase
  - A long term (average) operating current of 3mA/m2 of steel during the galvanic phase
  - A standard output capacity of 820Amp Hours per kilogram of zinc with an allowable utilisation factor of 90%
  - Anode zones to be designated with respect to (a) tidal ranges, and (b) restricted to a maximum number of 20 anodes per string

For the design and measurement of the Electrochemical chloride extraction system(s), the following parameters have been adopted in accordance with NACE SP0107-2021[5] standard for ECE:

- A maximum operating current density of 4A/m2 of steel reinforcement
  - Chloride content by weight of cement (Criterion A) <0.4% w/w of cement within 25mm of the steel reinforcement
- Alternative minimum end point criterion of 600Amp Hours/m2 of steel reinforcement, typical maximum end point criterion of 1,500Amp Hours/m2 (Criterion B)

System Description

Dual Phase Battery Hybrid Anodes

The Dual Phase Battery Hybrid Anodes combine the high-level performance of an ICCP system, with the long-term maintenance-free characteristics of galvanic cathodic protection. The single-unit system does not require complex wiring or an external DC power supply and can be installed in the concrete with direct connection to the embedded rebar.

When installed, the inbuilt impressed current component provides an initial phase of high charge density to provide cathodic protection and / or passivate active corrosion (phase 1). Then, the anode automatically switches to a galvanic cathodic prevention phase, where galvanic cathodic prevention current is provided to the reinforcing steel. This can maintain steel passivity and provide long- term, maintenance-free corrosion protection (phase 2).

For most structures it is the concrete cover that is the area of high chloride concentration. For optimum protection, the anode is designed for installation within the concrete cover. The spacing of the anodes is determined by the steel density, corrosion activity of embedded rebar, and the level of chloride concentration within the concrete.

Galvanic Anode System

Galvanic anodes with high zinc content and a relatively small diameter were utilised. As this anode type has been specifically designed for local / global galvanic anode protection systems and can improve the corrosion resistance of embedded rebar in concrete structures affected by chloride contamination.

The anode/backfill assembly is manufactured under strict quality control procedures and the anode is delivered to site ready to install. This eliminates major problems associated with backfill installation on site and problems associated with full encapsulation of the zinc anodes.

Electrochemical Chloride Extraction

The ECE system is designed to extract chlorides from the concrete at the depth of the steel reinforcement, while also promoting re-passivation through the application of a high-efficiency DC current. This system is temporary and can be reapplied to targeted areas as needed throughout the structure's service life.

Unlike traditional Cathodic Protection (CP) systems, ECE does not require extensive concrete breakout, complex cabling, or the installation of embedded components. The surface-applied nature of the system eliminates the need for physical embedment, making installation faster, less invasive, and more cost-effective. Additionally, the temporary nature of the system allows for modular and targeted deployment, further optimizing costs.

Access requirements for installation are significantly reduced when compared to traditional methods. For example, installations beneath wharf or jetty structures can typically be carried out from a workboat, minimizing disruption. The system also includes remote monitoring capabilities, allowing for regular checks without requiring physical access to the site. Figure 5 below provides a screenshot of the remote monitoring portal.

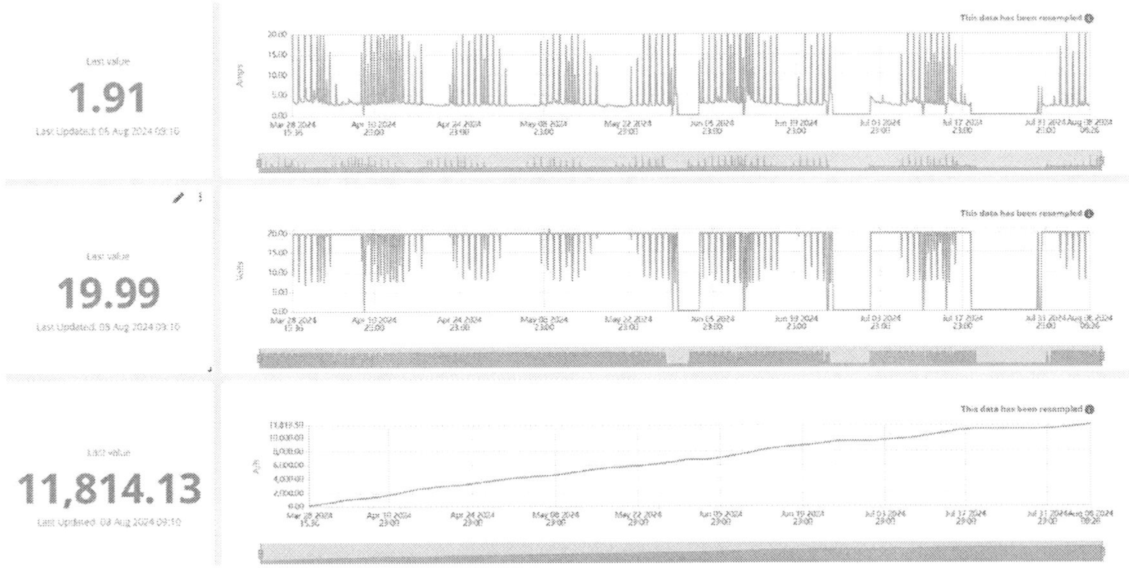

FIGURE 5 SCREENSHOT OF THE ECE ONLINE MONITORING PORTAL

The flexibility and scalability of ECE make it an ideal solution for progressively protecting large assets or multiple structures under term maintenance programs. ECE offers significant cost savings while providing the flexibility to protect specific elements or sections of a structure as needed.

The ECE treatment was implemented to extract chloride ions from the concrete and mitigate corrosion risk. As the system continues to operate, we are seeing a progressive reduction in chloride levels between the initial test points and subsequent progress test points. Table 1 shows the chloride ion concentrations at these stages, demonstrating the reduction achieved by ECE. While these early results are promising, ongoing monitoring will confirm the long-term effectiveness of the treatment.

TABLE 1 CRIB POINT ECE CHLORIDE ION CONCENTRATIONS

			Chloride w/w% cement+	
Structure	**Location**	**Depth (mm)**	**Initial Assessment**	**Progress Assessment**
Crib Point – Dolphin 4A	A – Soffit	0-55	2.901	1.735
		55-80	2.296	0.796
		80-100	1.611	0.333
		100-125	0.957	0.111
	B – Soffit	0-55	3.623	1.123
		55-80	2.420	0.611
		80-100	0.994	0.290
		100-125	0.698	0.204
Crib Point – Dolphin 4A	C – North Face	0-40	1.519	1.160
		40-65	0.827	0.420
		65-85	0.667	0.290
		85-110	0.315	0.130
	D – North Face	0-40	1.574	1.222
		40-65	1.611	0.870
		65-85	0.883	0.284
		85-110	0.420	0.148

+ *calibrated results using 16.2% of cement content from the laboratory results*

Access Arrangement & Staging

The primary access arrangement for works was achieved with a proprietary modular system of chain-suspended decking. The decks utilised varied in size subject to arrangement required onsite. They consist of steel frames with wire mesh infill, supported by chains from the superstructure so they were at a suitable height for working, which is approximately 1.8m below the deck soffit or beam.

Long Island Point works formed project critical path, and access decks were installed across all 3 locations to allow works continue in the event that shipping or weather did not allow access to a specific location. The staging of works was highly variable and required the crew to be extremely flexible to manage the frequent shipping movements and tidal conditions.

Project Challenges

Design Challenges

The remediation of the three structures at the Port of Hastings presented unique challenges that necessitated a customized approach for each site, focusing on environmental factors, operational limitations, and safety requirements. At Long Island Point, a comprehensive assessment revealed that the site's intrinsically safe exclusion zone and elevated chloride levels justified the use of a dual-phase anode system. This design choice effectively mitigates corrosion risks while adhering to stringent safety regulations. In contrast, the Electrochemical Chloride Extraction (ECE) system was implemented at Crib Point and Stony Point, where conditions required a minimally invasive, temporary solution that could be reapplied as necessary. These tailored strategies guided the decision-making process regarding the application of the ECE system and the use of galvanic anodes and hybrid systems.

ECE Application in Tidal and Splash Zones

A significant challenge was that ECE is most effective above the tidal zone, as the process requires a sustained electrochemical current which can be disrupted by continuous immersion in water. For both the Tug Berth and Crib Point Dolphin 4A, which have sections that extend into tidal waters, a different approach was required. In tidal zones, ECE could not be applied as effectively as in splash zones, where intermittent wetting and drying occur. This necessitated the adoption of hybrid solutions involving galvanic anodes, which can function more efficiently in submerged conditions, and the adjustment of anode placement to account for these varying exposure levels.

Tidal Zones vs. Splash Zones

In addition to ECE limitations, the constant immersion in tidal zones called for specialized design solutions. For example, the dynamic nature of the tidal areas required more durable and adaptable systems capable of coping with the varying chloride content and moisture exposure. This drove the decision to implement a multifaceted strategy combining ECE in splash zones and galvanic protection in tidal zones to ensure consistent protection across the structure.

Ongoing Power and Monitoring Requirements

Another critical consideration was the need for continuous power supply and ongoing monitoring to ensure the effectiveness of the cathodic protection systems. At sites like the Tug Berth and Crib Point Dolphin 4A, maintaining a reliable power source was essential to sustaining the ECE process over time. This requirement for continuous energy, combined with the challenge of ensuring ongoing access to monitoring equipment, introduced operational complexities, especially in remote or hard-to-access areas.

Intrinsically Safe Zones at Long Island Point

The Long Island Point wharf operates within a stringent exclusion zone designated for crude oil handling, which presents significant challenges due to safety regulations. This area is classified as an intrinsically safe zone, prohibiting the use of external power systems to mitigate the risk of igniting hazardous materials. Consequently, the design team opted for a dual-phase anode system, capable of functioning without an external power supply.

This choice was driven by the high chloride levels present and the infeasibility of implementing an impressed current cathodic protection (ICCP) system in this context. [6] This decision was crucial in ensuring both the structural integrity of the wharf and compliance with strict operational safety guidelines.

These varied challenges made it evident that a one-size-fits-all approach to concrete remediation would not be effective. Each structure required a unique combination of techniques, with adjustments made based on environmental exposure, operational requirements, and safety constraints.

PRACTICAL SITE RESTRICTIONS

As part of the contract conditions, works were unable to be conducted at Long Island Point whilst any vessel was in berth. Long Island Point has a significant increase in shipping volume as opposed to previous years, resulting in limitations to access. As such, consideration was given to limit the time required for onsite for both installation and ongoing maintenance of the cathodic protection system used. Combined with the required electrical exclusion zones due to being within a hazardous facility, a battery powered hybrid cathodic protection was specified. This met the practical site restrictions of the area, and provided an increased benefit during the initial energization and ongoing protection when the galvanic stage occurs.

Crib Point has significant adverse weather conditions along with also being a fuel and oil facility. ECE was ultimately chosen for this location, as it provided an easily installed and temporary system. This mitigated the effects adverse weather conditions that have previously had on the ongoing performance and condition of cathodic protection systems installed in the area.

INCREASED REPAIR QUANTITIES

The Scope of works incorporated a provisional quantity of concrete repairs based on a previous survey, which was to be measured and confirmed while works.

The most significant examples of increased repair quantities were

- Stony Point's quantity for concrete repair increased 3-fold, with a corresponding increase in quantities for steel augmentation and galvanic anode installation.
- Long Island Point's Berthing Dolphins required full depth concrete repair to the soffit as part of the scope of works, however once breakout was underway the steel reinforcement was in significantly worse condition than anticipated. Over 1000lm of reinforcement was replaced across the dolphins.

The additional repairs quantities were identified early in the project and resulted in significantly increased costs to the project. As part of the design, consideration was given as to systems which would provide the most cost-effective solution to the client and provide the most protection to the structures.

CONCLUSIONS

The project exemplifies the successful application of tailored concrete remediation strategies to address the unique challenges posed by different marine structures. Previous studies have highlighted the importance of customized approaches in mitigating corrosion in marine environments, particularly in structures subjected to chloride exposure and tidal conditions [1][5]. These studies provide evidence that a one-size-fits-all strategy is insufficient; instead, a multifaceted approach is necessary to adapt to specific environmental and structural conditions.

Each structure's specific environmental conditions and operational constraints necessitated a customized solution. At Long Island Point, the dual phase hybrid anodes addressed the high chloride content and restricted access conditions effectively, eliminating the need for external power sources and minimizing maintenance. In contrast, the ECE system, combined with silane treatments, provided a cost-effective and maintenance-friendly solution for the Crib Point Dolphin 4A and Stony Point facilities, where chloride levels and exposure conditions varied.

The project faced several challenges, including shipping and weather-related delays, increased repair quantities, and the need for flexible access arrangements. These were managed through meticulous planning, proactive problem-solving, and adaptive work strategies, ensuring project goals were met despite unforeseen difficulties.

In summary, the project successfully demonstrated that a well-considered, site-specific approach to concrete remediation can significantly enhance structural longevity and performance. While the strategies employed have addressed the immediate issues of corrosion and deterioration, it is important to recognize that the project is ongoing. Preliminary observations suggest that these approaches may contribute to the long-term sustainability of the Port of Hastings infrastructure. However, continued monitoring and data collection will be critical in validating the effectiveness of the remediation efforts and ensuring that they meet the anticipated performance criteria over time. This case underscores the importance of adapting remedial designs to the particular needs of each site and the value of employing a range of protection systems to achieve optimal results for each element.

ACKNOWLEGEMENTS

The authors wish to thank Port of Hastings Corporation for permission to submit this paper and to Esso Australia for providing access to Long Island Point. We also wish to acknowledge the verifying consultants Infracorr Consulting Pty Ltd for assistance on this project. Additionally, the authors thank Micronex who provided the portable power units and online portal for monitoring the ECE.

References

1. Moore A, Beushausen H, Otieno M, Ndawula J, Alexander M. Oxygen Availability and Corrosion Propagation in RC Structures in the Marine Environment—Inferences from Field and Laboratory Studies. *Corrosion and Materials Degradation*. 2022

2. Chess P.M (2005) Cathodic Protection of Steel in Concrete

3. AS 2832.5-2008, Cathodic Protection of metals, Part 5: Steel in concrete structures

4. Performance Specification for Concrete Remediation Works – Port of Hastings VIC, Infracorr Consulting – November 2021

5. NACE SP0107 – 2021, Electrochemical Realkalization and Chloride Extraction for Reinforced Concrete

6. Holloway L. Hybrid anode treatment for the management of corrosion to reinforced concrete bridge piles in an estuarine river. (In)Australian Small Bridges Conference, 9th, Apr 1-2, 2019, Surfers Paradise, Queensland, Australia, Paper 17.

AUTHOR DETAILS

Nate Berends, BEng (CivInfra)(Hons)

Nate is a Senior Technical Engineer at Marine and Civil Maintenance, a position he has held since the 2023. As part of the Engineering Services Team Nate supports MCM's engineers in the delivery of projects to ensure compliance with contract technical specifications, Australian and International Standards.

Dominic Branca, BAppSc (ConstMgt)

Dominic is a Project Management at Marine & Civil Maintenance with experience ranging in fields from civil and commercial construction to remediation of large-scale infrastructure. Dominic has delivered multiple projects including Civil Works, bespoke Cathodic Protection systems, Coatings, Carbon Fibre Strengthening and Structural concrete repair.

Blane McGuiness, BEng (Hons), MBA, RPEng, MIEAust, FIML

Blane is a Registered Professional Engineer (RPEng), and ACA-accredited Corrosion Technologist with over 15 years' experience in the remedial industry having extensive national and international experience in Cathodic Protection (Impressed Galvanic & Hybrid) system design, installation, commissioning & monitoring, structural remediation, structure condition assessments, durability consultation, and marine construction works.

Blane is an active member of various industry associations, holding key positions as Past NSW President, and current sitting National Director of the Australasian Corrosion Association.

CORROSION & PREVENTION
CAIRNS 2024

THE AUSTRALASIAN CORROSION ASSOCIATION INC.

Bi-modal Corrosion Behaviour of Magnesium Alloys

R. E. Melchers[1] *

[1]Centre for Infrastructure Performance and Reliability, The University of Newcastle, Australia 2308

**rob.melchers@newcastle.edu.au*

ABSTRACT

The potential advantages of the light weight of using magnesium alloys in practical applications has long been recognized but often out-weighed by the high rates of corrosion of most alloys. In most cases alloying with at least 3% (wt) of aluminium is required to obtain reasonably low corrosion rates, typically in the range 0.02-0.04 mm/y. However, a closer look at field data and some laboratory data indicates that the use of a 'rate' is a significant simplification of actual behaviour. In almost all cases for which the data is sufficiently discriminatory the trend in corrosion loss as a function of time can be modelled as 'bi-modal', a trend observed also extensively for steels, including stainless steels, copper alloys, aluminium alloys and nickel alloys. It implies that magnesium alloys follow a trend that commences with activation control on first exposure, followed by oxygen reduction and eventually by hydrogen evolution reaction (HER) as the controlling cathodic reactions. In the conventional literature, however, invariably it is assumed that HER is the only relevant cathodic reaction, with numerous attempts over many years to provide a justification for this reaction to be active from the very first exposure. This paper summarizes a considerable number of field results to demonstrate bi-modal behaviour and that it may occur over a very short period of time (days) for high purity magnesium alloys.

Keywords: Magnesium, corrosion, pitting, bimodal, mechanisms.

Introduction

Magnesium alloys are of much interest because of their light weight. They are used in various applications, including for biological implants, but their widespread use in commercial applications such as for automobiles and aircraft has been hampered by their inherent low strength and also concerns about their corrosion resistance. Both can be addressed by suitable alloying. Ultra-pure magnesium has a very high corrosion resistance [1, 2] but this can be depleted by even quite small amounts of alloying. Higher degrees of alloying, particularly with aluminium, have been found to improve corrosion resistance in a variety of exposure environments, both long-term and short-term, but usually not better than that of ultra-pure magnesium [3, 4]. There remains a need for improved understanding of the corrosion behaviour of practical magnesium [5]. The present paper focusses particularly on the relationship between shorter-term and longer-term corrosion response, for the mode common magnesium alloys.

Some of the earliest observations of the corrosion of magnesium was carried out by Beatz [6] in about 1866. He observed hydrogen evolving from the anode almost immediately after first exposure of magnesium (of undefined purity) in magnesium sulphate solution. For his experiments he used an impressed current system, rather than the open circuit conditions more akin to exposures in practical applications. He noted that the total amount of hydrogen evolved was proportional to the total amount of dissolution of magnesium. This observation has been used repeatedly in studies of the corrosion of magnesium alloy [7] to estimate corrosion rates, alongside electrochemical testing and has long been assumed to mean that the governing cathodic reaction for the corrosion of magnesium alloys is the hydrogen evolution reaction (HER):

$$2H_2O + 2e^- \rightarrow H_2\uparrow + 2OH^- \qquad (1)$$

Alongside, this observation has been used widely in experimental investigations estimating the rate of corrosion of magnesium alloys in laboratory settings [5, 8]. Results from both techniques have been used to make corrosion resistance comparisons between various alloys, typically, expressed as average corrosion rates. That there is a difference between such rates obtained from laboratory experiments average corrosion rates over longer term field observations (years) has been recognized [5, 7] but how these long-term rates might be related to short-term observations appears to have been given little attention.

Despite the wide acceptance that the hydrogen evolution reaction is the key cathodic reaction, it has been recognized since Beatz that 'all is not well' with relating the experimental observations with those of corrosion theory that typically assumed that initial corrosion in open circuit reactions and field exposures relies on the oxygen reduction reaction (ORR [9]):

$$O_2 + 2H_2O + 4e^- \rightarrow 4OH^- \qquad (2)$$

Attempts to assimilate theory based on the HER as the principal cathodic reaction with experimental observations have resulted in a number of (largely untested or untestable) suppositions, widely discussed in the literature over a long period of time [5, 7]. It is therefore highly relevant that evidence has emerged recently of oxygen reduction reaction (ORR) occurring as what was interpreted as a secondary cathodic reaction in observations over relatively short-term exposures (12 months or less) and in electrochemical tests. This includes ultra-high purity magnesium [], 10, 11]. This observation has suggested that magnesium alloys are not dissimilar from other metal alloys that have been shown to follow, for extended exposures, bi-modal behaviour [12, 13]. These alloys include steel, aluminium and nickel in a variety of exposure conditions including immersed, tidal, atmospheric [12]. Two recent papers have developed this theme [14, 15]. It is elaborated herein after first, in the next section, giving a brief overview of bi-modal behaviour. The section that follows summarizes a number of cases extracted from the (relatively limited) literature for corrosion of (mainly commercial) magnesium alloys. It will be seen that the development of corrosion for these can in all cases be interpreted in terms of the bi-modal model. The Discussion section briefly reviews the reasons for the observed bi-modal behaviour of magnesium alloys and considers why that was not previously elucidated. This includes the timing of the development of the bi-modal behaviour and also the implications it has for the laboratory observations, noting that hydrogen evolution usually is observed almost from very first exposure.

Bi-modal Corrosion Behaviour

A schematic view of bi-modal behaviour is shown in Fig. 1. Considerable experimental evidence for this trend was presented for steels (of different carbon contents) both in seawater and in de-ionized water [13] and has also been concluded from a variety of other observations for steels, stainless steels, copper alloys, aluminium alloys and nickel alloys under a variety of natural exposure conditions [12]. In Mode 1, corrosion is rate-controlled primarily by the oxygen reduction reaction (Eqn. 1). The rate control is by the rate of inward diffusion of oxygen from the exterior environment (atmosphere, water, etc.) to the cathodic regions on the metal surface. There, rusts build-up with continued exposure and increasingly inhibit the diffusion of oxygen. Not considered specifically in the original formulation [16], but clear from experimental observations [13], during this time there is also the development of pitting corrosion that after some period of exposure involves anaerobic conditions within the pits with associated production of a limited amount of hydrogen. This has been observed to cause hydrogen bubbles to evolve from pits [17]. It also is a critical (and classical) aspect of the transition from meta-stable to stable pitting of stainless steels, with rusts (rust caps) over the pits being ruptured as corrosion progresses [18]. The number of pits and the pit depths have been observed to increase [19, 13] with continued exposure, eventually, in Mode 2, becoming the dominant corrosion process with hydrogen evolution outwards through the increasing build-up of rusts being corrosion rate-controlling.

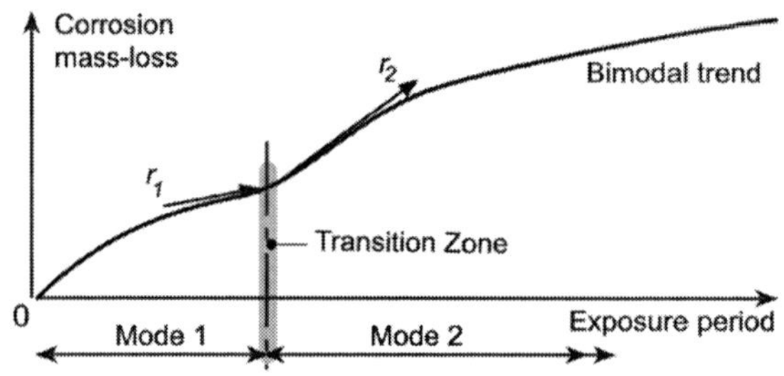

FIGURE 1. SCHEMATIC OF THE BIMODAL MODEL FOR CORROSION MASS LOSS SHOWING MODES 1 AND 2 AND THE TRANSITION BETWEEN THEM TOGETHER WITH THE CORROSION RATES r_1 AND r_2 BEFORE AND AFTER THE TRANSITION.

In the context of the corrosion of magnesium alloys pitting has been reported also, already from very early exposures [20, 21, 22 , 23]. It also has been reported for longer term exposures under field conditions [22]. Despite these observations, the potential contribution of such pitting to overall mass loss of magnesium alloys has received little attention in the magnesium alloy corrosion literature.

The change-over from Mode 1 to Mode 2 in the bi-modal model is shown (Fig. 1) represented by the corrosion rates r_1 before the transition to r_2 after the transition. The ratio between them has been argued as being theoretically $r_2 / r_1 = 4$ [24] based on (a) the rate of corrosion being directly proportional to the rate of inward oxygen diffusion (Mode 1) and outward hydrogen diffusion (Mode 2) and (b) the relative diffusivities of gaseous oxygen and hydrogen respectively through essentially the same thickness of rust products, i.e. for gaseous diffusion through a permeable barrier. For steel corrosion this ratio has been shown recently to apply (within experimental accuracy) in a variety of exposure conditions [24]. It follows that if the corrosion of magnesium alloys follows bi-modal corrosion behaviour, the ratio $r_2 / r_1 = 4$ should be applicable also. This aspect is examined in the next section.

DATA AND DATA TRENDS

For estimating comparative corrosion losses as a function of exposure time (and possibly corrosion rates) the 'gold-standard' remains the use of mass-(weight-) loss measurements [5]. This approach can be used for different alloys, various exposure conditions and also for extended periods of exposure. The latter is particularly important for infrastructure applications. The examples given below were all derived from mass-loss observations. As noted above, the present paper is concerned with the pattern of development of corrosion loss as a function of exposure time - other aspects such as the actual magnitude of corrosion losses, the topography and pitting of the corroded surfaces and the composition and the rate of formation of corrosion products as well as comparisons of corrosion losses for different alloys and environments are not considered. Information about most of those aspects is available in the original source materials and in comparisons for shorter-term and electrochemical experiments [7, 8].

For research purposes, electrochemical and hydrogen evolution measurement techniques have both been much used to estimate corrosion 'rates' of magnesium alloys, largely because they provide estimates in short periods of time. This gives them a major advantage over mass-(weight-)loss measurements since these may take many years for estimates of corrosion rates. The results of these accelerated techniques are not always consistent with mass-loss corrosion rates for reasons still not completely clear, despite decades of research and various propositions [5]. In consequence, electrochemical measurements and those using hydrogen evolution to estimate corrosion rates are not considered herein. However, some comments about how such observations might be reconciled with mass loss observations are available [15].

In the following, the examples are given in decreasing order of total period of exposure. It should be noted that the axes of the graphs are not necessarily in comparable units. Other cases are given in [14] and [15].

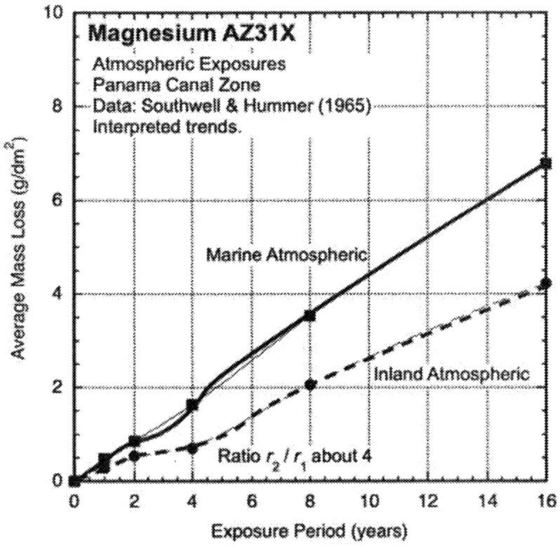

(A) DATA FROM [25].

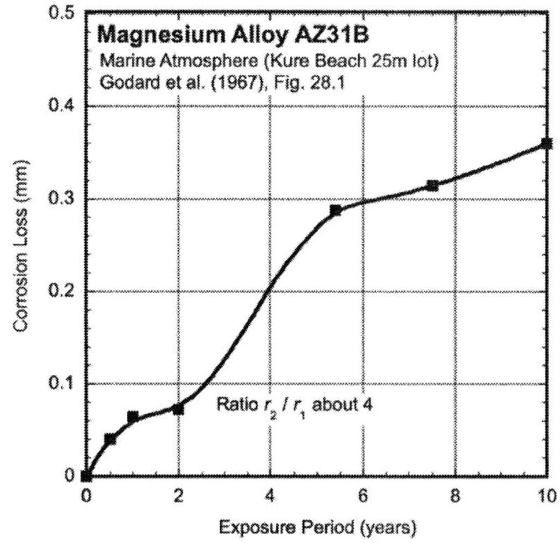

(B) DATA FROM [2].

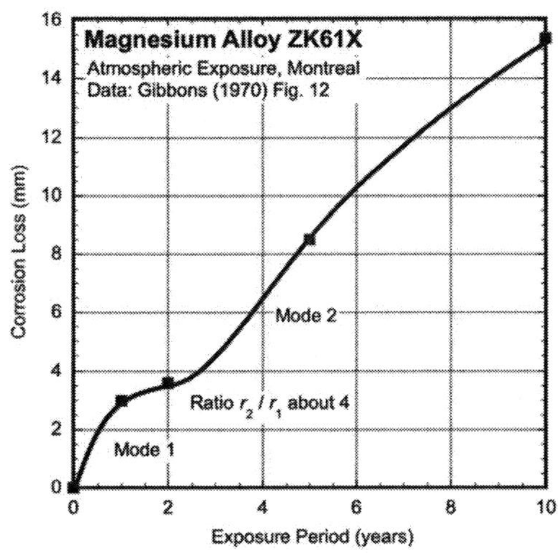

(C) DATA FROM [26].

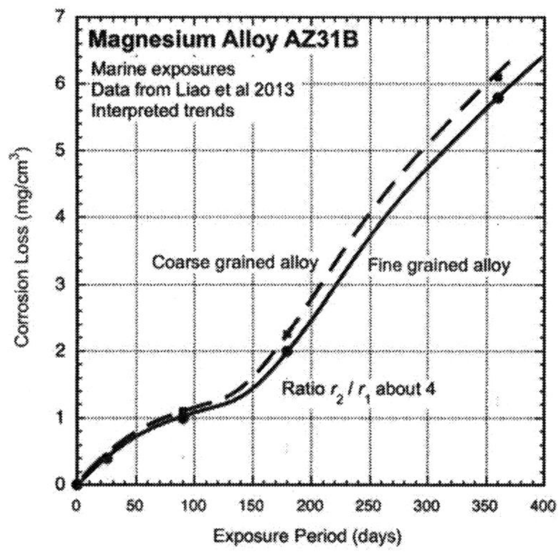

(D) DATA FROM [27].

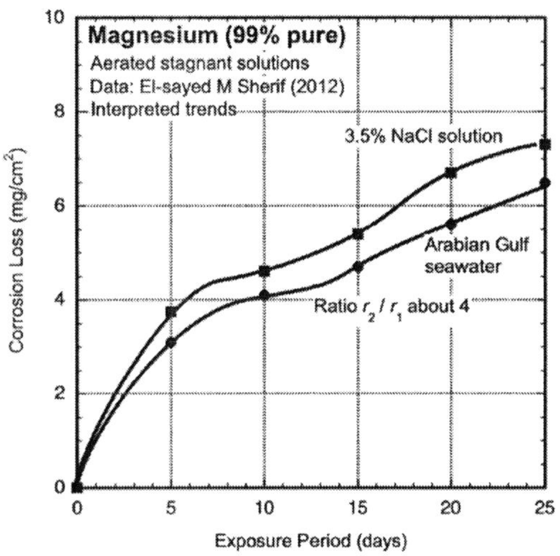

(E) DATA FROM [22].

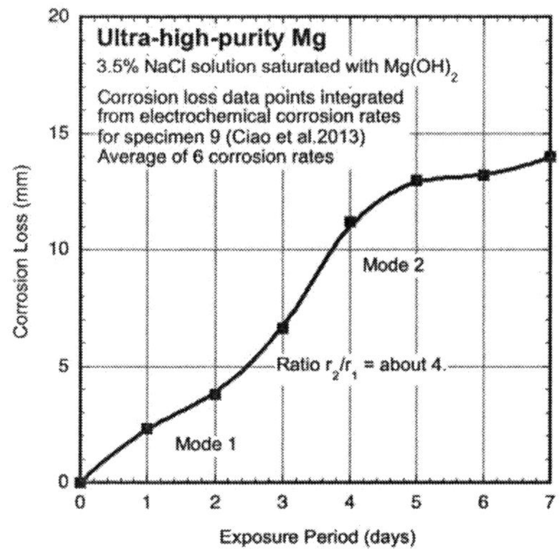

(F) DATA FROM [28].

FIGURE 2. EXAMPLES OF CORROSION MASS LOSS AS A FUNCTION OF EXPOSURE PERIOD FOR A NUMBER OF QUITE DIFFERENT MAGNESIUM ALLOYS, DIFFERENT EXPOSURE PERIODS AND DIFFERENT EXPOSURE ZONES.

In Fig. 2 the naming of the alloys employs the accepted convention that A = aluminium, Z = zinc and K = potassium, listed in order of composition and followed by the (rounded) value of that composition, e.g. AZ31B denotes magnesium alloy with about 3%(wt) aluminium, 1%(wt) zinc. In all cases the balance is magnesium. Any subsequent letters relate to minor components or physical treatment such as work hardening. The latter are ignored herein.

Each of the trends fitted to the data shown in Figs. 2(A-F) can be considered consistent with the overall notion that the trends are bimodal in character. For each of the data sets shown in Fig. 2 and the trend curves drawn through them, the slopes r_1 and r_2 can be estimated directly from the scaled plots. In all cases the ratio r_2 / r_1 is close to 4. This is additional support for the notion that the bimodal model applies for those magnesium alloys, irrespective of precise exposure conditions or alloy composition.

Discussion

The trends in Fig. 2 show that in most cases Mode 2 with HER as the dominant cathodic reaction with hydrogen evolution, develops relatively quickly, and for high purity magnesium within a few days. Apart from experimental and plotting accuracy, variations from the theoretical ideal $r_2 / r_1 = 4$ are likely to arise from departures from the theoretical assumption that the corrosion product layer (i.e. the diffusion barrier for oxygen and hydrogen respectively), does not change as corrosion progresses through the transition phase (Fig. 1). Evidently, an increase in corrosion product through the transition phase will permit only a slower rate of hydrogen diffusion and thus reduce the corrosion rate r_2 (Fig. 1). In turn this will somewhat reduce r_2 / r_1.

The above results and interpretations lead to the conclusion that the most significant difference between the development of corrosion for magnesium alloys and steels and at least some other alloys is the rate at which the respective cathodic reactions can occur and thus the time to the start of the transition zone (Fig. 1). In turn, this depends on the diffusivity of the corrosion products in the corrosion product layer, particularly in the transition zone, rather than directly on the alloy(s) themselves. Evidently, the time period to the transition zone is very short (days) for ultra-high purity magnesium (Fig. 2(F)). Conversely, alloying tends to extend that zone.

Much research on magnesium alloy corrosion has been carried out using electrochemical techniques. Their use extends back also to the observations by Beetz [6] who reported, for his electrochemical experiments, copious amounts of hydrogen bubbles from magnesium electrodes in magnesium sulphate solution. This occurred immediately after application of a potential across the electrodes. It also was noted as occurring for what are normally considered anodes. He termed this latter phenomenon the 'negative difference effect' (NDE). This term is still in use but proposed as more accurately termed 'anomalous hydrogen evolution' (AHE) [5]. The important difference between the present results for exposures in natural (open-circuit) conditions and those of the electrochemical observations lies in the use, in electrochemical tests, of a (large) cathode external to the magnesium alloy being tested. In effect this ensures, through the applied potential, sufficiently low local pH conditions for the immediate satisfaction of the thermodynamic conditions for the occurrence of the hydrogen evolution reaction. In other words, Mode 1 in Fig. 1, necessary to set up the conditions to allow corrosion in Mode 2, is essentially bypassed. Interestingly, once corrosion occurs under HER, irrespective of whether this was through the Mode 1 to Mode 2 process or directly electrochemically, the rate of loss of magnesium is closely correlated with the total release of hydrogen, including that attributed to being released from the anode (the anomalous part - i.e. that which might be attributed to AHE or NDE). This correlation underlies many estimates of corrosion loss through measurement of hydrogen evolution over extended periods of exposure, for example after 14 days [7].

The most likely reason for electrochemical test results under-estimating the (average) rate of corrosion loss of magnesium alloys is that electrochemical tests cannot deal well with localized corrosion including pitting [29]. Most electrochemical tests measure the rate of electron flow through the external circuit between the working electrode and the auxiliary electrode. However, pitting (and localized corrosion), involves local electron flows from the pit (anode) to the immediately neighbouring cathodic zone around the pit, there forming corrosion products. The corresponding electron flow is not measured in the external circuit, thus underestimating total metal loss.

It follows that pitting, as demonstrated (by others) to occur in the corrosion of magnesium alloys and as occurs also in the corrosion of alloys such as steels, is an important part of the overall corrosion process, even if not always isolated from 'general' corrosion or corrosion-loss obtained from mass-loss measurements.

As noted above, magnesium alloys display considerable pitting already soon after first exposure. Often that pitting consists of multiple pits in close proximity, with some that appear to have amalgamated, forming somewhat irregular plateaus in which, after further exposure, new pitting appears to have occurred. Similar observations have been made for pitting of steels [19, 13].

These observations can be interpreted as showing that some degree of anodic dissolution with the HER as the cathodic reaction can occur at the same time as dominant oxygen reduction reaction primarily responsible for Mode 1. With continued exposure the amount of pitting, and thus presumably hydrogen evolution, will increase and through lowering the pH at the metal surface permit the transition to Mode 2. As noted, corrosion during the latter period shows extensive pitting. Given these similar observations for steels and for magnesium alloys, it is reasonable to propose that a similar corrosion scenario holds for both, albeit with corrosion for magnesium alloys generally occurring at a faster rate, most likely owing to its higher electrochemical potential.

The present exposition has given a revised view of the progression of corrosion of magnesium alloys, based on the currently available experimental evidence for both short and long-term corrosion in natural environments. It remains to be verified by further experimental evidence, including that which takes into consideration the proposed mechanisms. This includes the role of corrosion pitting in magnesium alloys, a matter arising directly from invoking the bi-modal model. The interpretations given herein are likely to have an influence on practical issues, particularly for the early stages of exposure. That the ORR is involved soon after initiation of corrosion can be expected to have an influence on the selection and use of corrosion inhibitors since most of these reduce the availability of oxygen for the ORR and are intended to apply from very first exposure. There also are potential influences for protective coatings and possibly for cathodic protection.

More generally, the interpretation that bimodal behaviour is observed for magnesium alloys as well as a number of other alloys suggests an enhanced degree of unity in the literature on metallic corrosion.

CONCLUSION

1. The corrosion loss trends for magnesium alloys of various types show, irrespective of exposure environment, alloy surface condition, etc., bimodal behaviour, similar to that previously observed for steels, and some other alloys.

2. Compared to steels, the transition from Mode 1 to Mode 2 generally occurs much earlier in time for magnesium alloys. This likely reflects the higher electrochemical potential of magnesium alloys.

3. The corrosion rate immediately after the transition period (i.e. the start of Mode 2) is close to 4 times that near the start of the transition period (i.e. the end of Mode 1), similar to what has been observed previously for steels. This was interpreted to reflect the increase in diffusion rate between hydrogen and oxygen as the corrosion-rate-controlling cathodic processes for Modes 2 and 1 respectively.

4. Measurement of hydrogen evolution is consistent with Mode 2 and thus provides reasonable estimates of corrosion rate for the longer-term part of exposures, consistent with previously reported experimental observations.

4. Measurements of corrosion current in electrochemical experiments do not reflect the metal loss due to pitting corrosion and its development into localized corrosion. As a result, electrochemical measurements tend to under-estimate the total amount of corrosion loss.

ACKNOWLEDGEMENTS

Useful critical comments were made by Professors Nick Birbilis (Deakin University) and Andrei Atrens (University of Queensland) on a much earlier version of part of this manuscript. Review comments during the review processes of the preceding papers [14, 15] also are appreciated. No external funding was provided for this research.

REFERENCES

1. U.R. Evans, *The Corrosion and Oxidation of Metals: Scientific Principles and Practical Applications*, (London, Edward Arnold, 1960).

2. H.P. Godard, W.B. Jepson, M.R. Bothwell, R.L. Kane, *The corrosion of light metals*, (New York, John Wiley, 1967).

3. I.J. Polmear, Magnesium alloys and applications, *Materials Science and Technology*, 10 (1994) (1) 1-16. doi:10.1179/mst.1994.10.1.1.

4. M. Jonsson, D. Persson, C. Leygraf, Atmospheric corrosion of field-exposed magnesium alloy AZ91D, *Corros. Sci.*, 50 (2008) 1406-1413.

5 M. Esmaily., J.E. Svensson, S. Fajardo, N. Birbilis, G.S. Frankel, S. Virtanen, R. Arrabel, S. Thomass, L.G. Johansson, Fundamentals and advances in magnesium alloy corrosion, *Progress in Materials Science*, 89 (2017) 92-193.

6 W. Beetz, On the development of hydrogen from the anode. *Phil Mag.* 32 (1866) 269-278.

7 A. Atrens, Zhiming Shi, Syeda U. Mehreen, Sean Johnston, Guang-Ling Song, Xianhua Chen, Fusheng Pan, Review of Mg alloy corrosion rates, *J. Magn. Alloys*, 8 (2020) 989-998.

8 . Atrens, Guang-Ling Song, Fuyong Cao, Zhiming Shi, Patrick K. Bowen, Advances in Mg corrosion and research suggestions, *J. Magn. Alloys*, 1 (2013) 177-200.

9 D.A. Jones, *Principles and Prevention of Corrosion*, 2nd Ed., (Upper Saddle River, NJ, Prentice-Hall, 1996).

10. M. Strebl, M. Bruns, S. and Virtanen, Repirometric in situ methods for real-time monitoring of corrosion rates: Part 1. Atmospheric corrosion, *J. Electrochem. Soc.* 167 (2020) 021510.

11. Cheng Wang, Wen Xu, Daniel Hoche, Mikhail L. Zheludkevich, Sviatlana and V. Lamaka, Exploring the contribution of oxygen reduction reaction to Mg corrosion by modeling assisted local analysis, *J Magnesium and Alloys*, 11 (2023) 100-109.

12 R.E. Melchers, A review of trends for corrosion loss and pit depth in longer-term exposures, *Corros. Mater. Degrad.*, 1 (2018) 4; DOI:10.3390/cmd1010004.

13 R.E. Melchers and R. Jeffrey, The transition from short- to long-term marine corrosion of carbon steels, *Corrosion*, 78 (2022) (5) 415-436.

14 R.E. Melchers, Trends in longer-term corrosion loss of magnesium alloys, *Crystals*, 14 (2024) 172 *https://doi.org/10.3390/cryst14020172*

15 R.E. Melchers, Bimodal trending in corrosion loss of magnesium alloys, *Corrosion,* 80(2024) (9) 899-909. doi.org/10.5006/4548

16 R.E. Melchers, Mathematical modelling of the diffusion controlled phase in marine immersion corrosion of mild steel, *Corros. Sci.*, 45(2003) (5) 923-940.

17 H.W. Pickering, Important early developments and current understanding of the IR mechanism of localized corrosion, *J. Electro. Soc.*, 150 (2003) (5) K1-K13.

18 P.C. Pistorius, and G.T. Burstein, Metastable Pitting Corrosion of Stainless-Steel and the Transition to Stability. *Philosophical Trans. Royal Soc. Lond. Series A*, 341(1992) 531-559.

19 A.D. Mercer and E.A. Lumbard, Corrosion of mild steel in water, *British Corrosion Journal* 30 (1995) 43-55.

20 M. Jonsson, D. Persson, and C. Leygraf, Atmospheric corrosion of field-exposed magnesium alloy AZ91D, *Corros. Sci.*, 50 (2008) 1406-1413. See Figs 5-. 7.

21 Li-jing Yang, Yan-fang Li, Ying-hui Wei, Li0feng Hou, Ying-gang Li, and Yun Tian, Atmospheric corrosion of field-exposed AZ91D Mg alloys in a polluted environment, *Corros. Sci.*, 52 (2010) 2188-2196, See Fig. 4.

22. El-Sayed M. Sherif, Corrosion behavior of magnesium in naturally aerated stagnant seawater and 3.6% sodium chloride solutions, *Int. J. Electrochem. Sci.,* 7 (2012) 4235-4249.

23. Yong-gong Li, Ying-hui Wei, Li-feng Hou, Feng-ju Han Atmospheric corrosion of AM60 Mg alloys in an industrial environment, *Corros. Sci.*, 69 (2013) 67-76, See Fig 4.

24 R.E. Melchers and R. Jeffrey R., The transition from short- to long-term marine corrosion of carbon steels: 2. Parameterization and modeling, *Corrosion*, 78 (2022) (5) 427-436.

25 C.R. Southwell C.W. Hummer, Corrosion of metals in tropical environments - Aluminum and magnesium, *Mat. Prot.*, 4 (1965) (Dec.) 30-35.

26 E.V. Gibbons, Corrosion behaviour of the major architectural and structural metals in Canadian Atmospheres: Summary of ten-year results of Group 1, Technical Paper No. 328, Division of Building Research, National Research Council of Canada, Ottawa, Oct. 1970, NRCC 11630. https://doi.org/10.4224/20374352

27 Jinsun Liao, Makoto Hotta, Shin-ichi Motoda, Tadashi Shinohara, Atmospheric corrosion of two field-exposed AZ31B magnesium alloys with different grain size, *Corros. Sci.*, 71 (2013) 53-61.

28 Fuyong Cao, Zhiming Shi, Joelle Hofstetter, Peter J. Uggowitzer, Gangling Song, Ming Liu, Liu, Andrej Atrens, Corrosion of ultra-high-purity Mg in 3.5% NaCl solution saturated with $Mg(OH)_2$, *Corros. Sci.* 75 (2013) 78-99.

29. R.G. Kelly, J.R. Scully, D.W. Shoesmith and R.G. Buchheit, Electrochemical Techniques in Corrosion Science and Engineering, New York, Marcel Dekker, 2003.

AUTHOR DETAILS

Robert E. Melchers, Professor of Civil Engineering (Emeritus), The University of Newcastle, Australia, BE, MEngSC (Monash), PhD (Cambridge, UK), Fellow, Australian Academy of Technology and Engineering. Honorary Fellow, The Institution of Engineers Australia.

Awards include 2009 ACA Corrosion Medal, 2012 Jin S Chung Award (ISOPE), 2013 John Connell Gold Medal (The Institution of Engineers, Australia), 2014 Eminent Speaker, Structural College, The Institution of Engineers, Australia, 2006 and 2018 Guy Bengough Award (IOM3).

CORROSION & PREVENTION
CAIRNS 2024

THE AUSTRALASIAN CORROSION ASSOCIATION INC.

OBSERVATIONS OF MULTI-STEP PITTING CORROSION OF STEELS IN DE-IONIZED WATER AND SEAWATER

R. E. Melchers[1,*] R. Jeffrey[1]

[1]*Centre for Infrastructure Performance and Reliability, The University of Newcastle, Australia 2308*

**rob.melchers@newcastle.edu.au*

ABSTRACT

Some previous observations of the corrosion of steels in seawaters have shown that they pit and that the pitting may exhibit clearly identifiable 'steps' in depth. While 'pitting' is often associated only with alloys that form a passive film (e.g. stainless steels, aluminium) there is a long and clear history of pitting initiation and early development for mild steels, not only in chlorinated waters but also in triply distilled waters. Information about how such pitting develops with extended exposure was obtained in a major, extensive testing program. Three cohorts of similar-sized steel coupons (carbon contents 0.01, 0.04 and 0.06wt%) were immersion exposed, with 2 week observation intervals, for up to 1.8 years. One cohort was exposed in de-ionized water at 70°C and others in natural seawater at Taylors Beach (20°C average water temperature and 30°C constant temperature). Optical microscope observations of the corroded surfaces after cleaning showed, for the longer exposure periods, a small number of deep pits with multi-step characteristics. but the majority with simple geometry. The reasons for the step-wise behaviour are explored by invoking the competition between cathodic processes inside and outside a pit and the development and periodic removal of corrosion product over the oit mouth. Cross-reference is made to literature observations, predominantly for stainless steels. The overall outcomes will enhance the understanding of pitting for mild steels, relevant for water, oil and gas pipelines.

Keywords: steel, seawater, de-ionized water, pitting, progression.

INTRODUCTION

A previous paper [1] gave multiple images of the surface topography of steel samples as they were corroding in seawater and separately in de-ionized water. Surfaces were observed at 14 and 28 day exposures after standard non-invasive cleaning over a period of up to 511 days. The steels were low carbon (C content between 0.01 and 0.06 wt%) and all showed quite similar development of pitting corrosion. While the primary aim of the tests was to confirm that the overall development of corrosion followed a 'bi-modal' pattern with exposure period [2], irrespective of type of water or water temperature, the depth of pitting in seawater also tended to follow a bi-modal trend, it was noted that the depth of pitting began to show a step-wise development within weeks of initial exposure. Two example of corroded surface topography show evidence of step-wise morphology of the pit depth are shown in Fig. 1. It was earlier noted [2] that such step-wise development could be read into data reported by Mercer and Lumbard [3] for pit depth of mild steel exposed to triply distilled water. Some evidence of renewed pitting within earlier pits was reported by Wallwork & Harris [4]; Szauer and Jakobs [5] and Jeffrey [6] but not the potential for sequential step-wide pit development. For stainless steels observations have been reported recently [7, 8] that also can be interpreted as showing pitting within pits. There are similar observations for aluminium alloys [9]. However, no explanation was offered for these features or the possibility raised that there could be repetition in such pitting.

Some of the clearest evidence of multiple steps in the development of pit depth has emerged from the large amounts of pit depth data sourced from extreme value analysis of pipeline corrosion pigging after the data were filtered to remove all but the deepest pits in equal sections of pipeline. The steps in pit depth can be seen in the discontinuous pattern for those extreme value pit depths when plotted as extreme values (e.g. on Gumbel plots) [10, 11].

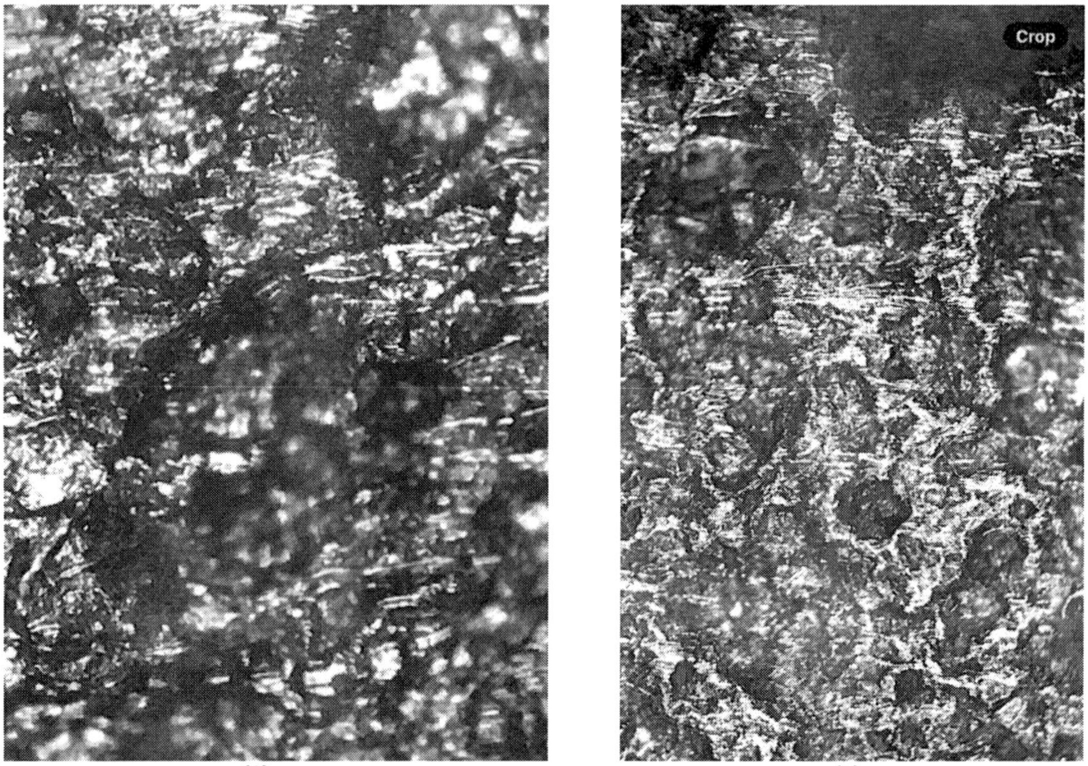

(A) (B)

FIGURE 1. OPTICAL MICROSCOPE IMAGES AS EXAMPLES OF PITTING TOPOLOGY THAT CAN BE INTERPRETED AS STEP-WISE IN DEPTH DEVELOPMENT (A) IN SEAWATER AND (B) IN DE-IONIZED WATER AFTER 28 DAYS IMMERSION. SCALE: WIDTH OF IMAGES = 0.9 MM.

Although there are some differences of opinion as to whether mild steels (i.e. alloys without a strong passivating film) pit or whether they show 'localized corrosion' [12], herein the classical Evans [13] view is taken that all alloys show pitting. Examples can be seen in Fig. 2 in Mercer & Lumbard [3] for steels in fresh- and in seawater and also in Butler et al. [14] for high-purity iron in a neutral chloride solutions. These were all cases of natural exposure conditions (open circuit), with rusts able to form freely under wet oxygenated conditions (immersed. atmospheric) and with the cathodic region(s) able to form in close proximity to the anodic (pit) zone.

The next section provides an overview of the experiments that were conducted for 3 different low carbon steels (0.01, 0.04 and 0.06 wt%C) in three different immersion exposure conditions - in seawater (at 20°C av.), in seawater at a constant temperature of 30°C (nom.) and in de-ionized water at constant 70°C (nom.). The latter exposure environment is similar to the triply distilled water experiments and results reported by Mercer and Lumbard [3] for low carbon steel. The following section summarizes the observations of the pit depth measurements and also compares the observed topography with the earlier observations, noting that the present observations extend over a much longer period of observation than those earlier works. Comparisons are made of the increments in pit depth for the three low carbon steels and for the three water environments used in the experiments. Also, the pit depth data for the carbon steels and the de-ionized immersion environment are compared with the bi-modal model trending for pit depth (and mass loss) that was earlier apparent for steel in seawater [1] and with pit depth data for steel in various waters [15]. The Discussion section considers the possible reasons for the incremental nature of the deeper pits and why this was not the case for all observed pitting. A model is proposed to explain the processes considered involved in step-wise pit depth development, including the crucial role of corrosion products. Comments are made about the applicability of the model for stainless steels and other alloys.

EXPERIMENTAL OBSERVATIONS

The experiments that provided the observations of steps in pit depth development consisted of exposing sets of coupons of three low carbon steels (Table 1) to three different exposure environments [1]: (a) quiescent natural seawater at 20°C (av., range15-25°C) temperature in a fibreglass tank open to the atmosphere but sheltered, located at Taylors Beach, NSW, with seawater replenished (at very low water velocity) every 12 hours for a 10 minute interval, (b) as for (a) but with the seawater heated to 30±1°C and (c) similar but with de-ionized water heated to 70±2°C. All coupons were 50 x 100 mm in size, nominal thickness 3 mm, slight thinner for the 0.06 wt% steel and slightly thicker for the 0.01 wt% steel to permit ready identification (cf. Table 1). The coupons were individually identified using a template system of four 2-mm diam. drilled holes. All coupons were briefly exposed to dilute, inhibited hydrochloric acid (16 wt% HCl with 2% hexamine) to ensure cleaning prior to exposure without damage to the coupon surfaces, verified with the use of blank coupons. The initial mass of each coupon was obtained to 0.1 mg accuracy prior to the coupon being placed in the relevant exposure tank, at a typical depth of at least 25 mm, suspended vertically and held in place with nylon cable ties. The coupons were all immersed at the same time. Further details are available in the preceding papers [1].

TABLE 1. COMPOSITION OF STEEL USED IN THE EXPERIMENTS (BALANCE FE) (WT%)

C	Mn	S	Ni	Cr	Mo	P	Comment	Coupon Thickness (nom)
0.01	0.05	0.004	0.02	0.02	<0.01	<0.01	Almost 'pure' iron	3+ mm
0.04	0.19	0.013	0.04	0.11	0.02	<0.01	Low carbon steel	3 mm
0.06	0.20	0.008	0.01	0.01	<0.01	<0.01	Typical 'mild' steel	3- mm

Si, Al below detectable limits

Throughout the test program coupons were recovered at 28 days (nom.) for the 20 and 30°C (nom.) tanks and at 14 days (nom.) for the 70°C tank. At each recovery the coupons were immediately cleaned of gross marine growth if present and the bulk outer rusts removed using gentle mechanical (wire-brush) cleaning followed by high-pressure water blasting. For the shorter-term exposures (about 100 days) this was sufficient to reveal the bare metal surface and pitting. For longer exposure periods the more resistant inner rusts were removed using dilute inhibited hydrochloric acid, followed immediately by rinsing in clean tap water and alcohol and drying, all consistent with ASTM procedures [16]. The cleaning process was repeated where microscope observations showed it has been insufficient to reveal the full depths of pits. All coupons were weighed to estimate mass loss. The depth of the deepest pits was obtained using the focus variation method and an optical microscope (Zeiss Axio Imager A1m at 100x magnification and sometimes higher). While laborious and time-consuming, this procedure provides a direct opportunity for detailed observations of any variation in the profile of each pit with depth (cf. Fig. 1). It also allows these variations to be measured relative to the original surface (where still observable). In other cases, it permits the measurement of steps, if any, in the pit depth profile and of depth of any steps in the corrosion plateaus, as described further below. As noted previously [1], it was found possible to set the microscope focus to a given pit depth and then move the microscope platen horizontally to quickly identify pits of similar depth as the pit bottoms came into focus.

For the longer exposures little evidence typically remained of the original surface of the coupon. In these cases only relative depths of pitting and pit depth increments could be measured, in some cases with sequential steps in pit depth. Examples of these measurements are reported in the next section. It is noted here that 3D optical and digital profilers provide surface topographies that tend to be too detailed for the present purposes and for obtaining a coherent impression of the overall topography of the corroding surface. This includes the detection of remnant corrosion products within the pits.

EXPERIMENTAL RESULTS

Optical microscope observations sequential in time of the overall, gradual, development of the surface topography of 0.06 wt% steel coupons corroding for up to 511 days in 20°C (nom.) seawater were reported earlier [1]. It was noted that very similar topographical corrosion surface development was observed for 0.01wt% and 0.02wt% C-steels and also for all three steels in 30°C (nom.) seawater and, at a somewhat faster time-rate of development for the three steels in 70°C de-ionized water. Examination of all the coupon surfaces showed that pitting was extensive in extent on all coupon surfaces and that this occurred from very early exposure onwards [1]. There also was visual evidence of step-wise pit depth profiles and what could be interpreted as amalgamation of pits to form small plateaus with non-uniform, undulating bases. For the longer exposures, the topography developed into what might be termed 'valleys' and 'islands' in the landscape of the corroded surface. These observations applied to all three steels and to all three immersion environments. Herein only the step-wise aspect is considered. Tables 2-4 give typical results from the many observations of absolute pit depth and of pit depth increments, including where the original surface was no longer identifiable. Because of the volume of observations, only a small number of exposure periods are reported here.

TABLE 2. PIT DEPTHS AND INCREMENTS IN NATURAL SEAWATER AT 20°C (AV.).

	0.60wt%C		0.04wt%C		0.01wt%C	
Exposure (days) (t)	Pit depths (microns)	Typ. incr. (microns)	Pit depths (microns)	Typ. incr. (microns)	Pit depths (microns)	Typ. incr. (microns)
28	40-60; 140-5.	~ 80	21; 26; 42; 50; 100<328<354; 282<416<467; 212,328<397; 122<158<274.	~ 80, 180	61<126; 49 (many); 100(many); 160.	~ 60 - 100
56	12; 0<x<531<819; 0<243<460; 0<373<646; 480.	~ 80, 200	132(many); 209(many); 228; 620; 773		356<614<815; 183<204; 0<177<339; 482; 864	~ 150-200
84	25-40 (many); 135 (many);148;183	~ 80	67<186; 89; 115	~ 90	140<342	~ 100 - 200
143	60-70 (many); 320-357 (many); 396; 462; 556; 639; 719; 993.	~ 70, 250	300; 326; 513-620; 654; 778; 937.	~ 200	299; 368 (most); 608; 735; 740; 790 (several); 1189	~ 70 - 140
256	307; 374; 475; 600; 747; 965; 354<570; 300<520; 506<644; 469<897; 395<738; 1068.	~ 350	376<571; 464<708; 319<500<731; 822-900.	250	613; 307<462<674; 341<700; 674; 460); 474; 611-650.	~ 350
364	-		0<x<779<1193; 469; 740.	300-350	140 (many); 629<1085; 300-310(many); 740(many); 593(many)	~ 400
446	257<438<538; 0<264<483<942; 0<203<349<673.	350	390<700; 241<657<846	350	306<672<919<1153 ; 310-330 (many)	390
511	412<714; 670<x<1026. NOS	300	-		327<521<752<973< 1166 269<390<780; 240<350; 603; 889; 1225; 1928; 1933; 2167.	90; 200-240

NOS = original reference surface not visible for pit depth measurement

TABLE 3. PIT DEPTHS AND INCREMENTS IN 30°C (NOM) SEAWATER.

Steel	0.06wt%C		0.04wt%C		0.01wt%C	
Exposure (days) (t)	Pit depths (microns)	Typ. incr. (microns)	Pit depths (microns)	Typ. incr. (microns)	Pit depths (microns)	Typ. incr. (microns)
28	40-60		30-70		20; 40; 156; 95<194	20; ~ 100
56	~ 50		65; 305-400	90	320	
84	20-30; 90-100		220; 230; 0<160<480	200	100-117; 232; 117<380<750<1185	250 - ~ 350
168	178; 223; 0<244; 0<133<458<768< 801	300	303; 365; 374	330	304; 333; 544	~ 300
256	344; 0<x< 425<726 0<280<664; 264.	300	348; 365; 401; 410	90, 320 (many)	NOS; <487	320
389	x<326<572<683	250-300	40; 50; 93<482	330 (many);	NOS	100, 340, 390

NOS = original reference surface not visible for pit depth measurement

TABLE 4. PIT DEPTHS AND INCREMENTS IN 70°C (NOM) DE-IONIZED WATER.

	0.06wt%C		0.04wt%C		0.01wt%C	
Exposure (days) (t)	Pit depths (microns)	Typ. incr. (microns)	Pit depths (microns)	Typ. incr. (microns)	Pit depths (microns)	Typ. incr. (microns)
14	387<769 <1051	350	380 (typ.)	308	80<430<800	350
28	340; 430	350	343 (typ.)		80<435<784	350
70	335; 340, 420; 733	~ 300	~ 394 (many)	~ 350	190-210; 300; 360	300
84	260-380	~320	386	~ 350	92<261<621; 342; 371; 380; 489; 585.	~ 360
154	248<564<828 334 (av)	~ 300	30(many); 42<243; 367	~ 330	102<252<500; 390<645; 340 (av.)	250
210	100<172	75 (est.),	329; 373	~ 350	212<369; 362<600	~ 380
238	20-40	50-80	371 (few)	~ 350	20-40; 342 (some)	~ 300
350	NOS	314; 321; 330 (many)	NOS	300; 321; 383 (many)	NOS	240-400 (many)
522	NOS	~ 360	NOS	300 (est.)	NOS	320; ~ 350

NOS = original reference surface not visible for pit depth measurement

INTERPRETATION AND COMPARISON OF RESULTS

The topographies shown in Fig. 1 and the data in Tables 2-4 show that despite the exposure for the 70°C environment being in de-ionized water rather than seawater, pitting did occur, and that it developed almost immediately from first exposure, an observation entirely consistent with those of Mercer and Lumbard [3] for steel in 70°C triply distilled water. Similarly, pitting occurred almost immediately after first exposure for the steels in seawater and that the higher temperature played a role in the rate of development of corrosion and of overall pit depth. Both increased with temperature, being faster in time for the 70°C environment than for the other two lower temperature environments. These observations run counter to the notion of pitting requiring an alloy with a passive film and also the classical notion, much questioned [17], and not consistent with the model for pitting developed by Wranglen [18] that chlorides are a necessary part of the pitting process. However, the present results also show that it is possible for some pits at least to have a repetition in their development of depth. This is consistent with earlier observations for pits in water-injection pipelines [10, 11].

As shown earlier for the three steels in 20°C seawater [1], the present results show that pit depth development has some degree of consistency with bi-modal behaviour (Fig. 2) but also that there are significant deviations from that trend, even at shorter periods of exposure and that, overall, the average pit depth shows bi-modal trending although at a faster rate as water temperature increases. It is also clear that pit depth development in non-quiescent waters results in considerably deeper pitting. Irrespective of these influences, however, Tables 2-4 show that the steps in pit depth development are very similar at similar relative periods of exposure.

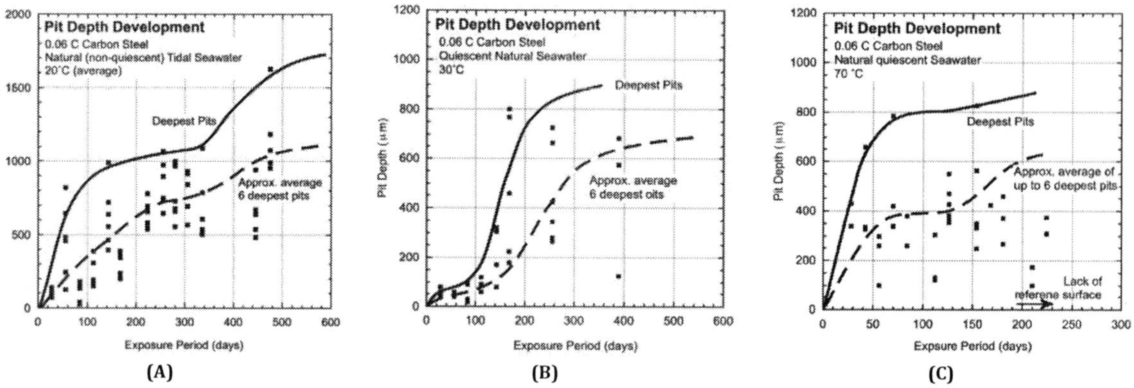

FIGURE 2. PIT DEPTH DEVELOPMENT TRENDS (MAXIMUM PIT DEPTH AND AVERAGE OF 6 DEEPEST PIT DEPTHS FOR 0.06%WTC CARBON STEEL IN (A) 20°C (AV.) NON-QUIESCENT (TIDAL) NATURAL SEAWATER, (B) IN 30°C QUIESCENT NATURAL SEAWATER AND (C) IN 70¬C DE-IONIZED QUIESCENT WATER. NOTE DIFFERENCES IN SCALES.

The data in Tables 2-4 can be interpreted as showing that for longer-term exposures the increment in pit depth typically is around 300 μm while for shorter exposure periods it is around 70 μm and although the data is sparse, the it depth increment is likely even less for very short exposure periods. These estimates are compatible with those obtained from extreme value plots of the variation of pit depth with probability of occurrence [10, 11].

MODEL FOR STEP-WISE PIT MECHANISM

A model for the development of repeated step-wise pitting will now be outlined. Initiation is assumed to follow the conventional notions of being caused by an inclusion or other means by which a local potential difference permits what is essentially differential aeration or galvanic action, thereby initiating the corrosion process. In the following interest lies with the development or progression of a pit once initiated. It is useful to commence with the classical model proposed by Greene and Fontana [19] model for open pits in which the oxygen reduction reaction governs until sufficient rusts have built-up to provide the essentially anaerobic model of Wranglen [18] for pit development governed by the hydrogen evolution cathodic reaction for much of the inner surfaces of the pit. This model was considered applicable to mild and low alloy steels as well as stainless steels (and aluminium [20]. As usual, pit initiation and development implies the release of ferrous ions (Fe^{2+}) by the anodic reaction within the pit:

$$Fe \rightarrow Fe^{2+} + 2e^{-} \quad (1)$$

Herein this reaction is assumed, as usual, to occur, initially, throughout the interior of the pit and does so under to electrochemical potential E between the ferrous component of the bare metal (Fe) within the pit and the first-formed rusts (corrosion products) $Fe(OH)_2$ exterior to the pit (Fig. 3(A)). Usually it is assumed that pits progress inwards relatively directly but closer examination of pit geometries obtained under open circuit or field conditions often show broadening of pits and in the case of stainless steels (and some aluminium alloys) what appears to be under-cutting [21]. This broadening effect can be considered a consequence of the electrochemical potential E since it applies to all metal surfaces within the pit, thereby permitting anodic reactivity everywhere within the pit, including 'sideways' as well as into the pit, causing what has been termed 'flask-like' shape of some pits [22].

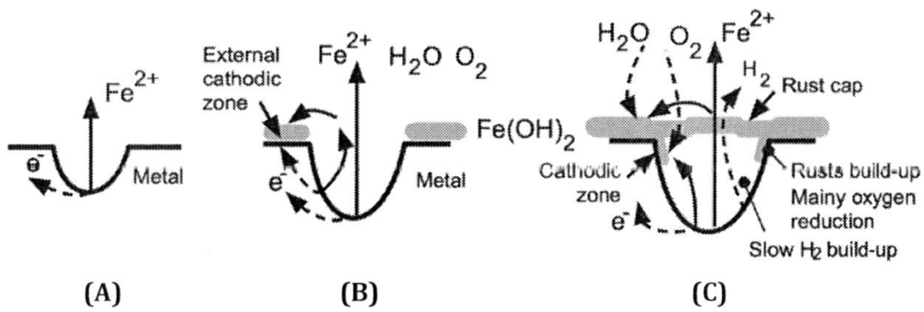

FIGURE 3. SCHEMATIC REPRESENTATION OF (A) INITIATION AND VERY EARLY GROWTH OF A PIT (B) EARLY PIT GROWTH UNDER OXIDATION REDUCTION AND (C) DEVELOPMENT OF RUSTS AND RUST CAP OVER THE PIT MOUTH.

As is well-known, the rate of pit depth penetration reduces as the pit deepens, attributed to the increasing difficulty of the ferrous ions to travel out of the pit as the pit deepens and as the layer of corrosion product over the pit mouth [21, 22]. The conventional assumption is that the ferrous ions that leave the pit through the pit mouth travel (mainly by diffusion) to external cathodic zones, where, under the action of electron availability through the metal and the presence of water and oxygen (for immersion corrosion as dissolved in water) they form hydroxide ions according to the cathodic oxygen reduction reaction (ORR):

$$O_2 + 2H_2O + 4e^- \rightarrow 4OH^- \quad (2)$$

With the combination of the ferrous ions that have migrated from the pit mouth the hydroxide ions produce the external corrosion products) $Fe(OH)_2$ usually observed on the exterior of the steel (Fig. 3(B))

$$2Fe + O_2 + H_2O \rightarrow 2Fe(OH)_2 \quad (3)$$

In the case of the absence of oxygen, and with sufficiently low pH, the cathodic reaction is the hydrogen evolution reaction (HER):

$$2H_2O + 2e^- \rightarrow H_2\uparrow + 2OH^- \quad (4)$$

but the overall corrosion reaction is still (Eqn. 3) but now with production also of gaseous hydrogen. The other reaction products, namely the $Fe(OH)_2$ remain the same.

According to Wranglen's model [18], the $Fe(OH)_2$ rust products, which are permeable, build-up in the regions immediately surrounding the pit mouth. This concept extends back to early observations summarized by Evans [13] and also has support in numerical simulation [23]. It includes the notion that these 'porous' rust products eventually cover ('cap') the pit mouth (Fig. 3(C))

In the classical model for pitting corrosion it is assumed that a annular region develops around an isolated pit and that this acts as a cathode, where, according to Eqn. (3), the positively charged ferrous ions interact with water (and oxygen if present) to deposit rusts in the form of $Fe(OH)_2$. With continued anodic activity this cathodic zone has been observed to increase in spatial extent and in thickness of the corrosion product deposited on it. However, this is not the only possibility for cathodic reactivity for pitting. It has been postulated, supported by mathematical modelling and experimental observations, that at least some of the ferrous ions transfer to cathodic zones located some distance away from the region immediately around the pit. This mechanism has been used to make estimates for the depth of pitting likely to occur for stainless steels, on the basis that for atmospheric corrosion at least this cathodic mechanism out-performs the cathodic zone around the pit [24, 25]. A similar approach has been employed for pitting of aluminium [26]. The theory has been developed mainly for pitting of stainless steels, for which isolated pitting is more common than for mild and low alloy steels. For the latter it is more usual for pitting to occur as 'fields' of pits [2]. The important point is that the mechanisms for pitting and the cathodic zones to support such pitting are essentially the same for stainless steels and for mild and low alloys steels [27]. In both cases a sufficiently large cathodic zone (or zones) is required - the precise shape of the cathodic zone(s) is not relevant [25]. This also means that the cathodic area could be somewhat removed from the pit (or multiple pits), the only effect being some (relatively small) 'ohmic drop' equivalent to the resistance suffered by the ferrous ions moving from the anodic zone (pit or pits) to the cathodic zones. For wet atmospheric conditions [25] and for immersion conditions [22] this has been shown to be the rate-limiting step for the rate of progression of pitting. The other potential rate-limiting steps are the rate of metal dissolution in the anodic process, the outward diffusion of ferrous ions from the pit through the pit mouth and the rate of the cathodic reaction itself. Usually these are all considered to be faster than the ionic transport process [22, 24]. As noted further below, these interpretations are all based on observations from electrochemical experiments.

While the exterior cathodic region(s) are recognized as the main contributors to pitting development [28] and have received most attention in terms of modelling to predict the depth of pitting, less attention has been given to the role of the cathodic region within the pits themselves.

After an inital period when oxygen becomes depleted within the interior volume of the pit oxidation of Fe by water to produce $FeOH^-$ and H^+ ions becomes the governing cathodic reaction [18]. The H^+ ions are reduced to gaseous hydrogen at the metal surface through the availability of electrons that have transmitted as an anodic current from the anodic zones (i.e. through the hydrogen evolution reaction, Eqn. 3) (Fig. 3(C)). Evidently, the $FeOH^-$ ions can oxidize to $Fe(OH)_2$ that then can deposit on the interior surface of the pit (including on any rusts deposited earlier). This deposition can occur only if those zones of deposition are cathodic and can occur further inside the pits as such cathodic zones develop. Thus, it has been observed that the build-up the rust-covered cathodic zone starts from the pit mouth [29] and slowly develops and builds-up towards the pit tip (Fig. 3(C)), consistent with theory and simulation [23]. The internal rust build-up also will tend to constrict the size of the opening at the pit mouth and eventually may cause the pit mouth to be covered.

As oxygen becomes less radily available within the pit, the HER becomes involved and thus the continued production of ferrous ions from the anodic areas within the pit will generate gaseous hydrogen. However, the build-up of hydrogen within the pit is tempered by its diffusion outwards through the rusts (if any) over the pit mouth [18] (Fig. 3(C), 4(A)), and may cause the occasional bursting or perforation of such rusts. This has been observed by various investigators [30, 31] (Fig. 4(B)).

With these preliminaries the proposed mechanism for the generation of step-wise pit depth development is as follows. It relies on two key aspects (a) the competitive process between the cathodic reaction exterior to the pit and the simultaneous cathodic reaction within the pit with associated generation of rusts within the pits, and (b) sufficient generation of gaseous hydrogen within the pit to cause loss or breaching of the corrosion products (pit cover) over the pit, with release of hydrogen. Depending on the rate of corrosion inside the pit and release of hydrogen and ferrous ions through the pit mouth, the deposition of rusts inside the pits may not completely envelop the pit interior and may leave, at the tip of the pit (Fig. 4(B)) a small anodic area for the development of new pitting (Fig. 4(C). It is this possibility that is responsible for the step-wise pitting behaviour. The sequence of events is illustrated in Fig. 4.

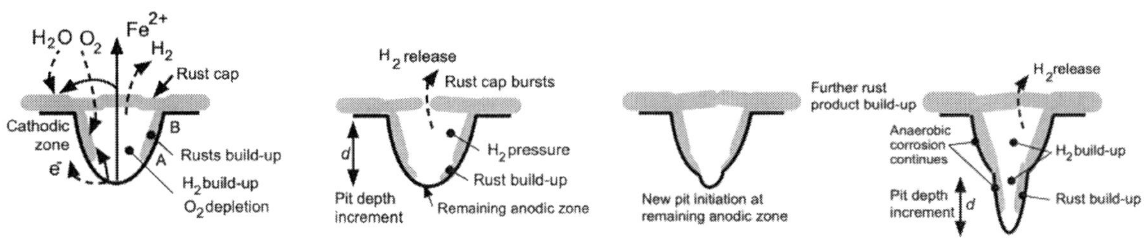

FIGURE 4. SEQUENCE OF STAGES IN THE DEVELOPMENT OF STEP-WISE CORROSION PITTING.

In summary, for the on-going process of pitting there will be ferrous ion loss at the anodic zones within the pit (Fig. 3 (B,C)), migration of ferrous ions to the exterior environment under the potential *E* and also ferrous ions that form rusts within the pit on its interior surfaces. In this process the $FeOH^+$ ions will form within the pit cavity, increase in concentration and when the solubility limit is exceeded deposit on the pit walls as $Fe(OH)_2$. Because of the pH gradient this occurs first near the pit mouth, both externally [29] and internally [32] and progressively moves further towards the pit tip, all the while building-up the corrosion products. The build-up of corrosion products over the pit mouth to form a permeable cover appears to have no effect on the development of pit growth [28]. Further, it should be clear that the process of rust deposition within the pit and outside does not directly affect the corresponding build-up of hydrogen and therefore the bursting of any rust covering over pit mouths. These pits, therefore, can show conventional pit cover perforation or bursting and stagnation of pit growth, as has long been observed for pitting of stainless steels [27, 22, 33]. In the past this slowing process has been described as 'passivation' of the pit process.

Provided the corrosion products from the interior cathodic reaction do not seal all of the interior of the pit, some bare metal remains as an anode. With sufficient cathodic reaction within the pit, hydrogen production can be sufficient to burst or perforate the corrosion products covering the pits [28]. Experimental observations for pitting of stainless steels, have reported that for pits that had passivated or nearly so, when the (perhaps 'lacy') cover over the pit mouth was removed, new pitting corrosion initiation of bare metal was immediately observed in locations near the pit tip. Similar observations have been reported by others [32]. In the context described above it can now be seen that the removal of the pit rust cover exposes the interior of the pit to the (almost) full potential E in the same way as if the already formed pit did not exist - it can be viewed as merely forming a 'tunnel' for ion transfer and for transfer of electrochemical potential (Fig. 4(D). The comment 'almost' is used here since some loss of potential can be expected through the mouth of the pit and through the already passivated region near the pit mouth.

It should be clear that the above situation involves a balance between the cathodic reaction within the pit, the corresponding degree of rust cover within the pit, the amount of hydrogen generated and the rate of loss through the porous pit cover and the presence of a cathodic reaction outside the pit. The latter must have sufficient potential to 'invade' the pit once the cover has been lost or damaged and to permit new initiation of pitting at the tip of the previous but now occluded pit (Fig. 4(D). This implies that the external cathodic region is sufficiently large. The experimental evidence referred to above clearly shows that this is the case for mild steels in seawater, in de-ionized water and in TDW. Other observations indicate that it also is the case for pitting of steel pipelines conveying injected water or oil and gas with a high 'water-cut' (i.e. a high percentage of water) [10, 11].

If the corrosion products seal all of the steel area within the pit, corrosion much reduces or ceases and pit depth essentially stops increasing. Evidently, this requires a relatively large amount of corrosion productsto occur within the pit and to deposit on the pit walls. In turn, this implies lesser ferrous ion transfer to the external cathodic zones, even though this likely remains the dominant cathodic reaction. Pits entirely filled with corrosion products are commonly observed after extended periods of exposure [34], suggesting that step-wise behaviour is a transient phenomenon.

If the external cathodic area is very large, most, if not all, of the ferrous ions from the anodic area within the pits will migrate out of the pit with only a small fraction remaining inside the pit. This can be envisaged as the scenario represented in most pit progression modelling approaches [24, 25], in which the pit can grow under the influence of E both into the metal and sideways, with little if any corrosion product build-up within the pit and hence little or no possibility for passivation. In this case there are no restraints for the pit depth to grow unabated whilst sufficient potential E remains to support such growth as pit depth increases. A similar scenario could arise if the environment outside the metal (and thus outside the pit mouth) is of sufficiently low pH to prevent the formation of corrosion products that are then free to migrate away.

Discussion

The above synthesis provides a set of mechanisms for the way step-wise pitting can arise, the condition under which deep pits sometimes form without obvious evidence of step-wise behaviour and why most pits tend to cease growing. The relative occurrence of these can be considered to relate to the availability of the cathodic region outside the pits to support each mechanism. As noted in the Introduction, for mild and low alloy steels pitting tends to occur in clusters over the metal surface, presumably because of the distribution of pit initiators such as inclusions.

For pits within such clusters the available external cathodic area is likely limited, and thus pit depth will be limited as noted above. This is likely to be the dominant cohort of pits given the density, pervasiveness and likely random nature of occurrence on the metal surface of pit clusters. On the other hand, statistically it can be expected that there will be some locations, randomly over the surface, for which initiated pits have access to larger areas of external cathodic surfaces. These pits are candidates for step-wise pitting.

Tables 2-4 provide some information about the step sizes for the 3 different steels in the 3 environments. Surprisingly, perhaps, is the consistency between the values - pit depth increments in the early corrosion stages of about 60-80 microns and about 250-300 for the longer-term exposures. Overall, the early exposures occur under the dominance of the oxygen reduction reaction for the production of rusts in Mode 1 of the bi-modal model [1, 2] while the pit depth increments under longer term exposures occur predominantly in Mode 2 in which hydrogen evolution is the dominant cathodic reaction rate-controlling the overall corrosion process. This suggests that the relativity between step-increments could be related to overall bi-modal behaviour and the relative rate-control of the respective cathodic reactions. This remains a matter for further research.

A related question is why pit depth measurements do not necessarily display clear bi-modal trending and that where they appear to do so on average the data show large excursions from the average trend. The first point is that while water (seawater or de-ionized or other) can be idealized as a homogeneous medium, this is not the case for the exposed surfaces of real alloys. These invariably display, and certainly so at a microscopic level, some degree of variability of grain structure. This implies variability in grain size and in grain boundary dimensions and, in turn, electrochemical potential from point to point on the metal surface, noting that such potential can result only from, perhaps only slight, variability in localized alloy content [35]. Most discussions of pit initiation recognize that such differences in composition must exist at a local level as a result of inclusions [18, 21] but the overall effect of variability across the surface itself is seldom addressed. Such variability immediately suggests that the sizes of the pits that form at initiation will not be the same in size or depth. Together with the then accepted development of corrosion from that initial pitting, leads immediately to differences in the depths of early pitting. It may be supposed that in turn such differences may lead to some pits increasing in depth through stepped behaviour at a rate faster than others. This clearly is a further matter for investigation, given the observations of very deep, stepped pitting already within 4 weeks or so of first exposure in seawater at 20°C (Table 2) and within 14 days for the de-ionized water at 70°C (Table 4).

A question that arises is why step-wise pitting is seldom observed in electrochemical test results. These almost always show what appears to be steady pit depth progression, without the obvious changes in topography and thus in profile such as in Fig. 1. The reason can now be seen to lie with the artificially large cathode invariably employed in such tests, in which most attention is given to the anodic processes. Using a large external cathodic area will overshadow any cathodic activity inside the pits and thus tends to negate or much reduce the protective effect any such rusts over the pit mouth and further into the pit (Fig. 4(A)).

Much discussion of pitting corrosion has presumed the presence of chlorides and indeed the widely-quoted Wranglen model [18] is almost always associated with their presence. However, the model is viable without chlorides. Similarly, the observations by Ernst and Newman [32] apply both for high chloride contents and for very low chloride contents [28]. Moreover, earlier observations [3] clearly show pitting of mild steel in the absence of chlorides, consistent with the even earlier observations by Heyn and Bauer [36] that for quiescent waters salt concentration (for a range of salt types) has no obvious effect on overall corrosion. The results reported earlier for mass loss in seawater and de-ionized waters support that proposition [1] while the results reported herein support that proposition for pit depth and for step-wise pit depth increments.

CONCLUSION

The following conclusions can be drawn from the observations presented herein and their interpretation:

1. Irrespective of precise carbon content and water temperature, low carbon and mild steels exposed in de-ionized waters or in seawater show clear visual evidence of pitting corrosion almost immediately after initial exposure.

With continued exposure pit depth (both mean and extremes) develops overall in a manner consistent with the bi-modal model.

2. The observed pitting occurred over the metal surface in areas that may be termed 'field of pits'. For exposures to around 100 days the majority of pits reached depths to around 60 microns and contained much corrosion product that was difficult to remove using ultrasonic corrosion coupon cleaning techniques.

3. With continued exposure much of the visible pitting exhibited steps in pit depth development, with early pit depth increments around 60-80 microns and later around 300-350 microns. In a model developed to explain this step-wise behaviour, this change is attributed to a change, within the pits, of the governing cathodic reaction from predominantly oxygen reduction to hydrogen evolution, the latter responsible for the known production of hydrogen within pits that also is known to cause bursting of pit covers.

4. Working from the starting point that the metal surface is not perfectly homogeneous and isotropic, these observations can be used to explain the classical notion of 'uniform' corrosion appearing to move around the corroded surface as corrosion progresses. They also can be used to explain why some isolated pits develop to greater depths than the others while showing step-wise development with depth.

5. The proposed model has corrosion products depositing initially inside and around the pit mouth and then gradually depositing further into the pit on cathodic zones as the anodic region contracts towards the pit base, leaving the possibility that some pits will fill completely with corrosion products while others will generate sufficient hydrogen gas pressure to burst the rust cap over the pit, in the process leaving only the pit base for further anodic activity and the further progression of the pit depth.

ACKNOWLEDGEMENTS

The authors acknowledge the continued support from the Australian Research Council (grant DP140103388), the Port Stephens Fisheries Institute at Taylors Beach, NSW, and the laboratory staff at The University of Newcastle, Australia.

REFERENCES

[1] R.E. Melchers and R. Jeffrey, The transition from short- to long-term marine corrosion of carbon steels: 1. Experimental observations, *Corrosion*, 78 (2022) (5) 415-426.

[2] R.E. Melchers, A review of trends for corrosion loss and pit depth in longer-term exposures, *Corros. Mater. Degrad.*, 1 (2018), 4; DOI:10.3390/cmd1010004.

[3] A.D. Mercer and E.A. Lumbard, Corrosion of mild steel in water, *British Corrosion Journal* 30 (1995) 43-55.

[4] G.R. Wallwork and R. Harris Localized corrosion in mild steel, (In) Localized Corrosion, (Ed.) R.W. Staehle, B.F. Brown, J. Kruger, A Agrawal, NACE-3, National Assoc. Corrosion Engineers, Houston, (1974) 292-304.

[5] T. Szauer and J. Jakobs The pitting corrosion of low alloys and mild steels, *Corrosion Science*, 16 (1976) 945-949, Figs. 3 and 5.

[6] R. Jeffrey and R.E. Melchers, The changing topography of corroding mild steel surfaces in seawater, *Corrosion Science*, 49 (2007) 2270-2288.

[7] G. Sander, V. Cuz, N. Bhat, and N. Birbilis, On the in-situ characterisation of metastable pitting using 316L stainless steel as a case study, *Corrosion Science*, 177 (2020) 109004.

[8] N. Ida and J. Tani, Dependence of pit growth rate on applied potential and chloride concentration under conditions with and without fluid flow, *Corrosion* 79 (2023) (2) 230-239.

[9] D. Sinclair, S. Niverty and N. Chowla, Understanding initiation and evolution of pitting corrosion in 7075-T6 aluminum alloy by four-dimensional X-ray microtomography, *Corrosion* 79(2023) (1) 4-16.

[10] R.E. Melchers, New insights from probabilistic modelling of corrosion in structural reliability analysis, *Structural Safety*, 88 (2021) (1) 102034.

[11] R.E. Melchers, Extreme Value Analysis for offshore pipeline risk estimation (In) *Engineering for Extremes*, (Ed.) D.V. Rosowsky and MG. Stewart, Springer (2021) pp. 355-372.

[12] G. T. Burstein, C. Liu, R. M. Souto, and S. P. Vines, Origins of pitting corrosion, *Corrosion Engineering Science and Technology,* 39 (2004) (1) 25–30.

[13] U.R. Evans, *The corrosion and oxidation of metals: scientific principles and practical applications*, London, Edward Arnold, 1960.

[14] G. Butler, P. Stretton and J.G. Beynon, Initiation and growth of pits on high-purity iron and its alloys with chromium and copper in neutral chloride solutions. *British Corros. J.* 7 (1972) (7) 168-173.

[15] R.E. Melchers, Progress in developing realistic corrosion models, *Structure and Infrastructure Engineering*, 14 (7) (2018) 843-853.

[16] ASTM Standard G1-03, *Standard practice for preparing, cleaning and evaluating corrosion test specimens*, American Society for Testing and Materials, West Conshohocken, PA, 2011.

[17] T.R. Foley The role of the chloride ion in iron corrosion, *Corrosion*, 26 (1970) (2) 58-70.

[18] G. Wranglen, Pitting and sulphide inclusions in steel, *Corrosion Science*, 14 (1974) 331-349.

[19] N.D. Greene and M.G. Fontana, A critical analysis of pitting corrosion, *Corrosion*, 15 (1959) (1) 41-47.

[20] G. Wranglen, *Introduction to Corrosion and Protection of Metals,* Butler and Tanner, London, UK, 1972.

[21] Z. Szklarska-Smialowska, *Pitting corrosion of metals*, Houston, National Assoc. of Corrosion Engineers, 1986.

[22] P.C. Pistorius, and G.T. Burstein, Metastable Pitting Corrosion of Stainless-Steel and the Transition to Stability. *Philosophical Trans. Royal Soc. Lond. Series A*, **341**(1992) 531-559.

[23] S. M. Sharland and P. W. Tasker, A mathematical model of crevice and pitting corrosion-I. The physical model, *Corrosion Science,* 28 (1988) (6) 603–620.

[24] Z.Y. Chen,, F. Cui, and R.G. Kelly, Calculations of the cathodic current delivery capacity and stability of crevice corrosion under atmospheric environments. J. Electrochem. Soc., 155 (2008) (7) C360-C368.

[25] J. Srinivasan and R.G. Kelly, On a Recent Quantitative Framework Examining the Critical Factors for Localized Corrosion and Its Impact on the Galvele Pit Stability Criterion. *Corrosion*, 73 (2017) (6) 613-633.

[26] S.T. Pride, J.R. Scully, and J.L. Hudson, Metastable Pitting of Aluminum and Criteria for the Transition to Stable Pit Growth. J. Electrochem. Soc., 141 (1994) (11) 3028-3040.

[27] H.S. Isaacs, and G. Kissel, J. Surface Preparation and Pit Propagation in Stainless Steels, J. Electrochem. Soc. 119 (1972) 1628-1631.

[28] P. Ernst and R.C. Newman, Pit growth studies in stainless steel foils. II. Effect of temperature, chloride concentration and sulphate addition, *Corrosion Science*, 44 (2002) 943-954.

[29] J.R. Galvele, Transport processes and the mechanism of pitting of metals, *J. Electrochemical Soc.* 123 (1976) 464–474.

[30] G.D. Bengough and F. Wormwell, The theory of metallic corrosion in the light of quantitative measurements. Part VI. The distribution of corrosion, Proc. R. Soc. Lond. *A*, 140 (1933) 399-425.

[31] H.W. Pickering, Important early developments and current understanding of the *IR* mechanism of localized corrosion, *J. Electrochem.l. Soc.*, 150 (2003) (5) K1-K13.

[32] P. Ernst P. and R.C. Newman, Pit growth studies in stainless steel foils. I. Introduction and pit growth kinetics, *Corrosion Science*, 44 (2002) 927-941.

[33] G.S. Frankel, L. Stockert, F. Hunkeler and H. Boehni, Perspective on Metastable Pitting of Stainless Steel, *Corrosion* 43 (1987) (7) 429-436.

[34] I. Comanescu, R.E. Melchers and C. Taxén, Corrosion and durability of offshore steel water injection pipelines, *Ships and Offshore Structures*, 11(2016) (4) 424-437.

[35] R.E. Melchers, Corrosion at the steel–medium interface. *Corros. Mater. Degrad.* 5 (2024), 52–72. https://doi.org/10.3390/cmd5010003

[36] E. Heyn and O. Bauer, Ueber den Angriff des Eisens durch Wasser und wässerige Losungen, *Stahl und Eisen*, 28 (1908) (44) 1564-1573.

AUTHOR DETAILS

Robert E. Melchers, Professor of Civil Engineering (Emeritus), The University of Newcastle, Australia, BE, MEngSC (Monash), PhD (Cambridge, UK), Fellow, Australian Academy of Technology and Engineering. Honorary Fellow, The Institution of Engineers Australia.

Awards include 2009 ACA Corrosion Medal, 2012 Jin S Chung Award (ISOPE), 2013 John Connell Gold Medal (The Institution of Engineers, Australia), 2014 Eminent Speaker, Structural College, The Institution of Engineers, Australia, 2006 and 2018 Guy Bengough Award (IOM3).

Robert Jeffrey, Conjoint Fellow, The University of Newcastle and Principal, Pacific Testing. He holds a BSc (Chem) and PhD from The University of Newcastle. He specialises in corrosion in marine, tidal and atmospheric conditions.

He is an honorary life member of the ACA (2011), been awarded the ACA Corrosion Medal (2012) is a past president of the ACA and has been presented with the ACA Kennet award and twice with the ACA's Marshal Fordham award.

Australasian Corrosion Association Conference 2024 – Paper number 0024-21588.

CORROSION & PREVENTION CAIRNS 2024

THE AUSTRALASIAN CORROSION ASSOCIATION INC.

Recent Progress in Ultra-low Thermal Conductivity Polymer Materials: A Mini-Review

Nicholas S L Tan,[1,*] Andrew B. Lowe,[2] Ibukun Olowoye,[1] Thunyaluk Pojtanabuntoeng,[1]

[1]Curtin Corrosion Centre, Western Australia School of Mines: Minerals, Energy and Chemical Engineering, Curtin University, Kent St, Bentley WA 6102

[2] Department of Chemistry, Khalifa University, PO Box 127788, Abu Dhabi, United Arab Emirates.

**Corresponding author email:* nicholas.tan1@curtin.edu.au

ABSTRACT

Effective thermal insulation plays a crucial role in various industries, offering advantages such as reduced energy consumption, improved thermal management, and enhanced worker safety. While polymer materials are commonly utilized for thermal insulation, there's a noticeable lack of research focusing on the development of polymer composites with ultra-low thermal conductivity (λ). This paper aims to address this gap by providing an overview of recent advancements in creating polymer materials with ultra-low thermal conductivity for thermal insulation applications. It also explores how insights gained from research on ultra-high thermal conductivity materials can be leveraged to inform the development of their low thermal conductivity counterparts. Additionally, the paper seeks to identify significant challenges and propose future research directions necessary for the advancement of materials with ultra-low λ for thermal insulation coating (TIC) application.

Keywords: ultra-low thermal conductivity, polymer, thermal insulation coating, TIC

Introduction

Thermal insulation materials are characterized by their low thermal conductivity (*k*) which is typically around 0.1 W/m·K (100 mW/m·K). To put it into perspective, one of the world's most thermally insulative material which are aerogels having a *k* value of around 0.013 W/m·K or 10 mW/m·K and pure copper having a thermal conductivity of around 400 W/m·K which is 40,000 times greater in magnitude than most polymers. Ultra-low thermal conductivity materials have garnered attention for their potential in enhancing energy efficiency and thermal protection where personnel protection is required.

Currently, the scientific community is largely focused on developing ultra-high thermal conductivity for thermal management applications in microelectronics. The highest thermal conductivity achieved by a polymer composite material has a k value measured at 695 W/m·K achieved by fabricating a polymer composite with vapor grown carbon fibre.[1] In comparison, there is a noticeable paucity in reports of ultra-low thermal conductivity polymers in recent years with the lowest k value measured at 0.030 W/m·K achieved by the incorporation of aerogel-based particles and the utilisation of a novel fabrication method which encompasses multi-step processes.[2] However, no breakthroughs in this class of materials have been reported since. Despite the scarcity of the reports of the lower limit of thermal conductivity in bulk polymers and their composites, significant understanding has been made since as knowledge gain in the development of ultra-high thermal conductivity polymers can also be applied to the advancement in research in ultra-low thermal conductivity polymers.

This review aims to provide an overview of the current state of ultra-low thermal conductivity polymer materials for thermal insulation applications. Firstly, the review will briefly cover the principles underlying thermal transport in polymer materials and the thermal metrologies available to understand the thermal properties of the polymer. Following that, the current understanding of how polymer morphology and topology influences these phenomena are discussed. Finally, the review will also provide a quick look at the various methods that has been used been employed in the scientific literature in the development of ultra-low thermal conductivity materials and their pitfalls. Finally, a future outlook will be provided to highlight the ongoing challenges and potential breakthroughs in the development of ultra-low thermal conductivity polymers and their composites by identifying key area for future research.

General Principles of Thermal Transport

In dielectric solid materials, heat is transported via quantised lattice vibrations called phonons.[3] The standard model in which phonon transport is based on is the phonon gas model (PGM) which readily predicts thermal properties in crystalline materials. However, in disordered materials like polymers, the PGM does not fit experimental results which led to development of a new thermal transport model proposed by Allen and Feldman[4, 5] which further breaks down phonon into propagons, diffusons and locons. Since then, attempts have been made to propose a unified model to predict thermal transport behaviour in both crystalline and amorphous solids which has been discussed elsewhere.[6, 7]

Phonon transport efficiency is dictated by its phonon mean free path (MFP) which is defined as the average distance a phonon travels between scattering events in a material. A shorter phonon MFP will result in a lower thermal conductivity while a longer phonon MFP results in a higher thermal conductivity material. In crystalline materials such as metals, phonon scatterers can be in the form of vacancies, voids and dislocations which lowers phonon transport efficiency. In disordered materials such as polymers, phonon transport efficiency is affected by polymer chain rigidity, crystallinity, orientation, molecular structure, and intermolecular interactions which includes but not limited to Van der Waals forces, Coulombs, and hydrogen bonding.[8]

Bulk Polymer Morphology, Topology and Chemistry

Polymer morphology is defined as the arrangement and physical structure of polymer chains within the material at different length scales. In simple terms, it describes how the polymer chains are arranged and organised in the material micro- and macroscopically. For the purposes of the current discussion, polymers can be divided into two types of polymers: thermoplastic and thermosets. Thermoplastic materials can be reshaped by heating and harden upon cooling and are composed of linear or branched molecular chains.[9] On the other hand, are polymers the cross-linked to form a polymer material of infinite molecular weight that is rigid, infusible, and insoluble three-dimensional network structure that cannot be reshaped via heating.[10]

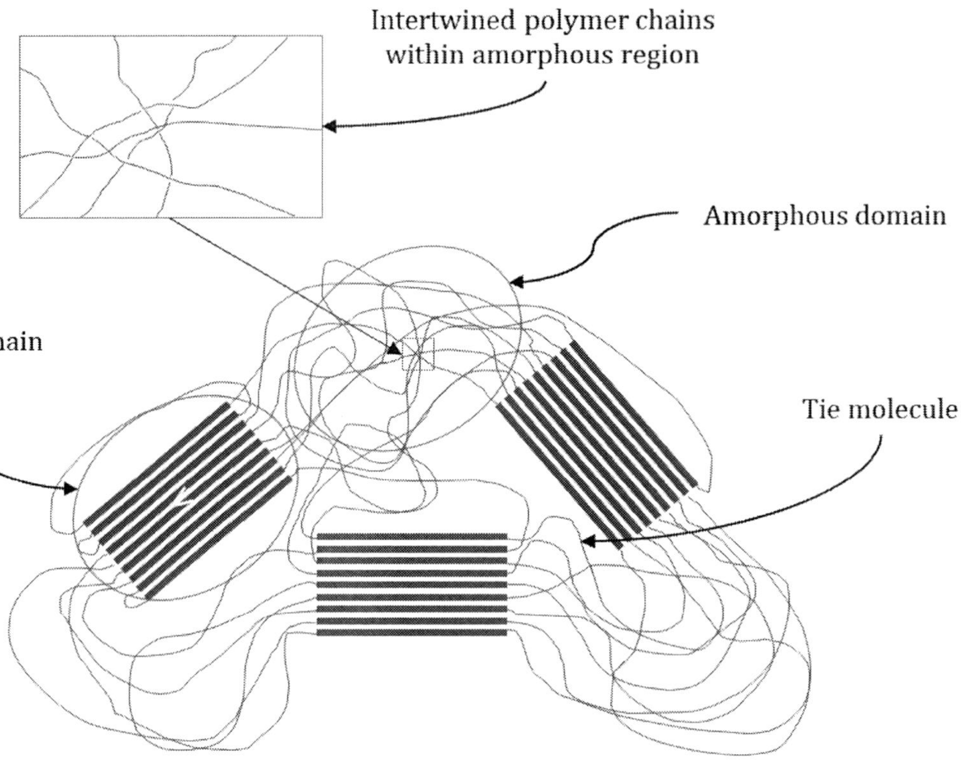

FIGURE 1 SCHEMATIC OF CRYSTALLINE DOMAINS AND INTERTWINED AMORPHOUS REGION IN A SEMI-CRYSTALLINE POLYMER.

Briefly, most thermoplastic polymers are typically semi-crystalline materials which are composed of crystalline and amorphous domains whose polymer chain, often, form intertwined regions at the crystalline-amorphous interfaces as illustrated in Figure 1. In this intertwined region, the entangled structure in the region contains "tie" molecules which are molecules that connects two lamellae from different crystalline domains.[11, 12] These are thought to be responsible for higher elongation at break, toughness and impact strength in polymer materials.[12] Thermoset polymers have different complex three dimensional structures compared to their linear thermoplastic counterparts, generally challenging to elucidate, however, recent molecular dynamic study have reveal the microstructural feature that underpins the mechanical properties of epoxy thermosets.[13]

The microstructure of polymer can be manipulated using various mechanical processes or by chemical modification, thereby, altering its thermal behaviour. In the context of thermal conductivity, phonon transport is affected by the degree of crystallinity and the orientation of this crystalline domain. The most well characterised material for structure-thermal conductivity relationship study is polyethylene (PE). Choy et al., demonstrated that with increasing crystallinity, the k value increases proportionally and phonon transport conducts preferentially along the lamellae of these spherulites. Choy et al.,[14] and recent reports[15-17] showed that with increasing alignment of this crystalline domain, PE can exhibit ultra-high k value with the highest recorded k of 104 W/m·K which was achieved on PE nanofibre fabricated with extrusion processes. The theoretical upper limit of 310 W/m·K as determined by an *ab initio* study[18] whose value is greater than metals such as aluminium (237 W/m·K) and comparable to metals such as copper (400 W/m·K) and gold (327 W/m·K).

Comparatively, amorphous polymer has comparatively lower thermal conductivity values due to the lack of long-range order which increases phonon scattering, thereby, lowering the MFP. Case-in-point, polysiloxanes are inorganic polymers having a Si-O-Si backbone, which have very low torsional barrier[19] and they are also amorphous with an associated k value of 0.092 W/m·K.[20] This dynamic flexibility decreases the MFP due to increasing phonon scattering, thereby, lowering the thermal conductivity. This can be rationalised by looking at the behaviour of molecules in response to heat for which the glass transition temperature (T_g) is an important and unique characteristic of a polymer, defined as the temperature at which a polymer material transition from a glassy state to the rubbery state. As the temperature increases, the molecular movement increases and in polymers, the molecular movements entail longitudinal motions and torsional rotations at the glass transition temperature which increases disorder; facilitating phonon scattering, thereby, culminating in a lower thermal conductivity. [21]

Polymer topology refers to the spatial arrangement and connectivity of the polymer chains within the amorphous material, in other words, it describes how polymer chains have connected each other. There are variety of polymer architecture namely linear, branched, star-shaped, comb polymers to name a few and has been discussed in detail elsewhere. [22] Recent in silico studies reveal that the key parameter that has been overlook for their contribution in phonon transport is the intermolecular interactions which encompasses intermolecular forces such as Van der Waals forces, Coulombs and hydrogen bonding to name a few.[8] The strength of this interaction can be affected by the polymer free volume which is defined as the volume of mass not occupied by the polymer chain i.e., the spacing between the chains. A decrease in polymer free volume either via cross-linking or increasing crystallinity, decreases the intermolecular distances, consequently, increasing the strength of intermolecular forces whose strength is distance-dependent property, resulting in a higher recorded k value for the bulk polymer material.

Most recently, the role of chemical structure, specifically, the type of bonding was shown to affect thermal conductivity significantly. For example, the carbon-carbon chemical bond in epoxy-based materials can be aromatic (sp^2 bond or C=C double bond) or aliphatic (sp^3 or C-C single bond) which was demonstrated to have an impact on thermal conductivity based on reports published by Lv et al.,[23, 24] in which materials with aromatic structure has a higher thermal conductivity compared to aliphatic structure due to the rigidity of the carbon-carbon bonding in aromatic structures. Additionally, the same author also studied the effect of cross-linker length and dynamic bonding between the borane-diol structure on the thermal conductivity, reported that the as the chain linker length increases, the thermal conductivity decreases due to decreasing rigidity and increasing dynamic bonding of the polymer network due to increasing temperature, likewise, increases phonon scattering, thereby, lowering the phonon MFP.[25]

Putting it all together, to fabricate a low thermal conductivity material, an ideal candidate would be an amorphous polymer with high torsional and longitudinal flexibility to encourage phonon scattering. The glass transition temperature of the polymer would also be key parameter to consider and should be lower than the operating temperature (i.e., material is in the rubbery state) to decrease phonon MFP to facilitate phonon scattering. The culmination of these features would be a bulk polymer with ultra-low thermal conductivity. A potential candidate that currently fits this description are polysiloxanes which also have incredibly high thermal stability.[26]

THERMAL INSULATIVE POLYMER COMPOSITES

Despite the inherent low thermal conductivity of polymers (~0.2 W/m·K), polymer composites fabrication or formulation is still essential in lowering thermal conductivity of polymers. This is because a lower k value material will result in the lower thickness material for insulation purposes which enhances practical applications. Theoretically, the temperature difference on the cold side can be calculated based on Fourier's law of heat transfer as shown in equation (1) where the Q is the heat transfer rate (W), k is thermal conductivity of the material, A is the cross-sectional area perpendicular to the path of heat flow, ΔT is temperature difference across the insulation material (°C) and d is the thickness of the material.

$$Q = \frac{k \cdot A \cdot \Delta T}{d} \qquad (1)$$

If the desired temperature on the "cold" side is 60 °C (a safe-to-touch temperature according to ASTM C1055), the required k value requirement can be estimated based on equation (1) on a 120 °C substrate temperature. With the assumption that Q value is ~420 W and the thermal insulation sample thickness is a maximum of 10 mm, then the proposed k is around 0.07 W/m·K. Indeed, a polymer material with baseline thermal conductivity of 0.2 W/m·K is required to decrease its *k* value to 0.35 % of its original value to meet the necessary thermal insulation performance requirements.

Such thermal insulation performance would not be achievable with the current generation of bulk polymer materials, and a polymer composite material is required for to achieve such low *k* values. The simplest method of decreasing thermal conductivity of the material is to introduce voids that contain air in the polymer matrices. This can be achieved by adding low density fillers such as hollow glass microspheres (HGMS) whose density is around 0.125 – 0.600 g cm^{-3}[27] to make a syntactic polymer foam or aerogel particles whose density is even lower than that of HGMS at 0.045 g cm^{-3}.[28]

Hollow Glass Microspheres-based Composites

HGMS-based TIC polymer composites, also known as syntactic foams, are well-characterised composite materials that incorporates controlled voids containing air. These HGMS materials are typically silica-based materials[29] but can be sometimes be manufactured using polymer materials.[30] Thermal insulation materials based on HGMS-based polymer composites at their highest inclusion rates can achieve a thermal conductivity range between 0.17 – 0.09 W/m·K depending on the density of the HGMS.[27]

The main drawbacks of this methodology is due to the fragility of these low density hollow fillers as they have lower compressive strength leading to issues during high shear processing which occurs during formulation e.g., high shear mixing or spray application causing these hollow fillers to shatter on impact with a hard surface.[31, 32]

Aerogel-based Composites

Aerogel-based polymer composites are materials made from the incorporation of aerogel particles whose density can be as low as 0.045 g cm^{-3} which results in ultra-low density and thermal conductivity material. An acrylic emulsion TIC formulation with hydrophobic aerogel microsphere particles was reported to have a thermal conductivity of around 0.06 W/m·K; similar reports of aerogel-based polymer composites fabricated using other methods have also been noted to be of similar *k* value range.[33, 34]

Indeed, aerogel-based particles are effective thermal insulation fillers due to their mesoporous structure lending itself ultra-low thermal conductivity of around 0.013 W/m·K.[35] However, in addition to its fragile nature similar to to HGMS-based TIC formulations, pore infiltration by the liquid resin or solvents in the formulation present significant formulation challenges as the thermal conductivity of the aerogel particles increases with higher degree of pore infiltration reducing the effective thermal conductivity that can be achieved with this technique.[36] As a result, various pore preservation techniques have been proposed such as exploiting the interface behaviour of hydrophobic surfaces and hydrophilic medium in coating formulation[37] or pore restoration techniques whereby the aerogel particles are filled with a low-boiling point blowing agent that evaporates, but requires elevated temperatures limiting its use in coating formulation.[38]

Self-Foaming Polymers

Polymers that are self-foaming utilises blowing agents which can be divided into two groups, that is chemical foaming agents (CFAs) and physical blowing agents utilising gaseous molecules which includes but not limited to CO_2 or H_2. For the purposes of this mini-review, chemical foaming agents (CFAs) are the focus of discussion due to the practical nature in the formulation of self-blown TIC formulations i.e., latent foaming action. However, an in-depth discussion on the foaming techniques and foam physics can be found elsewhere.[39, 40]

The most well-known self-foaming polymer foam are polyurethane foams which is the result of the thermal decomposition reaction of carbamates into amine and CO_2 (blowing agent) and the resulting foam can have a thermal conductivity as low as 0.013 W/m·K.[41] In general, there are two foam structures that can result with this technique which are open, closed cell structure or mixed of the two.[42] For TIC, a closed-cell structure would be ideal for barrier properties compared to an open-cell structure which has interconnected porous structure.

The advantageous of self-foaming TIC is apparent in that the thermal conductivity is lower compared to aerogel-based polymer composites and the pore morphology can be controlled. Additionally, due to the volume expansion of the foam, it requires less time per coating application due to the increased wet film thickness resulting in higher dry film thickness which reduces downtime with less consumption of materials compared to a regular TIC formulation that is dependent on hollow fillers such as HGMS-based or aerogel-based materials.

THERMAL METROLOGIES

The characterisation of novel polymer formulation thermal properties such as thermal conductivity is essential in the development of ultra-low thermal conductivity polymer materials. There are numerous thermal metrologies which can be broadly classified into two categories – steady state and transient methods. For steady-state methodologies, a constant and uniform heat flow is applied to the system, it is then allowed to reach thermal equilibrium i.e., temperature is relatively constant within a narrow standard deviation. In comparison, transient methods, the thermal behaviour is measured as a function of time during heating or cooling processes and temperature response is monitored over time. In this section, the most common thermal metrologies will be discussed in each category along with their benefits and draw backs.

STEADY-STATE METHOD

A common steady-state method is guarded hotplate (GHP) method and its derivatives. The setup is as shown in Figure 1 consists of two identical samples (usually in slab form) which sandwiches a heating plate with precise temperature controlled. A guard ring is placed in such a way that it surrounds the hot plate to minimize lateral heat flow and ensure a one-dimensional heat flow through the sample. Two cold plates flank the sample on the outside where it is constantly monitored for changes in temperature. Once a steady state is reached, the temperature difference is measured, and the amount of heat supplied is recorded with which the thermal conductivity can then be calculated.

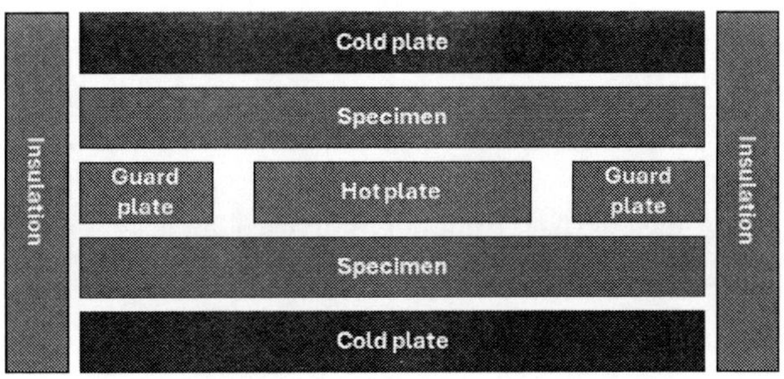

FIGURE 2 GENERAL SCHEMATIC OF THE GUARDED HOTPLATE METHOD SETUP

Although this method boasts accuracy in the determination of thermal conductivity, the method is time-consuming as thermal equilibrium can take days to weeks especially for ultra-low thermal conductivity material. The sample dimension requirement is large slabs which may not be feasible for TIC characterisation. The GHP setup is relatively expensive and requires stringent temperature control and measurement which may not be practical in conventional testing laboratories.

TRANSIENT METHODS

For transient methods, transient plane source such as hotdisk method developed by Gustafsson et al.,[43] is reliable method except for a few caveats (isotropic, sample size). In general, the sensor in the hotdisk method consists of a thin disk-shaped sensor made of a nickel foil which acts as both the heating element and a resistance-based temperature probe. This disk-shaped sensor is then sandwiched in between two identical samples with a good thermal contact between the interfaces as shown in Figure 2. A known amount of power (heat) is generated in the sensor which increases the temperature of the surrounding sample material and simultaneously, the temperature rise is recorded as a function time. In general, a material with low thermal conductivity will see a rapid increase in temperature due to the slow conduction of heat leading to exponential rise in temperature in the surrounding sample material whereby a high thermal conductivity material will see a lower temperature ramp in the same analysis time frame. The thermal transport properties such as the thermal conductivity is then calculated by fitting the temperature-time data to a theoretical model with a couple of assumptions such as geometry and heat transfer characteristics which provides a reasonable accuracy to within 5 % of the true value and with high precision.

FIGURE 3 TYPICAL HOT DISK SETUP (TRANSIENT PLANE SOURCE METHOD) WITH POLYMETHYLMETHACRYLATE (PMMA) SAMPLES FOR VALIDATION.

Laser flash diffusion analyzer is also another transient method in which a short laser pulse is directed onto a sample surface and the temperature on the opposite surface of the sample is measured via a thermocouple or infrared detector as a function of time from which the diffusivity (α) is calculated. Equation (1) describes the relationship between the thermal conductivity (k) and thermal diffusivity (α), specific heat capacity (C_p) and density (ρ). With the thermal diffusivity known, C_p can be derived from differential scanning calorimetry (DSC) and density can be obtained via Archimedes principle or gas pycnometry, the thermal conductivity of the solid can be calculated.

$$k = \alpha \cdot C_p \cdot p \quad (1)$$

The primary advantage of TPS method is that it is relatively faster analysis compared to GHP with high precision in measurements. It is effective method for a wide variety of materials ranging from ultra-low to ultra-high thermal conductivity materials such as polymer, metals and ceramics. It is also non-destructive technique with minimal sample preparation with the requirement of a flat surfaced sample. However, the limitations of TPS are that thermal properties calculated can be incorrect or introduce large errors in the value. This is because the interpretation of the results is subject to the expertise and experience of the user in analysing the numerical fit of the data to the proprietary numerical model.

CONCLUSION

The thermal conductivity of bulk polymers remains a significant constraint in the development of practical thermal insulation coatings (TICs). While notable advancements were achieved in the past, recent progress in ultra-low thermal conductivity polymers has stagnated, largely due to a predominant focus on enhancing thermal conductivity for microelectronics applications. Despite breakthroughs in understanding polymer thermal conductivity, the lower limits remain unexplored. The reliance on hollow and low-density thermal insulation fillers, such as hollow glass microspheres (HGMS) and aerogels, provides only a temporary solution to the growing demand for effective thermal insulation materials.

ACKNOWLEDGMENTS

The author acknowledges the contribution in terms of financial contributions by Petronas Research Sdn Bhd and Curtin Corrosion Centre and the various expertise and guidance contributed by Professor Andrew B. Lowe, Professor Thunyaluk Pojtanabuntoeng and Dr Ibukun Oluwoye.

References

[1] Y.-M. Chen, J.-M. Ting, *Carbon* **2002**, *40*, 359.
[2] S. Y. Kim, Y. J. Noh, J. Lim, N.-H. You, *Macromolecular Research* **2014**, *22*, 108.
[3] M. Nomura, R. Anufriev, Z. Zhang, J. Maire, Y. Guo, R. Yanagisawa, S. Volz, *Materials Today Physics* **2022**, *22*, 100613.
[4] P. B. Allen, J. L. Feldman, J. Fabian, F. Wooten, *Philosophical Magazine B* **1999**, *79*, 1715.
[5] P. B. Allen, J. L. Feldman, *Physical Review B* **1993**, *48*, 12581.
[6] L. Isaeva, G. Barbalinardo, D. Donadio, S. Baroni, *Nature communications* **2019**, *10*, 3853.
[7] M. Simoncelli, N. Marzari, F. Mauri, *Nature Physics* **2019**, *15*, 809.
[8] V. Rashidi, E. J. Coyle, K. Sebeck, J. Kieffer, K. P. Pipe, *The Journal of Physical Chemistry B* **2017**, *121*, 4600.
[9] P. International Union of, C. Applied, "thermoplastic elastomer".
[10] P. International Union of, C. Applied, "thermosetting polymer".
[11] S. H. Cho, **2022**.
[12] S. H. Cho, R. A. Register, *Macromolecules* **2022**, *55*, 3249.
[13] S. Ma, P. Chen, J. Xu, X. Xiong, *Journal of Materials Science* **2022**, *57*, 1123.
[14] C. L. Choy, Y. W. Wong, G. W. Yang, T. Kanamoto, *Journal of Polymer Science Part B: Polymer Physics* **1999**, *37*, 3359.
[15] T. Kim, S. X. Drakopoulos, S. Ronca, A. J. Minnich, *Nature Communications* **2022**, *13*, 2452.
[16] S. Shen, A. Henry, J. Tong, R. Zheng, G. Chen, *Nature Nanotechnology* **2010**, *5*, 251.
[17] J. Liu, Z. Xu, Z. Cheng, S. Xu, X. Wang, *ACS Applied Materials & Interfaces* **2015**, *7*, 27279.
[18] J.-W. Jiang, J. Zhao, K. Zhou, T. Rabczuk, *Journal of Applied Physics* **2012**, *111*.
[19] J. E. Mark, *Accounts of Chemical Research* **2004**, *37*, 946.
[20] C. Zhang, L. Qu, Y. Wang, T. Xu, C. Zhang, *RSC Advances* **2018**, *8*, 9901.
[21] W. N. dos Santos, J. A. de Sousa, R. Gregorio, *Polymer Testing* **2013**, *32*, 987.
[22] A. C. Rinkenauer, S. Schubert, A. Traeger, U. S. Schubert, *Journal of Materials Chemistry B* **2015**, *3*, 7477.
[23] G. Lv, E. Jensen, C. Shen, K. Yang, C. M. Evans, D. G. Cahill, *ACS Applied Polymer Materials* **2021**, *3*, 259.
[24] G. Lv, E. Jensen, N. Shan, C. M. Evans, D. G. Cahill, *ACS Applied Polymer Materials* **2021**, *3*, 1555.
[25] G. Lv, B. Soman, N. Shan, C. M. Evans, D. G. Cahill, *ACS Macro Letters* **2021**, *10*, 1088.
[26] J. Zhao, W. Luo, L. Qi, L. Yuan, G. Huang, Y. Huang, X. Weng, *Coatings* **2018**, *8*, 125.
[27] B. Zhu, J. Ma, J. Wang, J. Wu, D. Peng, *Journal of Reinforced Plastics and Composites* **2012**, *31*, 1311.
[28] M. J. Burchell, S. A. J. Fairey, N. J. Foster, M. J. Cole, *Planetary and Space Science* **2009**, *57*, 58.
[29] G. Chen, F. Yang, S. Zhao, K. Li, J. Chen, Z. Fei, Z. Yang, *Ceramics International* **2022**, *48*, 19367.
[30] A. H. Lu, W. C. Li, G. P. Hao, B. Spliethoff, H. J. Bongard, B. B. Schaack, F. Schüth, *Angewandte Chemie International Edition* **2010**, *49*, 1615.
[31] S. Shira, C. Buller, "Mixing and dispersion of hollow glass microsphere products", in *Hollow glass microspheres for plastics, elastomers, and adhesives compounds*, Elsevier, 2015, p. 241.
[32] I. S. Akgün, E. Demir, M. Işık, N. Ekmekçiyan, E. Şenses, C. Erkey, *Powder Technology* **2022**, *402*, 117331.
[33] C. Boonrawd, S. Yodyingyong, L. Benyahia, D. Triampo, *Gels* **2022**, *8*, 7.
[34] K. Maghsoudi, S. Motahari, *Journal of Applied Polymer Science* **2018**, *135*, 45706.
[35] S. Zeng, A. Hunt, R. Greif, *Journal of Heat transfer* **1995**, *117*.
[36] M. Kucharek, W. MacRae, L. Yang, *Composites Part A: Applied Science and Manufacturing* **2020**, *139*, 106108.
[37] H. M. Kim, Y. J. Noh, J. Yu, S. Y. Kim, J. R. Youn, *Composites Part A: Applied Science and Manufacturing* **2015**, *75*, 39.
[38] H. M. Kim, H. S. Kim, S. Y. Kim, J. R. Youn, *e-Polymers* **2015**, *15*, 111.
[39] F. Monie, T. Vidil, B. Grignard, H. Cramail, C. Detrembleur, *Materials Science and Engineering: R: Reports* **2021**, *145*, 100628.
[40] M. M. M. Shirvan, M. Famili, A. Golbang, *Plast. Polym. Technol* **2016**, *4*.

[41] C. S. Carriço, T. Fraga, V. E. Carvalho, V. M. D. Pasa, *Molecules* **2017**, *22*.
[42] C. W. Visser, D. N. Amato, J. Mueller, J. A. Lewis, *Advanced Materials* **2019**, *31*, 1904668.
[43] S. E. Gustafsson, *Review of Scientific Instruments* **1991**, *62*, 797.

DETAILS

<table>
<tr>
<td></td>
<td>Nicholas Sheng Loong Tan is the corresponding author of this paper. He is a research associate at Curtin Corrosion Centre since September 2021. His role entails research and overseeing higher degree research (HDR) student and undergraduate students. His current research is on ultra-low thermal conductivity polymer, coating degradation mechanisms and sustainable polymer research.</td>
</tr>
</table>

CORROSION & PREVENTION
CAIRNS 2024

THE AUSTRALASIAN CORROSION ASSOCIATION INC.

CHALLENGES IN CONDITION ASSESSMENT AND LIFE EXTENSION METHODS FOR AGED PIPELINES WITH CATHODIC PROTECTION & AC MITIGATION SYSTEM

*07102024

Bob Phang [1,*], Lifu Wang[2] , Anton Bardak[3]

[1]SMEC (A) Pty Ltd, Queensland, 4006, Australia

** Bob.Phang@smec.com*

ABSTRACT

The population growth in most modern cities demand more out of existing infrastructure servicing the needs of business and people by way of operational utilities. Most structures above ground are visually inspected as part of maintenance schedules, however, services transporting wastewater, potable water, gas by way of pipelines that transition underground and out of sight are out of reach for maintenance programs. This paper discusses the challenges encountered by organisations in preparing inspection methodologies used for condition assessments that can provide useful information of the asset in place, provide results for benchmarking each critical asset database with timelines and predictive outcomes. In addition, historical records and results provide another layer of defence to organisation and asset objectives for the whole of life considerations. The use of new inspection tools and techniques are present to inform strategic optimisation of the asset using remedial solutions. Where steel structures are involved, prevention of metal loss in subsurface situations using first principles in electrochemical treatment by implementing cathodic protection for remaining life extension is a sustainable option. This paper aims to provide the best practices for asset managers in defining root causes of mechanisms that compromise asset performance, selecting and optimising repair solutions with practical methods in present value that is sustainable.

Keywords: Condition Assessment, Utility Pipelines, WOL-whole of life, cathodic protection, critical assets, ILI -inline inspection, asset integrity management, CIPS -close interval potentials survey, GPS-global position system

INTRODUCTION

In present time, most built-up cities have ageing assets in place that provide primary utility services such as supply of potable water, gas, services including power and wastewater drainage maintained in good working order. The challenges arise when the aged infrastructure underground present inspection challenges, where special inspection tools are considered. This motivates the inspector surveyor with condition assessment methodologies that can provide meaningful results to strategically manage the asset condition and objectively advocate repairs to extend the operational life without excessive cost.

Background and First Principles

The experiences of the engineering team selected to provide asset integrity and inspections was the key in being engaged by the client to undertake the case study of the infrastructure, in this case a natural gas pipeline condition assessment which started operations in 2014.

Following successful international bid for the condition assessment program, a detailed program to perform the scope of work across the 110.9-kilometre (Km) buried pipeline of 508mm (20") diameter was developed. This is based on information retrieved by desktop review of existing documentation provided by the client GNGC in Ghana, West Africa.

Prior to embarking on the journey to conduct surveys from Takoradi in East Ghana to the Western township Atuabo, border of Ivory Coast. These tasks were actioned to take place in October 2019:

- Understanding the Scope and Client's Objectives
- Desktop review of the existing infrastructure design specification, materials suitability of operations
- Confirm historical operating, maintenance, repairs, leak record.
- Conduct condition assessment of the asset in service and report
  - Non-Destructive Inspection methods preferred for asset
  - Alignment PCM mapping, depth of cover surveys
  - Visual inspections, direct current coating defects survey
  - Any third-party disturbance or interference by other buried structures in close proximity
- Remaining life cycle studies based on findings to improvise remediation to inform operating threshold risks

A scope of work for the client have been customised to ensure that this important infrastructure that provide energy to few power stations in the west of the capital city of Accra are uninterrupted. The historical fact was that the 110.9Km pipeline experienced failure at age 5, based on a design life of 25 years. Commissioned in 2014, multiple leaks caused by pitting corrosion at 17 locations occured through the pipe wall thickness ranging from 10.31mm to 12.7mm. Hence, the management of the pipeline operating company committed a task force to analyse all aspects of the design specifications, construction process and operations in 2015. This pipeline have a three (3) layer PE barrier coating backed up with an impressed current cathodic protection (ICCP) system. To mitigate HVAC-high voltage alternating current interference and step voltage AC safety in design, there were grounding discharge mechanisms using decouplers at forty (40) locations along the pipe route.

The range of work on site included comprehensive audits of the ICCP system, DCVG-direct current voltage gradient survey to confirm the coating integrity after seven (7) years in operation across some newly established industrial plants along the pipeline route.

Of importance to the operations and maintenance of long pipelines is the risk associated with HVAC transmission towers that share the same pipeline easement. The HVAC study scope included calculations to AS/NZS4853:2012 or NACE equivalent standard to assess the expected induced voltage and earth potential rise (EPR) for the pipeline in close proximity with the identified overhead lines. Modelling some sections of the pipeline for potential interference by the HVAC line using CDEGS for EPR. Information required include GIS data, pipeline data, historical data on past induced voltages on pipelines and maximum fault current.

References

Asset Life Cycle Plans are part of the original design for infrastructure that requires established codes of practice and standards, therefore robust enough to go beyond the design life, with the maintenance recommended as part of the operations.

This paper details the need to perform condition assessment and integrity inspection work with knowledge of the engineering standards resulting in decision making for the outcome of life extension remediation work where necessary for the whole of life cycle.

These standards and codes are referred during our case study of the structure under our obligation when the asset was concerned.

AS/NZS 2832 -1	Cathodic Protection for buried cables and metallic structures
NACE SP 0169	Control of External Corrosion of Underground or Submerged Metallic Piping Systems
NACE SP-0572	Design, Installation, Operation and Maintenance of impressed Current Deep Groundbeds
NACE SP-0286	The Electrical Isolation of Cathodically Protected Pipelines
AS/NZS 3835	Earth Potential Rise-Protection of telecommunications network users, personnel and plant
BS EN 1361	British Standards – Corrosion Control of Pipelines
IPWEA N7v3	Condition Assessment and Asset Performance Guidelines-Water Supply and Sewer Assets
AS/NZS3000	Electrical Installations Wiring Rules
AS4827.1	Coating Defects Surveys for Buried Pipeline.

Table. 1 Codes of Practice and Standards

In the leading works, the IIMM 5th edition of International Infrastructure Manual was referenced as well.

Inspection Methodology

Planning to maximise the pipeline condition assessment benefits there was a methodical approach to optimise the output of the survey time of fifteen (15) days that was allocted on location in Ghana. The logistics of travel time from Brisbane to Accra was almost 24 hours.

1. Desktop Study

a) Review the Master Plan to confirm the long-term usage of the facilities,

b) Review aerial imagery and GIS mapping,

c) Source and review existing as-built documentation

d) Extract and review the existing Equipment Register

e) Extract and review maintenance history that included CIPS records,

f) Extract the key points from the ILI-in line inspection report comprising of 280 pages,

2. On Site Investigation
   a) Arrange access for the drive along the pipeline route with attendance by Operations Officers,
   b) Explore site, photograph all the assets and note any necessary amendments to the Equipment Register,
   c) Record the location of all assets,
   d) Identify access issues that need addressing to enable safe and effective execution of the surveys,
3. Asset Criticality Analysis
   a) Identify operational personnel with knowledge of the pipeline,
   b) Source and review the operational and environmental context of the access route,
   c) Facilitate a Criticality Analysis for each asset listed in the Equipment Register, accounting for all relevant consequences and categories,
   d) Submit the site report of the Asset Criticality Analysis for preview,
   e) Finalise and deliver the Asset Criticality Analysis for GNGC management,
4. Condition Assessment
   a) Training session with the GNGC operations on the findings of the PCA,
   b) Resolve any priority issues regarding classification of remediation works,
   c) Finalise and deliver the PCA Report for acceptance by GNGC,
   d) Deliver the database in Digital Space for GNGC.

Survey Equipment

Test equipment required for the execution of this 110.9 Km pipeline from East to West Ghana along the equator required a list of equipment with data capture software and hardware. The test equipment was packed and transported across the Pacific and Atlantic Oceans. Whilst on site, the weather forecast was rain, high humidity and short period of sunshine every day on site. Waterproofing all electrical and electronic connections was key to the success of the site inspection surveys.

The equipment specified for conducting this condition assessment work included two (2) sets of DCVG equipment with GPS synchonising current interrupters of 100A capacity, GPS tracker and data logging meters, Earth Resistance meters, Digital Voltmeters, Current Clamp and copper-copper sulphate electrodes.

Recording all the data capture included proprietary software preloaded into iPad tablets and smartphones equipped with local modems to upload to the cloud in Ghana, which are then processed simultaneously in Sydney, Australia by SMEC's team. This prevented any loss of data throughout the survey across the cross-country pipeline.

The digital engineering work included data capture sectioned into 'blocks' across the forest terrain. This allows for a systematic and targeted analysis of survey results acquired by the inspection tools used by survey crews on the pipeline. Mobile network in Ghana telecommunication enabling the devices used to measure and detect pipeline data sets which included DCVG coating defects IR gradient results, pipe to soil potentials versus CSE reference electrodes, soil ERT measurements that are critical data recordings. These overall sections of the 110.9 km were mapped and shown below.

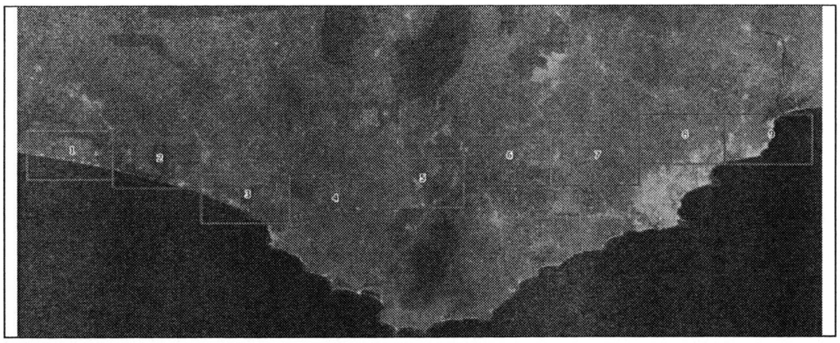

Figure 1.

In line Inspection

The desktop study included the review and comprehension of ILI – in line inspection by ROSEN intelligent pigging method using MFL-magnetic flux leakage tool that runs through the internal pipe with the flow of gas stream. Anomalies in the pipeline wall were detected and percentages of metal loss recorded in graphical format and translated for management of the condition found. In total, 350 metal loss anomalies were reported with a wall loss greater than 10% of the original pipe wall thickness. This ILI inspection was performed after the 17 leak locations have been removed by installing new sections of pipeline as replacements.

The pipeline parameters are provided in the table below.

nominal diameter	20 inch
type of pipe	SAWH / SWAL
steel grade	API 5L X65
nominal wall thickness	10.31 mm / 11.91 mm / 12.70 mm* / 14.27 mm
MOP	5.0 MPa
Design Pressure	10.0 MPa
SMYS	448 MPa
SUTS	530 MPa
minimum bend radius	6D
length (given)	110.9 km
length (recorded)	103.8 km
built in	2014
pipeline product	lean gas
inspection history	no previous ILIs performed

Table 2.

The characteristics of the findings provided an insightful angle towards the ICCP Audit and corrective decisions to ensure that any external coating defect found must be cathodically protected by achieving pipeline polarisation status to meet the NACE protection requirements.

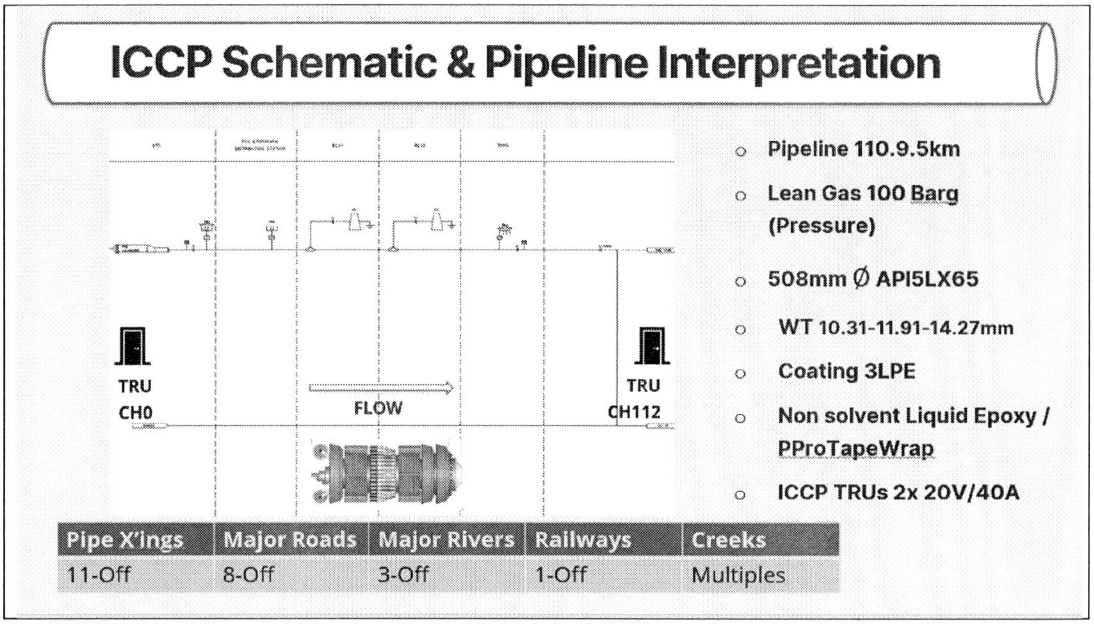

Figure 2. Impressed current CP stations and ILI pigging direction

Protection Criteria

The cathodic protection criterion adopted for this Pipeline is compatible with NACE standards as detailed below:

"All parts of the pipeline shall maintain a potential equal to or more negative than, -850mV (or -950mV at areas where sulphate reducing bacteria can thrive in marshes and swamps) with respect to a saturated copper/copper sulphate reference electrode with "IR" errors removed".

The same criterion will also apply to the buried pipework that connects to the AT Pipeline and thus also receiving cathodic protection.

Following discussions with GNGC dated 11th October 2019 during the kick-off meeting and review of the magnetic flux leakage tests (MFL) conducted by Rosen, severe pitting corrosion found at various sections of the pipeline.

Given the results from the MFL tests, it is prudent to increase the CP protection criteria to -950mV "OFF" potentials to arrest corrosion on pipeline with existing corrosion pits. Therefore, increasing the total current output of the CP systems to verified and recommended as an immediate follow up action before demobilising from site .

Excessive electronegative polarised potentials more negative than -1,200mV is avoided to prevent cathodic disbondment and hydrogen embrittlement on any buried high strength steel structure.

If pits are shallow and limited to millimetre size, corrosion prevention by re-alkalizing the pits through cathodic protection where new protective film will form on the pit surface.

However, if existing pits are deep and narrow in shape, it may be more difficult to arrest corrosion since acidity inside the deep pits and the bottom of the pits are shielded from CP current. From experience, application of adequate CP current will slow down the corrosion rate at these areas, however, achievement of full protection may require long polarisation periods or in some instances may not be possible without very high average current densities being provided.

Observations and Solutions

The DCVG coating defect survey found a host of anomalies along the pipeline when injection points for pipeline signal energized from both TRU-transformer rectifier units located at either extremity of the pipeline. The pipeline is protected by two (2) impressed current cathodic protection (ICCP) systems installed at both end of the pipeline within the AIS and TRMS stations. The survey detected 25 coating defects along the trunk gas mainline, of varying %IR severity. The surprising characteristics are of this calibre:

(I) Coating DEFECTS Exceeding 100m Long at 12 Locations,

(II) Established "Defects Signature" at offset of pipeline alignment route,

(III) Pipe cable to terminal burnt by "Arc" discharge to Earthing grid.

EPR – EARTH POTENTIAL RISE

Test results from the site locations at which pose potential hazards for earth potential rise and possible fault current discharge to the pipeline were obtained and used for the AC Modelling.

These features of the modelling:

1. Conduct Low Frequency Induction modelling at section of the pipeline that runs parallel to the HVAC power line and railway track crossing,
2. Conduct Earth Potential Rise modelling at Test Points,
3. Measure AC voltage at pipeline test posts parallel to the HVAC transmission lin,e
4. Measure current leakage to earth connections (within metering station),
5. Correlation of potential fault current leakage points, induction, capacitance and resistive coupling versus the installed pipeline,
6. Pipeline HVAC route soil logs study using CDEGS for Induced Capacitance Resistive Coupling for personnel safety.

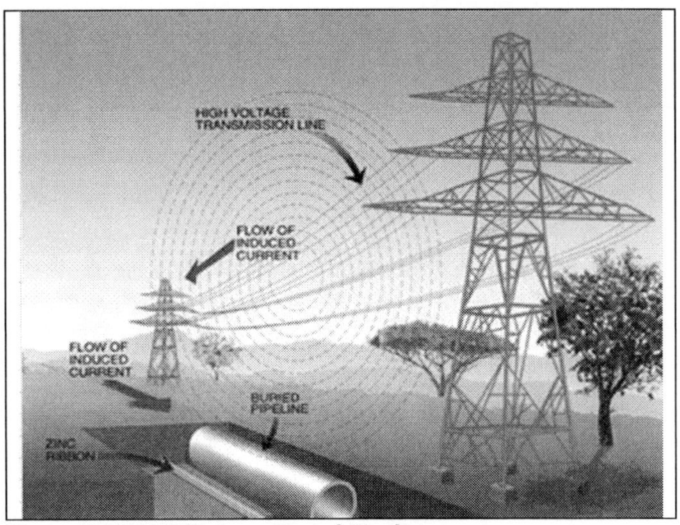

Figure 3. HVAC Effects on Buried Pipeline

Findings

The unexpected findings of coating defects with huge %IR severity over 100 metre length, was not in alignment with the ILI intelligent pig results along the 110.9 Km pipeline. These findings required in-depth discussions and evaluations into this case study.

It was essential to embark on analytic cathodic protection level across the full length of the pipeline which the following findings could explain the phoenomea and based on the following.

1. CP current 4.6A at CH(0) & 2.4A at CH(111) utilising shallow anode ground-beds,
2. Conduct Low Frequency Induction modelling at section of the pipeline that runs parallel to the HVAC power line and railway track,
3. Off take in AT (Atuabo City) 'Instant Off' -700mV to -848mV CSE indicated under-protection,
4. Coating Defects from 23% IR to 27%IR had anodic potentials – 0.398Volt and – 0.416Volt Vs CSE,
5. Mid point of pipeline 'Instant Off' -398mV to - 698mV CSE indicated anodic to under – protection,
6. Electrical Shorts of Pipeline to Anchor Blocks with steel rebars reducing CP level were identified,
7. Loss of CP current to foreign pipeline offtake at AT and Wankan located and electrical isolation,
8. Localised AC voltage discharge at JJ01 Test station with Vent Stack, inadequate earthing parallel to the HVAC transmission pipeline,
9. Shortfall of CP current density to well coated pipeline by compromised AC design to be corrected,
10. *Undetected* failed decouplers for AC Mitigation caused galvanic coupling by copper earth to pipeline,
11. Remediation to Cathodic protection level optimisation by replacing earthing beds with electronegative zinc electrodes and retest,
12. Leading up to Case Study II, after post remediation works.

LESSONS LEARNED

The need to redefine pipeline design quality in construction quality control and attention to HVAC, CP effective designs can certainly reduce the asset integrity risks in operation. These are the informed advantages and considerations undertaken.

In summarizing, this episode of over the line surveys to compare other conducted inspections provided:

1. Digital record of Condition Assessment combined Remedial Database,
2. Defects – Anomalies – Direct and Indirect Root Causes Located & Fixed,
3. Delineation of Conditions by Priority to Improve Pipeline Integrity,
4. Whole of Life Cycle Consideration in Real Time,
5. Asset Performance & Sustainability without Unscheduled Shutdown,
6. Unforgiving Pipeline Terrain can be tamed with well equipped survey team,
7. Targetted Performance 15 days including training by Certified AMPP Specialists,
8. Logistics and Delivery Management from Southern Tip of the World,
9. Digital Cloud Transfer off Site from Mobile Devices to SMEC Sydney Office for processing daily.

Figure 4. ICCP System Pipeline Signal Injection

Figure 5. DCVG Surveyor Over the Pipeline

ACKNOWLEDGMENTS

The authors wish to thank SMEC local and global management in Brisbane, Sydney and Ghana for their continuous support in encouragement to provide knowledge and experiences by the organisation to our young council of engineers in the Australasian Corrosion Association via these Conferences. Also, we thank GNGC of Ghana for their ground support and rich cultural experience during our stay in their country.

REFERENCES

1. CIP Survey Report No. PER0124 Rev. 2 Johannesburg
2. CDEGS software based on AS/NZS4853:2012
3. Little, R. G. (2012). *Managing the Risk of Aging Infrastructure.* Prepared for the International Risk Governance Council (IRGC) as part of the project on Public Sector Governance of Emerging Risks.
4. Noroznia, H., Gandomkar, M., Nikoukar, J., Aranizadeh, A., & Mirmozaffari, M. (2023). *A Novel Pipeline Age Evaluation: Considering Overall Condition Index and Neural Network Based on Measured Data. Machine Learning and Knowledge Extraction*, 5(1), 252-268. https://doi.org/10.3390/make5010016
5. Ma, Q., Tian, G., Zeng, Y., Li, R., Song, H., Wang, Z., Gao, B., & Zeng, K. (2021). *Pipeline In-Line Inspection Method, Instrumentation and Data Management. Sensors,* 21(11), 3862. https://doi.org/10.3390/s21113862
6. Cicek, V. (2013). *Cathodic Protection: Industrial Solutions for Protecting Against Corrosion.* Scrivener Publishing LLC.
7. Fan, L., Reis, S. T., Chen, G., & Koenigstein, M. L. (2018). *Corrosion Resistance of Pipeline Steel with Damaged Enamel Coating and Cathodic Protection. Coatings*, 8(5), 185. https://doi.org/10.3390/coatings8050185
8. Pipeline Risk Management Manual 3rd edition W.Kent Muhlbauer, Tx USA.
9. Field Guide for Investigating Internal Corrosion of Pipelines, Richard Eckert, NACE Press 2003

AUTHOR DETAILS

1. Phang. Bob -MEngSc (Materials and NACE Corrosion Specialist*) AMPP, Fellow MICorr. UK. MIEAust

Bob is a professional engineer based in Queensland for asset management, design of infrastructure with durability concepts that ensures realiable sustainable operations with corrosion mitigation approach. He is presently the Regional Director, Fuels for ANZ and SEA.

He is an ACA member since 1984 and AMPP CP / Basic Corrosion Instructor

2.Wang. Lifu PhD (Mechanical Engineering)

Wang is an Experienced Mechanical Engineer at SMEC, a position he has held since 2022. He is responsible for pipeline designs, network modeling, hydraulic simulations, and pump station design.

3. Bardak. Anton B.Eng Civil (Chartered Engineer and RPEQ)

Anton is an associate infrastructure civil engineer with over 11 years' experience. Over this period, he has specialised in condition assessment and asset management of water and water infrastructure. He has held key coordination, design manager and project management roles is multiple major programs with utilities in South East Queensland over the past 6 years. Anton also has strong technical background and has developed condition assessment techniques and asset management strategies for various water and wastewater assets, especially pressure pipelines.

CORROSION & PREVENTION
CAIRNS 2024

THE AUSTRALASIAN CORROSION ASSOCIATION INC.

Deep Well Impressed Current Cathodic Protection System For Steel Piles Of A New Bridge Over A River

S. Ali[1,*], F. Papworth[2]

[1]*Building and Construction Research and Consulting (BCRC)/NSW, 2066, Australia*

[2] *Building and Construction Research and Consulting (BCRC)/WA, 6005, Australia*

**Corresponding author email: s.ali@bcrc.com.au*

ABSTRACT

This paper presents the design and implementation of a deep well Impressed Current Cathodic Protection (ICCP) system for steel piles of abutments and piers supporting a new 700-meter-long bridge over a river. The decision to incorporate cathodic protection stemmed from geotechnical survey findings indicating extreme corrosion potential in the surrounding soil, posing a significant risk to the integrity of the bridge piles. The system was designed for 7 zones with requirements based on resistivity. The ground-bed resistance and ground potential rise were modelled using SES CDEGS software. This paper outlines the methodology and specifications employed in designing the cathodic protection system, including environmental protection using a cased and sealed anode system. Step and Touch voltage calculations were performed for all anode wells to ensure safe operations. By ensuring the durability and longevity of the bridge infrastructure, this study contributes to the advancement of sustainable engineering practices in bridge construction and maintenance. Comprehensive analysis and consideration of industry best practices were applied throughout the design.

Keywords: bridge, steel pile, ICCP, cathodic protection, design details, resistivity, SES CDEGS modelling

Introduction

The New Dubbo Bridge project is a landmark infrastructure development aiming to enhance connectivity across the Macquarie River in New South Wales. With a total length of 662.59 metres, the bridge is supported by 19 piers, featuring spans ranging from 24.5 metres to 34.1 metres. At its western end, the bridge incorporates reinforced abutment headstocks with a spill-through abutment, while a retaining wall marks its eastern terminus. The foundation of the bridge rests on driven tubular steel piles, with diameters varying between 900mm and 1200mm, and extending up to 60 metres underground. These piles are critical to the structural integrity and longevity of the bridge, particularly given their exposure to potentially severe environmental conditions. FIGURE 1 shows a schematic and a section of drawing indicating the locations of steel piles in the bridge [1].

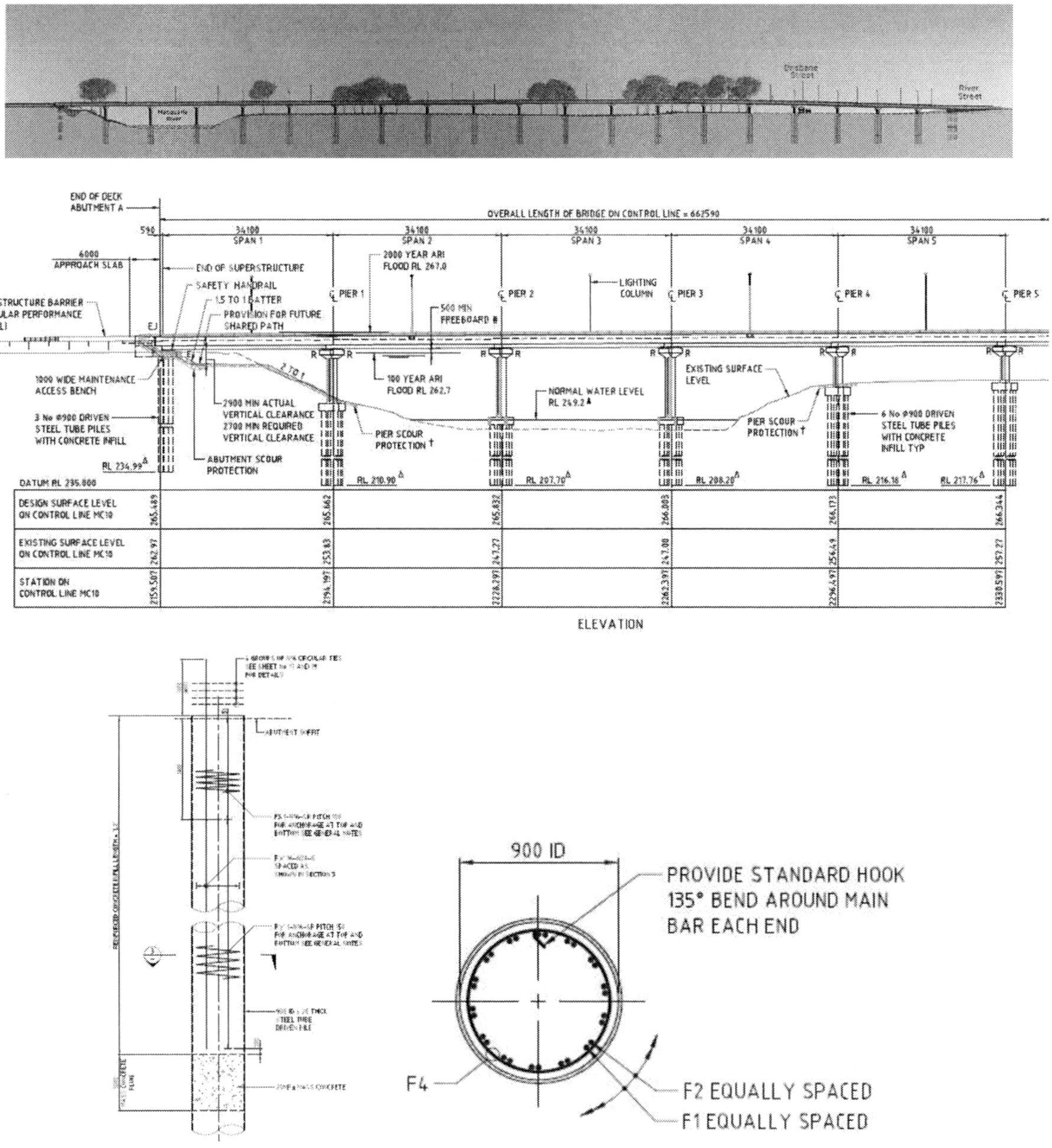

FIGURE 1: SCHEMATIC AND SECTION OF DRAWING OF THE BRIDGE SHOWING THE LOCATION OF PILES.

During the design phase, detailed geotechnical analysis revealed that the site's soil exhibited low resistivity, classifying the exposure of the steel piles as severe according to the Australian Standard AS 2159 [1-3,10]. This classification indicated a high risk of soil corrosion, necessitating robust protective measures to ensure the durability of the piles over the intended design life of 100 years. To address these concerns, the implementation of an Impressed Current Cathodic Protection (ICCP) system was identified as the optimum solution to mitigate the risk of corrosion and preserve the structural integrity of the piles.

Cathodic protection, particularly through deep well anodes, was chosen due to its effectiveness in combating soil corrosion for long, deeply embedded piles. This method is advantageous as it minimizes the risk of interference with existing utilities by ensuring that the current discharge occurs deep underground, rather than at the surface where most utilities are located. This consideration was paramount given the bridge's location in a developed area with numerous underground installations [4, 7].

METHODOLOGY FOR DETAILED DESIGN

The design and implementation of the Deep Well Impressed Current Cathodic Protection (ICCP) system for the steel piles of the New Dubbo Bridge involved a comprehensive, multi-step approach. This methodology section details the procedures, technical considerations, and modelling techniques employed to ensure effective corrosion protection for the bridge's structural foundation.

1. Site and Environmental Assessment

The initial phase of the methodology focused on a thorough site and environmental assessment to understand the conditions affecting the bridge's steel piles [1-4]. The site investigation consisted of three factors as below:

a) Geotechnical Investigation: Extensive soil testing was conducted to evaluate the soil conditions, including soil resistivity, and the existence of corrosive factors such as microbiological bacteria. This investigation was conducted using the bore holes at various depths and locations along the bridge alignment. Note that soil resistivity directly influences the effectiveness and design requirements of the ICCP system. Also, borehole drilling and soil sampling were performed to collect data on soil properties and moisture contents, which affect corrosion rates and cathodic protection efficiency [1].
b) Environmental Exposure Classification: Based on the data obtained, the exposure classification for the steel piles was determined according to AS 2159 standards, which was used as a guide to assess the severity levels for different environmental conditions. The site was classified as having severe exposure, necessitating robust protection measures to combat potential corrosion [1-5,10].
c) Utility Survey: A detailed survey of existing underground utilities was undertaken to identify any potential conflicts and to ensure that the ICCP system would not interfere with these utilities [1].

2. Design of Cathodic Protection Zones

To optimize the ICCP system, the bridge was divided into manageable cathodic protection zones. This zoning approach allows for tailored protection strategies based on localised conditions. The zones were selected based on the number of piles, and environmental conditions. The bridge's 19 piers and the associated spans were divided into seven distinct CP zones, each covering three spans. This segmentation was based on the ratio of the surface area of piles in each zone to the total surface area of all piles. This zoning facilitated targeted protection and ensured efficient distribution of the cathodic protection current across different sections of the bridge.

For each CP zone, the soil resistivity was measured and averaged. These values were critical for determining the electrical resistance and the required current output for effective cathodic protection in each zone [1].

3. Selection and Placement of Deep Well Anodes

Given the severe exposure classification and the presence of numerous underground utilities, deep well anodes were chosen for the ICCP system. This choice minimised the surface interference and ensured deep, uniform current distribution. Deep well anodes were designed to be installed at depths ranging from 30 to 60 metres below ground level, positioned centrally under the bridge between the piers. The well depth was selected to provide sufficient distance from the piles, reducing the risk of anodic interference while maintaining effective protection.

For the anode material and configuration, mixed metal oxide anodes (MMO) were chosen for their durability and conductivity, making them suitable for long-term use in the severe environmental conditions of the site. Each anode was housed within a perforated PVC casing filled with a conductive coke breeze backfill to enhance the current distribution and reduce the anode-to-soil resistance [3, 12].

4. Transformer-Rectifier Units (TRUs)

The ICCP system relies on transformer-rectifier units to convert AC power to the DC current required for cathodic protection. For this bridge, seven TRUs were specified, one for each CP zone, ensuring localised control and optimization of the cathodic protection current. The TRUs were designed with adjustable output settings to accommodate variations in soil resistivity and to maintain the required protection potential across the steel piles. The TRUs were planned to be installed within weatherproof enclosures mounted on the bridge structure, providing protection against environmental factors and ensuring ease of maintenance. Cabling from the TRUs to the deep well anodes and to the pile connections was planned to route through conduits to safeguard against mechanical damage and environmental exposure [11-12].

5. Modelling and Simulation with SES CDEGS

Advanced computational modelling using SES CDEGS software was employed to simulate the electrical behaviour of the ICCP system and to ensure optimal performance [4, 6, 12]. This is detailed below.

a) Ground-Bed Resistance and Ground Potential Rise (GPR) Modelling: The software was used to calculate the resistance of the ground bed around the anodes and to model the potential distribution in the soil. These simulations assisted to predict the ground potential rise and to design the system such that the piles were outside the critical 80% soil voltage gradient, minimising the risk of interference with other structures.

b) Remoteness and Attenuation Calculations: The attenuation of the protective current along the length of the piles was analysed to ensure uniform potential rise from the surface to the bottom of the piles. This analysis was considered to verify that the cathodic protection was effective across the entire length of each pile, particularly in zones with varying soil resistivity.

c) System Optimization: Based on the simulation results, adjustments were made to the anode placement, TRU output settings, and the overall configuration of the ICCP system to achieve the desired protection levels. The iterative process of modelling and adjustment ensured that the final design provided comprehensive protection while being cost-effective and efficient.

RESULTS – DESIGN DATA

The design of the cathodic protection system for this bridge involved segmenting the bridge into seven cathodic protection zones, each covering three spans. These zones were defined based on the surface area ratio of the piles within each zone relative to the total surface area of all piles. Soil resistivity was measured for each zone, and the average resistivity was assigned to guide the CP system design for the respective zones. Total CP current required was calculated to be 214.8A to achieve compliant CP on the piles, i.e. 89 piles with diameters varying from 900mm to 1200mm and buried depths between 25m to 42m. This current is based on the current density of 20 mA/m^2 for bare steel in the soil as per AS2832.1 [1, 3].

Table 1 presents the CP zoning detail of the bridge. CP zones were determined from the ratio of the surface area of each zone to the total surface area of all piles. Seven CP zones were identified, i.e. each zone covering three spans as per Table 1. The soil resistivity was measured for each zone at the proposed location of CP systems using 4-pin Wenner test up to the depth of 32m. The average resistivity was assigned to the zone covered by the corresponding CP system.

In the next stage of the design, calculations were completed to determine the details of the deep well anode groundbed for each TRUs. That consisted of anode requirement by consumption, minimum backfill quantity, actual backfill mass, groundbed resistance, resistance of positive and negative circuits. All data were measured to determine the CPU rating and configuration of groundbed [9-12]. Examples are presented in Table 2.

TABLE 1 - NODE GROUNDBED DETAILS.

Zone	Structure	No. of Piles	Current Req. (A)	Zone Resistivity (ohm.cm)
TRU-1	Abutment A	3	32.74	6891
	Pier 1	6		
	Pier 2	6		
TRU-2	Pier 3	6	40.25	6891
	Pier 4	6		
	Pier 5	6		
TRU-3	Pier 6	6	28.20	7948
	Pier 7	3		
	Pier 8	3		
TRU-4	Pier 9	3	23.58	7948
	Pier 10	3		
	Pier 11	3		

Zone	Structure	No. of Piles	Current Req. (A)	Zone Resistivity (ohm.cm)
TRU-5	Pier 12	3	26.01	3330
	Pier 13	3		
	Pier 14	3		
TRU-6	Pier 15	3	28.52	3524
	Pier 16	3		
	Pier 17	3		
TRU-7	Pier 18	3	35.55	3821
	Pier 19	3		
	Abutment B	11		

TABLE 2: EXAMPLE OF MEASUREMENTS AND CALCULATIONS FOR CPU RATING

1. The number of anodes (A_N) required to provide protection over the design life is calculated based on the average protective current.
   $A_N = I_t/I_s = 7$ No.
   Where:
   I_m= 5 amps (Manufacturer's rated output)
   T_m= 20 yrs (Manufacturer's design life)
   I_t= 32.7 amps (Total system protective current)
   t_d = 20 yrs (Design Life, DL)
   I_d= 5 amps (Max. individual anode current for DL)
2. The minimum backfill mass (BM_{min})required to achieve the design life and protective current.
   $BM_{min} = C_m \times I_t \times t_d = 654$ kg
   Where:
   C_m= 1 kg/A yr (Backfill consumption rate)
3. The actual backfill mass (BF_A)to meet design life and protective current requirements.
   $BF_A = BF_d \times (GB_v - (N \times A_v))=1523$ kg
   Where
   A_v= 0.001 m^3 (Individual anode volume)
   GB_v= 1.287 m^3 (Individual active column volume)
   N= 10 (Design number of anodes)
   BF_d= 1188 Kg/m^3 (Backfill density)
4. The resistance of one ground (R_{gb}) is calculated using Dwight's equation for a single vertical rod.
   $R_{gb}= 0.005\rho(\ln(8L/d)-1)/(pi)L = 1.251$
5. The total resistance of the groundbed (GB_t) including the header cable and parallel anode configuration.
   $GB_t= 1/(1/R_{h1}+1/R_{h2}+...1/R_{hn})+R = 1.293$ ohms
6. The total positive circuit resistance (R_{pos}) including the main positive cable and anode groundbeds.
   $R_{pos}=R_m +GB_t = 1.298$ ohms
   Where:
   R_m= 0.052 ohms (Main positive cable resistance)
7. The total negative circuit resistance (R_{neg})
   R_{neg}= Main Negative Cable Res. = 0.010 ohms
8. The total circuit resistance is (R_{total})
   $R_{total}= R_s+R_{pos}+R_{neg} = 1.758$ ohms
   Where:
   R_s= Structure: electrolyte resistance = 0.45 ohms
9. The CPU Rating (V_{min}) is then assessed.
   $V_{min}= I_t \times R_{total} +V_{EMF} = 59.5$ Volts
   Where:
   V_{EMF}= 2 Volts (Back EMF)
10. Confirm the ground bed is adequate to maximise the output from the power supply.
   $V_{min} \leqslant 0.8 \times V_{OO}$
   Where:
   V_{OO} = 60.0Volts (80% of CPU Output Voltage)
11. Confirm the minimum pile to groundbed separation (GB_s) as per AS 2832.1
   $S=I_t \times \rho /2 \times (pi) \times V$= 14m structure, 133m foreign object
   Where at D meters away from anode
   V_p= 2 Volts (Soil potential on a pile)
   V_f= 0.2 Volts (Soil potential on foreign structure)

Deploying deep well anodes strategically under the bridge, midway between the piers, allowed for efficient use of the existing right-of-way, thereby reducing costs associated with securing additional space for installation [5, 9].

The next step was to calculate the cable size and length based on AS/NZS 3008. The size of cables was determined in two categories, including bond cables and feeder cables [9]. The optimum size of cables was calculated to be 95mm² and 35mm² for bond and feeder cables, respectively. FIGURE 2 shows an example of this measurement with a step-by-step calculation as an example.

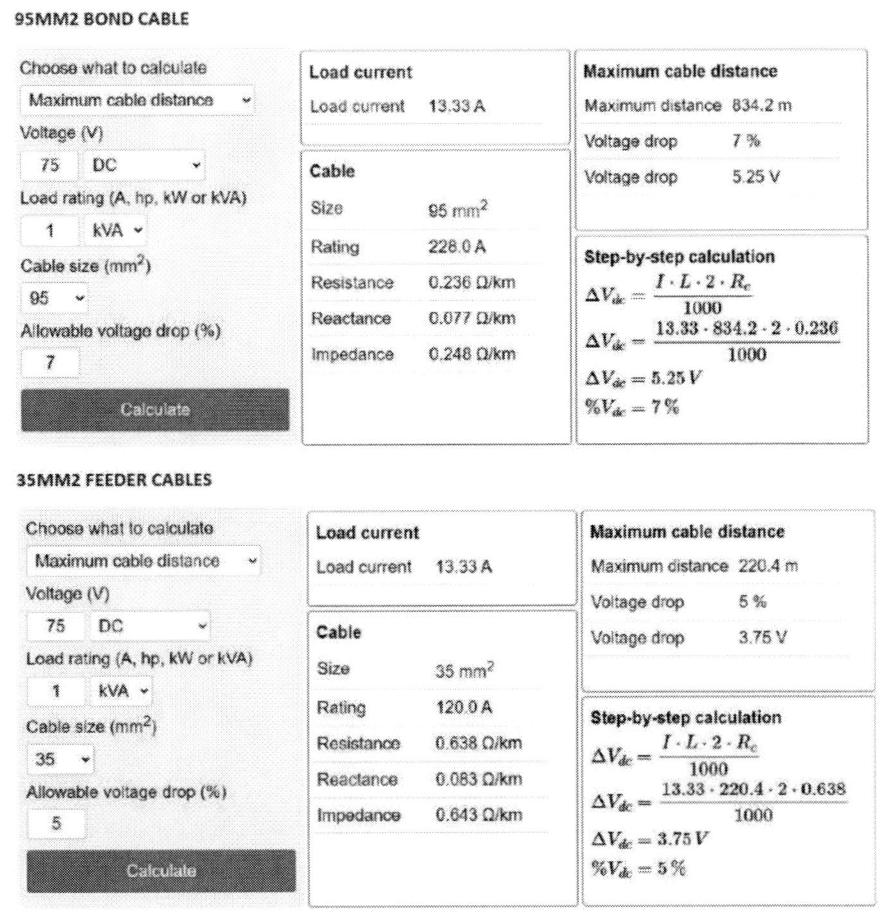

FIGURE 2: CABLE SIZE AND LENGTH CALCULATIONS AS PER AS/NZS 3008, FOR ONE LOCATION AS AN EXAMPLE.

Environmental considerations were taken into account in the design of the anode well. Three possible environmental risks with well drilling are surface water runoff, interchange between aquifers, and contamination by groundbed materials. To prevent the possibility of pollution by surface water runoff, the closed-hole system was deployed by considering the anode hole to be cased and sealed. A concrete slab (1500 mm x 1500 mm x 100 mm) and a covered pit were considered in the detailed design report for the installation of the holes to prevent pollutants from running off into the hole. To ensure the safety of the personnel and public, the ground potential riser surrounding anode wells was evaluated. Step and touch voltage calculations were performed for all anode wells and it was ensured that ground potential rise and the resulting step and touch potentials around the anode wells are within the safe limits for the personnel and public. FIGURE 3 shows an example of safety calculation. Note that the remoteness and attenuation calculations were performed to ensure the piles are outside an 80% soil voltage gradient around wells, and also the top and bottom of the piles receive a uniform potential rise. Advanced modelling techniques, including the use of SES CDEGS software, were employed for this simulation. These simulations ensured that the potential rise was uniformly distributed along the length of the piles, from the surface to the deeper sections.

This approach not only provided comprehensive protection against corrosion but also optimised the efficiency and effectiveness of the cathodic protection system.

```
REPORT 1
============< G R O U N D I N G    ( SYSTEM INFORMATION SUMMARY )
>===========

 Run ID.......................................: ANODE-WELLS 1& 2
 System of Units .............................: Metric
 Earth Potential Calculations.................: Single Electrode Case
 Type of Electrodes Considered................: Main Electrode ONLY
 Soil Type Selected...........................: Multi-Layer Horizontal
 SPLITS/FCDIST Scaling Factor.................:   1.0000
1
1

     MULTI-LAYER EARTH CHARACTERISTICS USED BY PROGRAM
```

LAYER No.	TYPE	REFLECTION COEFFICIENT	RESISTIVITY (ohm-meter)	THICKNESS (METERS)
1	Air	0.00000	0.100000E+11	Infinite
2	Soil	-0.999990	65.9700	1.00000
3	Soil	-0.179088	45.9300	1.00000
4	Soil	-0.364436E-01	42.7000	2.00000
5	Soil	0.424645	105.730	2.00000
6	Soil	-0.442133	40.9000	2.00000
7	Soil	0.318674	79.1600	2.00000
8	Soil	0.666745E-01	90.4700	2.00000
9	Soil	-0.132858	69.2500	4.00000
10	Soil	0.783257E-01	81.0200	4.00000
11	Soil	-0.129513	62.4400	4.00000
12	Soil	0.255781	105.360	4.00000
13	Soil	-0.470071	37.9800	4.00000
14	Soil	0.00000	37.9800	Infinite

```
1
 CONFIGURATION OF MAIN ELECTRODE
 ===============================

 Original Electrical Current Flowing In Electrode..:   40.000    amperes
 Current Scaling Factor (SPLITS/FCDIST/specified)..:   1.0000
 Adjusted Electrical Current Flowing In Electrode..:   40.000    amperes
 Number of Conductors in Electrode.................:     1
 Resistance of Electrode System....................:   1.2833    ohms

 SUBDIVISION
 ===========
 Grand Total of Conductors After Subdivision.:    13

 Total Current Flowing In Main Electrode......:   40.000    amperes
 Total Buried Length of Main Electrode........:   25.992    meters
```

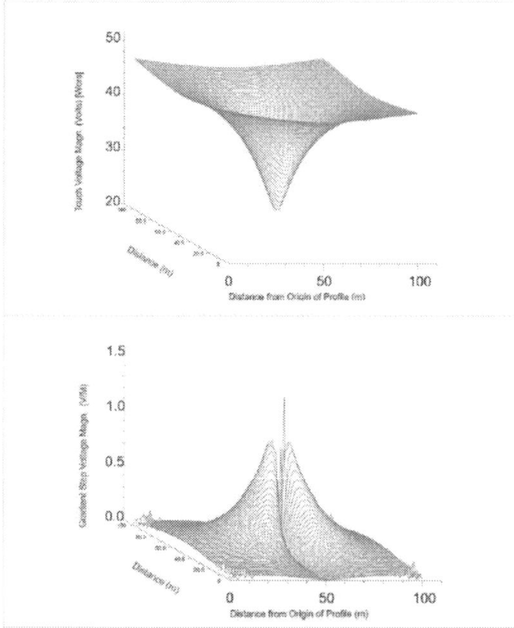

STEP AND TOUCH POTENTIAL AROUND ANODE WELL 1 (PASS IEC c1 curve)

FIGURE 3: CALCULATED STEP AND TOUCH, FOR ONE LOCATION AS AN EXAMPLE.

CONCLUSIONS

The design of the cathodic protection system for this bridge involved segmenting the bridge into seven cathodic protection zones, each covering three spans. These zones were defined based on the surface area ratio of the piles within each zone relative to the total surface area of all piles. Soil resistivity was measured for each zone, and the average resistivity was assigned to guide the CP system design for the respective zones. Deploying deep well anodes strategically under the bridge, allowed for efficient use of the existing right-of-way, thereby reducing costs associated with securing additional space for installation.

In summary, the detailed design of the ICCP system for this bridge demonstrated a significant advancement in protecting vital infrastructure from soil-induced corrosion. The project served as a case study in the effective application of cathodic protection techniques, offering valuable insights and methodologies that can be applied to similar infrastructure projects globally. As the bridge stands poised to facilitate improved transportation and connectivity in New South Wales, its foundation is secured against the ravages of corrosion, ensuring its longevity and serviceability for decades to come.

ACKNOWLEDGMENTS

The authors acknowledge that TfNSW and Jacobs were instrumental in determining the need for the CP system and establishing a review panel including GHD that gave valuable contributions to the final design.

REFERENCES

1. Ali S, Papworth F. – Detailed design report, 2022
2. Tomlinson M, Woodward J: Pile Design and Construction Practice, 6th Edition, CRC Press, 2015.
3. Cathodic Protection of Metals, Parts 1,2 & 3, AS2832, 2015.
4. Design, Installation, Operation, and Maintenance of Impressed Current Deep Groundbeds, NACE RP0572.
5. Electrical installations (known as the Australian/New Zealand wiring rules), AS/NZS 3000 - 2018.
6. Corrosion of metals and alloys – Basic terms and definitions, ISO 8044 - 2015.
7. What is Corrosion see: https://membership.corrosion.com.au/blog/what-is-corrosion/ Last accessed 26 March 2023.
8. Degrees of protection provided by enclosures for electrical equipment (IP Code) – Wall chart 2 (Supplement 2 to AS1939 - 1990), AS 1939 Supp2
9. Electrical installations - Selection of cables - Cables for alternating voltages up to and including 0.6/1 kV - Typical Australian installation conditions, AS/NZS 5000.1 – 2005
10. Australian Standard AS 2159-2009 Pilling – Design and installation
11. Bridge design Rehabilitation and strengthening of existing bridges, AS 5100.8 – 2017
12. NACE Cathodic Protection Specialist CP4 Course Manual 2013

AUTHOR DETAILS

	Saeed Ali is a senior materials and durability engineer at BCRC. He is a specialist in Corrosion and technical protection, and certified as a corrosion technologist by ACA. He is responsible for providing durability investigation and materials performance services to the industry.
	Frank Papworth has + 45 years' experience in concrete durability, inspection and repair, and corrosion. He has chaired CIA's durability committee, chaired fib COM8 on durability.

CORROSION & PREVENTION
CAIRNS 2024

THE AUSTRALASIAN CORROSION ASSOCIATION INC.

SOME MICROSTRUCTURAL INFLUENCES UPON LOCALISED CORROSION

N. Birbilis[1,*]

[1] *Faculty of Science, Engineering and Built Environment, Deakin University, Waurn Ponds, VIC, Australia.*

**Corresponding author email: nick.birbilis@deakin.edu.au*

ABSTRACT

The degradation of metallic alloys is a process that is, over extended time periods, basically assured. The 'engineering alloys' that are critical to essentially all facets of modern life, are highly engineered to 'cheat nature'. That is, immense amounts of energy are utilised to liberate ores (and specifically, oxides) to create metals and alloys. There is a considerable thermodynamic driving force for the reversion of metals and alloys to the ore-like state in atmospheric conditions – which gives rise to corrosion. It therefore follows that the composition, and the 'structure' of alloys will play a critical – and deterministic – role in how such reversion (i.e. corrosion) occurs. A foundational understanding of the role of alloy structure (specifically what metallurgists refer to as microstructure) is critical to understanding corrosion and its genesis at the functional length scale. Herein, some examples of microstructural effects on corrosion are touched up, including how they may impact early / localised corrosion.

Keywords: localised corrosion, alloys, microstructure, aluminium alloys, electron microscopy

INTRODUCTION

The impact of corrosion on all facets of our way of life is profound. Whilst occurring silently, corrosion costs the economy of most nations approximately 3-4% of GDP, annually[1]. Studies accounting for the cost of corrosion include the recent Impact Study (USA)[2], the recent Cost of Corrosion in China study[3], and a report released in June 2021 commissioned by the Australasian Corrosion Association ('Impact of Corrosion in Australasia report[4]). It was estimated up to $78 billion per annum is being spent on remediating assets affected by corrosion in Australia - along with the associated corrosion prevention measures. What is consistent across all studies regarding the impact of corrosion, is that the effects have a major influence on essentially every industry and most industrial processes.

There is a clear and ongoing (urgent) need for the development of more corrosion resistant alloys, with enhanced durability. But how? In order to begin to answer such a question, one must deterministically be able to rationalise the origins of corrosion – before having the 'tool kit' to be able to develop materials that can be more corrosion resistant by design. In other words, fundamental science needs to be the enabler that allows 'applied' outcomes – in service of society (where metallic alloys are now indispensable in energy, healthcare, communications and transportation).

One of history's finest examples of being able to link fundamental science to applied outcomes was a unique individual called Percival Faraday Thompson (PFT), (1885-1951), who was a metallurgist and analytical chemist – that worked a long career in the field of corrosion[5]. Much has been said regarding PFT, including his ability to be a thought leader and pioneer, but also a great scientist and educator. I will not repeat all such facets here, but whole-heartedly direct readers to wonderful relevant works[6,7]. One important statement that is attributed to PFT that rings true (and has, for over a century) is the following quote, "*Around no subject of technical interest have prejudice and wrong thinking in the past, so wrapped a web of obscurity as that of the corrosion of materials*". Such a statement can be interpreted thousands of ways, however what is still true, and remains true, is that there is ongoing obscurity regarding the deterministic origins of corrosion.

There are however, admittedly, good reasons why there has been ongoing obscurity regarding the deterministic origins of corrosion. As one example, advanced analytical microscopy was not readily available when many (if not, most) of the common engineering alloys were developed. The technology required to assess corrosion – and the factors controlling it – have emerged in earnest in the present millennium. Momentarily neglecting the consideration of corrosion, one may consider the development of other properties in engineering alloys. Key properties that are relevant to material usage include the yield strength of a material, the ductility of a material, and (mechanical) damage tolerance often assessed by quantities defined as tensile strength or toughness. The role of microstructural features on the development of such mechanical properties has been an endeavour of science that developed at a pace much more rapidly than say, the understanding of microstructural features on the development of corrosion properties of a material. Why is that? Perhaps PFT had some thoughts on why, but as a discipline, corrosion has (for reasons that defy all logic) remained an 'after thought'. In spite of nearly a century of studies on the cost of corrosion, the long-term economic perils of the corrosion menace have not been an adequate warning to much of global industrial practice (where, perhaps for reasons non-scientific, but driven by financial immediacy) to better understand corrosion. The good news is that things are changing. At the macro scale, practices such as durability management planning have been emerging across the infrastructure sector, which is very welcome. However, I will confine this paper to the scientifically interesting 'microscale'. So, what has been happening at the 'microscale' to better inform our understanding of corrosion? There are many things one could respond with, but I like a statement that was made by a professor in the USA by the name of Rudolph Buchheit, who after a seminal study in 1997[8] proclaimed that our understanding of corrosion is about to change forever. What Buchheit was referring to, was that the resolution from images captured by electron microscopes employing a field emission gun (FEG) as the electron source, were exceptional. Images from a FEG scanning electron microscope (SEM), were able to see localised corrosion occurring at the nano-scale, with unprecedented clarity previously not possible. In quick succession since then, the advent of analytical FEG-TEM (transmission electron microscopy), STEM (scanning transmission electron microscopy) and a method termed EBSD (electron backscatter diffraction) have permitted significant advances in our understanding and rationalisation of the genesis of corrosion; specifically, the role of microstructure on corrosion of engineering alloys. Some examples the author has had the pleasure of being involved with are concisely elaborated below.

Discussion

Examples herein are intended to provide a sample of experiments and studies that have contributed to deterministically explaining the role of microstructure on corrosion of engineering alloys. The author concedes the examples are a synopsis of their work, whilst noting there are exceptional studies by others that have been occurring over the past decade which merit exploration from interested readers (i.e. Kosari[9], Kosari[10], Gharbi[11], Williams[12], and Ryan[13]).

The role of grain size on corrosion

Engineering alloys undergo a range of processing conditions (thermo-mechanical processing) following their casting. These different processing conditions can result in a range of grain sizes in the material microstructure, as depicted for pure aluminium (Al) in Figure 1.

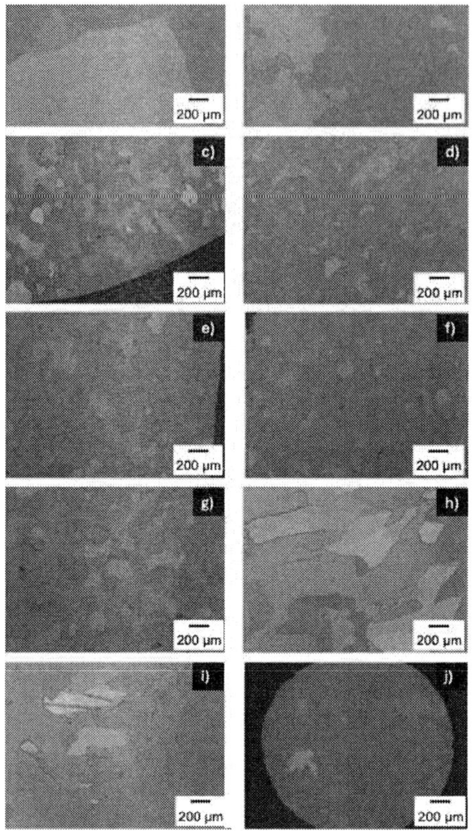

FIGURE 1: SAMPLE OPTICAL MICROGRAPHS OF HIGH PURITY ALUMINIUM AFTER DIFFERENT PROCESSING ROUTES AND HEAT TREATMENT TIMES AT 200°C: (a) AS-CAST 0 MIN, (b) ECAP 1-PASS 0 MIN, (c) ECAP 4-PASS 0 MIN, (d) ECAP 8-PASS 0 MIN, (E) SMAT 0 MIN, (f) SMAT 10 MIN, (g) SMAT 60 MIN, (H) COLD-ROLLED 0 MIN, (i) CRYO-ROLLED 0 MIN, AND (j) Al WIRE 10 MIN. IN ALL CASES THE SCALE BAR IS 200 MICROMETRES[14].

The ability to study the grain size of engineering alloys was rapidly advanced with the advent of electron backscatter diffraction (EBSD), over the past two decades. An early EBSD image from a magnesium specimen is shown in Figure 2, taken by the author[15]. The advent of EBSD not only allows the ready assessment of grain sizes, but has the advantages including: (i) being able to quantify very fine grains, (ii) being able to also (simultaneously) quantify grain orientation, and (iii) accurate determination of grain size for grains that are irregularly shaped or over a wide size-range distribution.

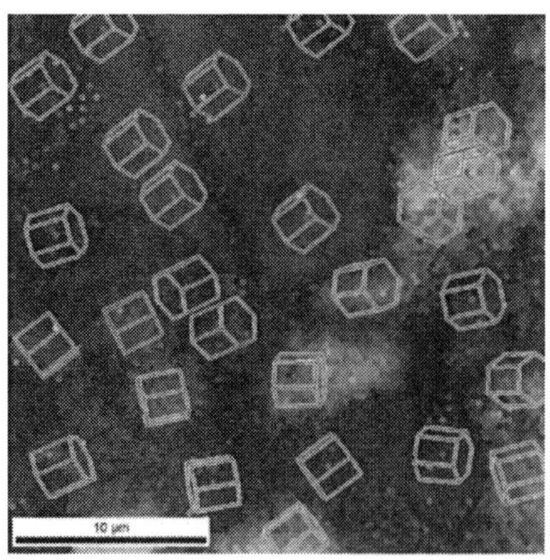

FIGURE 2: UNIQUE GRAIN EBSD MAP OF MAGNESIUM (99.9%) AFTER FOUR ECAP PASSES.

Through extensive review[16] and original works[17], a formalism relating the corrosion rate of engineering alloys (and pure metals) to grain size was developed. This is depicted by Eqn. 1, where A and B are constants that depend on the environment / test conditions, and *gs* is grain size.

$$i_{corr} = (A) + (B)gs^{-0.5} \qquad \text{Eqn (1)}$$

THE ROLE OF PRECIPITATE SIZE ON CORROSION

Through many works in the 1990's using scanning electron microscopy, the localised nature of corrosion (in the origins of corrosion) was being exposed. Through the study of light alloys, the works of Buchheit, Scully, Nisancioglu, and several other groups across the world – showed that localised corrosion was being defined as a 'deterministic' process. Whilst the concise paper herein will not elaborate on the extensive works carried out on stainless steels and brasses (and, there is much exceptional work to speak of), the case study of aluminium alloys is an important example to elaborate on. This is because (i) it was noted that 'particles' of varying types (and sizes) in aluminium alloys influence the localised corrosion process, and (ii) aluminium alloys develop their mechanical properties through aging processes, that stimulate the growth of hardening precipitates. This latter point is one that was exploited by the author, in order to study the critical 'precipitate' (i.e. particle) size that may trigger particle / second phase localised corrosion. By looking at the rate of pitting (as electrochemically measured from the metastable pitting frequency using potentiostatic transient testing[18]), as a function of aging time, one may determine at what critical aging time the alloy undergoes a transition from 'no localised corrosion', to, 'localised corrosion' prone. This is demonstrated in Figure 3, for a model Al-Cu-Mg alloy.

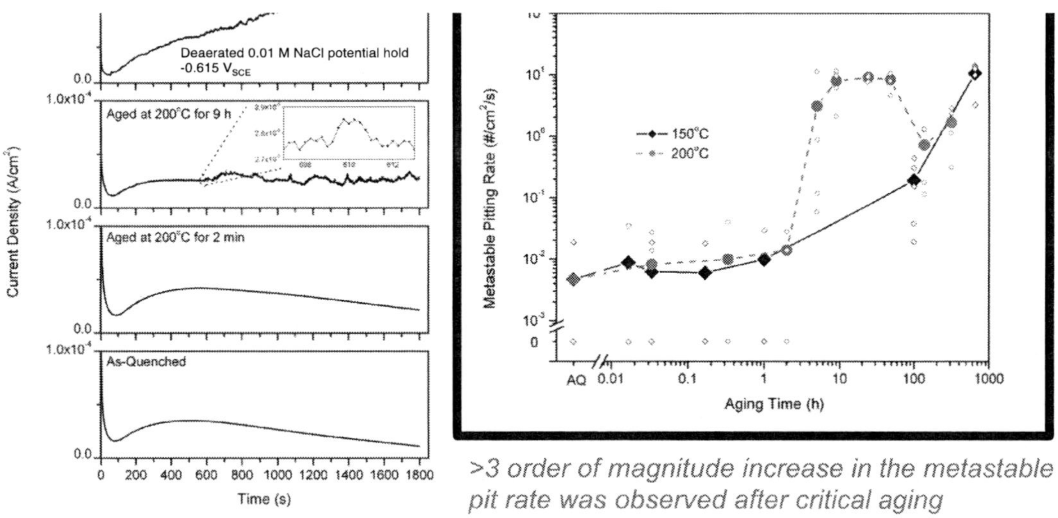

FIGURE 3: METASTABLE PITTING RATES FOR AL–1.1Cu–1.7Mg (at. %) AS A FUNCTION OF AGING TIME AT 150 °C AND 200 °C. RESULTS FROM INDIVIDUAL TRANSIENT EXPERIMENTS ARE DISPLAYED AS EMPTY DIAMONDS AND CIRCLES FOR 150 °C AND 200 °C, RESPECTIVELY.

Through correlative high-resolution scanning-TEM, it was possible to determine a critical precipitate thickness below which pitting did not occur, and above which pitting occurred[19]. This is depicted in Figure 4, where the images reveal the critical thickness for precipitate (i.e. particle) induced localised corrosion is at ~5 nm. This notion of a critical thickness below which localised corrosion did not occur, was subsequently demonstrated (again with advanced correlative microscopy) for other alloys, including Al-Zn-Mg (7xxx series) alloys[20], and Al-Cu-Li (2xxx series) alloys[21].

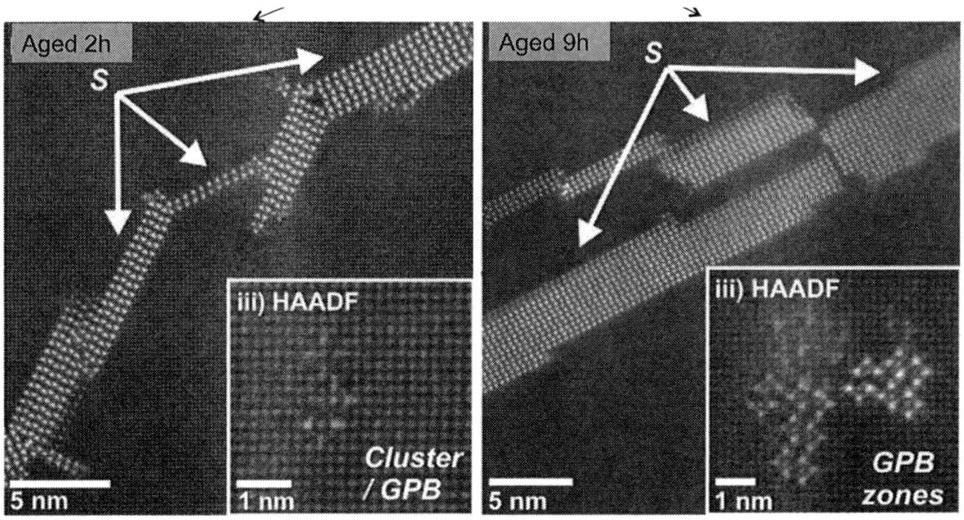

FIGURE 4: BRIGHT FIELD (BF) AND HIGH ANGULAR ANNULAR DARK FIELD (HAADF) SCANNING TRANSMISSION ELECTRON MICROGRAPHS (STEM) OF REPRESENTATIVE MICROSTRUCTURES IN SAMPLES AGED: (LEFT) 2 H AT 200 °C, (RIGHT) 9 H AT 200 °C. INSETS ARE HIGHER MAGNIFCATION IMAGES OF ATOMIC SCALE CLUSTERS THROUGHOUT THE ALLOY MATRIX.

Corrosion mechanisms at the nano/atomic scale

Ongoing developments in transmission electron microscopy allowed the simultaneous imaging and x-ray mapping through scanning transmission electron microscopy. The use of field emission electron sources also allowed imaging and analytical x-ray mapping to occur at high resolution. The aforementioned early works of Buchheit had revealed that localised corrosion occurred on Al-Cu-Mg alloys, also indicating that the process of localised (particle induced) corrosion was associated with the so-called 'de-alloying' process. That is, corrosion of particles did not occur uniformly in the alloys they populate. The particle itself would undergo a process of incongruent dissolution, which is directly akin to 'de-alloying'. Through meticulous experiments, Kairy developed a quasi in-situ process for tracking localised corrosion at the nanoscale as a function of time (i.e. time of exposure to a corrosion environment). The results of such testing are depicted in Figure 5 [22] for Al-alloys used in automotive applications.

What is seen from Figure 5, is that incongruent dissolution occurs for a precipitate, namely $Al_4Cu_3Mg_7Si_6$, at the nanoscale. Time dependent imaging reveals the preferential dissolution of aluminium; but also revealed the dissolution and redeposition of copper. A closer look at the redeposited copper is provided in Figure 6, revealing that essentially pure Cu is redeposited on the alloy surface following incongruent dissolution of $Al_4Cu_3Mg_7Si_6$. This is depicted in Figure 6, where it is also revealed that redeposited Cu forms with the inclusion of so-called 'nano-twins'. The broader works that relate to this system [23,24] (Al-Mg-Si-Cu) are also foundational in suggesting that all localised corrosion (in all alloys) occurs by a process of incongruent dissolution.

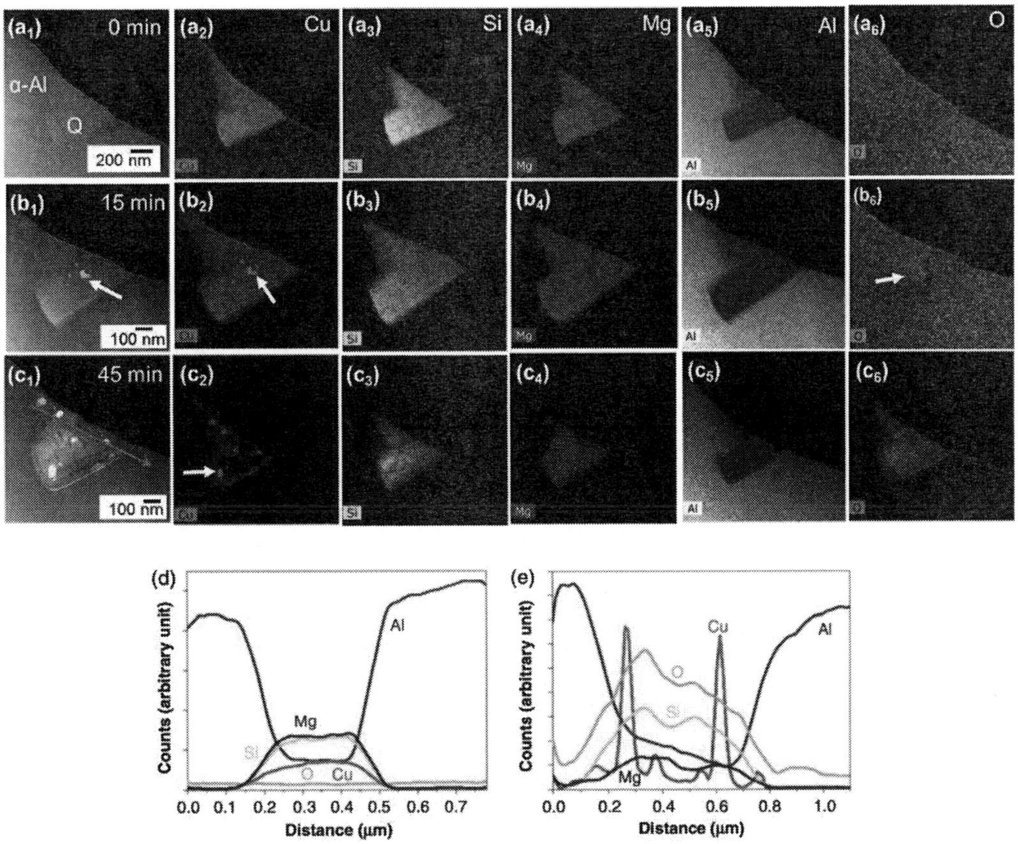

FIGURE 5: HAADF-STEM IMAGES OF Q-PHASE ($Al_4Cu_3Mg_7Si_6$) IN THE MATRIX OF Al-0.9Si-0.74Mg-0.84Cu-0.08Fe-0.14Mn (wt. %) SHEET AGED AT 400°C FOR 30 D, AS A FUNCTION OF IMMERSION TIME IN 0.1 M NaCl; (a1), (b1), AND (c1) ARE THE HAADF IMAGES PRIOR TO IMMERSION, AND AFTER 15 MIN AND 45 MIN IMMERSION, RESPECTIVELY, AND (a2) THROUGH (a6), (b2) THROUGH (b6), AND (c2) THROUGH (c6) ARE THE EDS MAPS CORRESPONDING TO (a1), (b1), AND (c1), RESPECTIVELY. (A1) THE NANO-Q-PHASE PRECIPITATE IN THE α-Al MATRIX. ARROW IN (b1) POINTS TO THE HIGH ATOMIC NUMBER CONTRAST NANOPARTICLES ON Q-PHASE, AND THE CORRESPONDING REGIONS IN (b2) AND (b6) SHOW Cu AND O ENRICHMENT, RESPECTIVELY. ARROW IN (c2) POINTS TO THE Cu ENRICHMENT IN Q-PHASE. (d) AND (e) EDS LINE ANALYSIS ALONG THE ARROWS IN (a1) AND (c1), RESPECTIVELY.

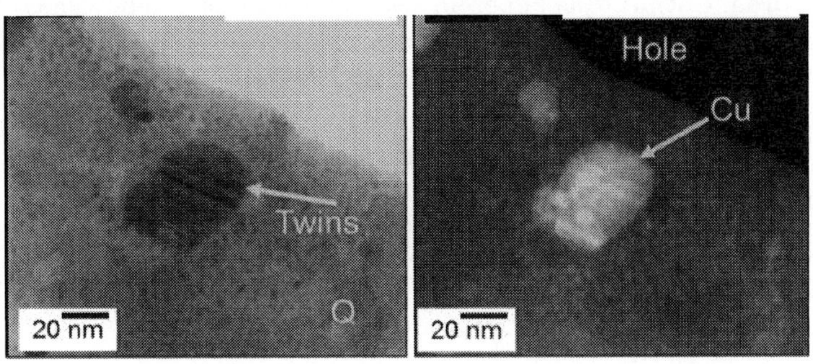

FIGURE 6: (A) AND (B) HIGH-MAGNIFICATION BF- AND HAADF-STEM IMAGES OF THE Q-PHASE ($AL_4CU_3MG_7SI_6$) PRECIPITATE, IN THE MATRIX OF Al-0.9Si-0.74Mg-0.84Cu-0.08Fe-0.14Mn (wt. %) SHEET AGED AT 400°C FOR 30 D, AFTER 45 MIN IMMERSION IN 0.1 M NaCl. ARROW IN (A) SHOWS TWINS ON THE Cu PARTICLE.

CORROSION OF COMPLEX ENGINEERING ALLOYS

The corrosion of engineering alloys is a deterministic result of the alloy microstructure, as noted from the examples provided above. However, there is still much to learn. As one example, the emerging class of additively manufactured alloys (i.e. produced by 3D printing), pose some unique microstructures and morphologic traits [25], as shown in Figure 7.

Understanding (deterministically) the corrosion of additively manufactured alloys is an area of enquiry that requires much future work. The same is true for the emerging class of so-called multi-principal element alloys[26].

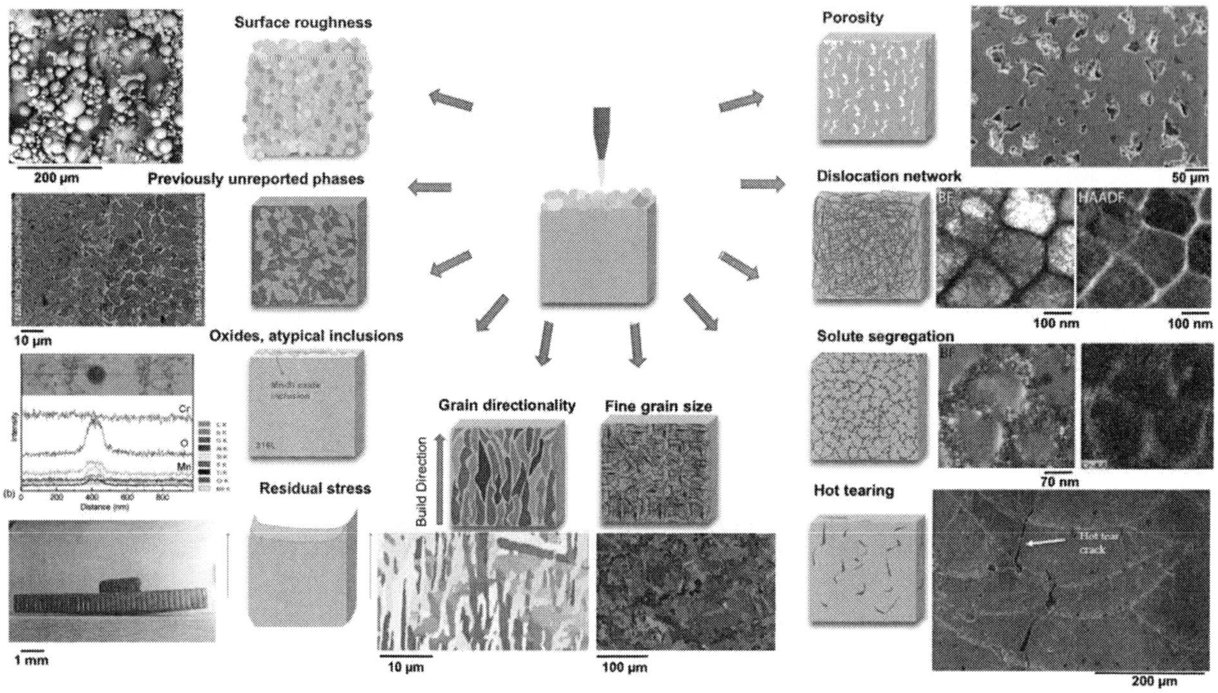

FIGURE 7: SCHEMATIC DIAGRAM SUMMARIZING THE POTENTIAL IMPACT OF AM PROCESSING ON THE OVERALL MICROSTRUCTURE OF THE PRODUCED COMPONENTS. THE POTENTIAL INFLUENCE OF EACH PROCESSING VARIABLE IS ILLUSTRATED BY DATA REPORTED IN THE LITERATURE.

CONCLUSIONS

Corrosion is a deterministic process, that occurs in a manner that is rational (based on the metal/alloy microstructure, and its environment). The paper herein has shown examples of how microstructure will influence corrosion. The examples herein are not exhaustive, and are admittedly confined to a short review of only some works done by the author. The study of microstructural effects on corrosion is (in spite of the wide use of engineering alloys) still rather nascent. There however remains much work to do, to understand the many facets of microstructure, and how they deterministically influence corrosion.

A more quantitative and definitive understanding of microstructural effects on corrosion is a necessary platform for predictive modelling of alloy corrosion (and degradation more generally). Corrosion models that can predict corrosion, its propagation and rate, will serve to inform durability management tools – but more importantly be the platform from which corrosion resistant alloys may be rationally designed.

ACKNOWLEDGMENTS

This concise paper has utilised the exceptional work of many researchers (now professional engineers and academics) over the past two decades. These include Dr. Kevin Ralston, A/Prof. Rajeev Gupta, Dr. Shravan Kairy, Dr. Oumaïma Gharbi and Dr. Victor Cruz de Faria. A special thanks to Professor Rudy Buchheit for many sustained discussions over the years, regarding localised corrosion.

REFERENCES

[1] Bhaskaran, R., et al. *Corrosion: Materials* 13 (2005): 621-628 (https://doi.org/10.31399/asm.hb.v13b.a0003968)

[2] http://impact.nace.org/ (accessed 3 Oct, 2024)

[3] Hou, B., et al., *npj Materials Degradation* 1.1 (2017): 1-10.

[4] https://membership.corrosion.com.au/blog/impact-of-corrosion-in-australasia-report/ (accessed 3 Oct, 2024)

[5] Potter EG, The corrosion scene in Australia, Australasian Corrosion Engineering, March, (1972) 21-29.

[6] Will, G., S. Denman, M. Rahman, and J. Davies. "Electrochemistry, Corrosion and Data Analytics: A Pathway to Education and Better Asset Integrity." Proceedings of The Australasian Corrosion Conference 2023 (available at: https://www.corrosion.com.au/electrochemistry-corrosion-and-data-analytics-a-pathway-to-education-and-better-asset-integrity/) (accessed 3 Oct, 2024)

[7] Green W, Hinton B, P F Thompson - Australia's pioneer in corrosion research and education (2015) p. (available at https://hdl.handle.net/10536/DRO/DU:30081795) (accessed 3 Oct, 2024)

[8] Buchheit, R. G., R. P. Grant, P. F. Hlava, B. Mckenzie, and G. L. Zender. "Local dissolution phenomena associated with S phase (Al2CuMg) particles in aluminum alloy 2024-T3." *Journal of the electrochemical society* 144, no. 8 (1997): 2621.

[9] Kosari, A., H. Zandbergen, F. Tichelaar, P. Visser, P. Taheri, H. Terryn, and J. M. C. Mol. "In-situ nanoscopic observations of dealloying-driven local corrosion from surface initiation to in-depth propagation." *Corrosion Science* 177 (2020): 108912.

[10] Kosari, A., F. Tichelaar, P. Visser, H. Zandbergen, H. Terryn, and J. M. C. Mol. "Dealloying-driven local corrosion by intermetallic constituent particles and dispersoids in aerospace aluminium alloys." *Corrosion Science* 177 (2020): 108947.

[11] Gharbi, O., D. Jiang, D. R. Feenstra, S. K. Kairy, Y. Wu, C. R. Hutchinson, and N. Birbilis. "On the corrosion of additively manufactured aluminium alloy AA2024 prepared by selective laser melting." *Corrosion Science* 143 (2018): 93-106.

[12] Williams, David E., Matt R. Kilburn, John Cliff, and Geoffrey IN Waterhouse. "Composition changes around sulphide inclusions in stainless steels, and implications for the initiation of pitting corrosion." *Corrosion science* 52, no. 11 (2010): 3702-3716.

[13] Ryan, Mary P., David E. Williams, Richard J. Chater, Bernie M. Hutton, and David S. McPhail. "Why stainless steel corrodes." *Nature* 415, no. 6873 (2002): 770-774.

[14] Ralston, K. D., Dan Fabijanic, and Nick Birbilis. "Effect of grain size on corrosion of high purity aluminium." *Electrochimica acta* 56, no. 4 (2011): 1729-1736.

[15] Birbilis, Nick, K. D. Ralston, Sannakaisa Virtanen, H. L. Fraser, and C. H. J. Davies. "Grain character influences on corrosion of ECAPed pure magnesium." *Corrosion Engineering, Science and Technology* 45, no. 3 (2010): 224-230.

[16] Ralston, K. D., and Nick Birbilis. "Effect of grain size on corrosion: a review." *Corrosion* 66, no. 7 (2010): 075005-075005.

[17] Ralston, K. D., Nick Birbilis, and C. H. J. Davies. "Revealing the relationship between grain size and corrosion rate of metals." *Scripta Materialia* 63, no. 12 (2010): 1201-1204.

[18] Ralston, K. D., Nick Birbilis, M. K. Cavanaugh, Matthew Weyland, B. C. Muddle, and R. K. W. Marceau. "Role of nanostructure in pitting of Al–Cu–Mg alloys." *Electrochimica Acta* 55, no. 27 (2010): 7834-7842.

[19] Ralston, K. D., Nick Birbilis, Matthew Weyland, and C. R. Hutchinson. "The effect of precipitate size on the yield strength-pitting corrosion correlation in Al–Cu–Mg alloys." *Acta Materialia* 58, no. 18 (2010): 5941-5948.

[20] Gupta, Rajeev Kumar, Alexis Deschamps, Mary K. Cavanaugh, Stan P. Lynch, and Nick Birbilis. "Relating the early evolution of microstructure with the electrochemical response and mechanical performance of a Cu-rich and Cu-lean 7xxx aluminum alloy." *Journal of The Electrochemical Society* 159, no. 11 (2012): C492.

[21] Yan, Y., L. Peguet, O. Gharbi, Alexis Deschamps, C. R. Hutchinson, S. K. Kairy, and N. Birbilis. "On the corrosion, electrochemistry and microstructure of Al-Cu-Li alloy AA2050 as a function of ageing." *Materialia* 1 (2018): 25-36.

[22] Kairy, S. K., P. A. Rometsch, C. H. J. Davies, and N. Birbilis. "On the electrochemical and quasi in situ corrosion response of the Q-phase (AlxCuyMgzSiw) intermetallic particle in 6xxx series aluminum alloys." *Corrosion* 73, no. 1 (2017): 87-99.

[23] Kairy, S. K., P. A. Rometsch, C. H. J. Davies, and N. Birbilis. "On the intergranular corrosion and hardness evolution of 6xxx series Al alloys as a function of Si: Mg ratio, Cu content, and aging condition." *Corrosion* 73, no. 10 (2017): 1280-1295.

[24] Kairy, Shravan K., Talukder Alam, Paul A. Rometsch, Chris HJ Davies, Raj Banerjee, and Nick Birbilis. "Understanding the origins of intergranular corrosion in copper-containing Al-Mg-Si alloys." *Metallurgical and Materials Transactions A* 47 (2016): 985-989.

[25] Sander, G., J. Tan, P. Balan, O. Gharbi, D. R. Feenstra, L. Singer, S. Thomas, R. G. Kelly, John R. Scully, and Nick Birbilis. "Corrosion of additively manufactured alloys: a review." *Corrosion* 74, no. 12 (2018): 1318-1350.

[26] Ghorbani, M., Z. Li, Y. Qiu, P. Marcus, J. R. Scully, O. Gharbi, H. Luo et al. "Current progress in aqueous corrosion of multi-principal element alloys" *Metallurgical and Materials Transactions A* (2024): 1-18.

AUTHOR DETAILS

Nick Birbilis is the Executive Dean of the Faculty of Science, Engineering and Built Environment at Deakin University, Australia. His research has sought to rationalise the behaviour of engineering alloys from a deterministic perspective. His interests also include computational studies of materials and machine learning tools to accelerate materials discovery. His awards include the H.H. Uhlig Award, U.R. Evans Award, ATSE Batterham Medal, ACA Corrosion Medal; along with recognition as a Fellow of ASM International, The Electrochemical Society, International Society of Electrochemistry, NACE/AMPP and Engineers Australia.

CORROSION & PREVENTION
CAIRNS 2024

THE AUSTRALASIAN CORROSION ASSOCIATION INC.

PORTS: ANCHORS OF THE ENERGY TRANSITION

C. Crawshaw[1,*], S. Barker[2]

[1]*Infrastructure Advisory Group, VIC, 3000, Australia*

[2] *Infrastructure Advisory Group, QLD, 4000, Australia*

Corresponding author email: Christine.crawshaw@infragroup.com.au

ABSTRACT

Australia's ambitious target to achieve net zero emissions by 2050 and integrate 82% renewable energy into its electricity grid by 2030 underscores the critical role of Offshore Wind in the nation's energy transition. With extensive wind zones identified along our coastline, offshore wind will be pivotal to achieve our ambition. However, this renewable future hinges on the readiness of Australia's ports, which will serve as essential hubs for construction, logistics, and long-term maintenance of offshore wind farms. Port infrastructure upgrades are necessary to accommodate the scale and complexity of offshore wind development, especially in managing the increased size of turbine components, specialised vessels, and operational demands. Additionally, the expansion of offshore wind will drive significant demand for corrosion technologists, not only for port infrastructure but also across the entire Offshore Wind asset lifecycle—from manufacturing and storage to installation and ongoing management. This paper explores the corrosion challenges inherent in offshore wind, including material selection, protective coatings, cathodic protection, and maintenance strategies to extend the life of wind structures. It further discusses how the corrosion industry can proactively prepare for this transition, ensuring resilience in both Port infrastructure and offshore wind operations. With international examples guiding the way, the Australian corrosion sector must now take active steps to meet the demands of this emerging energy landscape. How prepared are we for this transformation?

Keywords: Offshore Wind, Port Infrastructure, Corrosion Protection, Energy Transition, Renewable Energy

FIGURE 1: RENEWABLE ENERGY OFFSHORE WIND FARM [1]

INTRODUCTION

As Australia seeks to transition to a low-carbon economy, the nation's commitment to achieving net zero emissions by 2050 and increasing renewable energy generation to 82% of its electricity grid by 2030 has intensified the focus on offshore wind as a critical component of this energy transition. Offshore wind offers significant potential to harness vast marine wind resources and support decarbonisation efforts, making it a central pillar in the country's renewable energy strategy. With several offshore wind zones identified along Australia's extensive coastline, ports are set to become vital infrastructure hubs that will enable the construction, operation, and maintenance of this new energy frontier.

The readiness of ports is a crucial factor in enabling offshore wind development. Given the scale and complexity of offshore wind projects, including the transport and assembly of large turbine components, specialised vessels, and the need for long-term maintenance, Australian ports must undergo significant upgrades to meet the demands of this burgeoning industry. Furthermore, these developments will increase the need for corrosion management, both for new and existing port infrastructure and across the lifecycle of offshore wind assets. Corrosion technologists will play a vital role in ensuring the longevity and durability of these structures in challenging marine environments.

This paper will explore the intersection of port infrastructure and offshore wind, examining the upgrades necessary for ports to support the energy transition, with a particular focus on corrosion considerations. It will discuss the critical role of the corrosion industry in preparing for this transition and draw on lessons from international experiences in offshore wind development. The aim is to offer insights into how the Corrosion workforce in Australia can effectively prepare for the future of renewable energy.

Discussion

Australia aims to achieve net zero emissions by 2050 and 82% renewable energy in its electricity grid by 2030

The federal and state governments are investing heavily in renewable projects, with a significant focus on infrastructure that can support large-scale renewable energy integration. Australia has seen rapid growth in solar and onshore wind energy. In 2024, renewables generated nearly 40% of Australia's electricity, a major increase from previous years [2]. However, the country still faces challenges in reaching its 82% renewable target by 2030.

The growth of renewables needs to accelerate significantly to phase out coal and natural gas, this is where offshore wind will play a vital role. Offshore wind is considered a crucial solution for Australia to meet its renewable energy targets. Our vast coastline provides ample opportunity for large-scale offshore wind farms, which have higher energy generation capacity than onshore wind or solar [3].

Offshore wind will play a key role in achieving our renewable energy target

Australia's first offshore wind projects, such as the Star of the South in Victoria [4], are now in the planning stages, with more expected to follow, including the Great Eastern Offshore Wind Project [5] and Spinifex Offshore Wind Farm in South West Victoria [6]. Offshore wind will support energy generation during periods of low solar output and provide consistent, scalable energy to meet growing demand. Offshore wind will play a role in balancing the grid as other fossil fuel power plants are retired. In Victoria, fossil fuel plants are earmarked to retire in 2028 for Yallourn Power Station, followed by Loy Yang B power station in 2032 and for Loy Yang A power station in 2035, although there are considerations to bring forward its retirement to meet Victoria's climate goals [7].

Offshore wind zones in Australia

Six offshore wind zones have been designated for development in Australia, in key regions including Gippsland (Vic), Hunter (NSW), Illawarra (NSW), Portland (Victoria), Southern Ocean, Bass Strait. The Gippsland zone in Victoria is currently the most advanced, with projects like the Star of the South leading development. Currently, several large-scale projects are moving through the various approval stages. The map shown in Figure 2 shows the areas deemed suitable by Federal Government for Offshore Wind based on average wind speed estimates and proximity to infrastructure that can support offshore wind farms.

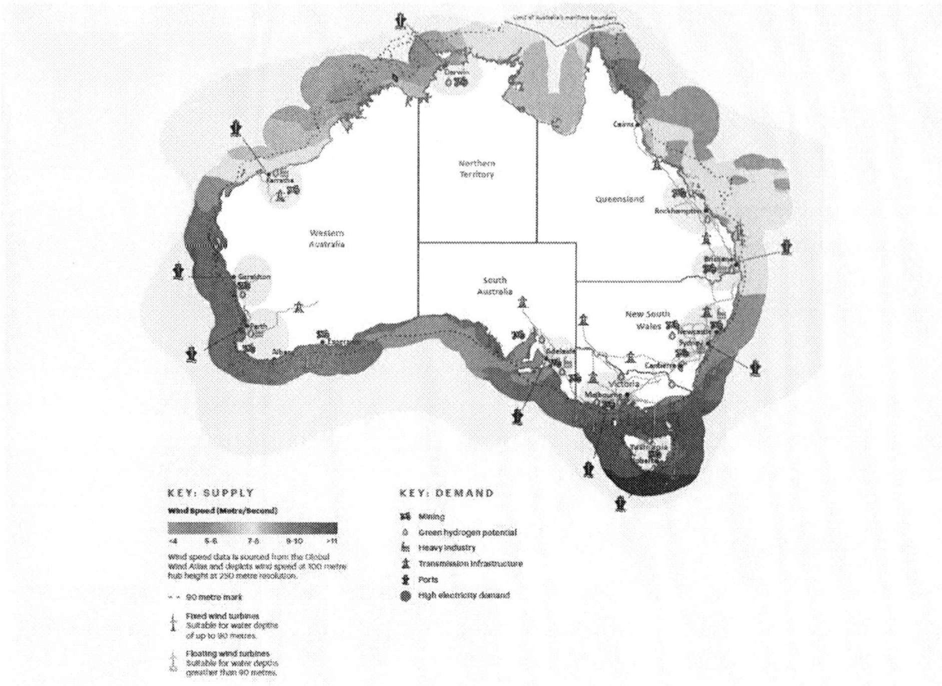

FIGURE 2: AREAS DEEMED SUITABLE BY FEDERAL GOVERNMENT FOR OFFSHORE WIND WITH WIND SPEED ESTIMATES [8]

Ports are the anchors to offshore wind development

Ports are essential for assembling, maintaining, and deploying offshore wind turbines. They will become hubs for offshore wind development, handling large-scale components and facilitating construction.

The Port of Hastings in Victoria has been identified as the most suitable site for the Victorian Renewable Energy Terminal, which will support offshore wind construction in the Gippsland zone. The proposed Victorian Renewable Energy Terminal (VRET) at the Port of Hastings will facilitate the construction of offshore wind farms by serving as a staging area for the assembly and transport of large wind turbine components. This development will be critical because Australia currently lacks ports equipped to handle large-scale offshore wind construction. The site is strategically advantageous due to its deep-water channels and proximity to the offshore wind projects in Gippsland. The VRET is currently undergoing an Environment Effects Statement (EES) process, including the establishment of a Technical Reference Group (TRG) to oversee the process [9].

Given the volume of expected offshore wind projects in Australia, a number of Ports like Port of Hastings will play a key role and will require significant upgrades to support offshore wind infrastructure, including heavy-duty quay areas, expanded storage, and enhanced corrosion protection.

Figure 3 shows an artists impression of a renewable energy Port, showing the extent of space required for laydown area of offshore wind components.

FIGURE 3: ARTISTS IMPRESSION OF RENEWABLE ENERGY PORT [10]

PORT UPGRADES WILL BE ESSENTIAL TO SUPPORT THE NEEDS OF OFFSHORE WIND CONSTRUCTION AND ONGOING MAINTENANCE

Significant port upgrades will be required to support the needs of Offshore wind construction and ongoing maintenance. These upgrades will include refurbishment and strengthening of existing port components, as well as the addition of capacity through new construction. Offshore wind turbines, especially their foundations and towers, can be up to 60-120 metres long and weigh 1,000-1,500 tonnes, depending on the size of the Turbine. Ports will require reinforced quays with heavy-duty pavements to withstand the weight of these components during assembly and staging.

The United Kingdom, Germany, China and Denmark made significant Port upgrades to support their globally leading offshore wind development.

1. UNITED KINGDOM

In the UK, the Port of Hull and Grimsby underwent significant upgrades to support offshore wind operations, particularly through the Green Port Hull project, completed in 2016. Siemens Gamesa, alongside Associated British Ports (ABP), invested £500 million to develop key infrastructure at Hull, transforming the port into a hub for wind turbine manufacturing and servicing. This included the construction of a 540,000 m^2 blade manufacturing facility, alongside new quayside infrastructure that supports roll-on-roll-off operations for easier and safer loading of turbine components.

Additionally, a new service and maintenance facility, spread across 12,300 m^2, was built to support turbine upkeep and pre-assembly phases. The project also involved in-filling wet docks to expand storage and operational space. These developments enabled Hull to serve as a major logistical centre for offshore wind farms like Race Bank and Dudgeon East, which Siemens was commissioned to supply with turbines [11].

2. CHINA

China has made significant strides in upgrading its ports to support the offshore wind industry, enabling the offshore wind capacity to grow to 45% of global offshore wind capacity [13]. Ports such as those in Jiangsu, Guangdong, and Shandong provinces have undergone extensive development to accommodate large-scale offshore wind projects. Coastal provinces in China have been focused on developing new offshore wind capacity. Guangdong aims to install

18 GW of offshore capacity by 2025, while Fujian, Zhejiang, and Jiangsu aim to install 13.3 GW, 6 GW, and 9 GW of offshore wind power projects by 2025, respectively [12].

These upgrades included expanding port areas, enhancing logistics facilities, and building specialised infrastructure to handle massive wind turbine components. New port berths and heavy-lift cranes were installed to manage these logistics. The Port of Sheyang and Yancheng have been pivotal in supporting nearby offshore wind farms and facilitating the logistics for wind turbine components. China developed ports to support both construction and long-term maintenance of offshore wind farms. This rapid development has positioned China as the largest offshore wind market in the world, supporting both domestic and international growth of the industry [13].

3. Germany

Cuxhaven Port in Germany has undergone significant upgrades to bolster its capabilities for the offshore wind industry. The most notable enhancement was the construction of a new berthing space and a roll-on/roll-off (RoRo) ramp, which was completed in 2017. This expansion included 1,340 meters of berthing space and depths of up to 11.6 meters, to allow for the heavy loads to be loaded directly onto RoRo ships [14].

In addition, a new heavy-lift berth with a total capacity of 8.5 hectares was constructed to support the storage and transhipment of monopiles and transition pieces for Northland Power's Deutsche Bucht windfarm, in the North Sea [15].

4. Denmark

Denmark's Esbjerg Port was transformed into a major hub for offshore wind assembly and deployment. Esbjerg Port serves as a logistical hub for several major offshore wind projects, including the Horns Rev and the Walney Extension. The port expanded its facilities to accommodate large offshore wind components, including turbine towers and blades. This included the construction of new quays and deep-water berths. The total investment in the port's development was approximately DKK 2.5 billion (approximately AUD 550 million).

Esbjerg Port increased its berth capacity to handle larger vessels associated with offshore wind logistics. This included over 10 dedicated berths for offshore wind operations. The port area allocated for offshore wind expanded to around 600,000 square meters, providing the required space for assembly, storage, and transportation. The port has been a leader in the offshore wind industry since they participated in the construction of the first large scale offshore wind farm in Denmark in 2002. Approximately 80% of Europe's current offshore wind capacity has passed through the port, together exceeding 4 GW of offshore wind capacity [16].

Further expansion projects are planned with up to DKK 7 billion (approximately AUD 1.5 billion) construction of offshore wind turbine production facilities, new terminals and development of a multi modal hub, to integrate sea transport with road, rail and air freight – due for completion by 2032. This investment is part of the broader goal to establish 134 GW of offshore wind capacity by 2030 and 300 GW by 2050 in the North Sea region [17].

Port upgrades and construction will increase demand for Corrosion technologists, but so too will Offshore wind design, construction and maintenance

Australia is in its early stages of developing an offshore wind industry. The design, construction, ongoing maintenance, and the associated establishment of a new supply chain, will demand the skills of corrosion technologists in Australia. Many offshore wind projects will be part funded by Government grants. These Government grants aim to stimulate local employment, creating an incredible opportunity for the local Corrosion workforce in Australia to work closely with international offshore wind construction companies, local consultants and local manufacturers to bring marine corrosion capabilities to a new industry for Australia. We will explore the role the corrosion industry can play in offshore wind further, however before we do, let's understand the basics of offshore wind structures.

Overview of Offshore wind structures

The size and scale of offshore wind structures is hard to comprehend, without a reference point. Much larger than their onshore counterparts, offshore wind rotor diameters are reaching over 200m and are projected to reach 300m by 2050. The larger the diameter, the greater the power output of the turbine.

FIGURE 4: RELATIVE SIZE OF OFFSHORE WIND TURBINES [18]

Key components include foundations (monopiles, jackets, and floating structures and their auxiliaries), towers, turbines (which can generate 7 to 20 MW depending on their rotor diameter), and electrical infrastructure including cables (dynamic cables for floating structures) and substations. The construction timeline for these structures can span 2 to 5 years, with installation taking 6 to 12 months post-foundation completion. As the offshore wind sector continues to expand, global advancements in technology and infrastructure will enhance efficiency and capacity [19].

FIGURE 5: DIFFERENT TYPES OF OFFSHORE WIND TURBINES [19]

FIGURE 6: COMPONENTS OF A MONOPILE OFFSHORE WIND TURBINE [20]

The monopile construction approach is a prevalent method used for offshore wind turbine foundations, particularly in shallow waters. This technique involves driving large, cylindrical steel tubes, or monopiles, deep into the seabed to provide a stable base for wind turbines. Typically ranging from 2 to 6 meters in diameter, monopiles can extend up to 30 meters or more below the seabed, depending on soil conditions and water depth. The monopile foundation is favoured for its relatively low cost and installation efficiency, making it suitable for offshore wind projects here in Australia. See Figure 7 for the monopile foundation construction approach.

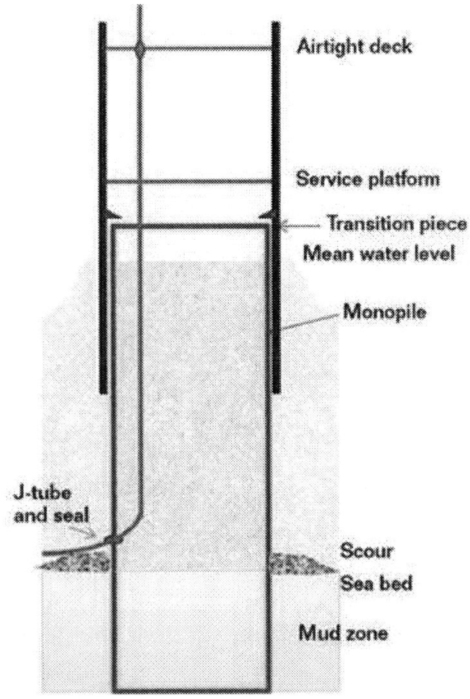

FIGURE 7: MONOPILE CONSTRUCTION [21]

Corrosion considerations over the offshore wind asset lifecycle

Corrosion is a major concern throughout the lifecycle of offshore wind turbines, given the harsh marine conditions and high wind zones. Consideration must be given to the durability of offshore wind turbines to secure Australia's energy future. Whilst Australia is lagging behind other countries in Offshore wind, it does come with the advantage of being able to take the learnings from the global offshore wind industry, with structures that have been in service for up to 10 years, to improve the design and construction methods. Let's explore the corrosion considerations throughout the lifecycle of an Offshore wind monopile structure and some of the learnings we can apply here in Australia.

1. Design and Manufacture

Selecting the right materials is critical to mitigate corrosion. Components such as the nacelle, blades, tower, and foundations are often made of steel, aluminium, composites, and have corrosion resistant coatings. The materials selection needs to take into account the exposure environment. The offshore wind structure corrosion zones are shown in Figure 8.

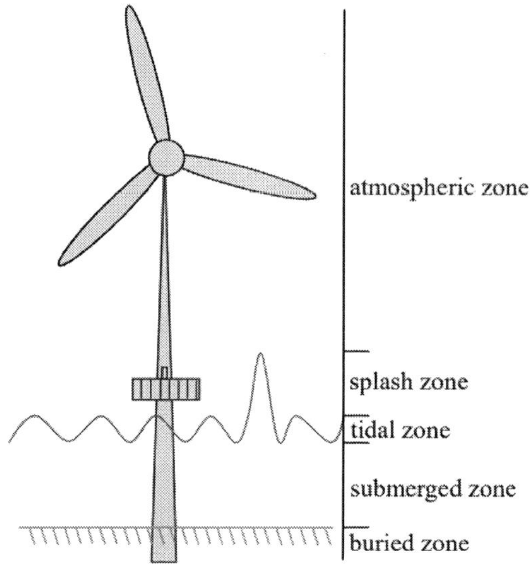

FIGURE 8: CORROSION ZONES FOR OFFSHORE WIND TURBINES [22]

External corrosion considerations for the foundation are similar to that observed in other offshore industries, and corrosion protection is prescribed in current industry guidelines based on experiences from the oil and gas industry. Cathodic protection (CP) of offshore structures with galvanic anodes is a well-established corrosion prevention technique, and DNV-RP-B401 provides requirements and guidelines for CP design, anode manufacturing, and installation of galvanic anodes. The design of an impressed current CP (ICCP) system is described in DNV-OS-J101, NACE SP0176, and BS EN 12495. Coating use is mandatory in the atmospheric and splash zones. External surfaces of the submerged zone have CP; and in the splash zone, CP may be assumed to be fully protective below the mean water level. In most projects, the CP system uses galvanic anodes placed on the transition piece, but ICCP systems are being applied in an increasing number of projects. A protective coating applied in the submerged zone is optional and primarily intended to reduce the level of CP required. General recommendations for coating systems to be used offshore are given in international standards such as EN ISO 12944, ISO 20340, and NORSOK M-501 [23]. Given the presence of sulfate-reducing bacteria in Australian waters potentially contributing to Microbial Induced Corrosion (MIC) in the tidal zone [24], ICCP, combined with coating application will provide the greatest protection against MIC.

Internal corrosion considerations must also be taken into account. Internal corrosion of the transition piece for example should include coatings and/or CP as part of the internal corrosion protection. Several wind farm owners have retrofit existing monopile structures with internal galvanic CP systems as well as include internal CP as part of the design for new projects.

Industry experience has shown, that it is difficult in practice to completely seal compartments and render them airtight. If the closed-compartment structure is not properly sealed, direct ingress of air is possible. Seawater and oxygen ingress have been detected in foundations that are two to 10 years old, which increased the rate of corrosion and localised corrosion attacks. Furthermore, in some cases, the increasing water level led to the internal aluminum ladder acting as a sacrificial anode (Figure 9). Today, air tight platforms are now much better controlled, the offshore wind industry has come a long way with improving the design to significantly reduce internal corrosion.

FIGURE 9: THE LOWER PART OF THE INTERNAL ALUMINUM LADDER, WHICH IS LOCATED ABOVE THE SERVICE PLATFORM, CORRODED AWAY DUE TO THE WATER LEVEL RISING ABOVE THE SERVICE PLATFORM. [21]

Design considerations must take these learnings into account, and design appropriate corrosion mitigation measures.

2. Storage and Pre-Assembly

As shown in the artists impression Figure 3 , Turbine components are stored at or near coastal ports. During storage, temporary corrosion protection measures such as temporary epoxy protective coatings or wrapping of sensitive components are used to shield them from the marine environment. Careful protection of the components during lifting and transport are also used, to reduce the risk of coating damage.

3. Construction and Installation

During the assembly of offshore wind structures, welded joints are areas prone to corrosion. Reapplication of corrosion protection after welding is essential. Any coating damage that occurs during transportation or installation must be addressed to maintain corrosion resistance. Touch-up coatings and sealants are applied to restore the integrity of the protection layers. The installation of foundation piles, often driven into the seabed, can wear away corrosion protection layers. Protective sheathings or coatings that are impact-resistant are applied to piles prior to installation. Even with these measures in place, studies have shown that 30% of corrosion caused on Offshore Wind structures is from mechanical damage [23]. Offshore wind structures are typically constructed in phases (called campaigns), using 3-4 month windows of opportunity where the weather is most suitable for construction. The first campaign focuses on foundation installation and the transition piece, with the second campaign focused on turbine installation. The period between the first and second campaign can be 6-12 months. During this period, (called the Standalone phase), the foundations are subject to bird droppings, as birds use the top of the transition piece as a resting place. Prior to the second campaign, a costly exercise is undertaken to remove the bird droppings and perform touch ups to the protective coating.

4. Ongoing Management and Maintenance

In offshore wind turbine maintenance, structures are generally inspected at regular intervals to prevent corrosion and structural damage. For instance, coating systems should be inspected every 4,200 hours, which typically means about two inspections per year. This frequent inspection schedule is crucial due to the harsh marine environment, where high moisture and salt levels can rapidly accelerate corrosion.

Maintenance challenges in these environments include costly repairs, particularly for on-site coating applications, which can be up to 7,000% more costly when compared to factory coatings. Surface preparation for repairs can also be difficult, given the constant exposure to moisture and salt. New developments in repair coatings are better suited for these conditions, improving protective performance despite poor surface conditions [25].

Additionally, the use of drones and quadcopters equipped with camera and laser systems for inspections is increasing. These unmanned aerial systems allow for more efficient inspections, particularly in high-wind environments where manual inspections might be impractical or dangerous. This technology is being integrated with plant monitoring systems (such as SCADA), further reducing the need for manual on-site inspections [25].

Recent developments in digital, automated corrosion management include drones that film affected areas, providing automatic detection and filtering of damaged sections. The programme generates the documentation and transfers it into a 3D model. The Hamburg-based offshore wind farm operator, Global Tech I, won the award for the Product Innovation of the Year. The team includes experts in digitalisation and artificial intelligence and is backed by the University of Aarlborg and the Energy Cluster Denmark. [26]

Submerged parts of wind turbines also require ongoing inspection and monitoring. Monitoring the effectiveness of the ICCP systems and ensuring their potentials are balanced to prevent any surface area from being under-protected will be essential to ensure the design is providing the intended protection. The condition of antifouling coatings will also form part of the inspection regime, to prevent marine biofouling from accelerating corrosion on submerged components. Routine underwater inspections, which have traditionally been performed by divers, will shift to being performed by ROV's (Remote Operated Vehicles) (Figure 10), to detect signs of early corrosion. This will improve the safety of these inspections and allow for longer periods between dive inspections. In Europe, companies are moving away from contracting professional divers due to safety concerns and cost considerations, making ROV's more prevalent for inspections. ROV's capability goes beyond visual inspection, to now include NDT as well as underwater welding, though this is still in its infancy.

FIGURE 10: EXAMPLE OF OFFSHORE WIND INSPECTION ROV [27]

Inspections focus on high-risk areas such as welds, submerged structures, and areas with mechanical stress or coating damage. ROV's can also serve other purposes, such as monitoring marine life and checking critical infrastructure for security threats.

The Corrosion industry can prepare now for the transition

The time is now to prepare for the energy transition. As Corrosion technologists, consultants and asset managers, there are a number of steps we can take to prepare:

1. Enhance knowledge of offshore wind
2. Develop, identify, and test new inspection and monitoring technologies suitable for offshore wind
3. Innovate and seek out sustainable corrosion prevention and management solutions suitable for offshore wind environments
4. Contribute to standards development and research

1. Enhance Knowledge of Offshore Wind

There are three ways Corrosion technologists interested in working in the Offshore Wind industry can build their knowledge; **gain international experience** directly on offshore wind projects, **partner locally with offshore wind developers**, such as Star of the South, or **undertake further research** and study in the field.

Gaining international experience on offshore wind projects will provide valuable and practical skills that can be applied to Australia's emerging offshore wind industry. Sean Barker (co-author) did this, as a Certification Manager for some of the first offshore wind projects in France. Sean led collaboration with certification bodies, regulatory agencies, engineering firms, and constructors to guarantee adherence to international standards, regulations, and industry best practices. The projects were governed by the internationally recognised IEC, IECRE and DNV standards with the site conditions, design (including load and limit-state analyses, offshore corrosion and underwater scour management), construction, transport & installation and commissioning of the turbine and its foundation and the offshore substation applicable for certification. By leveraging global best practices and lessons learned from international projects, Sean's background serves as a blueprint for overcoming the technical and logistical hurdles that Australia's fledgling industry will face. His global experience will play a pivotal role in ensuring that the local industry can develop efficiently, align with international standards, and rapidly scale to meet the country's renewable energy goals.

Supporting offshore wind companies locally, such as Star of the South, will provide exposure to the offshore wind industry, as they bring their in house global expertise to Australia.

Undertaking further research or study in the field will enhance knowledge in preparation to participate in the offshore wind industry in Australia. There are several ways this can be done, ranging from full tertiary degrees to short courses and professional certifications. International Universities like Strathclyde (UK) [28] and Cranfield (UK) [29] offer postgraduate programs in offshore renewable energy, while locally, UNSW provides undergraduate and masters courses in renewable energy engineering [30]. Short courses, such as those from DTU Wind Energy (Denmark) [31] and TU Delft (Netherlands) [32], focus on specific aspects of offshore wind, including technology, economics, and structural design. For hands-on skills, Global Wind Organisation (GWO) provides essential safety and technical training [33]. Certifications from institutions like Renewables Academy (RENAC Academy) in Germany [34] are also an option (with international online learning options available). Lastly, attending offshore wind conferences like WindEurope [35], Global Offshore Wind Conference [36], and APAC Offshore Wind [37] often includes workshops, masterclasses, and panels on the latest technologies and project developments. Locally, the Australia Wind Energy Conference [38] will provide a great opportunity to learn about the latest developments in offshore wind in Australia. Soon, we will start to see more options for training available in Australia, including the establishment of the Victorian Governments Wind Worker Training Centre to build the necessary skills to support the local workforce to transition to the offshore wind industry [39].

2. Develop, identify, and test new inspection and monitoring technologies

Offshore wind operations are using predictive maintenance systems to detect early signs of corrosion. Developing, identifying and testing new and improved remote Corrosion monitoring technologies, as well as developing data analytics models that can forecast corrosion failure and optimise maintenance schedules that can be deployed in an offshore wind context will support the ongoing management of corrosion in future offshore wind structures.

Inspection techniques such as robotic systems like drones and ROV's are increasingly used to inspect and monitor offshore wind infrastructure. Practicing using these technologies now for inspections of Port infrastructure and offshore oil and gas infrastructure assets and testing their accuracy against data gathered from dive inspections will enable Corrosion specialists become proficient in using and analysing data from these inspection tools.

3. Identify sustainable Corrosion protection solutions

Exploring Eco-Friendly Corrosion Inhibitors, such as bio-based coatings and low-VOC systems, and researching their effectiveness in similar environmental conditions will help lead the way to be able to utilise these coating technologies for offshore wind, contributing to more sustainable corrosion maintenance practices.

4. Contributing to Standards Development and Research

Contributing to the development of standards and research is essential for advancing corrosion management in offshore wind applications. Corrosion technologists can play a critical role by participating in organisations like AMPP, DNV, or ISO, helping to create and refine standards that address the unique challenges of offshore wind. In addition, technologists can engage in research projects through universities or industry partnerships, focusing on corrosion issues specific to offshore environments. Cross-disciplinary collaboration—bringing together materials science, engineering, and environmental science—can drive the development of innovative solutions for corrosion protection in these extreme environments.

Conclusions

Australia's energy transition towards net zero by 2050 relies heavily on the successful integration of offshore wind into its renewable energy mix. Ports will be the key enablers of this transformation, serving as essential hubs for the assembly, transport, and maintenance of offshore wind farms. However, this comes with significant infrastructure challenges. Port upgrades, particularly in corrosion management, are critical to ensuring that Australia's ports can handle the scale and complexity of offshore wind development.

As we've seen from international examples, robust corrosion protection strategies—spanning material selection, coatings, and cathodic protection—are vital to extending the life of both port and wind infrastructure. The Australian corrosion industry has an important role to play, not only in upgrading port assets but also in safeguarding offshore wind assets throughout their lifecycle. By engaging in research, contributing to standards, and developing new solutions for corrosion protection in extreme marine environments, the corrosion sector can help ensure the resilience of this emerging industry.

The energy transition presents an exciting opportunity for both the port, offshore wind and corrosion industries. With proactive planning, investment in research, and the adoption of best practices, Australia's ports and corrosion professionals can be ready to support the offshore wind boom, ensuring the nation's renewable energy targets are met.

The question remains: how will you prepare for the transition?

References

1. Australia's first offshore wind project: Star of the South see: https://cop.dk/star-of-the-south/ Last accessed 5th October 2024

2. Clean Energy Australia Report see: https://cleanenergycouncil.org.au/news-resources/clean-energy-australia-report Last accessed 5th October 2024

3 . Onshore and Offshore Windfarms: What's the difference? See: https://automation.honeywell.com/us/en/news/featured-stories/2023/onshore-offshore-wind-farms-difference Last accessed 5th October 2024

4. Australia's most advanced offshore wind project see: https://www.starofthesouth.com.au/ Last accessed 5th October 2024

5. Great Eastern Offshore Wind see: https://infrastructurepipeline.org/project/great-eastern-offshore-wind Last accessed 5th October 2024

6 . Parkwind and Alinta to Jointly Develop 1+GW Offshore Wind Farm in Australia see: https://www.offshorewind.biz/2024/04/24/parkwind-and-alinta-to-jointly-develop-1-gw-offshore-wind-farm-in-australia/ Last accessed 5th October 2024

7 . Cheaper, Cleaner, Renewable: Our Plan for Victoria's Electricity Future see: https://www.energy.vic.gov.au/__data/assets/pdf_file/0021/715008/our-plan-for-victorias-electricity-future-summary.pdf Last accessed 5th October 2024

8. Is Australia a great place for offshore wind farms? see: https://www.weatherzone.com.au/news/is-australia-a-great-place-for-offshore-wind-farms/1635485 Last accessed 5th October 2024

9. Powering Victoria's future, together see: https://renewableenergyterminal.com.au/ Last accessed 5th October 2024

10. Australian Port Making Room for Offshore Wind see: https://www.offshorewind.biz/2022/07/15/australian-port-making-room-for-offshore-wind/ Last accessed 5th October 2024

11 . Old dog, new tricks – the UK's Green Port Hull project see: https://www.windpower-international.com/analysis/old-dog-new-tricks-the-uks-green-port-hull-project-5715251/?cf-view Last accessed 5th October 2024

12. China wind energy market size and share analysis- growth trends and forecasts (2024 – 2029) see: https://www.mordorintelligence.com/industry-reports/china-wind-energy-market Last accessed 5th October 2024

13 . China reaches 45 per cent of global offshore wind capacity see: https://www.bairdmaritime.com/offshore/exploration-development/offshore-construction/china-reaches-45-per-cent-of-global-offshore-wind-capacity-new-report-reveals Last accessed 5th October 2024

14. Cuxhaven boosts offshore wind facilities see: https://www.maritime-executive.com/article/cuxhaven-boosts-offshore-wind-facilities Last accessed 5th October 2024

15 . Cuxhaven Port - Cuxport is base port for new German windfarm see: https://www.vesseltracker.com/en/Port/Cuxhaven/Dashboard.html Last accessed 5th October 2024

16. Port Esjberg, deploying offshore wind see: https://www.hbs.edu/environment/blog/post/IFC-Port-Esbjerg Last accessed 5th October 2024

17. Port of Esbjerg lines up EUR 780 Million investment in offshore wind turbine production facilities see: https://www.offshore-energy.biz/port-of-esbjerg-lines-up-eur-780-million-investment-in-offshore-wind-turbine-production-facilities/ Last accessed 5th October 2024

18. The what, why and how of Offshore wind, Australia Edition see: https://core-marine.com/news/what-why-how-offshore-wind-australia Last accessed 5th October 2024

19. Decode Offshore Wind challenges at every stage see: https://blog.dhigroup.com/decode-offshore-wind-engineering-challenges-at-every-stage/ Last accessed 5th October 2024

20. Design and construction considerations for offshore wind turbine foundations in North America see: https://www.researchgate.net/figure/Components-of-a-Wind-Turbine-System_fig1_267604619 Last accessed 5th October 2024

21 . Corrosion Risks and Mitigation Strategies for Offshore Wind Turbine Foundations see https://www.materialsperformance.com/articles/material-selection-design/2016/03/corrosion-risks-and-mitigation-strategies-for-offshore-wind-turbine-foundations Last accessed 5th October 2024

22. Recker L, State of the art: Corrosion protection for offshore wind turbines, FH M¨unster, Stegerwaldstraße 39, 48565 Steinfurt, Educational Journal of Renewable Energy Short Reviews (2024) Last accessed 5th October 2024

23 . Corrosion Risks and Mitigation Strategies for Offshore Wind Turbine Foundations see: https://www.materialsperformance.com/articles/material-selection-design/2016/03/corrosion-risks-and-mitigation-strategies-for-offshore-wind-turbine-foundations Last accessed 5th October 2024

24. Wade, S., & Blackall, L., Development of a laboratory test for microbial involvement in accelerated low water corrosion. MICROBIOLOGY AUSTRALIA, 39(3). 2018 https://doi.org/10.1071/MA18049

25. Plagemann P. and Momber A. "Corrosion protection of offshore wind energy constructions in Germany: Challenges and approaches". Journal of Telecommunication, Electronic and Computer Engineering (JTEC) 10.2-3 (2018), pp. 1–4.

26 . German renewables award hails great energy transition ideas see: https://hamburg-business.com/en/news/german-renewables-award-hails-great-energy-transition-ideas Last accessed 5th October 2024

27 . Autonomous Inspection Service for Offshore Wind to be developed in UK see: https://massworld.news/autonomous-inspection-service-for-offshore-wind-to-be-developed-in-uk/ Last accessed 5th October 2024

28 . Sustainable Engineering: Offshore Renewable Energy see: https://www.strath.ac.uk/courses/postgraduatetaught/sustainableengineeringoffshorerenewableenergy/ Last accessed 5th October 2024

29. Renewable Energy MSc see: https://www.cranfield.ac.uk/courses/taught/renewable-energy Last accessed 5th October 2024

30. Master of Engineering (Renewable Energy) see: https://www.unsw.edu.au/study/postgraduate/master-of-engineering-renewable-engineering Last accessed 5th October 2024

31. DTU Wind Department of Wind and Energy Systems Education see: https://wind.dtu.dk/education Last accessed 5th October 2024

32. European Wind Energy Master see: https://www.tudelft.nl/en/ewem Last accessed 5th October 2024

33. Training standards for a safer and more productive workforce see: https://www.globalwindsafety.org/ Last accessed 5th October 2024

34. RENAC Renewals Academy see: https://www.renac.de/ Last accessed 5th October 2024

35. Wind Europe Events see: https://windeurope.org/events/ Last accessed 5th October 2024

36. Global Offshore Wind see: https://events.renewableuk.com/gow24 Last accessed 5th October 2024

37. APAC wind energy summit see: https://gwec.net/apac-wind-energy-summit-2024/ Last accessed 5th October 2024

38. Australia Wind Energy 2025 see: https://www.windenergyaustralia.com/ Last accessed 5th October 2024

39 . Worker training centres see: https://www.energy.vic.gov.au/renewable-energy/victorian-energy-jobs-plan/worker-training-centres Last accessed 5th October 2024

AUTHOR DETAILS

C. Crawshaw is the corresponding author of this paper. As a Director at Infrastructure Advisory Group (IAG) since May 2024, she leads the Asset Management practice, helping clients overcome asset management challenges and achieve sustainable improvements. Christine specialises in operating model transformation and implementation, focusing on driving greater efficiencies and improved capabilities to deliver value for money outcomes. With a passion for sustainability, she is particularly excited about the opportunities within the offshore wind industry and its potential to shape a greener future.

	S. Barker is the Director at Infrastructure Advisory Group (IAG), a position he has held since 2024. Sean has 15 years of global experience in various roles within the Offshore Wind, Offshore Oil & Gas, Mining and Infrastructure industries. He has a diverse skillset in both design / engineering and construction including project management, regulatory compliance and certification management, interface management, resource management, technical document review and reporting, documentation for operation and associated coordination tasks.

CORROSION & PREVENTION
CAIRNS 2024

THE AUSTRALASIAN CORROSION ASSOCIATION INC.

CATHODIC PROTECTION AND ITS IMPLICATIONS FOR VESSELS IN THE ROYAL AUSTRALIAN NAVY

K. Dylejko*, W. Neil

Defence Science and Technolgy Group, Department of Defence, Fishermans Bend, Victoria, Australia

**Corresponding author email: kathryn.dylejko@defence.gov.au*

ABSTRACT

Cathodic protection systems on Royal Australian Navy vessels have unique challenges which need to be managed by Defence. Australian ships operate in warmer waters, berth for longer periods, and have operational profiles that can differ significantly from commercial ships. This can lead to issues such as interactions between systems, excessive calcareous deposits, and complex multifactorial problems that can be difficult to diagnose. In this paper we present some examples and case studies of problems that can arise relating to Navy cathodic protection systems and methods to address and manage them.

Keywords: cathodic protection, corrosion, ships, Royal Australian Navy

INTRODUCTION

The first practical application of cathodic protection (CP) occurred 200 years ago, in experiments conducted by Sir Humphry Davy on Royal Navy's ships [1-3]. Prior to Davy's experiments, during the 1760's and 1770's, copper sheathing had been introduced to the timber hulls of Royal Navy ships as it was found to prevent timber worms and the adherence of weeds to the hull [4]. By the early 1820's, corrosion of copper sheathing was perceived to be an issue and Sir Humphrey Davy was approached by the Navy Board for advice. Then followed a series of experiments run by Davy on Royal Navy warships demonstrating the first practical application of CP [1-3].

Davy's initial experiments were conducted using thin strips of zinc, malleable or cast iron attached to Navy ships moored in Portsmouth harbour [1]. The results were such a success, that the Royal Navy issued an order for all their ships to be fitted with Davy's 'protectors'. Davy had noted a problem with the formation of carbonates of lime and magnesia to which weeds had attached, which he was able to mitigate through sizing of the protectors [4]. Davy further refined the sizing of the protectors in open-sea experiments when he sailed on His Majesty's Steam Boat Comet in July and August 1824 [3-5].

Signs of problems started to arise with reports of increased fouling on the hulls of ships fitted with Davy's protectors. This included conflicting reports regarding the fouling on HMS Samarang (

Figure A) upon its return from a voyage to Bermuda [3]. Over the coming years, fouling was proven to be an issue on these protected ships, and the method was abandoned, despite its success in preventing hull corrosion.

FIGURE A: HMS SAMARANG , ONE OF THE FIRST SHIPS TO BE CATHODICALLY PROTECTED [6].

An attempt to protect ships from corrosion with CP briefly emerged during the 1890s, when Thomas Edison experimented with towing impressed current anodes behind a ship [7]. This "trailing anode" technique was unsuccessful due to the cost of electricity and lack of appropriate materials and techniques. However, it demonstrated the ongoing desire for a system to reduce the labour and time required to maintain ship hulls.

Despite the initial discontinuation of the CP methods demonstrated by Davy, his scientific insights were ahead of his time and he pioneered a technique that we see on ships around the world today. CP of active (seagoing) ships began to re-emerge in the late 1940s and 1950s [7-9] and is now essential to preventing corrosion of Navy ships around the world. The contribution of his successor, Michael Faraday was also instrumental in laying the scientific foundations of CP, in particular his 1834 discovery of the quantitative connection between corrosion weight loss and electric current.

Here we have only provided the briefest of introductions to the history of CP on Navy vessels. There are numerous books and papers detailing the history of electrochemistry, CP and the contributions of Sir Humphrey Davy, many of which are well worth a read for the corrosion or CP enthusiast [4, 5, 10, 11].

CATHODIC PROTECTION OF NAVY SHIPS

CP is commonly applied today to prevent corrosion of a wide array of submerged or buried metallic structures including pipelines, offshore platforms, boilers, concrete reinforcement, wharves and ships. In all systems, there is a metallic structure to be protected, anodes to provide CP current, and an electrolyte (usually water or soil). Systems may also include additional components such as reference electrodes and transformer-rectifier units.

On ships, the metallic structures typically include painted steel or corrosion resistant alloys (e.g. aluminium, copper nickel, nickel aluminium bronze (NAB), etc.). As many of these tend to be electrically connected, for example the propeller and the hull, galvanic couples exist that require CP in addition to protection of the individual materials themselves.

The electrolyte in a ship CP system is seawater, fresh water, or somewhere in between (brackish water). In the case of Navy ships, due to the nature of their operations, seawater is the most common electrolyte. Seawater is more conductive than soil or fresh water, which affects the protection current density requirements of the metals being protected. It also significantly reduces IR losses in the electrolyte, so potentials such as the "instant off" potential become irrelevant as the IR drop is negligible.

Whilst some smaller Navy vessels rely solely on sacrificial (galvanic) anode systems (Figure B, left), larger Navy ships in the Royal Australian Navy (RAN) tend to use impressed current cathodic protection (ICCP) on their hulls (Figure B, right). This is due to their ability to provide higher currents and protect large surface areas with only a few anodes, thereby reducing the ongoing maintenance costs. However, large ships with ICCP systems tend to also be fitted with sacrificial anodes to provide additional protection to specific areas such as sea chests, ballast tanks, propellers or shafts. One reason ICCP tends not to be used in tanks as there is a risk of hydrogen evolution, which creates an explosion risk when hydrogen gas accumulates in a confined space. This can occur with ICCP systems when potentials drop below around-1.1 V_{CSE} and hydrogen evolution becomes the dominant reduction reaction.

FIGURE B: HMAS ARUNTA SHOWING SACRIFICAL ANODES (LEFT) [12], AND AN ICCP ANODE ON A SHIP [13].

Ships tend to experience larger varations in CP current demand compared with other structures in seawater such as offshore platforms or wharves and piers. One of the main reasons for this is that when ships move through seawater, the protection current density is much higher than when they are stationary, due to the increased oxygen supply at the cathode. In addition, protection currents can change significantly over time due to the breakdown of paint coatings and the formation of calcareous deposits. Protection current density also changes slightly due to varying ocean conditions (temperature, salinity, dissolved oxygen). Because of the large variations in protection current required, ship ICCP systems tend to always be fitted with reference electrodes and a control system to ensure the appropriate current is being applied. Too little current can lead to corrosion, whilst too much can lead to overprotection, which is associated with problems such as hydrogen embrittlement, paint debonding, calcareous chalking of moving parts and excessive power usage.

CP of RAN ships comes with unique challenges not encountered in commercial shipping. One of the main challenges arises from the significant time spent in port compared to commercial ships. When alongside for weeks or months, the interactions with wharf CP systems become very important. Biofouling is also exacerbated, and without regular movement of parts such as bearings, calcareous chalk can accumulate and later cause excessive wear. This is further worsened in Australian waters, where many of our Navy bases are in tropical locations which are known for their harsher biofouling environments compared with more temperate climates [14]. Furthermore, their operational profiles including speeds and locations can differ from commercial ships, and their designs do not always follow the same standards. Another unique challenge is the electromagnetic fields that arise from CP systems, which can be detected by adversaries and may need to be reduced for certain platforms and operating environments.

To summarise, RAN ships are complex systems that often operate under unique conditions. In identifying and addressing problems that occur in these systems, it is always important to consider the system as a whole, including not just the CP components but all other possible interactions. In the following sections of this paper, we discuss some examples of CP issues faced by the RAN and the approaches used to diagnose and address these issues.

CALCAREOUS DEPOSITS: FRIEND OR FOE?

When CP is applied to structures in seawater (as is the case for Navy ships), the reactions occurring at the cathodic surface cause a local increase in pH, which in turn causes the precipitation of calcium carbonate (usually as aragonite) and magnesium hydroxide (as brucite). Even Davy noted such deposits during his 1824 experiments, when he described the copper sheathing as being "coated with carbonate of lime and magnesia" [1]. These deposits reduce the required CP current by acting as a barrier to oxygen and other cathodic depolarisers, as well as increasing the circuit resistance and interfacial pH.

Calcareous deposits on ships can have significant benefits. On ships using ICCP, calcareous deposits result in a lower load on the ship's power system and the ability to protect a larger hull area without causing excessively negative potentials near the anodes. The main benefit for sacrificial anode systems is that fewer anodes are required, and they will need to be replaced less frequently. Overall, Navy ships cost less to maintain and may have increased availability thanks to calcareous deposits.

Despite their benefits, calcareous deposits can be troublesome if they occur in unwanted locations, such as moving parts including shaft bearings. Calcareous deposits are an ongoing problem in water-lubricated propeller shaft applications causing accelerated bearing wear [15]. Figure B shows the appearance of a calcareous deposit on a propeller shaft bearing liner with CP at the macro scale and as seen using a scanning electron microscope (SEM). The deposit was also analysed using Fourier transform infrared spectroscopy and found to be primarily aragonite (calcium carbonate), with some brucite (magnesium hydroxide), and possibly calcite (calcium carbonate). The crystalline structure of the deposit has a "spikey" appearance which can be typical of aragonite. Because of this, and the fact that aragonite is harder than the bearing liner materials (which tend to be polymeric), its action on the bearing liners can lead to increased wear, and costly premature replacement.

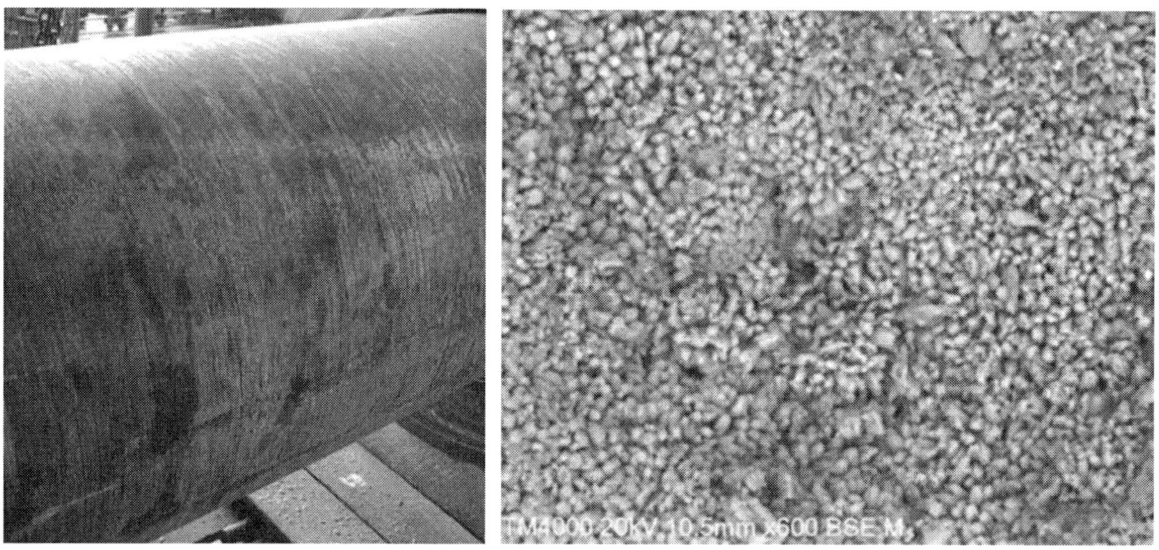

FIGURE B: PROPELLOR SHAFT BEARING WITH A CALCAREOUS DEPOSIT (LEFT) AND AT X600 MAGNIFICATION (RIGHT).

Other areas on ships that can be affected by calcareous deposits are any large, bare metallic components – such as hull valves that are exposed to seawater and CP. Figure C shows an example of this, where a hull valve on a Navy ship had seized closed due to the formation of calcareous deposits whilst under the protection of the wharf ICCP system. The inoperability of hull vales could have a catastrophic outcome if there was need for an emergency opening of these hull valves to allow seawater in (for firefighting purposes) or closing (if there was an uncontrolled flood on board).

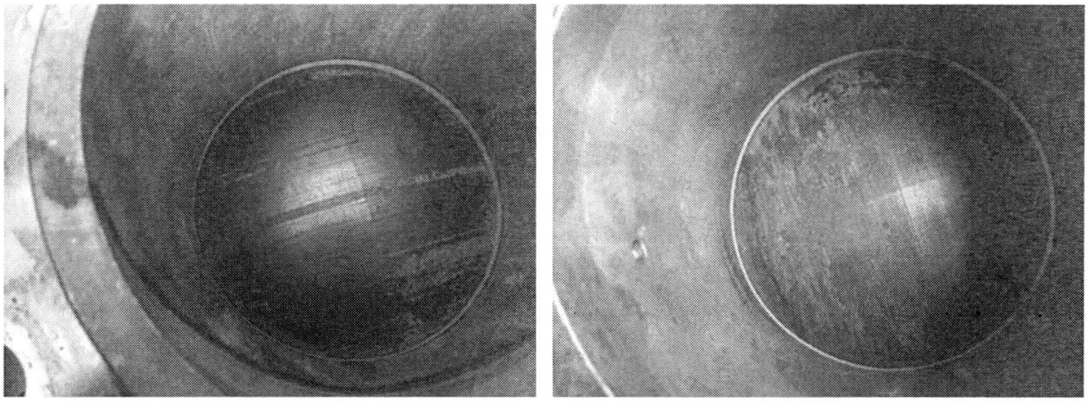

FIGURE C: HULL VALVES – CALCAREOUS DEPOSIT FROM ON VALVES WHICH WERE SEIZING CLOSED.

There are several solutions to problematic calcareous deposits, and the most appropriate will depend on the cause. Problems associated with calcareous deposits in shaft bearings can be exacerbated for Navy ships which tend to spend more time alongside/berthed than commercial vessels. In this instance, a typical treatment is to turn the shaft regularly to remove any early deposits before they become too thick.

If the formation of problematic calcareous deposits is shown to be due to overprotection from an CP system, the problem may be solved by reducing the set point of an ICCP the system (and therefore the anode currents), or by removing sacrificial anodes near the shaft. However, this may not be possible as it may no longer prevent corrosion, and other solutions may be required. They may also be caused by overprotection when bonded to a wharf with its own CP system—this type of interaction will be discussed in the next section.

Another possible solution currently being investigated for shaft bearings is to introduce additional zinc anodes near the bearings. Zinc ions have been shown to inhibit the formation calcareous deposits on steel [16-18]. Furthermore, deposits found in Navy bearings with zinc anodes and where zinc was found in the crystal structure have tended to be associated with reduced wear. Experiments by the authors and their colleagues have confirmed that zinc ions reduce calcareous deposits on bearing liner materials. Figure D shows SEM images of deposits formed on gunmetal held at –900 mV vs Ag/AgCl/KCl(sat.) for one week in artificial seawater. The left image shows the deposits without zinc ions, whilst the right image shows the deposit formed with 10 mg/L of zinc ions added. The deposit on the right is sparser, has a different structure, and was found to rub off more easily. These results have led to larger scale trials which are currently underway.

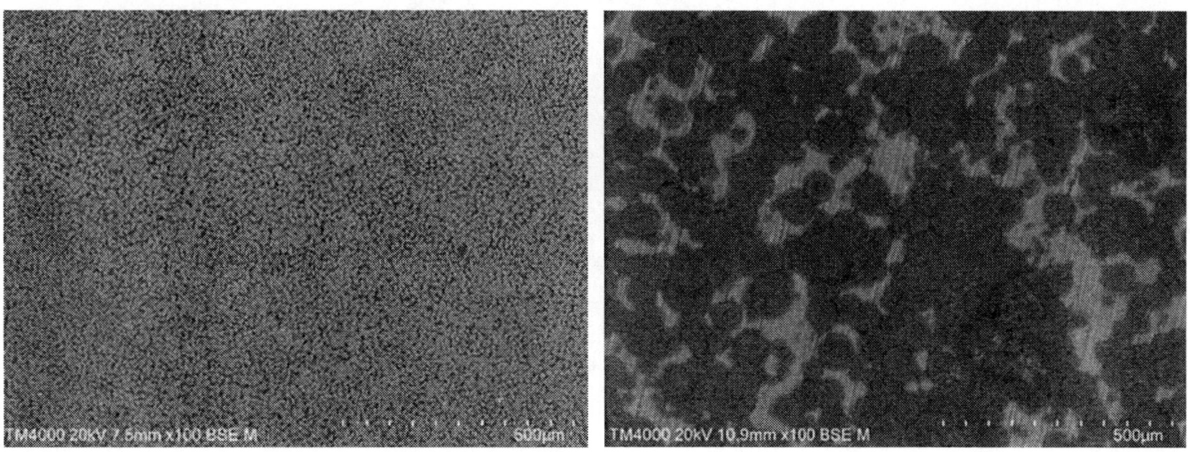

FIGURE D: SEM IMAGES OF CALCAREOUS DEPOSITS FORMED ON GUNMETAL IN ARTIFICAL SEAWATER (LEFT) AND ARTIFICAL SEAWATER WITH 10 MG/L ZINC IONS (RIGHT).

INTERACTIONS WITH WHARF CP (OVER PROTECTION, UNDER PROTECTION ETC.)

The interaction between ship and shore CP systems can be complex in nature. Wharves and piers to which Navy ships are docked are often made from reinforced concrete or steel, which will usually have CP systems installed. When a ship is docked, it needs to be bonded to the shore to provide a current path to prevent stray current corrosion of the ship or the dock. Various issues can arise because of this arrangement, which can be exacerbated for Navy ships which tend to spend more time docked than commercial vessels. In this section we discuss some examples of such issues.

When a ship with ICCP bonds to a wharf, its ICCP system is turned off to prevent overload of the ship's ICCP system. The ship will instead be protected by the wharf CP. There are several problems that can arise.

The wharf CP system may cause over-protection near anodes or under-protection away from the anodes. This may not be a problem for short durations but for Navy vessels docked for long periods can cause problems. For example, it can lead to the development of excessive calcareous deposits as seen in the previous section.

If the wharf does not have CP or the system is not working, the ship and its sacrificial anodes will end up protecting the wharf. This can lead to accelerated corrosion of the ship's sacrificial anodes and under-protection or corrosion of the ship's hull. It can also lead to stray current corrosion of the wharf as shown in Figure E. If the ship ICCP were turned on, the high currents required to protect the wharf could cause overprotection near the ICCP anodes and under-protection in other areas.

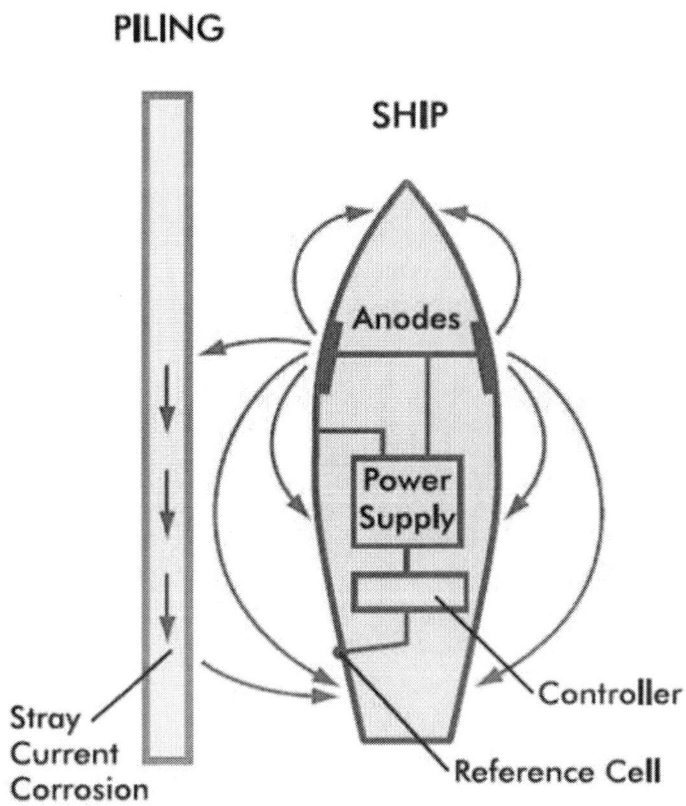

FIGURE E: STRAY CURRENT CORROSION OF THE ICCP SYSTEM OF THE SHIP TO THE DOCK [15].

There are several ways in which we can minimise the occurrence of issues related to ship and wharf CP interactions. Firstly, proper maintenance of wharf CP and training of Navy personnel is essential. Wharf CP interactions should be investigated if issues such as excessive galvanic anode consumption or excessive calcareous deposits are encountered without an obvious cause. If problems are suspected, interactions can be identified by checking the direction of the current flowing through the bond cable – current flowing from the shore to the ship indicates that the ship is protecting the wharf, and vice versa. A diagram of the correct current flow direction is shown in Figure F. As we show in the following sections, a consideration for the ship system as a whole, rather than individual systems, is essential for diagnosing and solving CP problems.

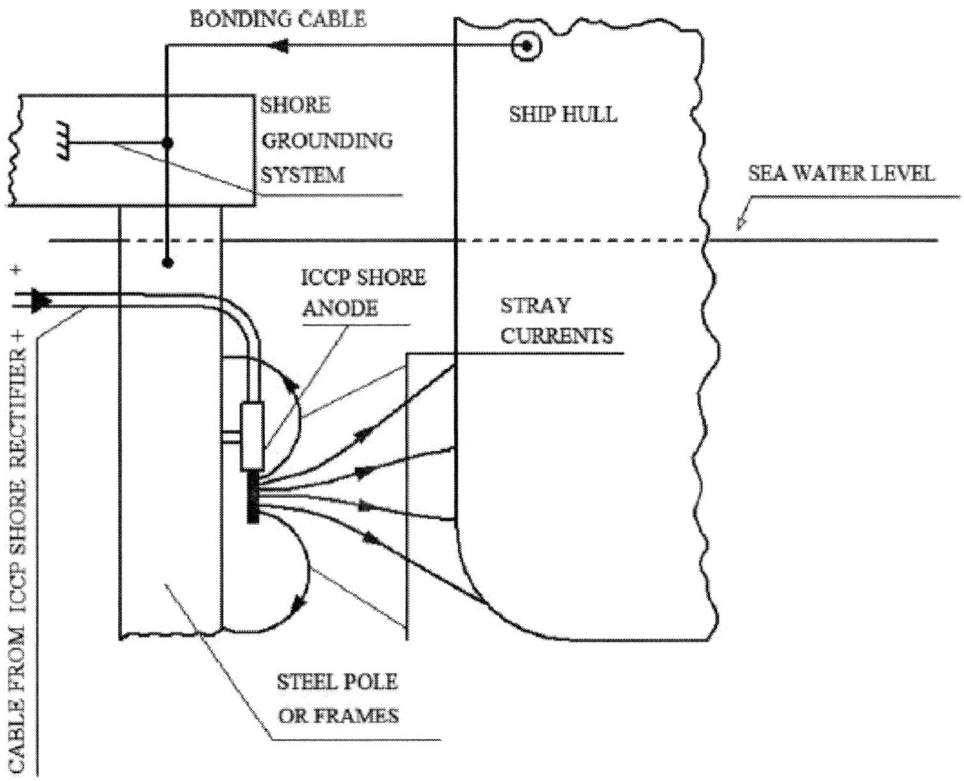

FIGURE F: CORRECT DIRECTION OF CURRENT FLOW WHERE THE WHARF ICCP IS PROTECTING THE BONDED SHIP [19]. NOTE THE DIRECTION OF THE CURRENT IN THE BOND CABLE IS GOING FROM SHIP-TO-SHORE.

Marine Growth Prevention System interactions

As we saw in Davy's experiments 200 years ago, the presence of copper ions in seawater inhibited marine growth (and in Davy's case, the lack thereof encouraged such growth!). This principle is still used to our benefit today on many ships in their marine growth prevention systems (MGPS). These systems use copper anodes in areas such as sea chests or strainers to release copper ions into seawater piping systems, which prevents the adherence of marine growth. They also include a ferrous or aluminium anode to prevent corrosion within the piping. A typical setup is shown Figure G.

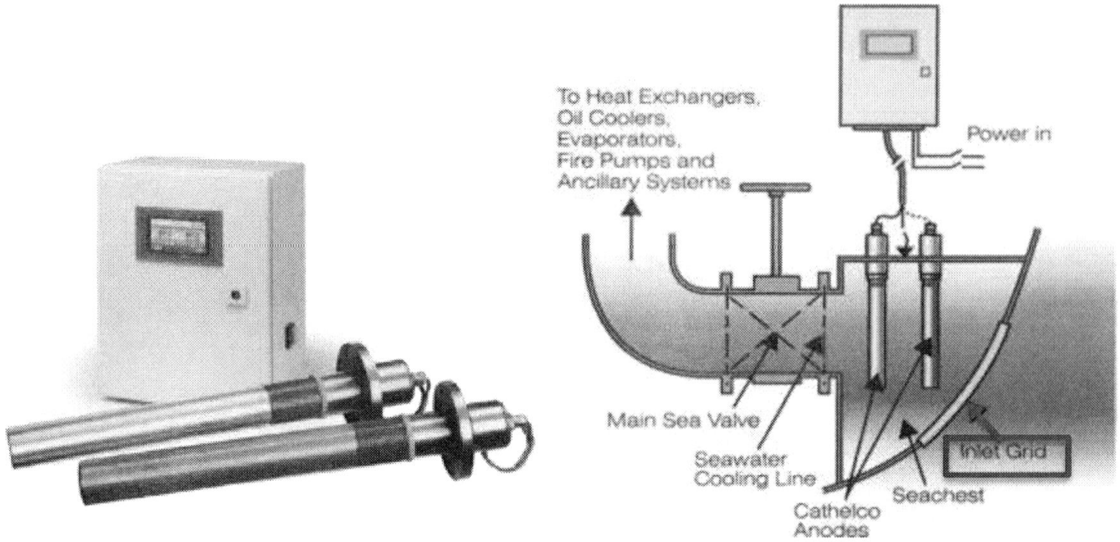

FIGURE G: A TYPICAL MGPS SYSTEM [19].

As with any corrosion cell, the MGPS system must also have a cathode, which is either a dedicated electrode alongside the copper and aluminium/ferrous anode, or the ship's hull (this is the case in Figure G). When the ship's hull is used, problems can arise when this internal system interacts with the external CP system designed to protect the hull from corrosion. For example, if the inlet grid (Figure G) is not earthed to the hull, current can flow to the external hull (or wharf, if things are particularly bad), quickly consuming the MGPS anodes. This requires untimely maintenance to replace depleted anodes, and excessive dissolution of copper from the MGPS system may even cause downstream issues such as plating out of copper in heat exchangers. In addition, the potentials in this situation can be as negative as –1000 mV vs. Ag/AgCl/seawater which can cause issues such as cathodic disbondment of paint.

The scenario above can also cause a problem if the MGPS is close to the ICCP reference electrodes. The ICCP system will assume the hull is protected by the ICCP anodes and reduce the current output accordingly. However, this will not change the potential at the reference electrode significantly, as it is protected by the MGPS. The ICCP current will continue to reduce, possibly even to zero. As a result, metallic areas which aren't near the MGPS and reference electrodes, can become under-protected.

In this scenario interactions between systems can be assessed by monitoring outputs from the MGPS, ship CP and wharf CP with various combinations of systems switched on and off. Solutions may include adding a dedicated cathode to the MGPS or earthing the inlet grid to the hull, resolving any problems with ship or wharf CP systems, or crew education and training.

The example discussed here demonstrates a case of a component designer not considering the whole system approach of how one component (the MGPS system) can interfere with other systems (ship and wharf CP) and cause greater issues. In the next section, we will discuss an example where multiple systems failed due to a simple design issue with the root cause not discovered until the system was assessed as a whole.

WHAT HAPPENS WHEN SHIPBOARD ICCP GOES BAD?

Here we describe a case of a whole-system issue which lead to catastrophic failure of components, that was caused by a simple problem of inadequate shaft grounding. On a ship, the drive shaft which transfers power to the propellors is earthed with silver/carbon brushes, which allow the shaft and propellors to be earthed back to the ship's hull. If this earth is not satisfactory, there can be electrical discharge damage in the bearings, leading to micro-arc welding of the rolling components, quickly resulting in an unserviceable bearing. Furthermore, it is essential to the shipboard ICCP system as it provides a return path for ICCP currents protecting the propeller and shaft. Without it, the ICCP system will not "see" the large exposed metallic propellor and will not provide enough current for adequate protection. A diagrammatic representation of a shaft grounding system is show in Figure H.

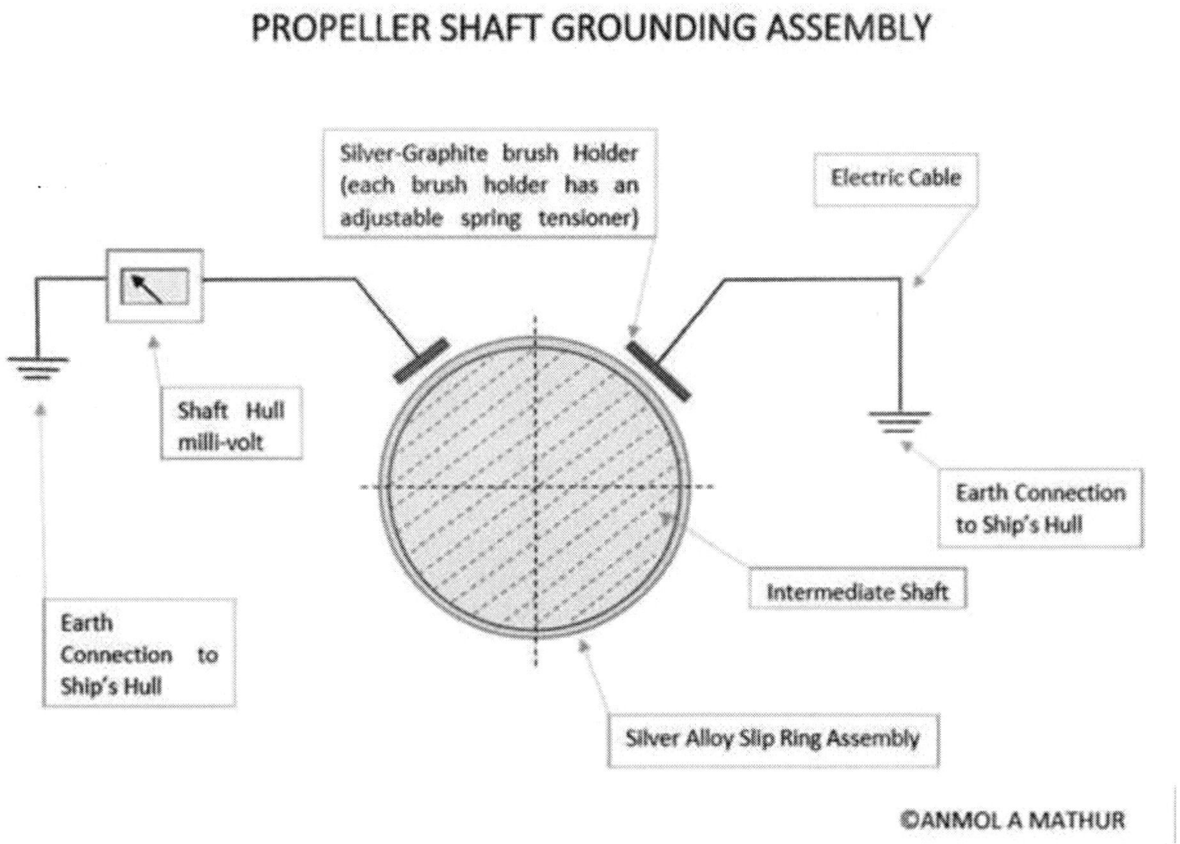

FIGURE H: DIAGRAMMATIC REPRESENTATION OF A SHAFT GROUNDING SYSTEM [19].

In the scenario presented here, the lack of grounding on the shaft, hence lack of CP on the propellors, caused the NAB propellors to be unprotected in the seawater. To add to the corrosion demand on the unprotected NAB propellors, the nuts used to connect the propellor blades to the hub were made from grade 410 stainless steel, which is not an appropriate grade for constant immersion in seawater. When the shipbuilder saw the rapid corrosion on the nuts, and without yet an awareness of the inadequate shaft grounding, they assumed it was due to the low grade of stainless steel. As a solution, the ship builder replaced the nuts with an appropriate marine-grade stainless steel: super duplex 2507.

Unfortunately, by changing the nut material, they had inadvertently changed the galvanic system completely. The original stainless steel 410 nuts were anodic to NAB and had therefore been protecting the propeller (cathode) from corrosion. Now the NAB propeller blades were the anode, and the 2507 nuts were the cathode. So, without any ICCP or accidental sacrificial anode CP, these NAB propeller blades corroded. Typically, a ship's propellor is not a consumable item, but the blades on this vessel had to be completely replaced after only 2 years in service. The images in Figure I show the excessive degree of pitting and general corrosion seen on these blades, with the super duplex nuts remaining in pristine condition.

FIGURE I: IMAGES OF NAB BLADES, SHOWING GENERAL AND PITTING CORROSION. FINGER FOR SCALE. THE SUPER DUPLEX NUTS CAN BE SEEN IN PRISTINE CONDITION IN THE LEFT-HAND IMAGE.

After the propeller was replaced, and once the entire system was considered, including the galvanic material interactions, CP system and shaft grounding, it was finally determined that the shaft grounding was inadequate to return CP currents from the anodes to the hull via the propeller. This meant that the propeller/nut galvanic system was completely isolated from the ship hull and CP system, and the most active of the pair was destined to corrode.

So, what caused this shaft grounding to be so inadequate? There appears to be multiple factors at play, all relating to poor installation and workmanship. The silver/carbon brushes were not aligned correctly, so only the carbon part of the brush (not the more electrically conductive silver part) was in contact with the shaft. The shaft brushes were then connected to the hull through a very long run of small gauge wire. The termination point was installed on a painted structure via the use of crimped connectors (Figure J). All these factors would have caused a high resistance path back to earth and were very easily avoidable mistakes that ultimately lead to the replacement of several very expensive components. Once repaired, the system functioned correctly and has not had a problem since.

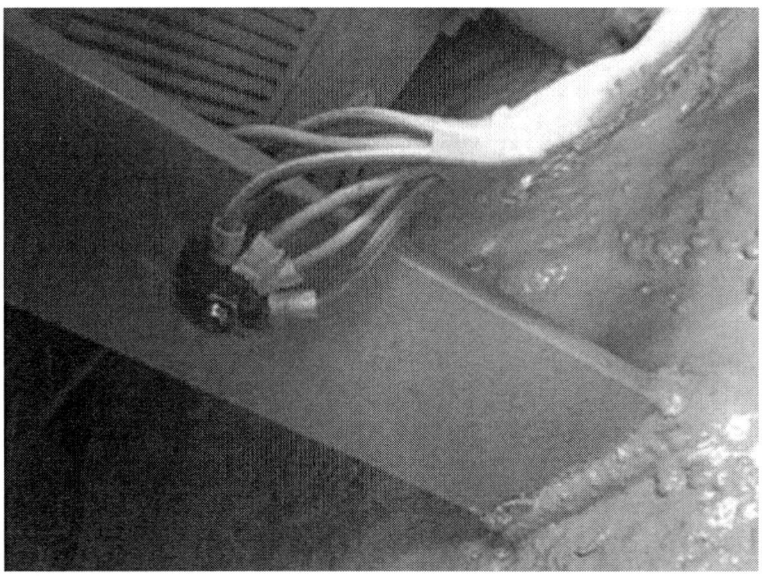

FIGURE J: EARTH CABLE TERMINATION POINT – NOTE THE SMALL GAUGE WIRE AND CRIMPED CONNECTORS.

Conclusion

CP of Navy ships has existed for 200 years and is an incredibly effective method for preventing corrosion of ship hulls and appendages. Despite this, problems with these systems can still occur. The Royal Australian Navy faces unique challenges in managing CP systems on their vessels, primarily due to the extended periods spent in port compared to commercial vessels and their unique operational profiles. The complexity of the interactions between various systems, such as galvanic couples, wharf CP, MGPS can often make diagnosing and solving issues with CP systems challenging. In this paper we have presented some examples of problems relating to CP systems that have been observed in Royal Australian Navy vessels, and the methods that can be used to mitigate them.

ACKNOWLEDGMENTS

We would like to thank Dr Bruce Ackland, Dr Markus Büchler and Warren Green for sharing papers and providing insights into the fascinating history of CP and Sir Humphrey Davy. Also, our friends in the Immersed Corrosion team at DSTG for making our work fun and for their contributions to the various case studies presented here.

References

1. Davy, H., *XII. Additional experiments and observations on the application of electrical combinations to the preservation of the copper sheathing of ships, and to other purposes.* Philosophical Transactions of the Royal Society of London, 1824. **114**: p. 242-246.
2. Davy, H., *IV. On the corrosion of copper sheeting by sea-water, and on methods of preventing this effect; and on their application to ships of war and other ships.* Philosophical Transactions of the Royal Society of London, 1824. **114**: p. 151-158.
3. Davy, H., *XV. Further researches on the preservation of metals by electrochemical means.* Philosophical Transactions of the Royal Society of London, 1825. **114**: p. 328-346.
4. James, F.A.J.L., *Davy in the Dockyard: Humphry Davy, the Royal Society and the Electro-chemical Protection of the Copper Sheeting of His Majesty's Ships in the mid 1820s.* Physis, 1992. **29**: p. 205-225.
5. Paris, J.A., *The Life of Sir Humphry Davy*. Cambridge Library Collection - Physical Sciences. Vol. 2. 2011, Cambridge: Cambridge University Press.
6. Ackermann, R. and T.G.D.S. Dutton, *H.M.S. Samarang 26 Guns In a gale crossing the Bay of Biscay on her passage home on the 23rd December 1846*. 1846, Image sourced from https://rmg.co.uk/collections/objects/rmgc-object-140767.
7. Morgan, J.H., *THE ECONOMICS OF SHIPS CATHODIC PROTECTION.* Anti-Corrosion Methods and Materials, 1964. **11**(11): p. 53-55.
8. BARNARD, K.N., *Cathodic Protection of an Active Ship In Sea Water. Part II.* Corrosion, 1951. **7**(4): p. 114-117.
9. BARNARD, K.N. and G.L. CHRISTIE, *Cathodic Protection of an Active Ship in Sea Water★*. Corrosion, 1950. **6**(7): p. 232-234.
10. Ackland, B., *Cathodic Protection – Black Box Technology?*, in *Corrosion & Prevention*. 2005: Queensland, Australia.
11. Baeckmann, W., W. Schwenk, and W. Prinz, *Handbook of Cathodic Corrosion Protection: Theory and Practice of Electrochemical Protection Processes*. 1997: Gulf Publishing Company.
12. Allanson, L. *HMAS Arunta Ships Company in the dry dock, Garden Island, Sydney*. 2024 [cited 2024 4 September 2024]; Available from: https://images.defence.gov.au/assets/Home/Search?Query=20240730ran8642974_0003.jpg&Type=Filename.
13. Fast Marine Services. *Cathodic Protection*. 2024 4 September 2024]; Available from: https://www.fastmarine.net/technical-repair-services/cathodic-protection/.
14. Floerl, O., S. Wilkens, and C. Woods, *Temporal development of biofouling assemblages. Report prepared for the Australian Department of Agriculture, Fisheries and Forestry*. 2010.
15. Sheridan, T., *Effects of Calcareous Deposits on Water-Lubricated Propeller Shaft Operations*. 2023.
16. Ahmadzadeh, M., et al., *Calcareous scales deposited in the organic coating defects during artificial seawater cathodic protection: Effect of zinc cations.* Journal of Alloys and Compounds, 2019. **784**: p. 744-755.
17. Banslimane, S., et al., *Thermodynamic study of Zn2+ inhibition properties and mechanism on calcium carbonate precipitation by chemical and electrochemical methods.* Desalination, 2016. **398**: p. 114-120.

18. MacAdam, J. and S.A. Parsons. *The Effect of Metal Ions on Calcium Carbonate Precipitation and Scale Formation*. in *Sustainability in Energy and Buildings*. 2009. Berlin, Heidelberg: Springer Berlin Heidelberg.
19. Sulligoi, G., et al., *Shore-to-Ship Power*. Proceedings of the IEEE, 2015. **103**: p. 1-20.

AUTHOR DETAILS

	Kate Dylejko is a corrosion engineer at the Defence Science and Technology Group, specializing in minimizing and controlling corrosion on Navy platforms. With roots as a mechanical engineer in armoured vehicles since 2006, she transitioned to corrosion 11 years ago, leveraging computational modelling to optimize cathodic protection systems for Navy vessels. Kate's expertise spans various facets of corrosion, with a keen focus on cathodic protection systems for maritime platforms. Her research includes electrochemical studies on calcareous deposits and corrosion potential monitoring across Navy vessels and wharves. Actively contributing to the Victorian branch of the Australasian Corrosion Association, Kate recently achieved certification as an AMPP CP2 Cathodic Protection Technician, reinforcing her commitment to advancing corrosion control technologies.
	Wayne Neil is a corrosion scientist in the Defence Science and Technology Group, where he leads a team specialising in maritime corrosion. His research focuses on technologies to mitigate and prevent corrosion on a broad range of areas within vessels for the Royal Australian Navy. With 15 years of experience in Defence, Wayne specialises in microbially influenced corrosion; corrosion of marine alloys including marine grade aluminium, copper nickel, stainless steels and high strength steels; environmentally friendly corrosion inhibitors; and corrosion of magnesium alloys. He is the current ACA Victorian Branch President.

EVOLUTION OF FOULING RELEASE COATINGS

D. Harvey

PPG Industries Australia

**Corresponding author email: dharvey@ppg.com*

ABSTRACT

The challenge of keeping a vessels' underwater hull free of macro fouling has been with us for thousands of years, ever since man invented boats for transport, commerce and leisure. The drive to improve the protection of the underwater hull led to the development of coatings containing more and more effective biocides, eventually leading us to a conundrum; "How do we continue to keep the hull clean and at the same time reduce the negative impact these developments may have on the environment in which we work and play?" To answer this question this paper looks at the development of biocide free, foul release coatings, how they were imagined, their development, their impact on the use of conventional antifouling coatings and their new found importance in the modern era of transport and our focus on the reduction of CO2 emissions.

Keywords: foul release, antifouling, biocides,

Note on keywords: Examples such as steel, seawater, adhesive, protective coating, etc. are example words that denote relevancy to the specific part of the industry and the contention of the paper. Keywords should be complementary instead of cosmetic to the Abstract and can be used by other academics or researchers to locate and cite your paper if necessary. Abbreviations are appropriate if they are renown industry-wide; otherwise, they are to be avoided where possible. The maximum number of keywords allowed is 5.

INTRODUCTION

This paper seeks to understand the need for underwater hull protection, the development of various options from copper and lead sheeting to coatings containing biocides and ultimately the development of niche coatings that contain no biocides, improve vessel performance, reduce maintenance costs and assist in protecting the environment and the reduction of CO2 emissions.

.

HISTORY OF FOULING AND PROTECTION

As far back as 410BC and probably beyond there are records of man's attempts at underwater hull protection. Lead sheeting was one of the ways they tried to keep the hulls clean. Around the 1500's copper sheeting came into use over the timber hull combined with tar or grease. Around the 1700's the iron vessels, when coated with copper sheeting, developed severe corrosion often leading to such degradation of the hull that the vessel was unusable. This became such a problem that the English Admiralty almost completely stopped building iron ships in favour of timber. In 1850 the first of a breed of new coatings was developed containing copper sulphide and later, mercury and arsenic were added. These two were later excluded in favour of copper in the form of cuprous oxide and in the early 70's a breakthrough came with the development of SPC (Self Polishing Copolymer) combining the new resin technology with cuprous oxide or cuprous thiocyanates and Tributyl Tin (TBT). These combinations proved extremely effective with 95% predictability outcomes, but it was later discovered TBT had a very detrimental effect on oysters which led to a banning of TBT in antifoulings in 1989 for vessels under 75 feet.

The IMO issued a directive stating that in 2003 a total ban of any further application of antifoulings containing TBT with an agreement that from 2008 there would be no vessels using an antifouling containing TBT.

This resulted in a loss of performance of the antifouling coatings and combined with rising fuel prices and an emerging awareness of the effect on the environment, new ways were being explored to keep shipping costs viable. This led to renewed focus on a technology that had been developed back in the late 70's, biocide free, low surface energy coatings, fouling release.

Evolution of fouling release

In 1977, Alexander Milne secured a patent on a biocide free coating product combining silicone elastomers and silicone polymer or oils. This formed the basis of what we see today. During the 80's a lot of work went into solving some of the issues with this technology such as:

- Suitable priming systems that would protect the substrate and allow adhesion of the silicone coatings
- Toughness and durability for the marine environment
- Practicality of use. Pot life/shelf life/ spray and roller application
- Prevention of sagging when using airless spray
- Maintenance and repair procedures.
- One of the first applications was in 1987 on a US Sealift Command container vessel.

It was determined that this technology was suitable for vessels constantly in use and operating at higher speeds to allow the surface to self-clean. It was trialled on recreational craft but proved ineffective as these vessels spent 80%-90% of their time static on moorings or marina berths, allowing the growth to settle on the hull and the resultant excess cleaning damaged the coatings, further reducing its effectiveness.

With the advent of the SPC antifouling technology and the subsequent reliable high performance, interest in the biocide free coatings waned. There seemed little need for the silicone technology as fuel costs were controlled and dockings were less expensive with the vessels being cleaner at the 5 year survey period. When the TBT component was banned and fuel costs rose a new focus came into place.

Development of biocide free systems

Originally the systems were cumbersome:

Anticorrosive epoxy primer- Epoxy intermediate- Silicone tie coat- silicone colour coat- Silicone clear finish coat

Anticorrosive epoxy primer- Epoxy intermediate- Silicone tie coat- Silicone coloured finish coat

They were then refined further:

Anticorrosive epoxy primer- Silicone tie coat- Silicone colour finish

BENEFITS

The advent of this technology brought many benefits to the vessel owner/operator over conventional antifouling

- Extended docking periods
- No need to apply full coat at next docking
- Hull is cleaner longer
- Extended and safer manoeuvrability
- Smoother finish results in reduced fuel usage
- Lower risk of transporting marine organisms to new and possibly vulnerable environments
- No biocides being leached into the water

Major paint companies all have an offer in the biocide free sphere. Some are 3 coat application, some 4 and others have combined biocides with the silicone to help improve idle time performance.

WHAT IS THE FUTURE FOR THIS TECHNOLOGY

The initial motivation for this technology was fuel saving and then protection for the marine environment from possible harmful biocides and it could begin to replace conventional antifoulings as more pressure comes on globally to ban copper and other biocides. This of course would mean the maintenance of vessels and particularly the annual slipping of pleasure craft, would have to be changed dramatically.

The new and immediate impact on the industry has been the declaration in January 2023 by the IMO that all commercial vessels of certain size would have to be measured and rated relative to their carbon emissions or, carbon footprint.

The goal of the IMO: Decarbonisation

Reduction in power (fuel usage) Energy Efficiency Existing Ship Index (EEXI) and

Annual reduction in CO_2 emissions. Carbon Intensity Indicator Rating (CII)

The IMO has set a target of 40% reduction in CO_2 emissions (2008 v 2030)

This requires operators to first determine what their rating is by measuring their GHG (Green House Gas) output based on miles travelled, speed etc and then put in place a plan with the IMO that outlines how, if they are rating low, they are going to improve their rating. This will then allow organisations wanting to engage freight carriers to, if they desire, award contracts based on lower GHG emissions meaning some carriers may lose business unless they comply.

As we have seen, fouling release coatings provide the mechanism to improve operators' ratings by reducing the friction of the underwater hull coating. Fouling is minimised and, due to the slipperiness of the coating, the vessel requires less power to achieve the same speed resulting in reduction in fuel, ergo reduction in CO_2 emissions.

SUMMARY

The need for ships as a form of global transport for people and goods continues to grow. This brings into greater focus the importance of protecting the marine environment while, at the same time aiming to stay in touch with the concerns over CO_2 emissions. The challenge of protecting the underwater hull from infestation of marine growth while looking after our environment has been ongoing for more than a thousand years and new technologies continue to emerge to meet these needs.

Silicone coatings certainly offer an alternative to conventional antifoulings and work hand in hand to provide the solutions for a more sustainable shipping fleets.

CONCLUSIONS

As continued focus comes on the inclusion of biocides in our coatings and the on going demand for the reduction of these we will see an even greater push for biocide free coatings and particularly foul release technology to replace conventional antifouling.

ACKNOWLEDGMENTS

AUTHOR DETAILS

Photos are to be submitted in 600 DPI. One of the authors should be nominated as the corresponding author in case of feedback or any other communication required between the author and the Conference. The corresponding author and contact details should also be clearly stated in list of authors at the start of the paper.

D. Harvey is the corresponding author of this paper. He is Marine Manager for ANZ at PPG Industries, a position he has held since 2008. He is responsible for the strategy, direction and implementation of the PPG marine business in Australia, PNG and the Pacific.

SUSTAINABLY LED ASSET MANAGEMENT: A PACIFIC CASE STUDY QUEENS WHARF, LAUTOKA FIJI

B. McGuiness[1,*]

[1]*Marine & Civil Maintenance, NSW, 2111, Australia*

**blanem@marineandcivil.com.au*

ABSTRACT

The presence and ingress of chlorides, particularly in marine environments, is one of the main factors responsible for the deterioration of steel reinforced concrete structures. Across the pacific, where the conditions are optimal for chloride induced corrosion, and substantial maintenance funds are limited, innovative and preventative approaches are required to ensure optimal value, and long-term durability. Through this case study, the reemergence of Electrochemical Chloride Extraction (ECE) and the retrofitting of water base anode Impressed Current Cathodic Protection (ICCP) Systems are two ways preventative intervention have been utilised to curb the progression of deterioration across Queens Wharf, Lautoka, Fiji.

Keywords: Electrochemical Chloride Extraction, Cathodic Protection, Durability, Sustainability, Asset Management

INTRODUCTION

Lautoka Port is Fiji's second largest commercial port and is one of the two primary import/export ports of the country, along with Suva. The main wharf comprises two berths. The southern berth was constructed in 1959 and a northern berth was added in 2005 (refer Figure 1). The two different wharf sections are of a different construction form which is described below.

The primary objective of the project was to (a) determine the current condition of the structures, (b) develop remediation plans for the rehabilitation of the port, and (c) execute the works via an option engineering design amendment and verification process.

FIGURE 1: QUEENS WHARF, LAUTOKA, LOCALITY PLAN

Condition inspections were conducted by Cardno in 2018/2019 in accordance with the Ports Australia Wharf Structure Condition Assessment Manual (WSCAM) [1]. Following a detailed design stage, the works were tendered initially in 2020 via a prequalification panel, and then by invitation in 2022 following the easing of travel restriction following the COVID-19 pandemic.

This case study will focus on the utilisation of sustainable innovations and option engineering to maximise the long-term durability for Pacific based infrastructure, in an economic environment where environmental protection and value for money are paramount considerations under a limited funding model.

It should be noted that the systems that are discussed following were installed in addition to the completion of traditional concrete repairs to already delaminated and spalled locations. However, discussions around traditional concrete repairs do not form a focus of this case-study as presented.

Construction Description

Queens Wharf (1959 Section)

The older section of the wharf was constructed in 1959 (see Figure 2). This section typically consists of a reinforced concrete deck integral with reinforced concrete deck beams. The deck is supported by circular concrete piles with a large (1.07 m diameter) concrete cap at the top of each pile. The concrete cap typically extends from deck level about 1.5 m down. There are three rows of square concrete piles around the outside of the circular piles. The concrete deck is typically 305 mm thick. The southwest corner of the wharf has been strengthened with an insitu deck thickening for the purpose of crane loading. The west side of the wharf is the only large vessel berthing face of this wharf and it has 'V' fenders spaced at approximately 10 m. The eastern side of the wharf has tyre fenders for the berthing of small and/or recreational vessels.

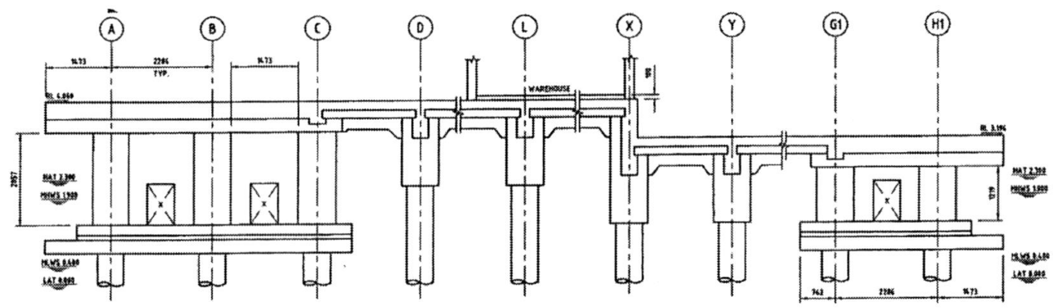

FIGURE 2: QUEENS WHARF (1959 SECTION) CROSS SECTION

QUEENS WHARF (2005 SECTION)

The new section of the wharf was constructed in 2005 (see Figure 3). This section typically consists of a reinforced concrete deck supported by vertical steel circular hollow section (CHS) piles. Piles are typically 900 mm diameter and are covered with a 20 mm thick HDPE sleeve. The steel piles have a concrete cap which creates a moment connection into the deck. The centre row of piles consists of 20 pairs of raker piles and the two pile rows of berthing faces include 6 pairs of raker piles. The fenders in this section are cone fenders with large HDPE face plates and upper and lower chain supports. There are deck thickenings on the three piles rows with rakers to provide additional strength for berthing.

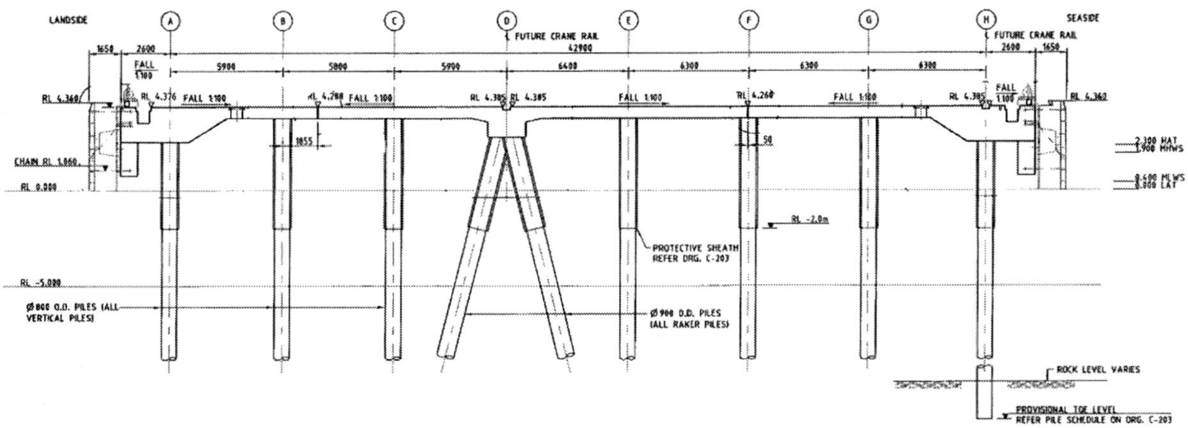

FIGURE 3: QUEENS WHARF (2005 SECTION) CROSS SECTION

ASSESSMENT REGIME

During the 2018 inspections, engineers completed on-water and on-deck inspections in accordance with the Wharf Structures Condition Assessment Manual (WSCAM) condition rating scale.

The inspection was primarily visual in nature, however, delamination investigation (i.e. hammer tests) on abutments was also carried out. Representative photos and video were taken during the inspections to show defect types recorded and allow for cross referencing to previous reports and diving inspections.

The WSCAM condition rating scale utilised gives a comparative structural component rating of 1-7 as detailed below in Table 1.

TABLE 1: WSCAM GENERIC CONDITION RATING SCALE

CONDITION STATE	GENERIC DESCRIPTION	EXPECTED REM. LIFE (% of original design life)	ACTIONS
1	New with no visible defects/damage.	100	No repairs required. Inspection at next scheduled inspection.
2	No or very minor defects which do not affect the overall integrity or durability of the element or component	55-100	No repairs required. Inspect at next scheduled inspection.
3	Limited defects present may affect the long term durability of the element or component. Minor deterioration of protective coating or parent material is evident.	40-55	Inspect at next scheduled inspection Continue planned and preventative maintenance. No repairs required.
4	Defects present may have minor impact on integrity. The short term durability of the element of component may be affected. Localised areas of moderate to advanced deterioration may be present.	25-40	Further testing is recommended and mostly reactive maintenance and some minor upgrades. Priority of repairs is low.
5	Advanced deterioration present. Defects present may have a moderate impact on integrity. The immediate durability of the element or component may be affected.	15-25	Maintenance; upgrade or rehabilitation works are required within 5 years or as dictated by expected remaining life. Structural assessment is recommended.
6	Advanced deterioration Defects present likely to have major impact on integrity. Further deterioration will compromise the safety of the structure.	0-15	Rehabilitation or renewal required immediately. Structural assessment is recommended.
7	Very advanced deterioration present. Defects present likely to have an extreme impact on integrity and may constitute failure of the element.	0	Rehabilitation required immediately or replace component/asset. Structural assessment is recommended where rehabilitation works are to be undertaken.

In addition to the visual inspection, a dive inspection was conducted across selected representative piles, whereby selected piles were cleaned using hand tools to remove localised marine growth. During the same inspection, existing pile repair and protection systems were inspected with reference to condition, construction techniques, and levels.

Assessment Findings

Queens Wharf (1959 Section)

A comprehensive structural assessment was conducted on the concrete piles and beams of the aging wharf. Circular concrete piles were generally in fair condition (WSCAM 3-4), though some showed significant cracking, especially near the pile caps, which were occasionally found to be severely corroded and spalled. Marine growth was prevalent below the waterline. In certain areas, such as pile row 5, significant defects were observed, with many piles showing extensive cracking and delamination, possibly due to poor construction materials.

The surrounding square piles were mostly in good condition (WSCAM 3), with some minor to moderate cracking. The seaward concrete walls between these piles were generally defect-free, whereas the shoreward walls exhibited cracking and delamination. The platform between square piles was in good condition but showed minor corrosion. Notably, square piles connecting the old and new wharf sections had been repaired, presenting minimal defects.

Longitudinal and transverse beams under the deck were in good condition (WSCAM 3), though areas such as pile rows X-H1 showed severe cracking and delamination. Several beams had undergone repairs, including shotcrete applications, though some repairs had begun to fail due to continued reinforcement corrosion, leading to further delamination and spalling.

The deck soffit displayed localised defects, especially on the eastern section, where large areas of spalling with corroded reinforcement were found. In some areas, previous repairs were also failing, contributing to further cracking and delamination. Notably, significant deterioration was observed around piles D3, D7, G33, I21, S31, T9, and T47, attributed to altered structural loading.

The wharf deck was generally in good condition, though some recent repairs and impact damage were noted. The 'V' fenders were in excellent condition (WSCAM 2/3), having been recently replaced. Bollards on the western side were used for larger vessels, while smaller bollards on the eastern side, typically used for smaller vessels, were mostly in good condition, although one had failed.

The approach bridge deck and soffit were in good condition, but the beams showed significant cracking, despite recent repairs with shotcrete.

FIGURE 4: TYPICAL UNDER WHARF INSPECTION OBSERVATIONS OF THE ORIGINAL 1959 SECTION

Queens Wharf (2005 Section)

This assessment evaluated the condition of steel piles, the concrete deck, and the fender system at the wharf. Steel piles are sleeved with HDPE from the connection point to below the water level, and above-water sleeves are in good condition with minimal marine growth. However, gaps at the top of the piles expose the steel to seawater, and the underlying condition of the piles was unknown due to the sleeve. Below the HDPE, divers visually observed Accelerated Low Water Corrosion (ALWC), which is an aggressive form of corrosion near the low-water line. ALWC was found on all piles, with corrosion depths ranging from 2-5 mm.

The concrete deck soffit was generally in a good condition (WSCAM 2), though minor cracking, delamination, and spalling were observed near construction joints. Reinforcement in these areas was often found close to the surface. Erosion of the concrete was noted near deck penetrations, causing localised damage. Concrete deck thickenings showed minor rust staining, likely from tie wires. Some cracks were observed between piles on deck-thickening beams, extending through the full depth of the beam.

The fender system was inspected at low tide to observe the condition of the panels and supports. While the concrete panels were in good condition, the steel fender frame was in poor condition (WSCAM 6), with significant corrosion and surface rust. Some fender panels displayed corrosion, and several fender chains had failed due to severe corrosion and compression forces. Many fender dampers were buckled or failing, compromising their functionality. Fender chains, which showed multiple failure modes, were attributed to the improper alignment of fender panels during berthing.

The wharf deck was found to be in average condition, with debris accumulation contributing to potential wear. Areas where cranes and machinery operate were coated with a thick oil and rubber film, obscuring the deck's condition. Significant kerb damage was noted along the seaward side, likely caused by forklift impacts. The service pit on the western side was also in poor condition, which may lead to accelerated deterioration of the utilities within.

Overall, while many areas of the wharf's structure remain in serviceable condition, the steel piles' exposure to ALWC, fender system failures, and kerb damage highlighted the required attention to maintain the structural integrity and functionality of the wharf.

FIGURE 5: TYPICAL UNDER WHARF INSPECTION OBSERVATIONS, AND OBSERVED PILE CONDITION (SHOWING ALWC) OF THE NEWER 2005 SECTION

INITIAL TENDER SCOPE

Based on the inspections and recommendations of the Cardno Inspection Report, the Fiji Port Authority issued tender documents with a prescribed scope of works, itemised across the original 1959 section, and the newer 2005 section as detailed below in Tables 2 and 3.

Under the contracted scope of work, tenderers were invited to submit conforming tenders, but also encouraged to submit alternatives based on value for money, cost reductions, advantages regarding ongoing maintenance and durability, and sustainability and environmental protection.

These alternatives were further detailed and explored during an extensive post tender interview and negotiation stage whereby the preferred contractor worked collaboratively with the client, and client-side engineer in a detailed optioneering process of scope and contract refinement.

TABLE 2: TENDER SCOPE – 1959 SECTION

Description	Qty	Unit
REMEDIATION (1959 SECTION)		
Pile / Pile Cap (Circular) Concrete Patch Repair. This item is to include concrete breakout and repair to original profile (including replacement of reinforcement and installation of patch anodes).	34	No.
Pile Cap (Square) Concrete Patch Repair. This item is to include concrete breakout and repair to original profile (including replacement of reinforcement and installation of patch anodes).	30	No.
Beams Concrete Patch Repair. This item is to include concrete breakout and repair to original profile (including replacement of reinforcement and installation of patch anodes).	75	m^2
Deck Soffit Concrete Patch Repair. This item is to include concrete breakout and repair to original profile (including replacement of reinforcement and installation of patch anodes).	200	m^2
Bridge Beam Concrete Patch Repair. This item is to include concrete breakout and repair to original profile (including replacement of reinforcement and installation of patch anodes).	5	m^2
Install new stainless-steel ladders	4	No.

TABLE 3: TENDER SCOPE – 2005 SECTION

Description	Qty	Unit
REMEDIATION (2005 SECTION)		
Wrapping of steel piles with protective jacket from seabed level to existing protective sheath level (approximately RL -5.0 to RL -3.0)	248	No.
Install new stainless-steel ladder in accordance with details on PM-3050.	5	No.

Option Engineering and Executed Scope

Shortly after award of the contract, the project team undertook a detailed Whole of Life (WOL) cost and option analysis whereby the following factors were considered:

- The identified repair quantities based on the 2018/2019 condition assessment, which also formed the basis (including contingency areas) for the 2022 tendered scope
- The "as-surveyed" repair quantities, based on 2022 onsite visual and delamination assessment
- Overall deterioration growth rates, and projected repair expenditure required over a 30-year service life
- Preventive maintenance interventions to reduce the overall 30-year expenditure, and maximise project value for money
- Alternative methodologies to achieve specified project outcomes in a more innovative and sustainable way

The comparative analysis of deterioration growth rates showed an approximate growth of required repair areas from 2% of the total wharf surface area in 2018, to approximately 6% of the total wharf surface area in 2022, and approximate growth rate of 1% per annum.

Using the initial contract sum and rates as a basis, and allowing for deterioration growth and CPI adjustments, a value was calculated across the 30-year service life, that accounted for treatment/repair across only 36% of the total wharf area (assuming a repair regime based on reactive concrete repairs only).

However, given the project teams ambition to implement sustainable and innovative technologies in a preventative manner, options including implementation of Electrochemical Chloride Extraction, and retrofitting of an existing, yet non-operational water anode Impressed Current Cathodic Protection (ICCP System) allowed the client to rationalise a maintenance plan that would treat/repair ~80+% of the total wharf area for approximately 40% of the calculated 30-year expenditure.

Implementation of Electrochemical Chloride Extraction

The primary post project award Value Engineering Proposal for the Queens Wharf Rehabilitation project focused on the application of an Electrochemical Chloride Extraction System, designed to halt and remove the causing agent of reinforced concrete corrosion. ECE offers a more comprehensive solution than the initially nominated traditional galvanic patch anode systems by protecting all concrete elements identified (via the presence of repair quantities) affected by corrosion, not just localized areas adjacent to repairs. This innovative approach, previously and successfully implemented across Australia, presented significant life-cycle cost savings for Fiji Ports Corporation Limited (FPCL).

In general terms Electrochemical Chloride Extraction (ECE) is a technique used to mitigate the damaging effects of chloride-induced corrosion in reinforced concrete structures. Chlorides, often from saltwater, penetrate concrete and initiate corrosion of the embedded steel reinforcement, leading to structural damage over time.

In ECE, an electrical current is applied between the reinforcing steel (acting as a cathode) and an external temporary, and reusable applied surface mounted anode system. Using a conductive electrolyte, high output DC current is applied across the structural elements for a duration from 10-20 weeks. This process helps to migrate the chloride ions away from the steel reinforcement and toward the external anode (see Figure 6), with performance assessed against compliance with *NACE/AMPP SP0107-2021: Electrochemical Realkalisation and Chloride Extraction for Reinforced Concrete* [2].

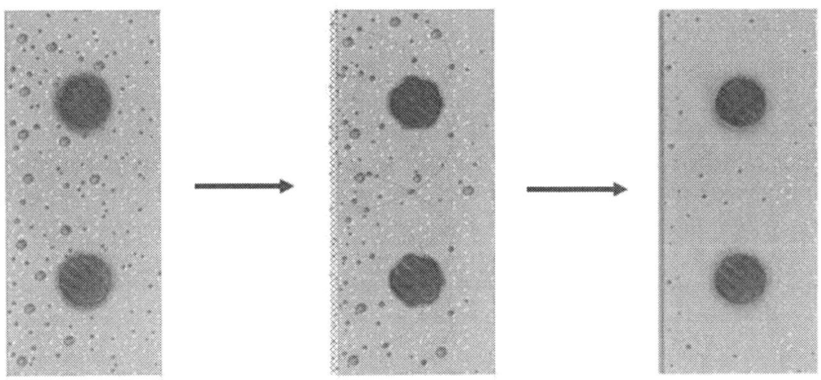

FIGURE 6: SCHEMATIC DIAGRAM ILLUSTRATING THE PROCESS OF ECE: (A) PRE-TREATMENT INDICATING THE PRESENCE OF CHLORIDES THROUGHOUT THE CONCRETE MATRIX, AND CORROSION OF THE STEEL REINFORCEMENT. (B) SURUFACE APPLIED TREATMENT INDICATING THE CURRENT FLOW, THE EXTRACTION OF CHLORIDES, AND THE TRANSITION TO REPASSIVATION. (C) TREATMENT COMPLETION, AND THE APPLICATION OF A BARRIER COATING.

The ECE implementation led to both immediate contract savings and long-term reductions in maintenance costs over the nominated 30 years. Specifically, the implemented solution capitalised on the potential savings through the substitution of the need for galvanic anodes and priming rebar during repairs, whereby these funds were reinvested into wider ECE application across not only the identified repair areas, but across the entire element nominate for treatment, including both repaired and non-repaired areas.

The life-cycle cost analysis predicted a substantial reduction in ongoing repair costs, ultimately extending the service life of the Queens Wharf structure and minimizing the risk of future repairs. Fiji Ports Corporation Limited (FPCL) adopted a recommendation for a phased application of the ECE system, beginning with areas heavily trafficked by machinery, followed by a broader roll-out across other structural elements, depending on the observed scope of repairs.

The proposal as implemented enable the project delivery team to balance immediate savings with long-term structural protection, aligning with FPCL's operational needs, whereby an initial treatment area of 810 m^2 was extended to 1,650 m^2 (an increase of over double), with a nominal expenditure increase of less than 10%.

Retrofit and Upgrade to Water Anode ICCP System

During the tender period, and prior to award, Marine & Civil Maintenance (MCM) identified significant cost-saving opportunities across the project, most notably an option that involved the elimination of the proposed encasement/jacketing of piles for corrosion protection by patch welding of corroded areas, and upgrading and recommissioning an existing, yet non-operational Impressed Current Cathodic Protection (ICCP) system, as part of the 2005 section scope of work.

This approach directly addressed the issue of Accelerated Low Water Corrosion (ALWC), which was at the time reducing the pile thickness in sections identified for repair. Commonly adopted as an industry-standard solution for managing corrosion in such cases, ICCP has be demonstrated successfully through various implementations, notably at the Port of Melbourne, Australia.

Under the alternative scope of work, the follow work activities were detailed and executed.

1. **Comprehensive System Audit**: Inspection and auditing of all existing ICCP components, including Transformer Rectifier Units, junction boxes, anodes, cables, and reference electrodes.
2. **System Repair and Upgrade**: Restoration and replacement of any damaged components in the existing system.
3. **Enhanced Monitoring**: Installation of additional reference electrodes to improve data collection and transparency in system performance.
4. **Advanced Control System**: Installation of a new Transformer Rectifier Unit equipped with remote monitoring and control capabilities for both on-site and off-site operation.
5. **Long-Term Maintenance**: Provision of a 10-year comprehensive maintenance and monitoring plan to ensure the system's ongoing operational effectiveness, which provides enhanced protection and transparency compared to the nominated pile jacket system.

This proposal was externally assessed by the client and their engineers and agreed to be a more efficient and cost-effective alternative to traditional pile encasement, ensuring the client long-term protection from ALWC and offering FPCL a robust, verifiable, sustainable, and cost effective solution for corrosion management.

Key to this solution was the removal of the single output, 1000 A Oil Cooled manual control system, and replacement with a project specific developed, seven (7) zone output Transformer Rectifier Unit (TRU). This enhanced system was manufactured complete with remote and local monitoring and control functionality, additional reference electrodes for remote monitoring, and installed within a temperature controlled server room.

As part of this process, the single zone system was rezoned into a total of seven (7) independent outputs (see Figure 7) across the 2005 section of the wharf (six zones), and a single additional single zone to protect the approach bridge, adjacent to the 1959 section, allowing for great transparency, control and assessment of the system performance.

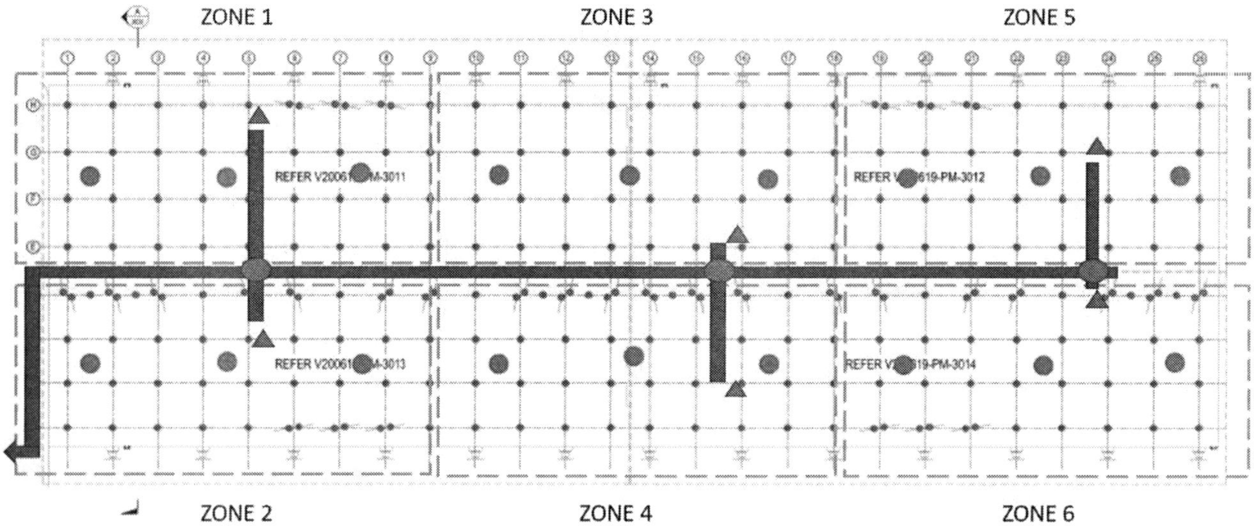

FIGURE 7: AMENDED SYSTEM ZONING AND MONITORING POINTS

Project Sustainability – Cost and Environmental Outcomes

The advantages of innovative maintenance over complete rebuilding of concrete wharf structures extend beyond qualitative benefits, with significant reductions in carbon dioxide emissions being a critical factor. On average, maintenance activities produce 183% fewer greenhouse gas emissions per square meter compared to new construction [3,4]. Furthermore, it has been found that for every square meter maintained, approximately 3.3 tons of CO_2-equivalent emissions are avoided [5]. In practical terms, this means that repair and remediation efforts drastically cut down on the carbon footprint of infrastructure projects, highlighted by global metrics whereby concrete production is responsible for 8% of global anthropogenic greenhouse gas emissions [6].

As an example, eliminating the need for HDPE pile jackets contributed to these reductions, as traditional pile jackets require manufacturing and installation processes that emit additional carbon. By reutilising the existing ICCP system, the project team was able to eliminate approximately 2.5 kg of carbon dioxide per kg of HDPE avoided, and a further 2-3 kg of carbon dioxide per man hour of avoided labour.

With regards to the concrete structural elements, by utilizing advanced techniques such as Electrochemical Chloride Extraction and Realkalisation, the lifespan of the existing structure is extended without resorting to high-emission material-based reconstruction, resulting in a 91%+ reduction in carbon emissions, as compared to a rebuild scenario that was possible without intervention.

Electrochemical Systems Assessment

At the time of writing this paper, a total of six sub-stages (Stages, 1, 2A, 2B, 2C, and Stages 4A, and 4C), corresponding to 60 unique locations, and 600 chloride content samples have been taken and assessed, analysed as %w/w cement.

From this data set, all 60 locations have achieved Criterion A in accordance with *NACE/AMPP SP0107-2021: Electrochemical Realkalisation and Chloride Extraction for Reinforced Concrete* [2], corresponding to a chloride concentration below 0.4% w/w cement within a radius of 25mm of the steel reinformcent, compared to an average initial baseline concentration at the level of the steel reinforcement of 0.7-0.9 % w/w cement, and a concrete cover concentration typical ranging from 1-2.5% w/w cement.

With regards to the ICCP system operating and protecting the steel piles of the 2005 section, this system was completed and commissioned in February 2024.

With an average natural potential measured across all piles in the range of -592 to -628mV (measured against a Ag/AgCl sea water reference electrode), all piles are now fully protected in accordance with *AS2832.3 Cathodic Protection of Metals Part 3: Fixed Immersed Structures,* section 2 [7], with a measured potential at all test points, more negative than -800 mV (with respect to a Ag/AgCl sea water reference electrode), and closely approaching the suggested protection level for ALWC of -850mV (with respect to a Ag/AgCl sea water reference electrode) suggested in literature [8].

DISCUSSION

The adoption of innovative techniques like Electrochemical Chloride Extraction (ECE) and the retrofitting of an Impressed Current Cathodic Protection (ICCP) system, reflects a significant shift in the Pacific towards sustainable asset management and value engineering.

The observed chloride concentration reductions, and the subsequent predicted reduction in corrosion activity, enabling an extension of structural lifespan, underscores the importance of preventive measures over reactive repairs. These technologies not only align with environmental goals by reducing greenhouse gas emissions and material usage but also provide long-term financial savings.

The optioneering process, which encouraged value engineering and collaboration between contractors and the client, played a pivotal role in delivering these sustainable solutions. This model of engagement demonstrates the potential for innovative engineering to not only solve current issues but to pre-emptively address future deterioration, thus maximizing the lifespan and functionality of vital infrastructure like Queens Wharf.

CONCLUSION

The Queens Wharf rehabilitation project exemplifies the potential for integrating sustainable technologies in asset management. The application of ECE and ICCP has demonstrated both environmental and economic advantages by reducing future repair costs and extending the service life of the wharf. This case study highlights the importance of adopting preventive maintenance approaches, particularly in regions where resources are limited, and environmental protection is crucial.

Moving forward, the continued success of this project will depend on the robust monitoring of these systems and their adaptability to emerging challenges. Nevertheless, the outcomes of this project suggest that similar approaches could be effectively applied across other aging infrastructures, particularly in coastal and marine environments. This proactive strategy not only safeguards structural integrity but also supports broader goals of sustainability and resource efficiency, offering a template for future infrastructure projects in the Pacific and beyond.

ACKNOWLEDGMENTS

The author would like to thank Marine & Civil Maintenance's Project team members, Mr Vajira Piyasena (CEO – Fiji Ports Corporation), Mr Krishnit Charan (Manager Projects – Fiji Ports Corporation), and Mr Scott Keane (Principal Industry Director: Ports and Marine – Cardno) for their support and confidence throughout the project delivery, and their embrace of innovative and alternative approaches to project delivery.

Additionally, Marine & Civil Maintenance wishes to thank and acknowledge our partners in delivering these works, Infracorr Consulting, for providing third party verification services, and Remedial Technology/MicroNex for the design support throughout the ICCP System upgrades.

REFERENCES

1. Ports Australia. (2017) *Wharf Structure Condition Assessment Manual*. 1st edn. Sydney: Ports Australia.
2. NACE International. (2021) *SP0107-2021: Electrochemical Realkalisation and Chloride Extraction for Reinforced Concrete*. Houston, TX: NACE International.
3. Zhou, Z., Alcalá, J. & Yepes, V., 2020. Bridge carbon emissions and driving factors based on a life-cycle assessment: Case study of a cable-stayed bridge over Hun He River in Liaoning, China. *International Journal of Environmental Research and Public Health*, 17(16), p.5953.
4. Saravia de los Reyes, R., Fernández-Sánchez, G., Esteban, M.D. & Rodríguez, R.R., 2020. Carbon footprint of a port infrastructure from a life cycle approach. *International Journal of Environmental Research and Public Health*, 17(20), p.7414.
5. Dequidt, T.C.E., 2012. *Life Cycle Assessment of a Norwegian Bridge*. Master's thesis, Norwegian University of Science and Technology
6. Belaïd, F., (2022). How does concrete and cement industry transformation contribute to mitigating climate change challenges? *Resources Conservation and Recycling Advances*, 15
7. Standards Australia, 2005. *AS 2832.3: Cathodic Protection of Metals - Part 3: Fixed Immersed Structures.* Sydney: Standards Australia.
8. Khoi, T., & Cottis, RA., 2009, Controlling the Components Of Calcareous Deposit To Protect Steel Structures From ALWC With Dc And Pulse Current, *NACE International Corrosion Conference & Expo*, CORR09, Paper No. 09008

AUTHOR DETAILS

Blane McGuiness is a Registered Professional Engineer (RPEng), Australasian Corrosion Association (ACA) accredited Corrosion Technologist, and senior management professional with extensive national and international experience in Electrochemical Protection Systems, structural remediation, and marine/civil infrastructure construction works.

With vast industry experience, Blane is a well-regarded technical leader and an active member across numerous associations and institutes. His recent roles include serving as a Director of the Australasian Corrosion Association (2019-2022) and a visiting research supervisor with the University of New South Wales (UNSW), Sydney, Australia.

Blane's contributions to the field of corrosion technology and structural remediation are widely recognized, and his research and expertise continues to influence best practices and advancements in the industry.

AUTHOR INDEX

Ali, S. 332
Bardak, Anton 321
Barker, S. 349
Berends, N. 262, 277
Birbilis, N. 340
Branca, D. 277
Brodribb, Richard 17
Brown, Matthew 1
Chaves, I. 51, 79, 126, 237
Chen, Jiarui 63
Crawshaw, C. 349
Cukierski, D. 214
De Prazer, S. 237
Dylejko, K. 367
Edelman, D. 9
El Sarraf, R. 106
Eliasson, B. 146
Fortune, K. 9
Fosdike, D. 176
Francis, R. A. 106, 137
Gillogly, S. 126
Glass, G. K. 29
Harvey, D. 379
Hossain, M. A. 51, 79, 237
Jeffrey, R. 299
Johnson, Peter 117
Karajayli, P. 214
Lam, C. Y. 51, 79
Lichtendonk, B. 247
Lowe, Andrew B. 312
Lu, A. 146
Madaleno, Liliana 195
Mandeno, W. L. 106
Masia, M. 51, 79, 126, 237
McDonald, D. 146, 176
McGuiness, B. 277, 383
McHendrie, R. 176
McKeown, D. 247
Melchers, R. 91, 204, 290, 299
Middleton, C. 237
Morton, David 195
Nearhos, S. 163
Neil, W. 367
Nicholas, D. 204
O'Brien, David 17
Odlin, S. 214
Olowoye, Ibukun 312
Panafieu, M. 262
Papworth, F. 332
Petersen, R. B. 91
Phang, Bob 321
Pojtanabuntoeng, Thunyaluk 312
Richardson, P. 204
Rigby, J. 40
Sellars, K. 176
Smith, B. 237
Sterling, A. 226
Stone, C. 29
Tan, Nicholas S. L. 312
Terry, L. 126
Van Woensel, L. 237
Vazey, J. 51
Wang, Lifu 321
Williams, Phil 117
Young, David J. 63
Zhang, Jianqiang 63

Australasian Corrosion Association Inc.
30a Jessie Street
Preston
VIC 3072
Australia

ISBN 979-8-3313-2966-2

9 798331 329662

CURRAN ASSOCIATES INC.
proceedings.com